P9-BYK-077

INSTRUMENTS AND MEASUREMENTS FOR ELECTRONICS

INSTRUMENTS AND MEASUREMENTS FOR ELECTRONICS

CLYDE N. HERRICK
San Jose City College
San Jose, California

McGRAW-HILL BOOK COMPANY
New York San Francisco St. Louis Düsseldorf
Johannesburg Kuala Lumpur London Mexico
Montreal New Delhi Panama Rio de Janeiro
Singapore Sydney Toronto

This book was set in Times Roman by Holmes Typography, Inc., and printed and bound by The Maple Press Company. The designer was Janet Bollow; the drawings were done by Gary V. Baird. The editors were Alan W. Lowe and Eva Marie Strock. Charles A. Goehring supervised production.

INSTRUMENTS
AND
MEASUREMENTS
FOR
ELECTRONICS

Printed in the United States of America.

Library of Congress catalog card number: 79–152004

34567890 MAMM 7987654

07–28367–2

CONTENTS

I

BASIC MEASURING INSTRUMENTS

VI ELECTRONIC COUNTERS AND FREQUENCY METERS

PREFACE

Electronics is a new and expanding field of scientific endeavor with many diverging areas, each changing rapidly, so that the devices we study and use today may soon be replaced by more sophisticated devices tomorrow. The electronic instrumentation used to make today's measurements will be replaced eventually with newer and more sophisticated instruments. However, there is a common core of knowledge threading through these areas, devices, and instruments.

It is the purpose of "Instruments and Measurements for Electronics" to provide the electronics technology student with a sound background in the basic theory and common-core concepts of measurements and electronic measuring instruments. To accomplish this purpose, the student is led through graduated measuring concepts and techniques using the measuring instrument as the supporting vehicle.

The student is also presented with a basic core of

measuring concepts and techniques through the study of electronic instruments. These instruments can be applied as the basis in every facet of measurement in electronics. Textual treatment includes both intuitive and descriptive approaches, as well as conceptual and analytic approaches.

The text provides the student with an understanding of the logic behind the selection of a specific type of instrument for a measurement and the accuracy to be expected from this instrument. The student will also learn the importance of proper care and application of each type of measuring instrument and the purpose of calibration and maintenance of this instrument. Careful distinction is made between accuracy and precision. The student will also obtain an understanding of the probability of error analysis for electronic instruments and measurements, and the limitation of each type of measuring instrument.

The foregoing objectives are realized in "Instruments and Measurements for Electronics" through the presentation of materials under six topics: Basic Measuring Instruments (Meters); Bridge-Type Instruments; Electronic Display Instruments; Generating Instruments; Tube and Semiconductor Device Testers; and Electronic Counters and Frequency Meters.

CLYDE N. HERRICK

INSTRUMENTS AND MEASUREMENTS FOR ELECTRONICS

BASIC MEASURING INSTRUMENTS

PROBABILITY AND ERROR ANALYSIS

1.1 PRECISION OF MEASUREMENTS

Although beginning students tend to confuse the *precision* of a measurement with the *accuracy* of a measured value, there is a basic distinction between these terms. For example, a battery or cell such as illustrated in Fig. 1.1 has a terminal voltage (strictly an electromotive force) that we call its true or actual voltage. Furthermore, this actual voltage value is not measurable, although we can approximate this value by careful measurement. The accuracy of a measurement denotes the extent to which we approach this actual value. Even the most careful measurements can establish an actual voltage value only within certain limits of accuracy. Of course, new and improved measuring techniques can narrow these accuracy limits.

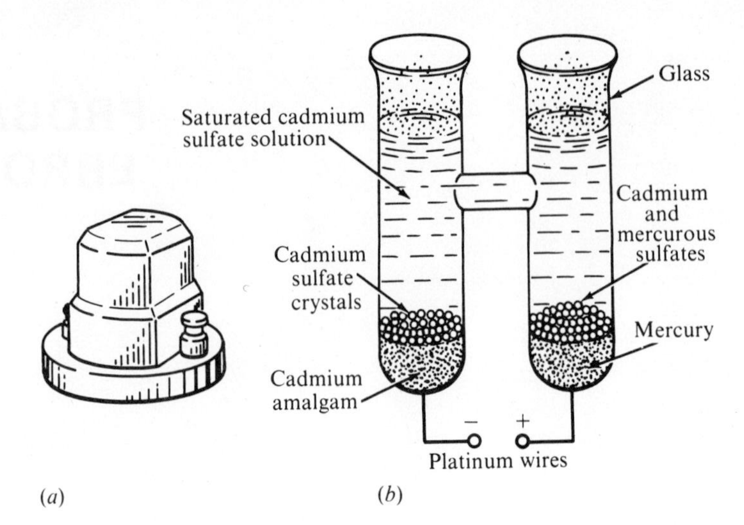

FIG. 1.1 *Appearance (a) and construction (b) of a Weston standard cell.*

On the other hand, the precision of a measurement denotes its departure from the average of a number of measured values. For example, suppose that we carefully measure the terminal voltage of a dry cell six times. Since an observational error is inevitably present in any voltmeter reading (except nonanalog types), we may take the precaution of asking five other observers to repeat this measurement and thus we have six separate measured values:

1.49	1.49
1.51	1.52
1.50	1.50

In this example, we are using the most accurate voltmeter available and accordingly are concerned only with the precision of the foregoing measurements. We proceed as follows: The *sum* of the measured values is 9.01 V. We divide this sum by 6 to find its *average value* of 1.50+ V. Since the remainder in the quotient is less than 5, we round off 1.50+ to 1.50 and thereby determine its *most probable value*. In turn, we conclude that the third and sixth measurements were the most precise within the group of six measurements.

Note that the first measurement has a precision of approximately 99.3 percent; this precision can also be stated as approximately −0.7 percent deviation from the mean. The third measurement has a precision of 100 percent, and so on. Next, if we obtain a voltmeter with a higher accuracy rating, we will measure a different group of values, and this new group will have another average or most probable value.

1.2 OBSERVATIONAL ERRORS

There is more than one source of observational error. As an example, the pointer of a voltmeter rests slightly above the surface of the scale. We thus will incur a *parallax* error unless our line of vision is exactly above the pointer. To minimize parallax error, highly accurate meters are provided with mirrored scales, as depicted in Fig. 1.2. When the pointer's image appears hidden by the pointer, the observer's eye is directly in line with the pointer. Although a mirrored scale will minimize parallax error, a residual error is necessarily present, though it may be very small.

In the first analysis, we find that various observers are not in complete agreement concerning whether the pointer image is exactly hidden by the pointer. Some observers have sharper eyesight than others. Some are more critical and more patient. Indeed, no two persons observe the same situation in exactly the same way where small details are concerned. On the second point, we recognize that a mirror cannot be mounted perfectly parallel with a scale; a mechanical tolerance—however small—is invariably present. Moreover, no mirror is perfectly flat, although it may closely approximate a perfect plane.

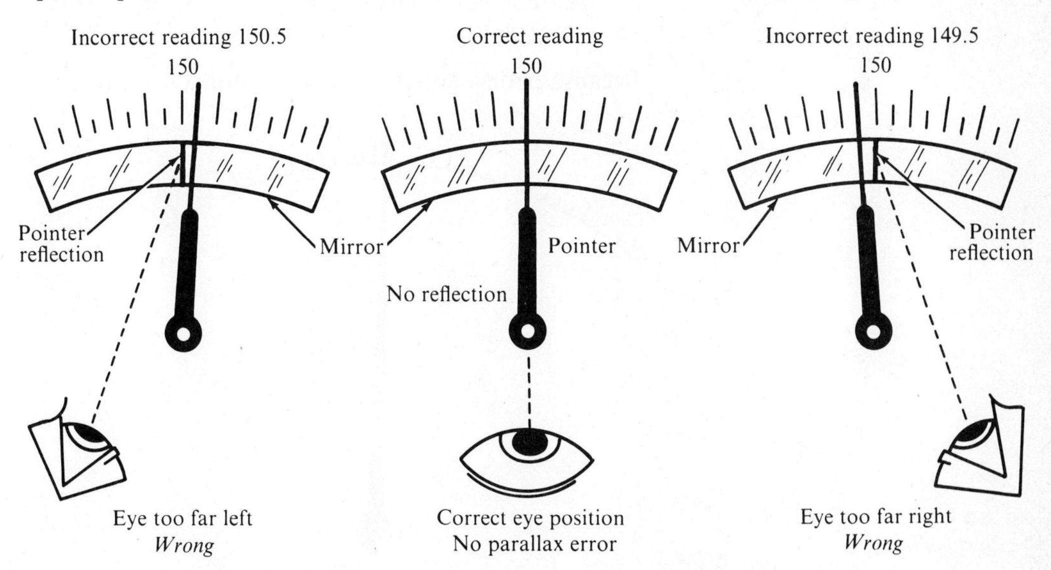

FIG. 1.2 *Use of antiparallax mirror. (Courtesy, B & K Manufacturing Co.)*

1.3 ERRORS OF ESTIMATION

All scale readings entail estimations of indicated values. For example, after parallax error has been minimized (Fig. 1.2), various observers will disagree concerning the exact position of the pointer. One observer may maintain that the pointer rests exactly over the 150 calibration mark. Another observer will protest that the pointer is very slightly to the left of the mark, and still another will state that the pointer is very slightly to the right of the mark. This difference of opinion is a simple example of estimation error.

Estimation error is more prominent when the pointer clearly rests between calibration marks on a scale, as depicted in Fig. 1.3. In this example, the observer must estimate the fraction of a scale division by which the pointer position exceeds the nearest calibration mark. Evidently, exact agreement among various observers cannot be expected. Errors of observation and errors of estimation are not arbitrary, but tend to follow a law of normal distribution. This law is visualized in Fig. 1.4, and is often called the "bell curve" because its shape resembles a bell. Technically, it is termed a gaussian curve and is formulated by

$$y = A\epsilon^{-bx^2}$$

where y = number of times a certain value is observed
x = value
ϵ = base of natural logarithms
A, b = constants

Because errors of observation and estimation tend to fall

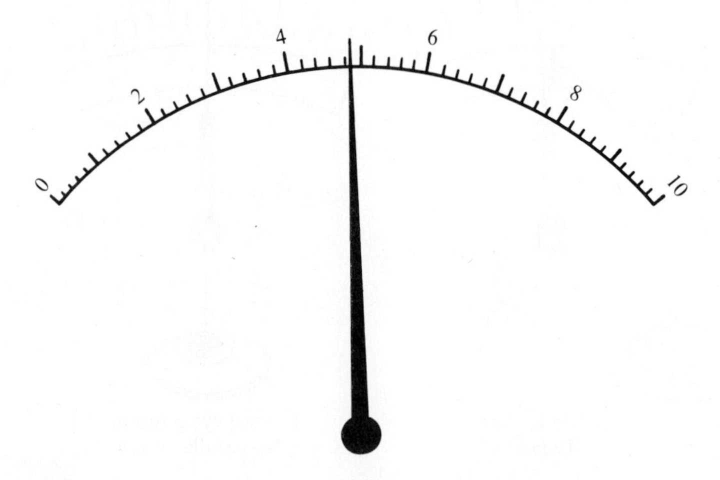

FIG. 1.3 *Scale indication subject to estimation error*

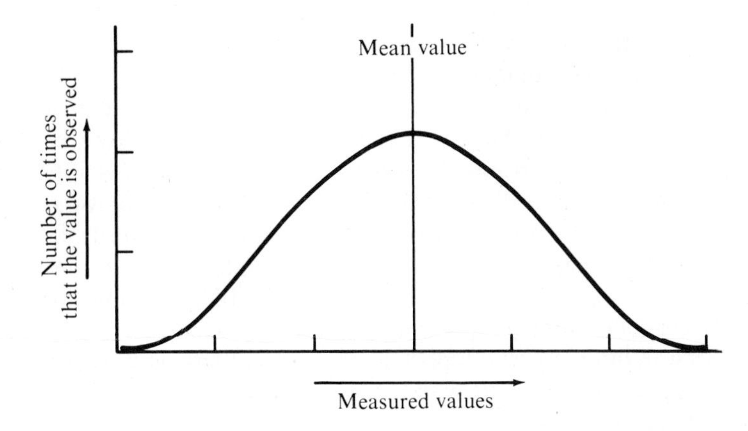

FIG. 1.4 *Normal distribution curve, or normal curve of error*

on the curve shown in Fig. 1.4, we are justified in taking the average of a group of observed values as the most probable value. Note that if a large number of observers read a scale indication as accurately as possible, the values they report will tend to cluster toward the mean value depicted in Fig. 1.4. Very few of the reported values will fall toward the ends of the normal curve of error.

1.4 SYSTEMATIC ERRORS

Every student eventually encounters systematic errors. Let us consider a practical example. Anyone who has used a vacuum-tube voltmeter (VTVM) knows that the zero setting tends to drift, and that a voltage measurement should not be made unless the zero setting of the pointer is first checked carefully. If, say, a dozen students in a class make successive measurements of a voltage value, the zero setting should be systematically checked before each measurement. It is obvious that if the zero setting drifts by 0.1 V between the time of the first measurement and the time of the last measurement, the experimental data will contain a systematic error in addition to the errors of observation and estimation.

The foregoing example of systematic error results in a skewed error curve; that is, a sufficient number of reported values will reveal that the curve is not symmetrical. For example, if the VTVM is gradually drifting toward higher scale indication, the early observers will report lower values on the average than are reported by the later observers. In turn, the curve of error will rise more rapidly than it falls and, in effect, will be skewed. This is a simple example of systematic error. Various other sources of systematic error may also be encountered.

Electrical instruments are compensated to minimize errors caused by changes in ambient temperature. However, temperature compensation is inevitably less than perfect, and any instrument will tend to indicate incorrect values when used at other than standard temperature. In other words, an instrument has maximum accuracy at standard temperature. If a room temperature changes substantially during the course of observing an indicated value, a systematic error will occur. Observers who report at low ambient temperature will disagree systematically with those who report at high ambient temperature.

1.5 ACCURACY RATINGS OF INSTRUMENTS

Although no instrument can be perfectly accurate, its accuracy can be stated within certain limits with respect to established standards. For example, the Weston cell shown in Fig. 1.1 is the basic standard of voltage. The National Bureau of Standards (NBS) maintains banks of Weston cells under carefully controlled conditions, and, in turn, establishes the volt unit as closely as possible in the prevailing state of the art. The departure in indication of a particular voltmeter with respect to the established volt unit is stated as the *accuracy* of the voltmeter.

Lab-type voltmeters have hand-drawn or custom scales, and therefore provide comparatively high accuracy. For example, the voltmeter illustrated in Fig. 1.5 has a rated accuracy of ± 0.1 percent of the *scale reading* at 25°C, or an accuracy of ± 0.25 percent at 15°C or 35°C. In other words, if the scale reading is 100 V, the actual voltage falls within the range from 99.9 to 100.1 V, at an ambient temperature of 25°C. At 15°C ambient temperature, the actual voltage in this example will fall within the range from 99.75 to 100.25 V. The same accuracy limits apply in this example to a measurement made at an ambient temperature of 35°C.

Most instruments have printed scales, and their indication accuracy is usually less than for instruments with hand-drawn scales. This wider tolerance on accuracy is due to manufacturing tolerances on the instrument mechanism, which can be fully corrected only by hand-drawn scales. A

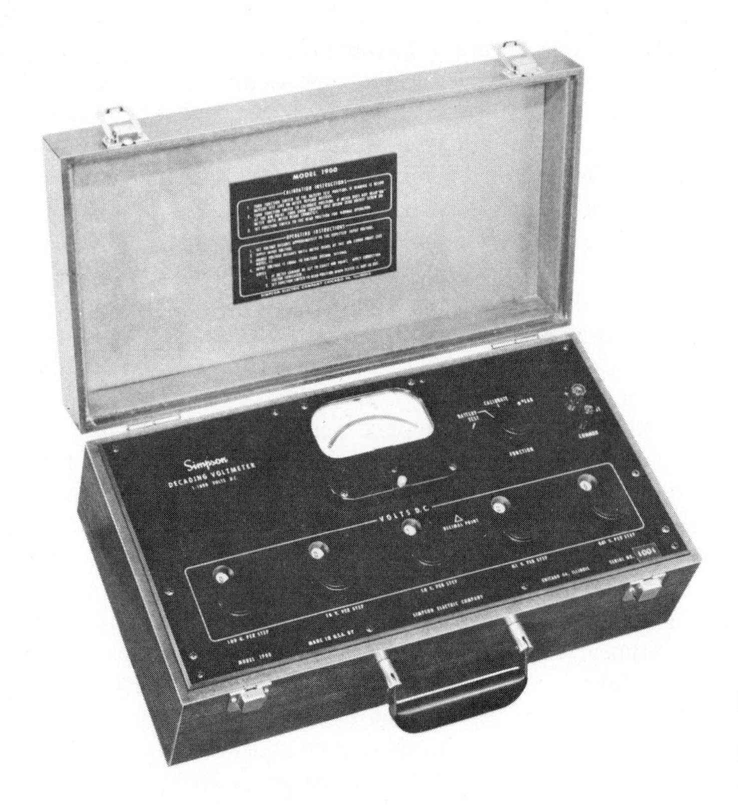

FIG. 1.5 *A voltmeter with a rated accuracy of ± 0.1 percent of scale reading.* (*Courtesy, Simpson Electric Co.*)

voltmeter with a printed scale is illustrated in Fig. 1.6. Its rated accuracy is ±0.5 percent of full scale at 25°C. For example, to measure a 100-V source, we choose the 150-V range. At 25°C, the rated accuracy will be ±0.75 V. Accordingly, a scale indication of 100 V denotes an actual voltage value that falls within the range from 99.25 to 100.75 V.

A utility-type voltmeter, such as that illustrated in Fig. 1.7, has an accuracy rating of ±3 percent of full scale at normal room temperature. For example, if we are measuring a source voltage of 150 V, we employ the 250-V range. In turn, the tolerance on the indicated value will be ±7.5 V. If the scale reading is 150 V, the actual voltage value will fall within the range from 142.5 to 157.5 V. Note that best indicating accuracy is obtained when the selected range is as low as possible. For example, if we measure 150 V on the 500-V range, the accuracy tolerance becomes ±15 V. In turn, the actual voltage value will fall in the range from 135 to 165 V.

FIG. 1.6 *A voltmeter with an accuracy rating of ±1/2 percent of full scale. (Courtesy, Simpson Electric Co.)*

FIG. 1.7 *A voltmeter with an accuracy rating of ±3 percent of full scale. (Courtesy, Simpson Electric Co.)*

The same accuracy considerations apply to ammeters and wattmeters. In principle, the same accuracy considerations apply also to ohmmeters. However, since an ohmmeter scale is nonlinear, as seen in Fig. 1.8, the tolerance on a resistance reading must be stated in terms of *arc of error*. In this example, the basic instrument has an accuracy rating of ±3 percent of full-scale value. Thus, on the 250-V range, the indication accuracy is ±7.5 V, corresponding to an arc of error of ±3°. Therefore, the tolerance on a resistance reading falls between the dotted lines in Fig. 1.8. If we read a value of 12 Ω on the resistance scale, the actual resistance value falls within the limits of 10.8 and 14.5 Ω.

Since the resistance scale depicted in Fig. 1.8 is progressively expanded toward the right-hand end, it is evident that best indication accuracy will be obtained when the selected resistance range provides indication above half scale. Resistance measurements below half scale are evidently subject to comparatively large error. The same principles apply to measurements made on the decibel scale, since this scale is nonlinear in the same general manner as the resistance scale. Note also that rated accuracy can be provided by an ohmmeter only when the internal batteries are fresh. Old batteries develop substantial source resistance that impairs the accuracy of resistance measurements.

1.6 OTHER ACCURACY FACTORS

Most electrical instruments are designed to operate properly in either a vertical or a horizontal position. Certain types of instruments, such as the student galvanometer illustrated in Fig. 1.9, are operable only in a vertical position and should be carefully leveled to obtain maximum accuracy. To anticipate subsequent discussion, this type of instrument employs a phosphor-bronze ribbon suspension, whereas instruments such as that illustrated in Fig. 1.7 utilize a movement that rotates in jeweled bearings, similar to the balance wheel in a watch.

Students of physics often conduct experiments—such as measuring the strength of the earth's magnetic field—with precision tangent galvanometers, shown in Fig. 1.10. This type of instrument is operable only in a horizontal position, and must be carefully leveled for maximum accuracy. The movement is basically a high-quality compass-needle assembly.

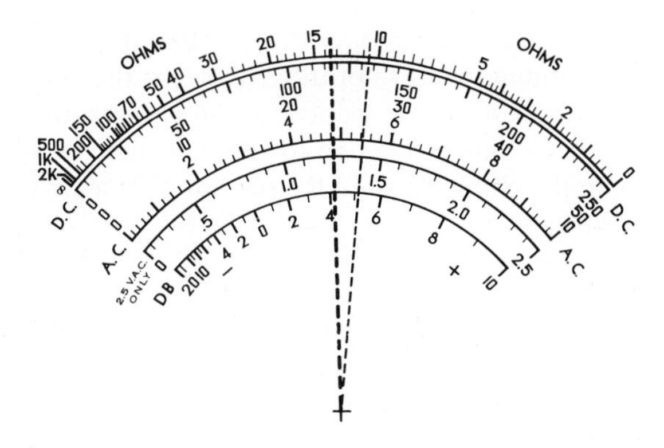

FIG. 1.8 *An example of arc of error.*

Many instruments are constructed with glass faceplates, and others are provided with plastic faceplates. We know that glass acquires an electric charge when rubbed by another substance, such as cloth; plastics may acquire stronger charges than glass. When dust is wiped from a faceplate, a charge may be produced that impairs the indication accuracy of the instrument; that is, the pointer is attracted toward the area of the faceplate that has the strongest charge. Antistatic solutions are available to minimize this source of inaccuracy; the solution is a partial conductor that assists in dissipating charges on faceplates produced by friction.

Instrument accuracy is often impaired by rough handling. Mechanical shock tends to demagnetize the permanent magnets contained in most instruments. As explained in greater detail subsequently, bearing jewels can be nicked or chipped by jarring, and delicately pointed pivots may be blunted, thus impairing the accuracy of the instrument. In school laboratories, the most common cause of impaired accuracy is accidental overload. When excessive current is passed through an instrument, its internal resistance is usually changed, even if burnout does not occur. Movements may be strained, pointers bent, and delicate springs may lose their temper as a result of overload.

The foregoing accuracy factors concern inherent instrument accuracy. These sources of error must not be confused with measurement inaccuracies caused by faulty application techniques. For example, if instrument voltage drops, instrument current drains, and circuit-loading factors are ignored, inaccurate measurements will result, even if we employ the instruments that have good inherent accuracy. Let us consider the sources of measurement error that may be significant in

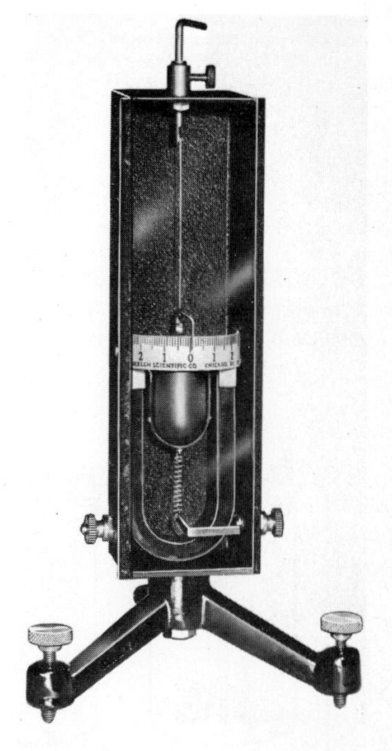

FIG. 1.9 *A student galvanometer. (Courtesy, Sargent-Welch Co.)*

FIG. 1.10 *A precision tangent galvanometer. (Courtesy, Sargent-Welch Co.)*

Fig. 1.11. It is not necessarily true that the voltmeter indicates the terminal voltage of the battery, or that the milliammeter indicates the current drain of the resistor, or that the ratio of the voltmeter reading to the milliammeter reading agrees with the ohmic value of the resistor. Our analysis proceeds as follows:

1. A typical milliammeter has an internal resistance of 30 Ω. If the resistor in Fig. 1.11 has a value of 30 Ω also, insertion of the milliammeter in series with the circuit obviously reduces the current flow to one-half its nominal value.
2. A typical voltmeter has an internal resistance of 5000 Ω. If the resistor in Fig. 1.11 has a value of 5000 Ω also, it is obvious that connection of the voltmeter across the resistor reduces the load to one-half its nominal value.
3. With both the voltmeter and the milliammeter connected into the circuit, the milliammeter indicates the sum of the currents drawn by the resistor and the voltmeter.
4. If the instruments are connected as shown in Fig. 1.12, the voltmeter indicates the terminal voltage of the battery; however, the current drain of the circuit is reduced as before by the internal resistance of the milliammeter.

The conclusions to be drawn from the foregoing examples are that maximum accuracy will be obtained if a current meter has a low value of internal resistance, compared with the load resistance. Maximum accuracy will be obtained if a voltmeter has a high value of internal resistance compared to the load resistance. Whenever these test conditions cannot be realized in practice, instrument readings must be corrected by calculations that take into account the conditions of the

FIG. 1.11 *Measurement (a) and schematic diagram (b) of voltage applied to a resistor and of current flow through a resistor.*

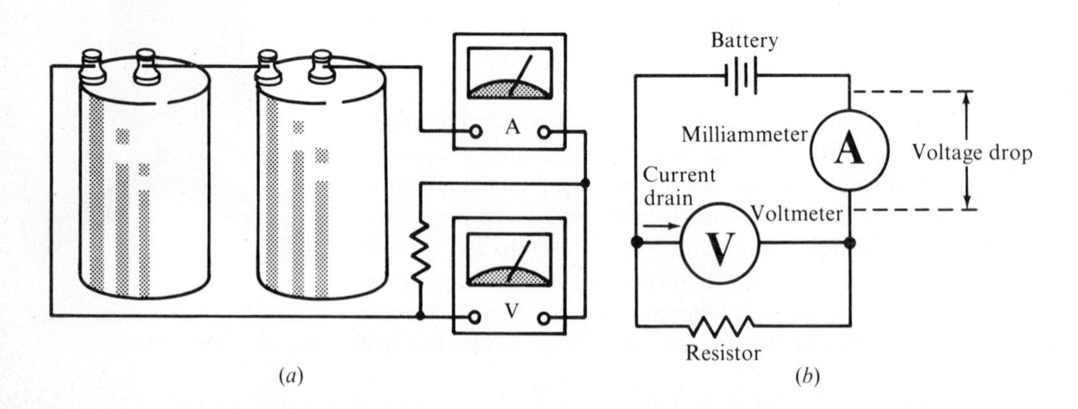

test circuit. It is left as an exercise for the student to calculate the true current and voltage values in Fig. 1.11*b* if (1) the resistor has a value of 5000 Ω and is shunted by a voltmeter that has an internal resistance of 5000 Ω and indicates 50 V, and (2) the current value indicated by a milliammeter with 30 Ω of internal resistance is 20 mA.

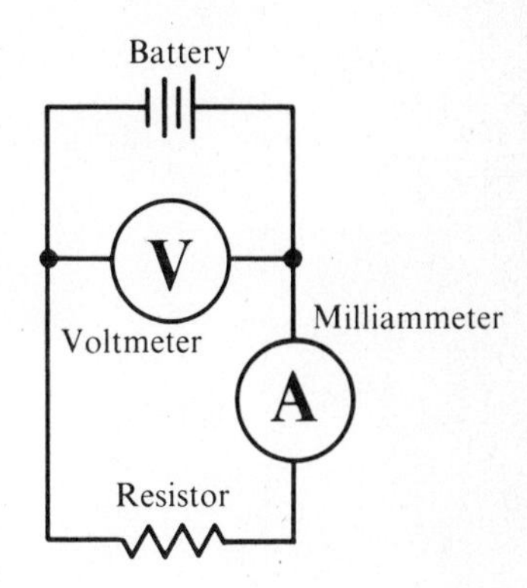

FIG. 1.12 *Alternate connection arrangement.*

1.7 APPLICATIONS

An example of the application of the theory of probability is shown in Fig. 1.13. It is sometimes possible to make a highly probable conclusion concerning the accuracy of a pair of voltmeters by making a comparison check. Note that three fundamental possibilities are involved:

1. The readings of the two voltmeters are exactly the same, within the range of observational error.
2. The readings of the two voltmeters agree within the accuracy rating of the instruments.
3. The readings of the two voltmeters are in disagreement by an amount that exceeds the accuracy ratings of the instruments.

In turn, our logical conclusions are as follows:

1. It is highly improbable that the indication errors of the two voltmeters selected at random would have exactly the same magnitude. Therefore, it is highly probable that both voltmeters are accurate.

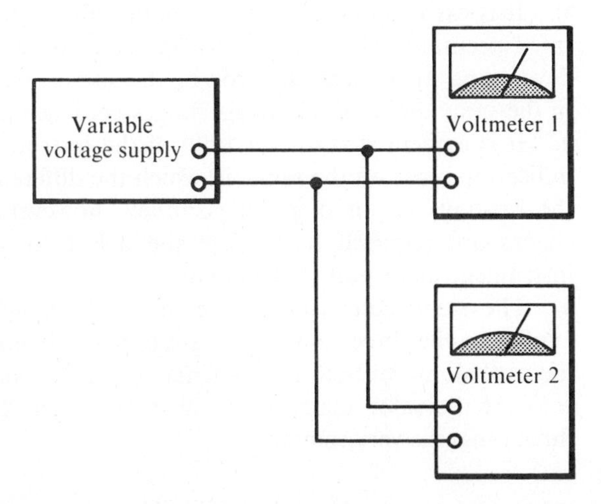

FIG. 1.13 *A comparison check of two voltmeters.*

2. This is the most common situation, for example, if each instrument has a rated accuracy of ± 2 percent of full scale. It is improbable that two voltmeters selected at random would be out of tolerance by the same amount and in the same direction. Therefore, it is probable that both instruments are within rated accuracy.
3. When this situation is encountered, we can conclude only that at least one of the voltmeters is out of tolerance. Supplementary checks are required to arrive at any useful conclusion.

Another example of the application of the theory of probability is provided by the comparison check depicted in Fig. 1.13 when successive voltage ranges are cross-checked. Several possibilities are as follows:

1. The readings of the two voltmeters agree within the accuracy ratings of the instruments on all ranges. (Six ranges will be stipulated in this example.)
2. The readings of the two voltmeters agree within the accuracy ratings of the instruments on all but one range.
3. The readings of the two voltmeters agree within the accuracy of the instruments on three ranges, and are out of tolerance on the other three ranges.
4. The readings of the two voltmeters agree within the accuracy ratings of the instruments on only one range.
5. The readings of the two voltmeters are out of tolerance on all ranges.

In turn, our logical conclusions are as follows:

1. It is extremely probable that both voltmeters are within rated accuracy. That is, the probability of both instruments developing an indication error of the same magnitude and in the same direction on all six ranges is exceedingly remote.
2. It is evident that one or both of the instruments has an indication error on the range in which the difference between the readings is out of rated accuracy; however, additional checks are required to localize the defect to a particular instrument or to both instruments.
3. The same conclusion made in the foregoing situation applies to the three ranges that are out of tolerance. That is, the probability of both instruments developing an indication error of the same magnitude and in the same direction on three ranges is very remote.

4. Since the probability of disagreement is very much greater than the probability of agreement, it is logical to conclude that both voltmeters are within rated accuracy on the range that is within tolerance.
5. The only logical conclusion in this situation is that at least one of the instruments is out of tolerance on all ranges.

QUESTIONS AND PROBLEMS

1. Explain the difference between the precision of a measurement and the accuracy of a measurement.
2. Explain *observational error* and give three possible causes.
3. Define *parallax error* as it applies to a meter and explain how it can be minimized.
4. Define the term *residual error* as it applies to a meter.
5. What is the purpose of the mirror on a meter scale?
6. Explain the term *estimation error*. Can it ever be eliminated from reading of electronic instruments?
7. What is the law of normal distribution?
8. Explain the term *systematic error* and give several reasons for its existence.
9. What are two precautions that may be taken to reduce the magnitude of the systematic error when using a VTVM?
10. Explain the term *accuracy rating of an instrument* as it applies to a voltmeter.
11. Why do laboratory-type voltmeters have hand-drawn scales?
12. On which part of a resistance scale is the most accuracy possible?
13. In what terms is the accuracy of an ohmmeter stated? Why?
14. What factors should be considered when choosing an operating position for an electrical measuring instrument?
15. What precautions should an operator take when cleaning the face of a meter?
16. Give three reasons why an electrical measuring instrument should never be dropped or jarred.
17. List at least five precautions that should be taken when using a voltmeter or ammeter.
18. List four precautions that should be taken when an ammeter or

voltmeter is used to make a measurement on an electrical circuit.

19. Seven students made the following readings with the most accurate voltmeter available in the laboratory; 4.45 V, 4.41 V, 4.51 V, 4.52 V, 4.5 V, 4.49 V, and 4.47 V. What is the most probable value of the voltage?

20. Five students made the following readings on a very accurate voltmeter: 3.15 V, 3.12 V, 2.97 V, 2.99 V, and 3.10 V. What is the most probable value of the voltage?

21. What are the precision and the deviation of each of the measurements in Prob. 19?

22. What are the precision and the deviation of each of the measurements in Prob. 20?

23. Suppose a voltmeter has an accuracy rating of ± 0.1 percent at 25°C. What is the range of the actual voltage reading (at 25°C) if the instrument reads 225 V?

24. A utility-type voltmeter with an accuracy of ± 3 percent of full scale is used on the 300-V scale to measure 230 V. What is the possible error?

25. A utility-type voltmeter with an accuracy of ± 2 percent of full scale is used on the 3-V scale to measure 1 V. What is the possible error?

26. Suppose the circuit shown in Fig. 1.11*b* has the following values: $R = 100\ \Omega$, $E = 1.5$ V, the resistance of the milliammeter $= 120\ \Omega$, and the resistance of the voltmeter $= 60{,}000\ \Omega$. What values of current and voltage will the instruments indicate if they are 100 percent accurate? How do the readings compare to calculated values?

27. Suppose the circuit in question 26 is changed as shown in Fig. 1.12. What are the new instrument readings and how do they compare to calculated values?

28. Suppose the circuit shown in Fig. 1.11*b* has the following values: $R = 1$ M, $E = 100$ V, the resistance of the microammeter $= 50\ \Omega$, and the resistance of the voltmeter $= 2$ M. What value of current and voltage will the instruments indicate if they are 100 percent accurate? How do these values compare to calculated values (no instruments in the circuit)?

29. Suppose the circuit in question 28 is changed as shown in Fig. 1.12. What are the new values of current and voltage? How do these values compare to the calculated values (no instruments in the circuit)?

DC VOLTMETERS, AMMETERS, AND OHMMETERS

2.1 INTRODUCTION TO METERS

A *voltmeter* is defined as an instrument for measuring voltage; an indicating scale is provided which may be calibrated in volts, millivolts, or kilovolts. A few voltmeters are calibrated in microvolts. Note that a millivolt is equal to 0.001 V, a kilovolt is equal to 1000 V, and a microvolt is equal to 10^{-6} V. An *ammeter* is defined as an instrument for measuring electric current. Its scale may be calibrated in amperes, milliamperes, or microamperes. A milliampere is 0.001 A, and a microampere is 10^{-6} A. An *ohmmeter* is defined as an instrument for measuring electrical resistance. Its scale may be calibrated in ohms, kilohms, megohms, or milliohms. Note that a kilohm is 1000 Ω, a megohm is 10^6 Ω, and a milliohm is 0.001 Ω.

When an instrument combines voltage, resistance, and current functions, it is called a *volt-ohm-milliammeter* (VOM). Various trade names for this type of instrument are multitester, multimaster, and volometer. A *volt-ohmmeter* measures voltage and resistance values, but does not measure current values. A *voltammeter* measures voltage and current values but does not measure resistance values. Direct current (dc) instruments measure dc values only. Alternating current (ac) instruments measure ac values only. Most VOMs measure dc and ac voltage, dc current, and dc resistance values. All ac instruments may be calibrated to indicate effective or root-mean-square (rms), peak, or peak-to-peak values of sine waveforms, as will be explained subsequently.

2.2 D'ARSONVAL METER MOVEMENT

Conventional VOMs utilize a meter mechanism or movement of the permanent-magnet moving-coil type, and are called D'Arsonval instruments. A permanent-magnet moving-coil assembly depends for its operation on the reaction between the current in a movable coil and the field of a fixed permanent magnet. This principle of operation is depicted in Fig. 2.1. The mechanical arrangement is termed an *external magnet*

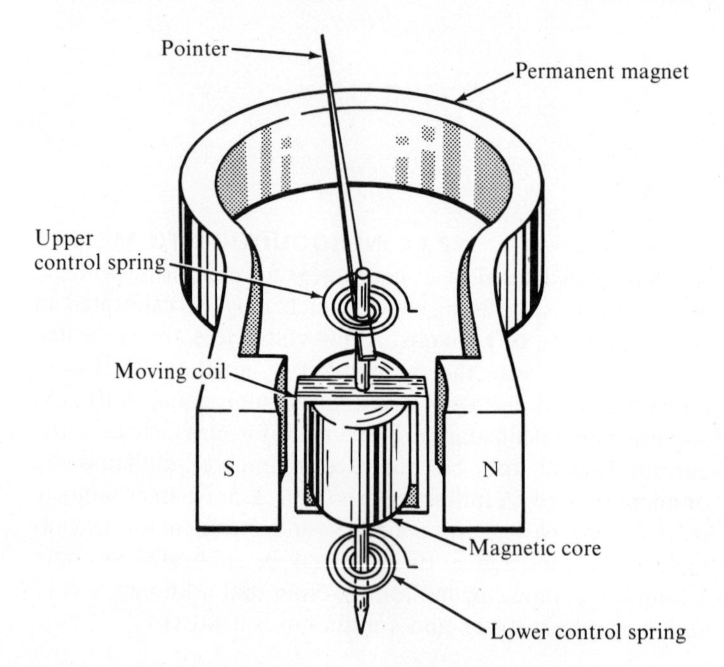

FIG. 2.1 *Magnet and coil of voltmeter.*

mechanism because the permanent magnet is located outside of the moving coil. The reaction of permanent-magnet flux with coil current establishes a torque (twisting force). This induced electromagnetic torque is balanced by a mechanical torque provided by control springs attached to the movable coil. In turn, the angular position of the movable coil is indicated by a pointer that is deflected along a calibrated scale. The external appearance of a VOM is illustrated in Fig. 2.2.

FIG. 2.2 *A volt-ohm-milliammeter. (Courtesy, Simpson Electric Co.)*

To calculate the torque developed by the moving coil in a D'Arsonval movement, we write

$$T = \frac{BAIN}{10} \tag{2.1}$$

where T = torque in dyne-centimeters
B = flux density in air gap measured in gauss
A = effective coil area in which $\text{cm}^2 = 2 \times$ radius of coil side from pivot center $\times$ length of active coil side
I = current in coil measured in amperes
N = number of turns on coil

The coil area in Formula (2.1) is usually from 0.5 to 2.5 cm^2, and the flux density may range from 1500 to 5000 G. A moving coil may be wound with a very small number of turns of large-diameter wire, or up to several thousand turns of wire that is smaller in diameter than a human hair. Coil resistances may range from a few ohms to a million ohms. One ampere-turn produces one line of magnetic flux. One gauss is equal to one line per square centimeter, or 5.452 lines per square inch. Note that the intensity of the earth's magnetic field is approximately 0.5 G.

External magnet design (Fig. 2.1) permits the use of a maximum-size magnet in a given space, and is employed when maximum flux in the air gap is desired. It is common practice to grind one side of an alnico V block of magnetic material and to solder or cement to the faces a pair of soft-iron pole pieces to form a U-shaped magnetic structure. This construction is depicted in Fig. 2.3*a*. Another common construction is called the *core magnet*. With the advent of alnico and similar magnetic materials, it became possible to design a magnetic system in which the magnet has the location of a core in an older-type movement. This construction is illus-

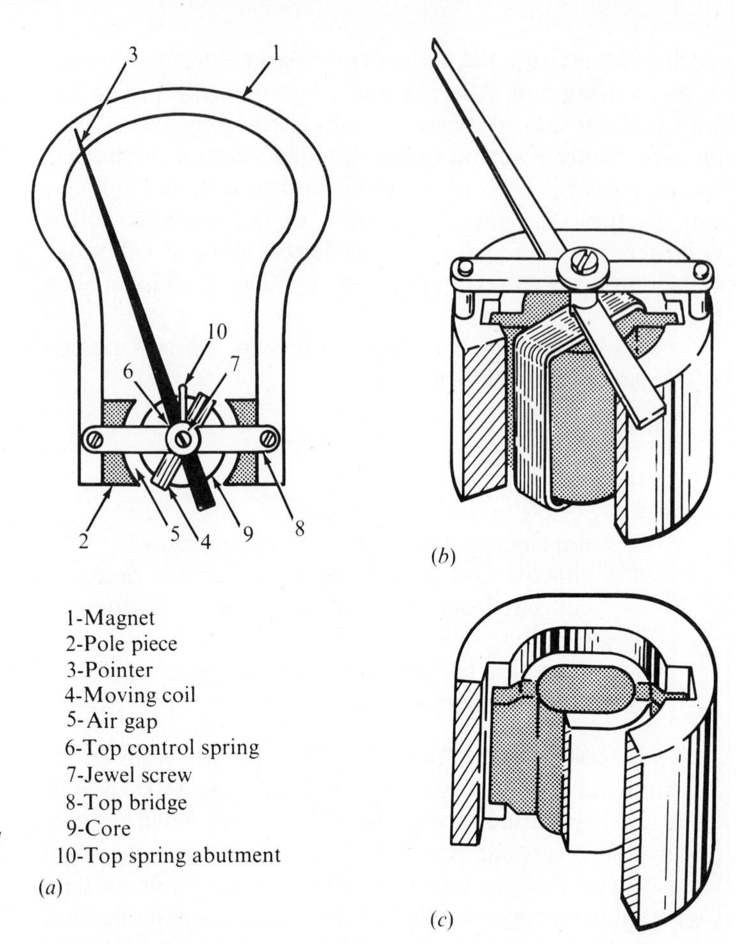

FIG. 2.3 *(a) External magnet and pole pieces; (b) magnetic-core mechanism; (c) magnetic core and surrounding soft-iron ring.*

trated in Fig. 2.3*b* and *c*. Such magnets operate with maximum energy and minimum length, and provide a compact mechanism. Another advantage of the magnetic-core mechanism is its self-shielding feature which makes the indication practically unaffected by external stray magnetic fields. Figure 2.4*a* shows the direction of a magnetic-core field, and an exploded view of a core-type movement is shown in Fig. 2.4*b*.

Practically all D'Arsonval mechanisms are provided with jeweled bearings. A V jewel bearing, as depicted in Fig. 2.5, is commonly used. The pivot may have a tip radius from 0.0005 to 0.003 in., depending on the weight of the coil assembly. Note that the radius of the jewel tip is somewhat greater, so that contact is made in a circle with a radius of less than 0.001 in. A moving element weighing 300 mg resting on a circular

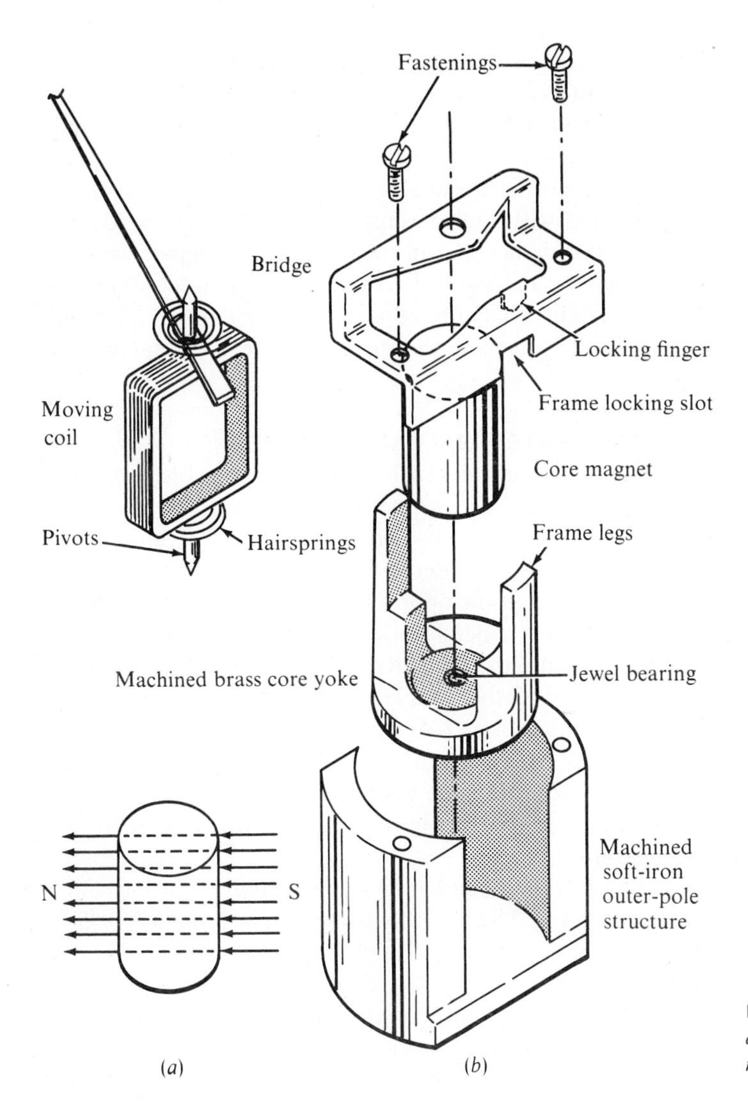

FIG. 2.4 *(a) Magnetic-core field direction; (b) exploded view of magnetic-core meter mechanism.*

area of 0.0002 in. diameter produces a bearing stress of 20,000 lb/in.^{2} or 10 tons/in.^{2} Delicate meter mechanisms must be protected from shock and vibration. The spring-back jewel bearing depicted in Fig. 2.6 is widely used in the better instruments to provide shock absorption. Two spiral phosphor-bronze springs are provided to maintain the reference position of the moving coil and exert a mechanical torque against electrical torque.

Instead of spiral phosphor-bronze springs, many modern

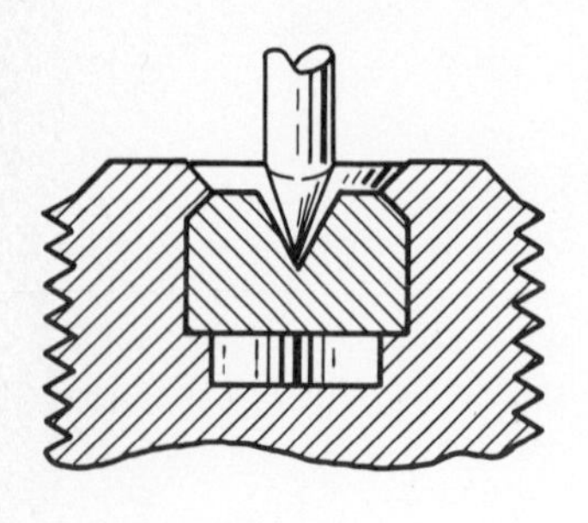

FIG. 2.5 *A V jewel bearing.*

meter mechanisms employ a taut-band design. As seen in Fig. 2.7, this suspension-type mechanism dispenses with jewels, thereby minimizing friction. The ribbons are kept under tension by a tension spring, so that the mechanism can be used in any position. Taut-band suspension instruments can be fabricated with higher sensitivity than those using pivots and jewels. Any mechanism must include a balance weight to counteract the mechanical torque of the pointer. When the balance weight is exactly adjusted, a meter mechanism indicates accurately in either the horizontal or the vertical position.

2.3 CALIBRATED GALVANOMETERS

A simple galvanometer, such as the lecture-table instrument illustrated in Fig. 2.8, is provided with a *zero-center scale.* When direct current is passed through the galvanometer in one direction, the pointer deflects to the left; when direct current is passed in the opposite direction, the pointer deflects to the right. The scale is not calibrated, in the strict sense of the term, because the scale divisions do not indicate standard current units directly. Simple galvanometers are used chiefly to indicate the presence or absence of current, and the zero (null) point is generally sought in the adjustment of associated equipment. To anticipate subsequent discussion, the simple galvanometer is commonly used as an indicating instrument in a resistance bridge.

Any galvanometer has a basic sensitivity. The instrument illustrated in Fig. 2.8 has a sensitivity of approximately 0.35 mA per scale division. If the pointer deflects to the 5 mark on the scale (in either direction), the current through the galvanometer has a value of approximately 1.75 mA. Accordingly, the galvanometer can be used as a current meter by calculating the current value that corresponds to a given pointer deflection. With an internal resistance of 5 Ω, this instrument has a voltage sensitivity of approximately 1.75 mV per scale division. If the pointer deflects to the 5 mark on the scale (in either direction), the voltage across the galvanometer has a value of approximately 8.75 mV. In turn, the galvanometer can be used as a voltmeter by calculating the voltage value that corresponds to a given pointer deflection.

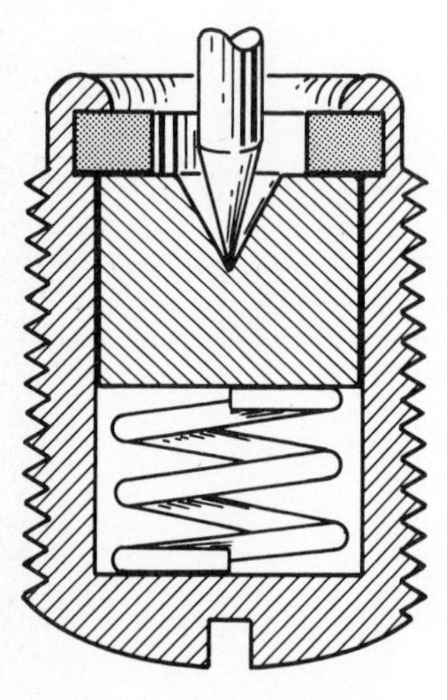

FIG. 2.6 *A springback V jewel bearing.* (*Courtesy, Weston Instruments, Inc.*)

If the galvanometer is connected to an ac source, the pointer does not deflect. Although a torque is developed by the moving coil, this torque reverses alternately as the applied

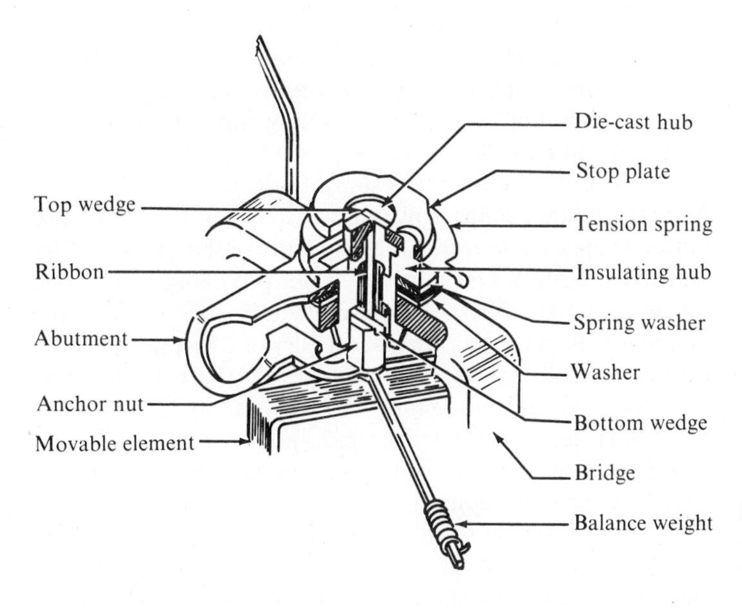

FIG. 2.7 *Suspension-type mechanism. (Courtesy, Weston Instruments, Inc.)*

voltage changes from positive to negative. Since the inertia of the movement is comparatively large, no perceptible vibration of the pointer occurs unless the frequency of the applied ac voltage is quite low. Beginning students should note that a galvanometer may be burned out when connected to an ac source although no pointer deflection occurs. In other words, a galvanometer is suitable for use only in dc circuits.

When a galvanometer is to be used for measurement of current or voltage values, it is advantageous to employ an instrument that has a scale calibrated in standard electrical units. For example, the galvanometer illustrated in Fig. 2.9 is calibrated for a sensitivity of 1 mV per scale division. If the pointer deflects to the 20 mark on the scale (in either direction), the applied voltage has a value of 20 mV. This instrument has an internal resistance of 100 Ω; hence, each division on the scale represents 10 μA. If the pointer deflects to the 20 mark on the scale (in either direction), the current through the instrument has a value of 200 μA.

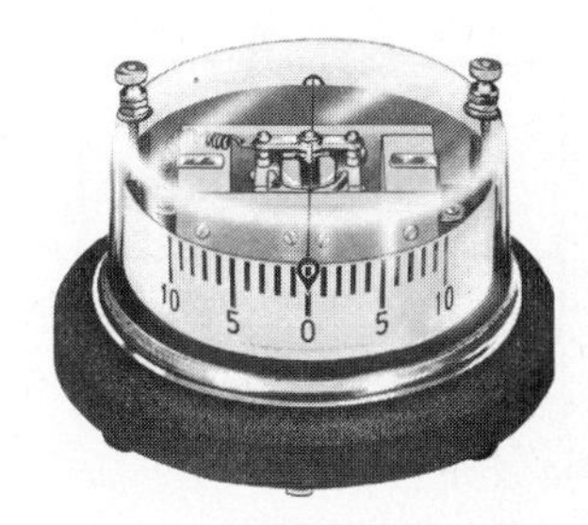

FIG. 2.8 *A typical lecture-table galvanometer. (Courtesy, Sargent-Welch Co.)*

2.4 AMMETERS, MILLIAMMETERS, MICROAMMETERS, AND SHUNTS

If a calibrated galvanometer is used with a *shunt*, the instrument can be used to measure comparatively high current values.

FIG. 2.9 *A calibrated galvanometer. (Courtesy, Sargent-Welch Co.)*

Figure 2.10 shows how a shunt is connected across a galvanometer. A calibrated galvanometer with a matching shunt is illustrated in Fig. 2.11. When used without a shunt, this instrument has a voltage sensitivity of 0.2 V per scale division. When used as a voltmeter, it has a full-scale range of ± 5 V. A second voltage range is also provided (when the right-hand button is pressed on top of the instrument); the galvanometer sensitivity is then 4 mV per division, providing a full-scale range of ± 100 mV. If the left-hand button is pressed, the instrument then has a current sensitivity of 20 μA per division, providing a full-scale current range of ± 500 μA. In this case, the upper scale indication must be multiplied by 5.

Next, let us see how the galvanometer operates as an ammeter when a suitable shunt is connected across the instrument terminals, as depicted in Fig. 2.10. The shunt is designed to provide a voltage drop of 100 mV when the current flow is 1 A. When the left-hand button is depressed (or locked in place by a half turn), the galvanometer indicates 40 mA per scale division, providing a full-scale current range of ± 1 A. In this case, the upper scale indication must be divided by 100.

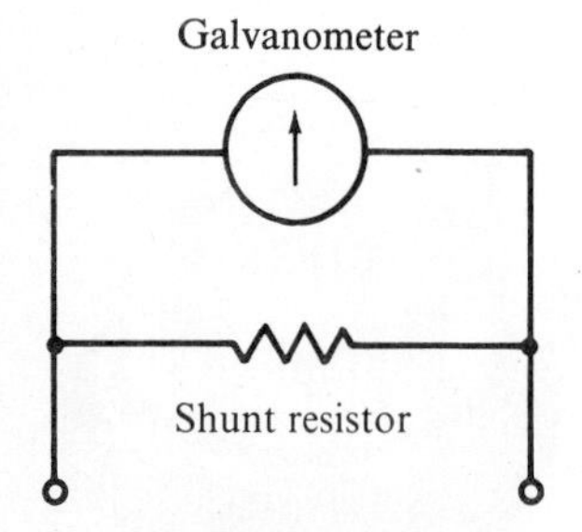

FIG. 2.10 *Connection of a shunt to a galvanometer.*

Other shunts provided with the instrument described in this example provide full-scale current ranges of 0.05, 0.5, 5, 10, and 25 A. High-current shunts must be designed to maintain a fixed resistance value when heated. When power is dissipated by a shunt, it is converted into heat. Therefore, shunts are constructed from alloys, such as manganin, that have a low temperature coefficient of resistance. It is also advantageous to provide considerable surface area to facilitate heat radiation. A shunt that develops a 100-mV drop at a current value of 25 A dissipates 2.5 W of heat.

FIG. 2.11 *(a) Calibrated galvanometer and matching shunt; (b) galvanometer scale. (Courtesy, Sargent-Welch Co.)*

When a shunt must dissipate large amounts of heat, it is designed with several rectangular resistance elements connected in parallel, as illustrated in Fig. 2.12. This construction exposes a comparatively large surface area to the surrounding air, and thereby increases the heat-radiation ability of the shunt. Note in Fig. 2.12 that the shunt elements are embedded in heavy blocks of metal. These massive conducting blocks ensure that the applied current will divide equally among the shunt elements. If the current division were not uniform, one of the shunt elements would be required to dissipate more power than the other elements.

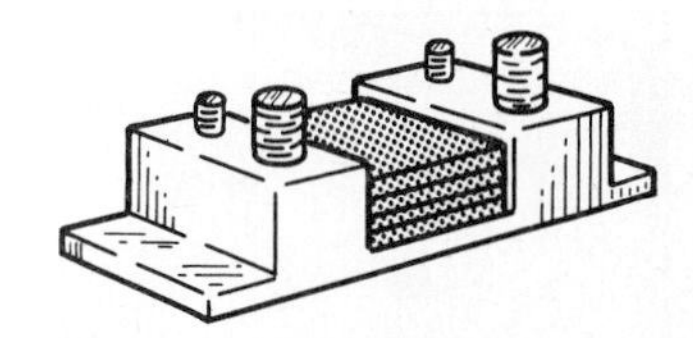

FIG. 2.12 *A 200-A shunt. (Courtesy, Simpson Electric Co.)*

Another basic consideration in shunt design is that of uniform current distribution in the metal blocks. In Fig. 2.12 two sets of binding posts are provided for the shunt: the outer pair of large binding posts connect to the circuit under test; the inner pair of small binding posts connect to the meter. Current enters and leaves the shunt at the large binding posts, and the meter indicates the voltage drop between the small binding posts. Note that the voltage drop between the large binding posts is slightly greater than the voltage drop between the small binding posts, indicating that although heavy conducting blocks are used, a block nevertheless has a small resistance value. Let us see how this bulk resistance affects current distribution.

Current enters the shunt in Fig. 2.12 at the upper surface of the block. As the incoming electrons spread out in the metal, the upper surface of the block has a higher current density than the lower surface. This current density rapidly equalizes as electrons flow toward the shunt resistance elements, the current dividing with practical equality among the several shunt elements. The rapidity with which the current density equalizes depends on the current value. When operated at maximum current capacity, the current density equalizes more slowly than when operated at a low current value. Therefore, maximum indication accuracy requires that the meter be connected near the shunt elements and not at the points of current entry and exit.

The problem of obtaining uniform current density at still higher current values requires larger conducting blocks and larger binding posts for connection to the external circuit. A shunt for current values up to 7000 A is illustrated in Fig. 2.13. Note the massive construction, and the placement of the binding posts. The meter posts are placed near the shunt elements and are also mounted at the center plane on the

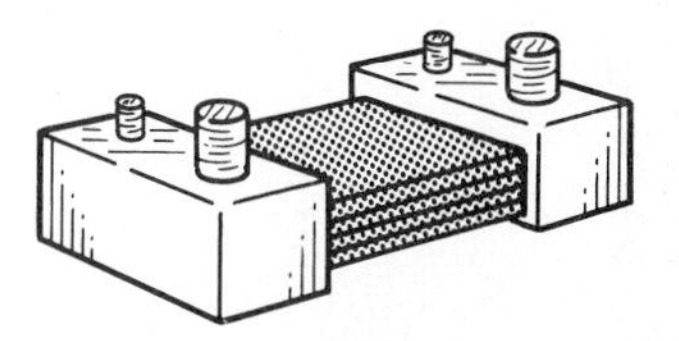

FIG. 2.13 *A 7000-A shunt. (Courtesy, Simpson Electric Co.)*

FIG. 2.14 *A panel-type ammeter. (Courtesy, Westinghouse Electric Co.)*

sides of the blocks. Placement of these posts centrally between the four shunt elements minimizes indication error due to residual nonuniformity of current distribution.

Large shunts are commonly used with *panel meters*, such as illustrated in Fig. 2.14. A panel meter differs from a D'Arsonval galvanometer chiefly because of its scale and the resting position of its pointer. A galvanometer scale is commonly marked with numbers that may be associated with either current or voltage units (Fig. 2.11). A panel meter scale is marked in numbers that correspond to a single electrical unit, such as amperes (Fig. 2.14). Nearly all galvanometers provide center-scale zero indication, whereas nearly all panel meters provide zero-left indication. Laboratory-type galvanometers are designed to have extremely high sensitivity and may indicate 0.0001 μA per division. Few panel meters have a sensitivity greater than 15 μA full scale, and may indicate only 0.3 μA per division.

2.5 VOLTMETER MULTIPLIERS

Next, let us return to the calibrated galvanometer illustrated in Fig. 2.11. It was noted that when neither of the top buttons is depressed, the full-scale voltage indication is ± 5 V. However, when the right-hand button is depressed, the full-scale voltage indication is ± 100 mV. These two voltage ranges are provided by a *multiplier resistor*, as depicted in Fig. 2.15. When the right-hand button is depressed, the multiplier resistor is short-circuited. Let us see how a multiplier operates.

We will stipulate that the meter movement in Fig. 2.15 provides full-scale deflection at 100 mV, or at 500 μA. It follows from Ohm's law that the internal resistance of the movement is 200 Ω. If we connect a 9800-Ω multiplier resistor in series with this movement, the total voltmeter resistance becomes 10,000 Ω, and the full-scale range becomes 5 V. This fact follows from Ohm's law:

$$I = \frac{E}{R} = \frac{5}{9800 + 200} = 500\ \mu\text{A} \tag{2.2}$$

It is left as an exercise for the reader to calculate the value of multiplier resistance required to provide a full-scale indication of 1000 V with the meter movement stipulated in the foregoing example. Multipliers are commonly used with

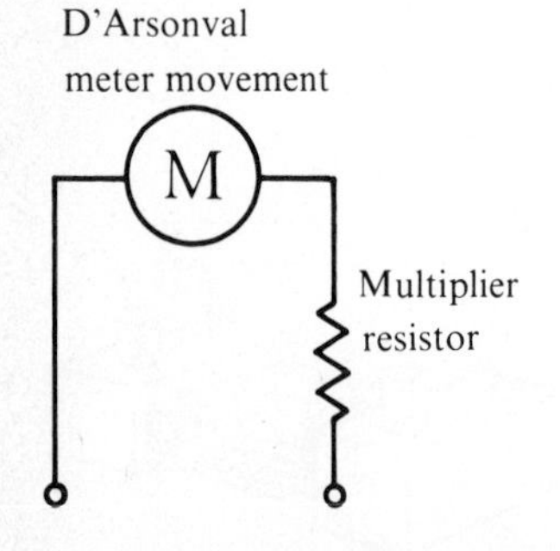

FIG. 2.15 *Basic voltmeter circuit.*

panel meters, as illustrated in Fig. 2.16. They are designed with precision resistors having a typical tolerance of ± 1 percent. When a multiplier must dissipate appreciable power, it is designed with a comparatively large surface area and is mounted in a ventilated case. Tolerance considerations are explained in greater detail in the next chapter.

Temperature compensation of a measuring instrument involves several factors. An increase in temperature tends to reduce the force of the springs in the movement, and tends to reduce the number of flux lines that pass through the moving coil. These are conflicting factors, but their net effect is to produce an *increase* in scale indication as the temperature increases when a D'Arsonval movement is used as a current indicator. Practical temperature compensation can be obtained in a voltmeter by winding the multiplier with resistance wire that has a negligible temperature coefficient, such as manganin. Moving coils are usually wound with copper wire, which has a positive temperature coefficient of resistance. Let us see how suitable proportions of manganin and copper wire can minimize the indication error of a voltmeter due to temperature changes.

A typical D'Arsonval movement has an effective temperature coefficient of approximately 0.02 percent/°C. This is the indication error on a constant-current basis. On the other hand, since copper has a temperature coefficient of about 0.4 percent/°C, the indication error on a constant-voltage basis would *decrease* by $0.4 - 0.02 = 0.38$ percent/°C in this example. This negative indication error of the movement in a voltmeter application can be compensated by using an amount of manganin wire in the multiplier to make the net temperature coefficient in the system equal to 0.02 percent/°C. Thus, the decrease in current with temperature is exactly compensated by the increase in scale indication on a constant-current basis.

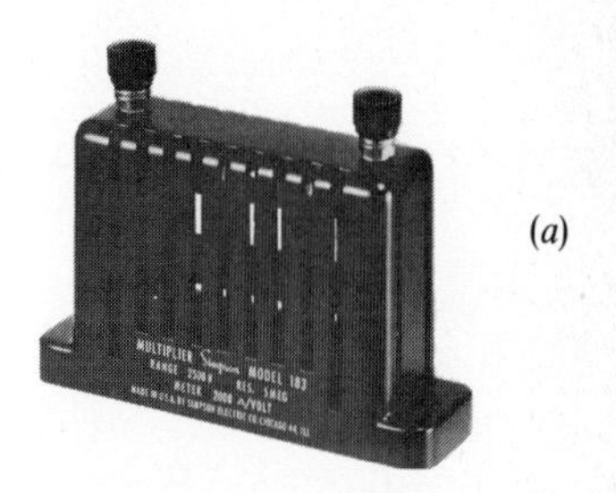

(*a*)

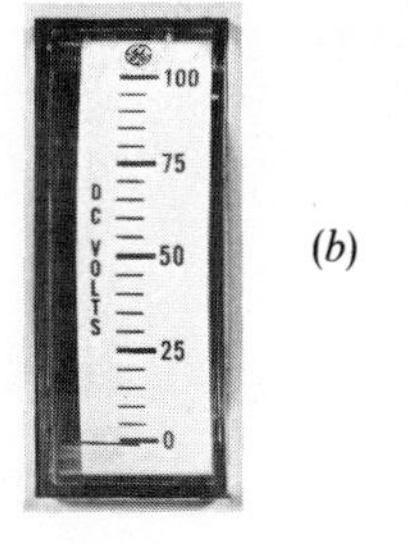

(*b*)

FIG. 2.16 (*a*) *An external multiplier;* (*b*) *a panel-type voltmeter.* (*Courtesy, Simpson Electric Co.*)

2.6 THE BASIC OHMMETER

We noted previously that an ohmmeter is an instrument for measuring resistance values. Resistance is defined as a voltage/current ratio, in accordance with Ohm's law:

$$R = \frac{E}{I} \tag{2.3}$$

An ohmmeter must be provided with a voltage source

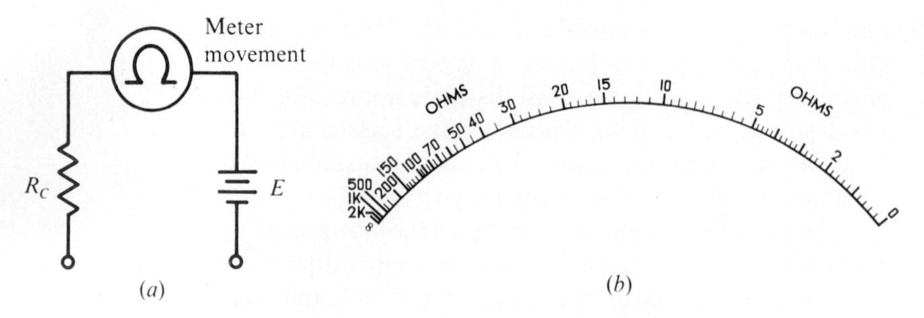

FIG. 2.17 (*a*) *Basic ohmmeter configuration*; (*b*) *an ohmmeter scale*.

to provide a corresponding current in the resistance under test. The ohmmeter is calibrated in resistance values; its basic configuration is depicted in Fig. 2.17. We observe that an ohmmeter scale is highly nonlinear. It is instructive to calculate the component values required in a simple ohmmeter. With reference to Fig. 2.17, R_C is a calibrating resistor that operates in combination with E to provide full-scale deflection when the test terminals or leads are connected together.

It is evident in Fig. 2.17 that half-scale indication will occur if we connect a resistance with a value equal to the internal resistance of the ohmmeter between the test terminals. To calculate the current in the circuit when an unknown value of resistance is connected between the test terminals, we write

$$I = \frac{E}{R_C + R_X + R_M} \tag{2.4}$$

where R_X = value of unknown resistance
R_M = internal resistance of meter movement

It follows from Formula (2.4) that an ohmmeter scale will be nonlinear, because the scale is calibrated in terms of R_X and the denominator has a value equal to the sum of a constant and R_X. Students who have taken a course in analytic geometry will recognize that an equation of this form does not plot as a straight line. In the example of Fig. 2.17, the internal resistance of the ohmmeter is 8 Ω. We will stipulate that the cell E has a voltage of 1.6 V, and that the meter movement has an internal resistance of 5 Ω and a full-scale current of 200 mA. In turn, the value of R_C is evidently 3 Ω:

$$0.2 = \frac{1.6}{5 + 3} \tag{2.5}$$

When a resistance of 8 Ω is connected between the ohmmeter terminals, the current value is 100 mA. The pointer deflects to half scale, and the scale indication is 8 Ω. It will be found that other measured resistance values produce scale indications as depicted in Fig. 2.17. The useful range of this ohmmeter configuration is from 0 to 1000 Ω. In practice, refinements are included to compensate for variation in the battery voltage, and to provide several ranges of resistance indication. These features are explained in the next chapter.

2.7 AGING OF PERMANENT MAGNETS

To stabilize the characteristics of a permanent magnet used in a D'Arsonval movement, some method of aging is customarily employed. For many years, magnets were artificially aged by various processes of heating and/or vibration. For example, after initial magnetization to saturation, a magnet might be boiled for a specified length of time. More recently, an aging

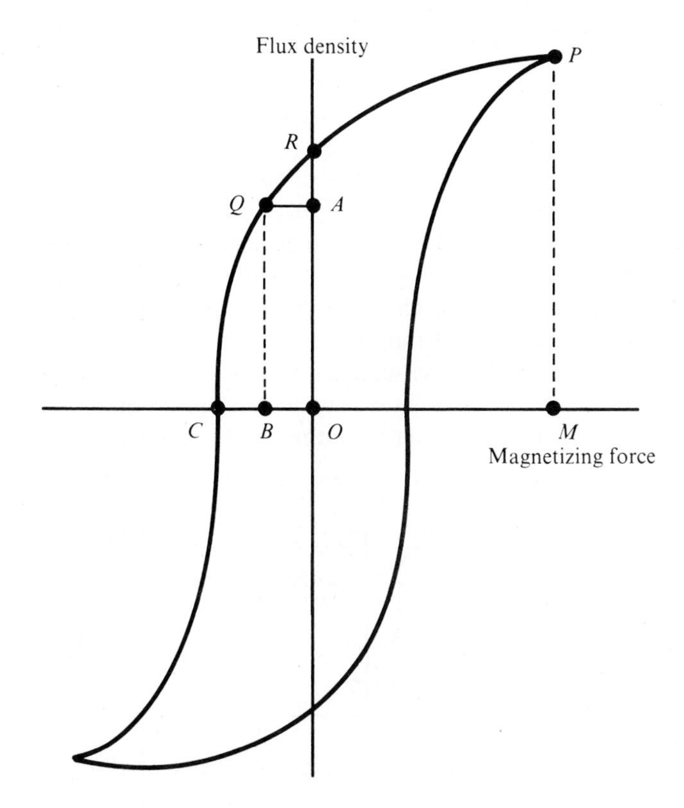

FIG. 2.18 *Hysteresis loop diagram of the Sangamo method of aging a permanent magnet.*

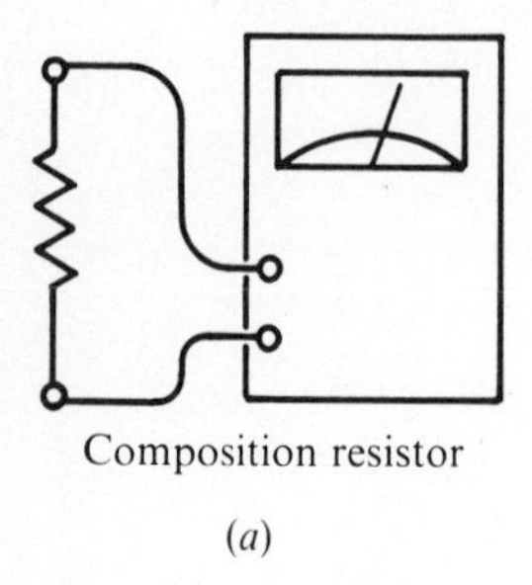

(a)

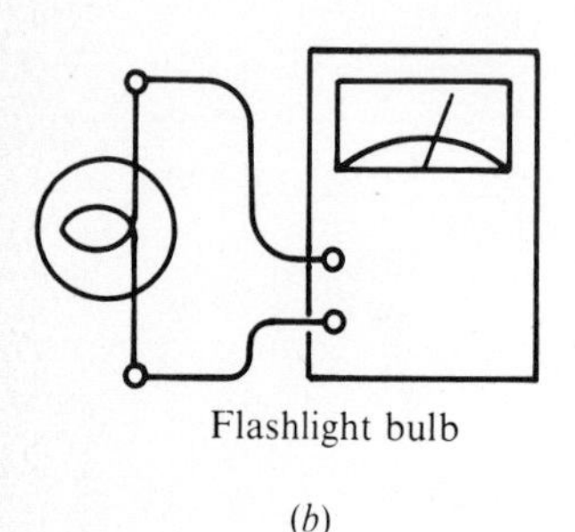

(b)

FIG. 2.19 *Ohmmeter comparison. (a) Readings agree on successive ranges; (b) readings do not agree on successive ranges.*

process was developed which provided controlled demagnetization by alternating flux from an initial saturation value to a certain lower value bearing a definite relation to the saturation value. There has been a definite trend toward the use of square-wave flux fields, instead of sinusoidal flux fields for this purpose. The general principle is illustrated in Fig. 2.18. If *OM* represents the initial magnetizing force, and *MP* is the resultant magnetization for the particular magnet, then *OR* may be considered as the magnetization after the removal of the magnetizing force. The aging process consists in the application of a demagnetizing force *OB*, which reduces the magnetization to *OA*. At this point on the hysteresis loop, the magnet will resist external or stray fields, the maximum value of which does not exceed *OB*. The value *OB* is chosen so that it is considerably greater than any stray field encountered in service. After this aging process, a permanent magnet becomes quite stable, and will not weaken appreciably during extended service.

2.8 APPLICATIONS

If an ohmmeter is within rated accuracy, its readings on successive ranges agree when a fixed resistor such as a composition resistor is under test. For example, with reference to Fig. 2.19*a*, the resistance value of a 10-k composition resistor may be measured on the $R \times 100$ range and on the $R \times 1000$ range of an ohmmeter. The two readings will agree within the rated accuracy of the ohmmeter unless the instrument is defective. On the other hand, if the filament resistance of a small flashlight bulb is measured on successive ranges of an ohmmeter, the two readings will be in substantial disagreement, even if the ohmmeter is within rated accuracy. The

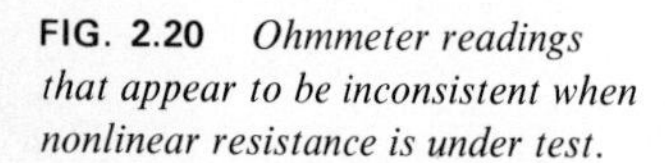

FIG. 2.20 *Ohmmeter readings that appear to be inconsistent when nonlinear resistance is under test.*

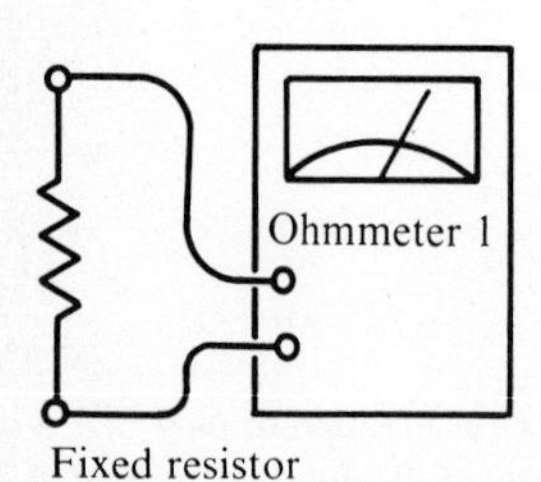

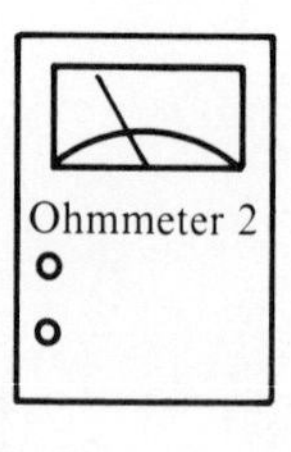

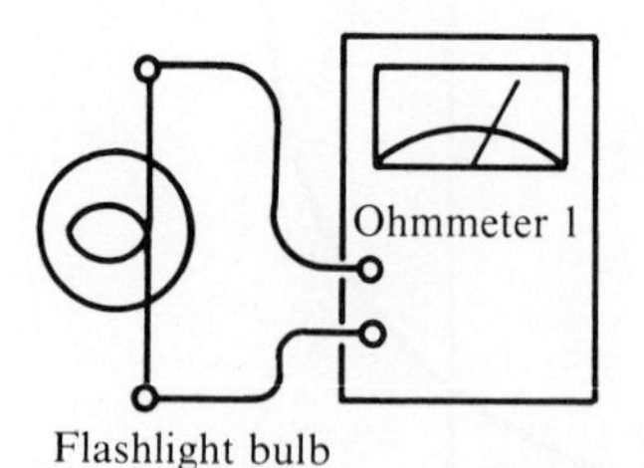

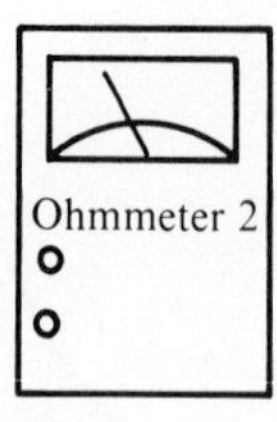

reason for this discrepancy is that the lamp filament is a nonlinear resistance, and its effective resistance changes when the applied voltage of the test is changed. This characteristic of the ohmmeter is illustrated in Fig. 2.20.

The foregoing characteristic of a nonlinear resistance also applies to checks of different types of ohmmeters. For example, even though the ohmmeters are operated on the same resistance range, the internal battery voltages may not be the same. That is, one brand of ohmmeter might employ a 1.5-V battery, and another brand might utilize a 3-V battery. If both instruments are in good condition a cross-check of the ohmmeter readings will agree (within the rated accuracy) when a fixed resistor is under test. When a nonlinear resistance is under test, the cross-check readings will not agree. This is true when the front-to-back ratio of a semiconductor diode or other semiconductor junction devices are under test, as semiconductor materials have a nonlinear E/I characteristic.

Although it is impractical to obtain complete data when checking nonlinear resistances with an ohmmeter, we are often concerned with questions such as the presence or lack of continuity or general orders of magnitude. For example, if we check a filament with an ohmmeter, we are usually concerned only with the possibility that the filament might be burned out. In other words, we are merely making a continuity check with the ohmmeter. If we are checking a semiconductor diode, we are usually concerned only with the question of whether a substantial front-to-back resistance ratio is present. This test is made as shown in Fig. 2.21. Since the polarity of the ohmmeter leads is known, the test also serves to identify the cathode and the anode of the diode. If a small transistor is in good condition, the resistance relations in Fig. 2.22 will be observed. Note that the ohmmeter test serves to identify a transistor as a PNP or NPN type.

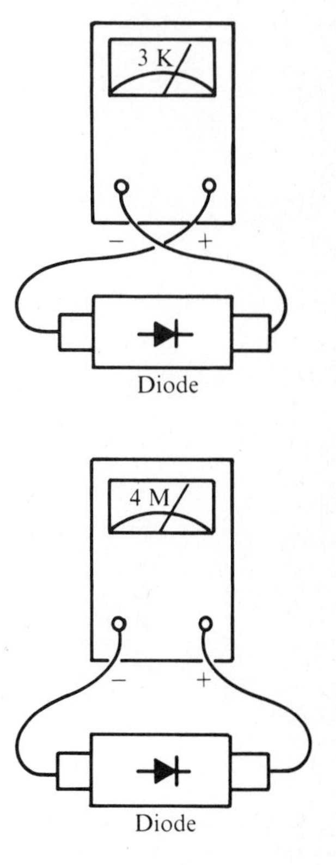

FIG. 2.21 *Checking a semiconductor diode for front-to-back ratio.*

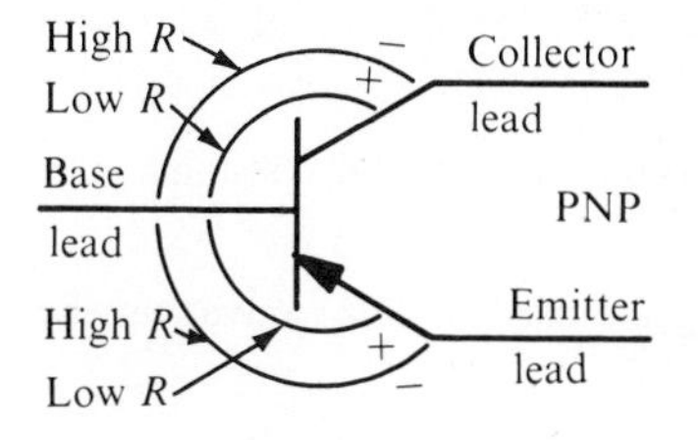

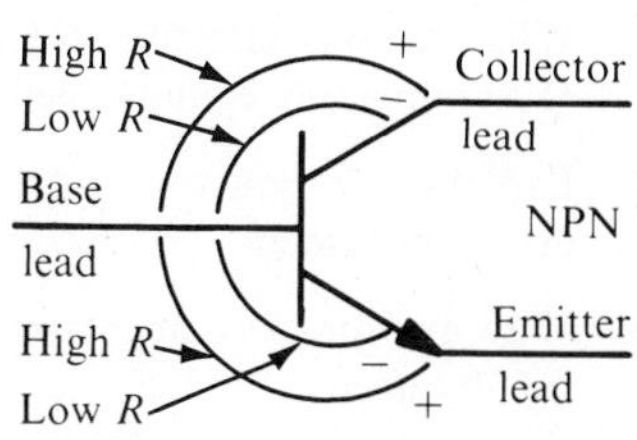

FIG. 2.22 *Comparative ohmmeter readings for PNP and NPN transistors.*

QUESTIONS AND PROBLEMS

1. What quantities can a VOM generally be used to measure?
2. What is the advantage of using an external magnet design in a D'Arsonval meter?
3. Why are meter mechanisms very delicate?
4. What is the purpose of the two spiral phosphor-bronze springs used with the moving coil of a D'Arsonval meter movement?
5. What is the purpose of the balanced weight on a meter pointer as shown in Fig. 2.7?
6. What is the purpose of the spring in the meter bearing shown in Fig. 2.6?
7. Why isn't the scale of the galvanometer, shown in Fig. 2.8, calibrated?
8. Explain how the galvanometer shown in Fig. 2.8 can be calibrated to read current or voltage.
9. Why doesn't the pointer of a galvanometer respond to an ac voltage?
10. How can the range of a galvanometer be extended to measure a large value of current?
11. Why should a galvanometer shunt be constructed of a material with a low temperature coefficient of resistance?
12. Why is the meter shunt shown in Fig. 2.12 constructed of parallel strips?
13. Why does the meter shunt shown in Fig. 2.12 have two sets of binding posts?
14. What is the basic difference in the appearance of a panel-type meter and a galvanometer?
15. What are the basic electrical differences between a panel-type meter and a galvanometer?
16. What is the purpose of a multiplier resistor?
17. What is the effective temperature coefficient of a typical D'Arsonval meter movement?
18. Explain how you would determine the internal resistance of a meter movement.
19. Why are the permanent magnets used in D'Arsonval meters usually aged?

20. Calculate the torque that is developed by the moving coil in a D'Arsonval meter movement when: $B = 4000$ G, $A = 2$ cm^2, $I = 50$ A, and $N = 5000$ turns.

21. Calculate the torque that is developed by the moving coil in a D'Arsonval meter movement when: $B = 5000$ G, $A = 2.5$ cm^2, $I = 1$ mA, and $N = 1000$ turns.

22. Calculate the necessary value of a multiplier resistor that would allow the galvanometer shown in Fig. 2.15 to read 500 V full scale.

23. Calculate the value of the shunt that would allow the galvanometer shown in Fig. 2.15 to read 10 A full scale.

AC VOLTMETERS, AMMETERS, AND WATTMETERS

3.1 CLASSES OF METERS

Various classes of ac voltmeters, ammeters, and wattmeters are used in the laboratory and in industry. Many ac voltmeters are classified as *sine-wave* instruments; that is, this type of meter indicates correctly only when a sine waveform is applied, as shown in Fig. 3.1*a*. Other ac voltmeters are classified as *true rms* instruments, and will indicate correctly whether the waveform is sinusoidal or complex. Figure 3.1*b* depicts a complex waveform called a *square wave*. Note that a square wave can be synthesized from a fundamental sine wave and its odd harmonics.

All ac ammeters and wattmeters are basically classified in this same manner. We will find that a majority of the ac

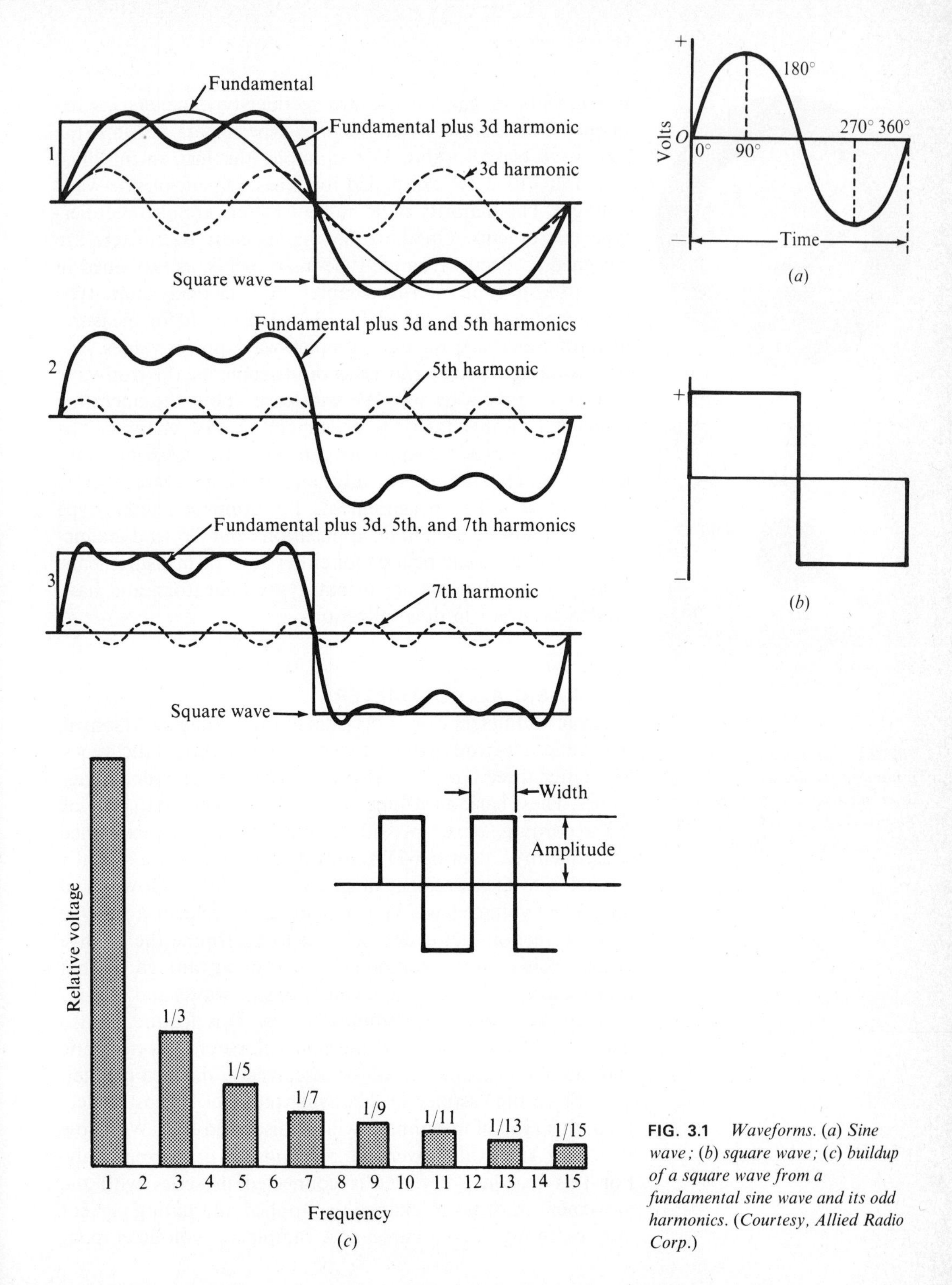

FIG. 3.1 *Waveforms. (a) Sine wave; (b) square wave; (c) buildup of a square wave from a fundamental sine wave and its odd harmonics. (Courtesy, Allied Radio Corp.)*

(*a*)

(*b*)

FIG. 3.2 (*a*) *A rectifier-type instrument for sine-wave application.* (*Courtesy, Simpson Electric Co.*); (*b*) *an electrodynamic-type instrument suitable for complex-wave application.* (*Courtesy, Westinghouse Electric Co.*)

instruments in general use are rectifier-type meters, as explained in the next topic. Rectifier-type meters fall into the sine-wave classification. For example, the instrument illustrated in Fig. 3.2*a* is designed for measurement of sine-wave voltages. The majority of ac current meters are also rectifier-type instruments. On the other hand, most wattmeters are designed as electrodynamic-type instruments, as explained in the latter portion of this chapter. An electrodynamic-type wattmeter, as illustrated in Fig. 3.2*b*, is suitable for measurement of either sine-wave or complex-wave power values.

Among the true rms class of instruments, the iron-vane type is in fairly wide use. We will learn about the operating principles of this type of instrument in this chapter. The foregoing types of instruments can be further classified into those that respond to ac only, and those that are equally useful for dc and ac measurements. For example, rectifier-type meters cannot be used in dc applications, but electrodynamic-type instruments can be used for either ac or dc measurements. Iron-vane instruments are primarily ac indicators, and have limited accuracy in dc applications.

3.2 BASIC AC VOLTMETERS

Basic ac voltmeters can be classified into rectifier-D'Arsonval, iron-vane, electrodynamic, thermal, and electrostatic types. The latter three types can also be used to indicate dc voltage values. These basic ac voltmeters may be elaborated to provide *XY* recording, or to provide digital readout, as explained subsequently. Rectifier-D'Arsonval voltmeters are usually restricted to measurement of sine-wave voltages. However, if an applied voltage has a known complex waveform, a suitable scale-correction factor can be used to determine the voltage value. Iron-vane ac voltmeters have an advantage in that they indicate correct voltage values of sine waves and reasonably correct values of complex waves. This feature is also true of electrodynamic and thermal voltmeters. Electrostatic and thermal instruments will be discussed in the next chapter.

Since the rectifier-D'Arsonval voltmeter is most widely used, this type of instrument will be discussed first. We know that a D'Arsonval movement responds to dc current only. For this reason a rectifier is connected in series with the movement in order to change the applied alternating current into pulsating direct current. A simple ac voltmeter configuration is shown in Fig. 3.3*a*. A semiconductor rectifier is used in this example (some dc voltmeters employ vacuum-tube rectifiers). A rectifier has a low forward resistance and a high reverse resistance. The circuit current flows in half cycles when a sine waveform of voltage is applied, as depicted in Fig. 3.3*b*. This is a pulsating dc current waveform. Pulsating dc is defined as a varying voltage (or current) that has one polarity only—that is, a pulsating dc waveform does not cross the 0-V axis. A meter movement responds to pulsating dc because the current is always flowing in the same direction, although its amplitude varies periodically.

Next, let us consider the response of a moving coil to half-sine waves of current. The force on the coil is changing periodically, but the pointer does not vibrate unless the applied frequency is quite low. That is, the inertia of the moving-coil assembly damps out the variations in force, and the coil responds on the basis of the *average value* of the half-sine waves. This average value is depicted in Fig. 3.3*c*. Note that area A_1 is equal to area A_2; it can be shown that the average level in the waveform is equal to 0.318 of the peak value. Therefore, when a sine-wave voltage is applied to the

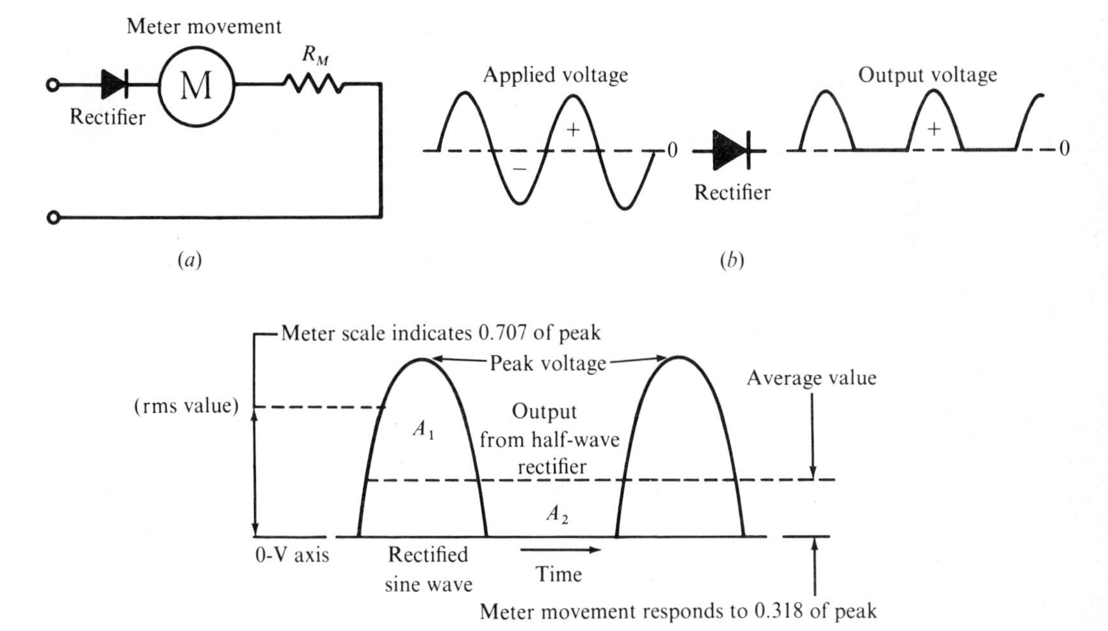

FIG. 3.3 (*a*) *Basic ac voltmeter configuration;* (*b*) *principle of rectification;* (*c*) *voltage relations in a half-sine wave form.*

ac meter circuit, the moving coil responds to the average value of the rectified waveform, and this average value is 0.318 of peak.

Although the scale of an ac voltmeter can be calibrated in average values, this is done only for special applications. For general applications, the scale is calibrated in terms of rms values. An rms value of an ac voltage is also called its *effective voltage.* This is an important unit of ac voltage measurement, because it is equivalent to dc units insofar as power values are concerned. That is, a lamp will glow just as brightly when energized by 117 rms V ac as when it is energized by 117 V dc. Thus, an ac voltmeter of the type depicted in Fig. 3.3 responds to the average value of half-sine waves, but its scale is ordinarily calibrated in rms values. It can be shown that the rms value of a sine wave is equal to 0.707 of peak, as shown in Fig. 3.4.

Note that the ratio of rms voltage to average voltage is 2.22. This means that if we apply dc voltage to the ac voltmeter configuration shown in Fig. 3.3*a*, the pointer will indicate 2.22 times the value of applied dc voltage.

$$\frac{\text{rms value}}{\text{Average value}} = \frac{0.707}{0.318} = 2.22 \tag{3.1}$$

If we apply 1.5 V dc to this type of ac voltmeter, the pointer will indicate 3.3 V. Of course, the dc voltage must be polarized in the forward-resistance direction of the rectifier; otherwise, the pointer will indicate zero. Note carefully that the foregoing relation between rms volts and average volts is true only for sine waves. In other words, an ac voltmeter such as depicted in Fig. 3.3*a* will not indicate correctly unless the input voltage has a sine waveform.

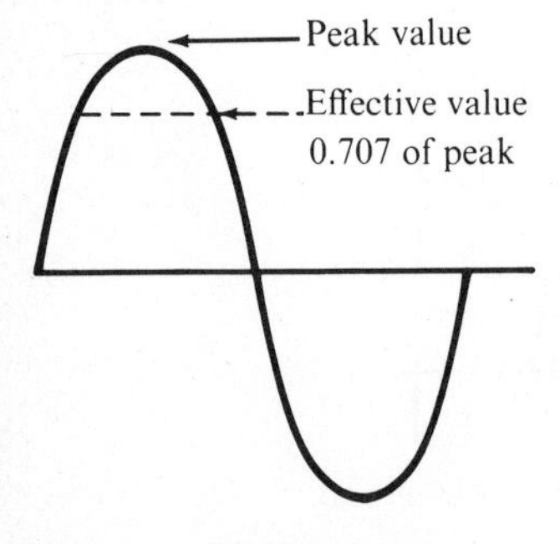

FIG. 3.4 *Effective value of a sine wave.*

Resistor R_M in Fig. 3.3*a* is a multiplier resistor. Its value is chosen to provide the desired full-scale voltage indication. The total internal resistance of the instrument is equal to the sum of R_M, the resistance of the meter movement, and the resistance of the rectifier. Although the sum of the first two resistance values is constant, the rectifier resistance depends upon the rectifier current, as exemplified in Fig. 3.5. Copper-oxide rectifiers are in wide use as instrument rectifiers; therefore, we will consider this type of rectifier in our discussion of basic ac voltmeters.

The forward resistance of a rectifier is equal to its E/I ratio at a given point on its characteristic. Let us calculate

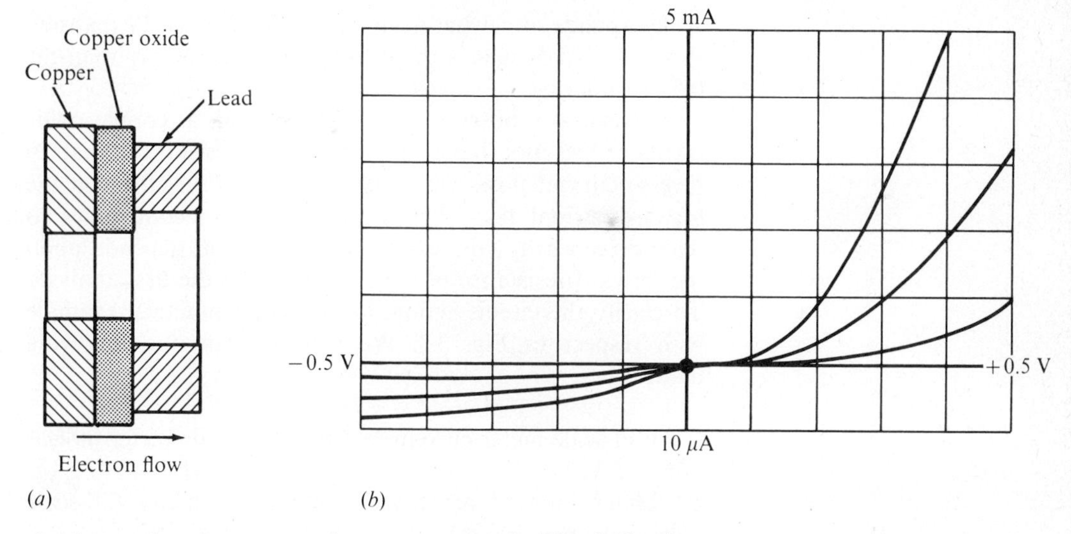

FIG. 3.5 (*a*) *Construction of a copper-oxide rectifier*; (*b*) *voltage-current characteristics of three copper-oxide instrument rectifiers.* (*Courtesy, Weston Instruments, Inc.*)

the forward resistance on the upper curve in Fig. 3.5*b* when 0.4 V is applied. The corresponding current value is 5 mA. The resistance at this point is calculated by applying Ohm's law:

$$R = \frac{0.4}{0.005} = 80\ \Omega \tag{3.2}$$

Next, let us calculate the forward resistance when 0.2 V is applied. The corresponding current value is approximately 1 mA. In turn, we write

$$R \approx \frac{0.2}{0.001} = 200\ \Omega \tag{3.3}$$

Note that the forward resistance increases as we proceed down the rectifier characteristic. At 0.1 V, the rectifier resistance is very high. Therefore, we draw two important conclusions: (1) The scale of the voltmeter will be nonlinear, and will be particularly cramped in the vicinity of the zero point, (2) it is impossible to obtain useful scale indications at applied voltages less than 0.1 V. A typical scale is depicted in Fig. 3.6. We observe that the 0.1-V mark is very close to zero. The

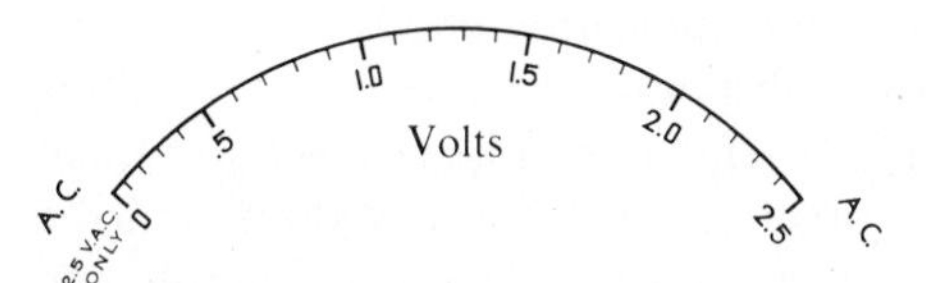

FIG. 3.6 *Nonlinear ac voltmeter scale.*

scale expands somewhat from 0 to 0.5 V; increased expansion from 0.5 to 1 V makes the 0 to 0.5 interval shorter than the 0.5 to 1 interval.

Next, we observe in Fig. 3.5*b* that a copper-oxide rectifier does not have infinite reverse resistance; that is, reverse current flows which has a cancellation effect on the forward-current flow. This cancellation also contributes to scale nonlinearity; the extent of cancellation depends upon the forward resistance of the instrument, in the first analysis. To clarify the circuit action, let us take a practical example with respect to Fig. 3.3. We will stipulate the following parameters:

1. Full-scale meter current = 1 mA; full-scale meter indication = 5 V.
2. Meter internal resistance = 100 Ω, or meter full-scale voltage drop = 100 mV.
3. Rectifier resistance = 200 Ω at 1 mA, or full-scale rectifier voltage drop = 0.2 V.
4. Multiplier resistance = 4700 Ω in accordance with Ohm's law.

The voltage drop across the rectifier is 0.2 V when 5 V are applied in the forward direction in Fig. 3.3. We will find that the voltage drop across the rectifier is *much greater* when 5 V are applied in the *reverse direction.* In other words, the rectifier resistance is much higher in the reverse direction, and *most of the applied voltage drops across the rectifier.* For example, the rectifier reverse resistance is about 10,000 Ω in the range of 3- to 4-V reverse potential. It follows from Ohm's law that the reverse voltage drop across the rectifier is 3.4 V in this example.

To summarize briefly, scale nonlinearity in the basic ac voltmeter configuration results both from the nonlinearity of the rectifier's forward resistance and from the nonlinearity of its reverse resistance (Fig. 3.7). Next, we will find that there is a problem of rectifier breakdown when the value of R_M is increased in Fig. 3.3. That is, if we increase the value of R_M in order to obtain a full-scale indication of 20 V, it is evident that the rectifier must withstand a reverse voltage of this order of magnitude. However, the rectifier *cannot withstand this reverse voltage* in practice—it will break down and be destroyed. Note that the forward voltage drop across the

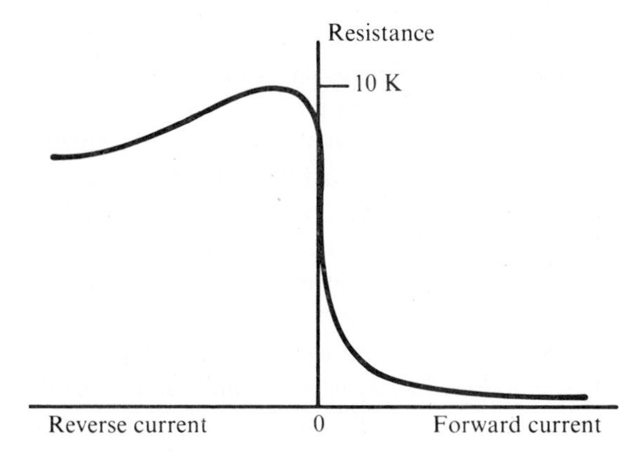

FIG. 3.7 *Comparative forward and reverse resistance curves for a copper-oxide rectifier.*

rectifier is the same, because the full-scale current demand of the meter remains unchanged.

The basic ac voltmeter circuit depicted in Fig. 3.3 must be elaborated to minimize the reverse voltage drop across the rectifier whenever we choose a value of R_M that is sufficient to cause the rectifier to operate out of its reverse-voltage rating. In practice, this is done as depicted in Fig. 3.8. Diode D_2 conducts on the reverse half cycle of applied voltage, and shunts a low value of forward resistance around D_1 and the meter. Neither D_1 nor D_2 is called upon to withstand appreciable reverse voltage. We will find that additional design considerations are imposed by production tolerances on copper-oxide instrument rectifiers. However, these minor problems are reserved for discussion in the next chapter.

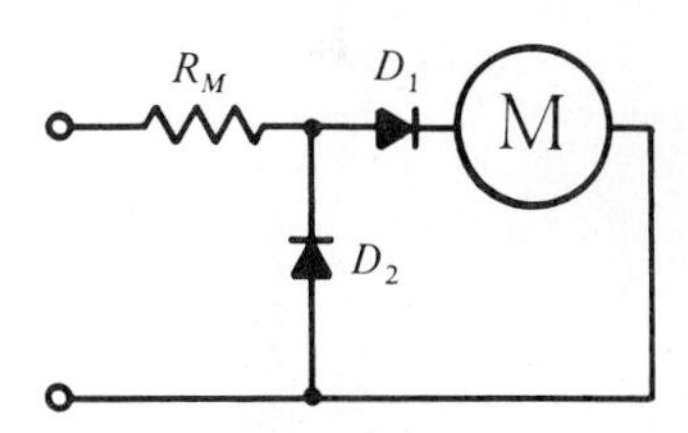

FIG. 3.8 *A protective diode, D_2.*

3.3 IRON-VANE AC VOLTMETERS

Iron-vane mechanisms are widely used in industry, frequently in ac panel meters. The forerunner of the iron-vane mechanism is the tangent galvanometer, with which most students make an acquaintance in the physics laboratory. A diagrammatic drawing of a tangent galvanometer is seen in Fig. 3.9. The response of the instrument is formulated:

$$I = \frac{10\,rH}{2\pi N} \tan\theta \qquad \text{amperes} \tag{3.4}$$

where r = radius of coil
H = horizontal component of earth's magnetic field in gauss
N = number of turns

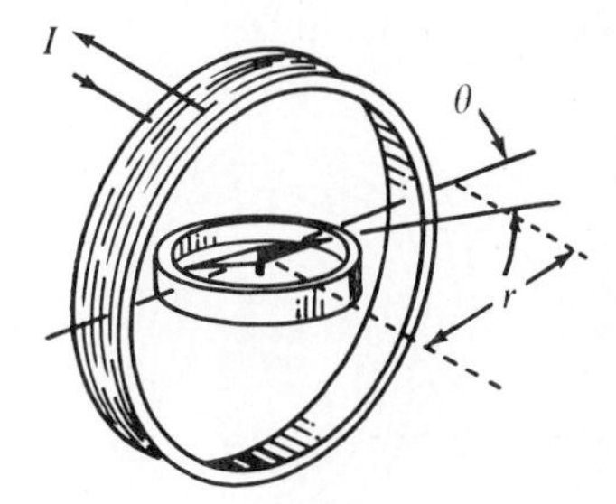

FIG. 3.9 *The tangent galvanometer. (Courtesy, Weston Instruments, Inc.)*

Although all the early instruments of this type employed permanent magnets of one form or another, disadvantages were encountered in applications other than measurement of the earth's magnetic field strength. When used as an ammeter (or voltmeter in combination with a multiplier resistor), the instrument depicted in Fig. 3.9 is very susceptible to error from stray magnetic fields. To overcome this disadvantage, Kelvin developed the astatic galvanometer depicted in Fig. 3.10. This suspension-type instrument with opposed magnetic needles is comparatively unresponsive to stray magnetic fields, and is very sensitive. It was used to indicate the feeble currents transmitted by the early transatlantic cables, for example.

The mirror in this type of galvanometer increases the effective sensitivity of the instrument, particularly when the scale is placed at a distance from the mirror. In a lab-type galvanometer, the scale is often mounted 1 m from the mirror.

Unfortunately, any type of polarized iron-vane instrument that has high accuracy is costly and comparatively

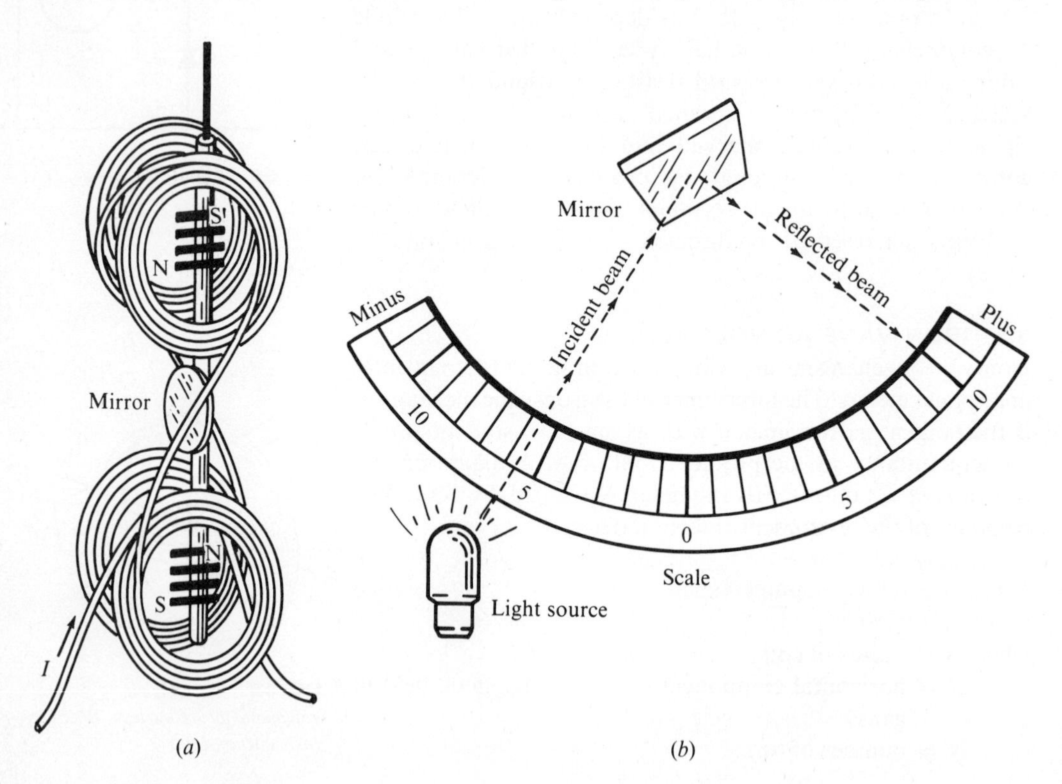

FIG. 3.10 *(a) The Kelvin astatic galvanometer. (Courtesy, Weston Instruments, Inc.); (b) effective sensitivity of a galvanometer increased by a reflecting mirror.*

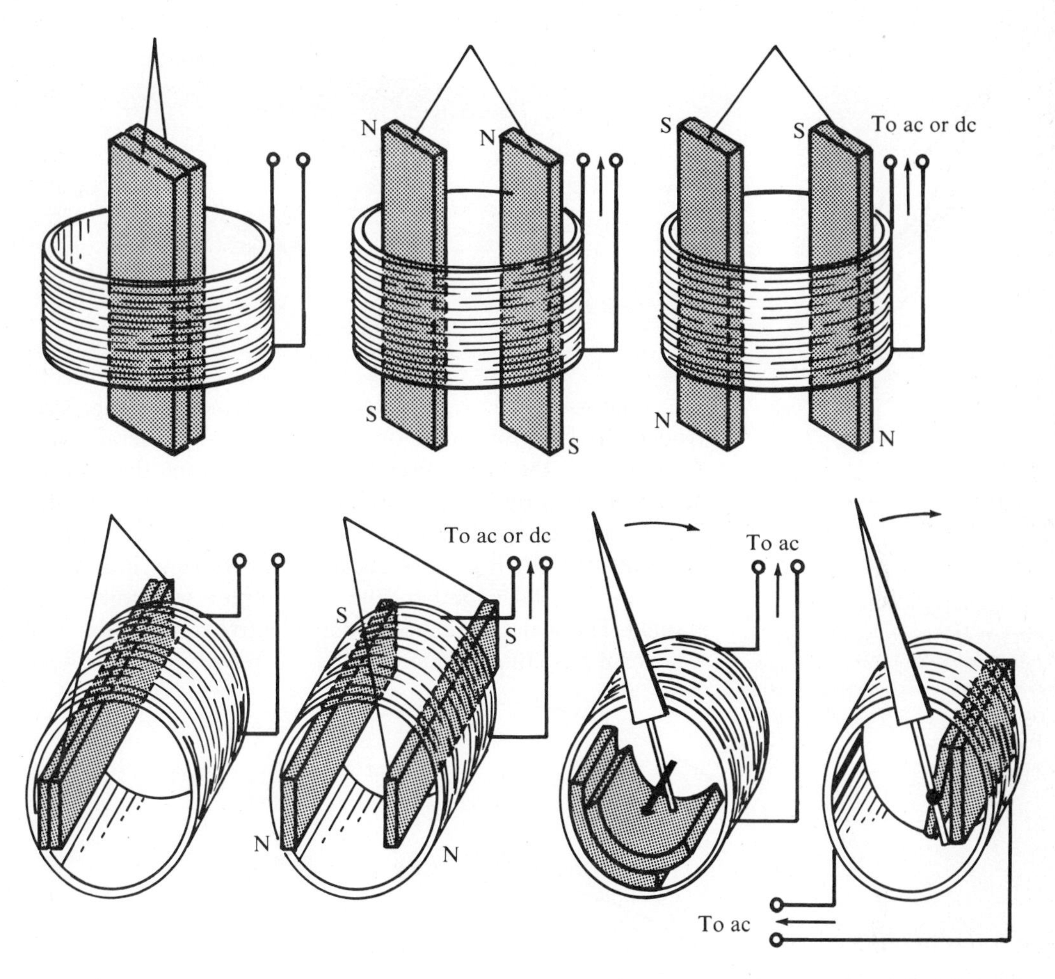

FIG. 3.11 *Principle of the moving iron-vane mechanism. (Courtesy, Weston Instruments, Inc.)*

difficult to manufacture. Hence most iron-vane instruments are now produced in unpolarized forms, and contain no permanent magnets. Figure 3.11 shows the principle of the moving iron-vane mechanism. If two similar adjacent iron bars are similarly magnetized, a repelling force is developed between them which tends to move them apart. In the moving-vane mechanism, this principle is used by fixing the position of one bar, and pivoting the second so that it will tend to rotate when current passes through the surrounding coil. A spring attached to the moving vane opposes the magnetic torque and permits the scale to be calibrated in terms of current or voltage values. As a voltmeter, a suitable value of multiplier resistance is connected in series with the coil.

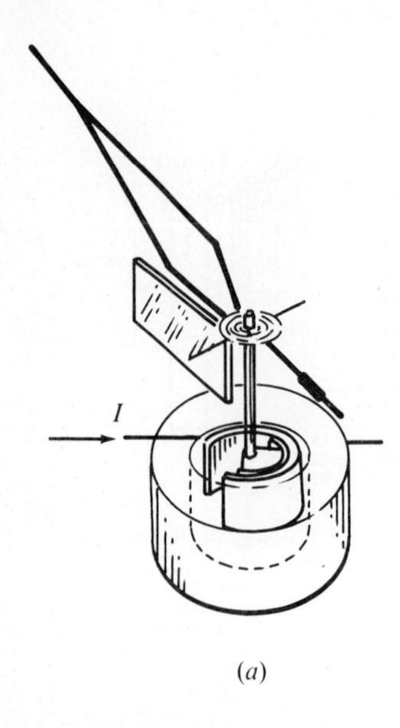

(a)

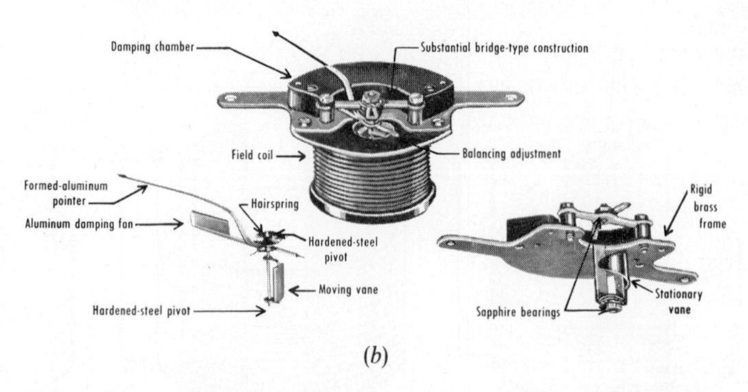

(b)

FIG. 3.12 *(a) The concentric vane mechanism. (Courtesy, Weston Instruments, Inc.); (b) photograph of repulsion-vane movement and components. (Courtesy, Sargent-Welch Co.)*

The concentric vane mechanism is depicted in Fig. 3.12. Under repulsive magnetic forces, the vanes slip laterally and the pointer is thereby deflected. In its basic form, the pointer deflection is proportional to current squared; that is, the instrument has square-law characteristics. Square-law scales are illustrated in Fig. 3.13. As noted previously, indication accuracy is good, whether sine or complex waveforms are applied. The scales are calibrated in effective (rms) values. If a waveform has components of different frequencies and effec-

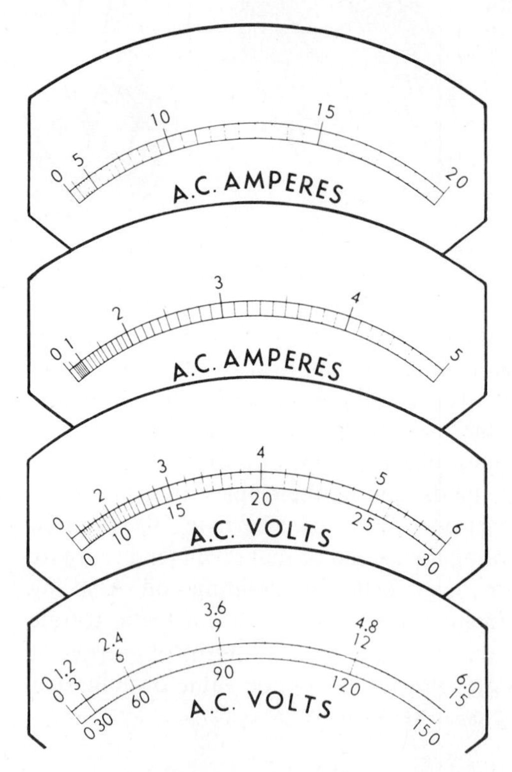

FIG. 3.13 *Square-law scales for an iron-vane instrument. (Courtesy, Sargent-Welch Co.)*

tive amplitudes I_1, I_2, I_3, etc., a square-law instrument gives the same deflection as a sine wave that has the effective value expressed by the formula:

$$\text{Effective value} = \sqrt{I_1^2 + I_2^2 + I_3^2 + \cdots} \qquad (3.5)$$

It can be shown by substitution in Formula (3.5) that a large harmonic amplitude, such as 20 percent, will change the scale indication by only 2 percent. It can also be shown that the indication of a square-law instrument remains the same when the phase of the harmonic component is shifted, as depicted in Fig. 3.14. Residual magnetism produces indication error if the iron-vane mechanism is used as a dc meter; this error can be minimized by reversing the dc input and taking the average of the two readings. Short magnetic vanes may be used to minimize the reversal error. This type of instrument has moderate sensitivity, compared with other forms. The vanes can be specially shaped to partially linearize the scale.

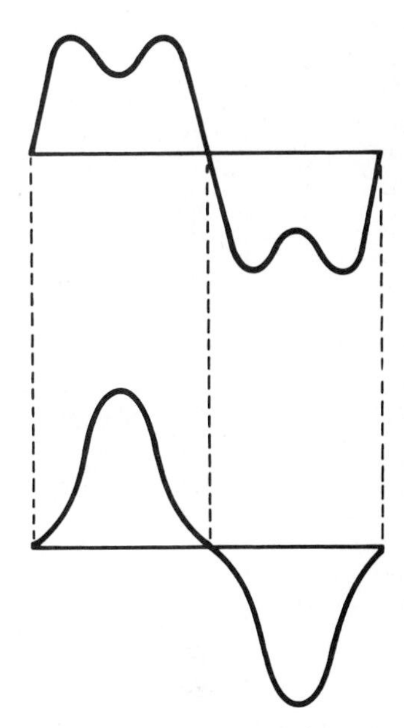

FIG. 3.14 *Effect of phase shift on wave shape.*

The radial vane mechanism depicted in Fig. 3.15 opens up like a book under the influence of magnetic repulsive forces. This construction has comparatively high sensitivity and provides an approximately linear scale. An aluminum damping vane is attached to the shaft of an iron-vane instrument just below the pointer. This damping vane rotates in a close-fitting air chamber, designed to damp out pointer oscillation and bring the pointer to rest on the scale quickly. Radial-vane instruments must be carefully designed and must utilize high-grade magnetic material to provide good accuracy.

3.4 ELECTRODYNAMIC AC VOLTMETERS

Electrodynamic mechanisms are the most fundamental of all the indicating devices now used. Like most of the basic meter movements, this mechanism is current sensitive; that is, the pointer is deflected on the basis of current flow. The electrodynamic mechanism is the most versatile of those that have been discussed, since its single-coil movement can be used to indicate current, voltage, or power values, with either ac or dc applied, and regardless of waveform. Crossed-coil movements can be used to measure power-factor, phase-angle, and frequency values. Chart 3.1 provides a survey of electrodynamic mechanisms.

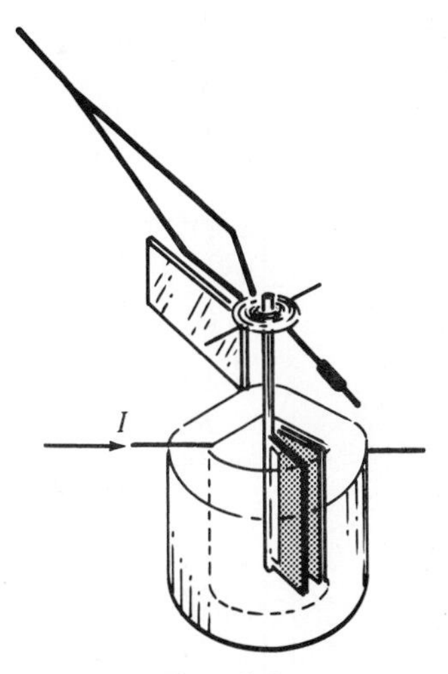

FIG. 3.15 *The radial vane mechanism. (Courtesy, Weston Instruments, Inc.)*

We observe that the basic electrodynamic mechanism is made up of a fixed coil winding and a rotating coil winding.

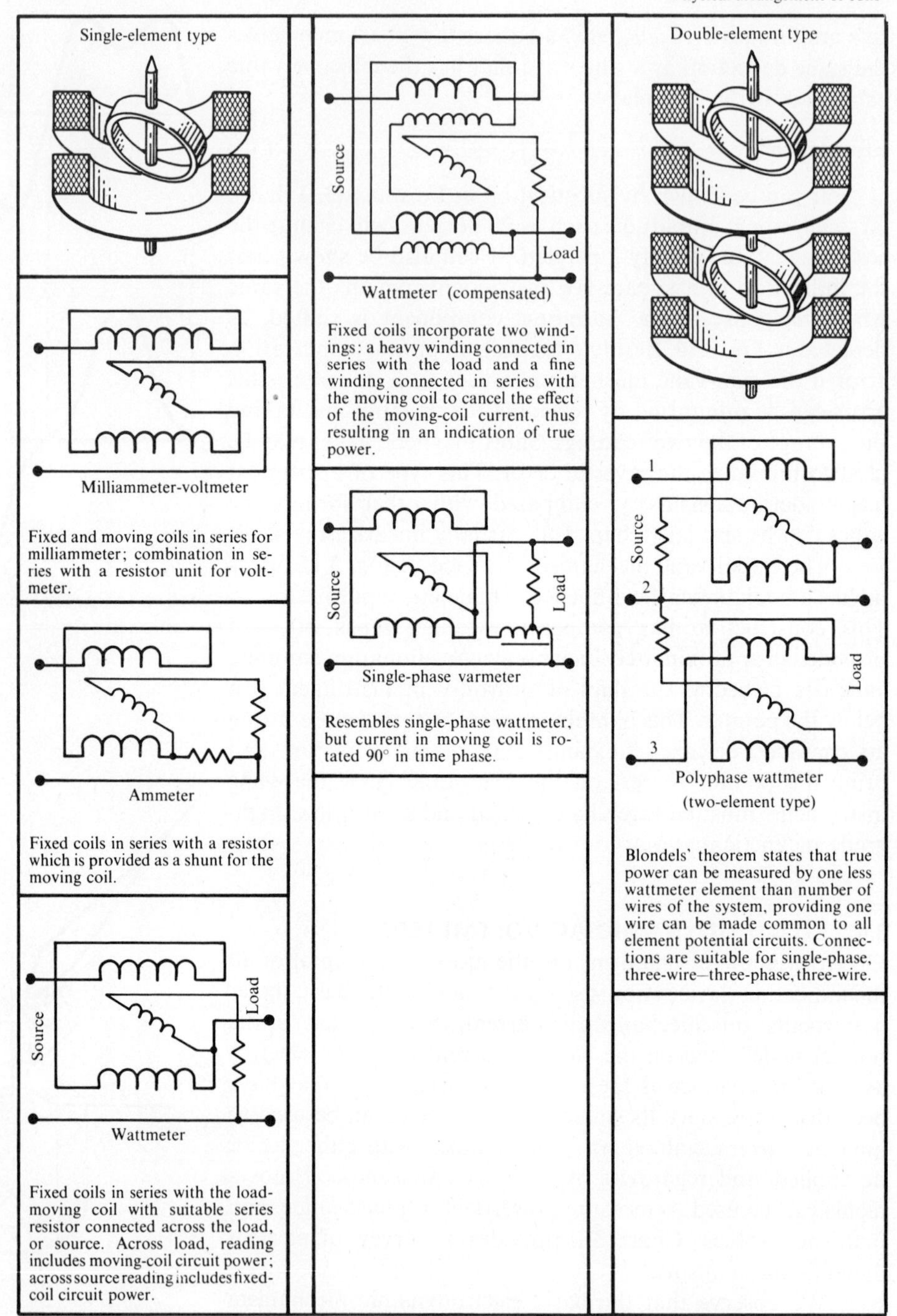

CHART 3.1 *Coil arrangements in electrodynamic and similar mechanisms.* (*Courtesy, Weston Instruments, Inc.*)

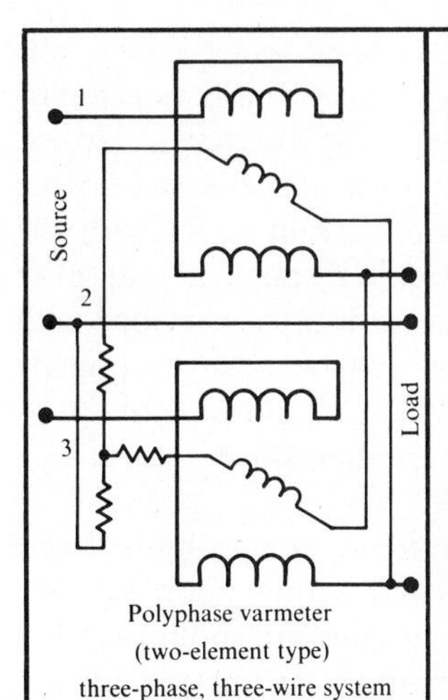

Polyphase varmeter
(two-element type)
three-phase, three-wire system

This and the four-wire version below resemble the polyphase wattmeters except moving-coil connections produce currents 90° rotated in time phase.

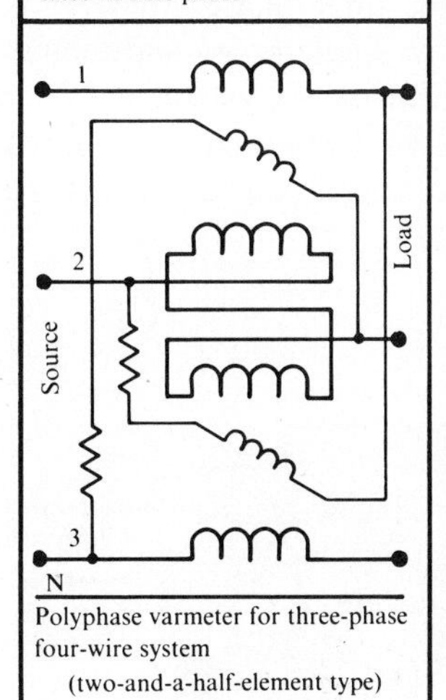

Polyphase varmeter for three-phase four-wire system
(two-and-a-half-element type)

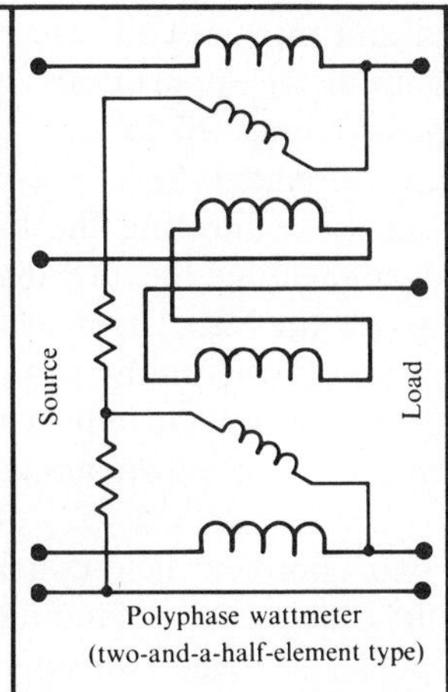

Polyphase wattmeter
(two-and-a-half-element type)

Above arrangement constitutes a compromise on Blondels' theorem for measurement of three-phase, four-wire system power—will indicate true power providing voltages are balanced. Called two-and-a-half element because of split connection of fixed coils. Note current in line 2 passes through both upper and lower element, in reverse direction, but is displaced in phase by 60°—its effect is thus half. Summation of two elements equals third element required.

Crossed-coil type

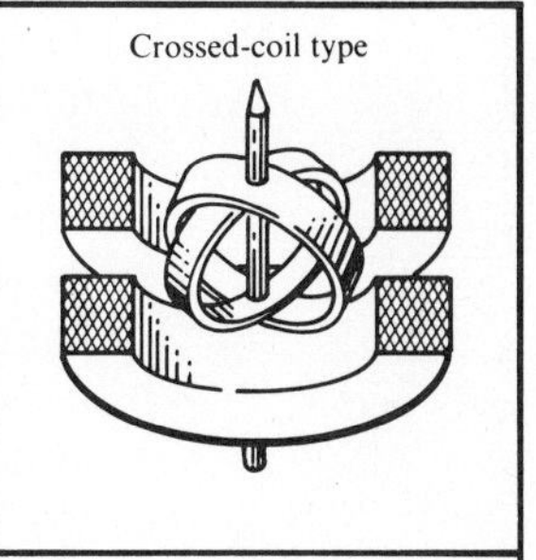

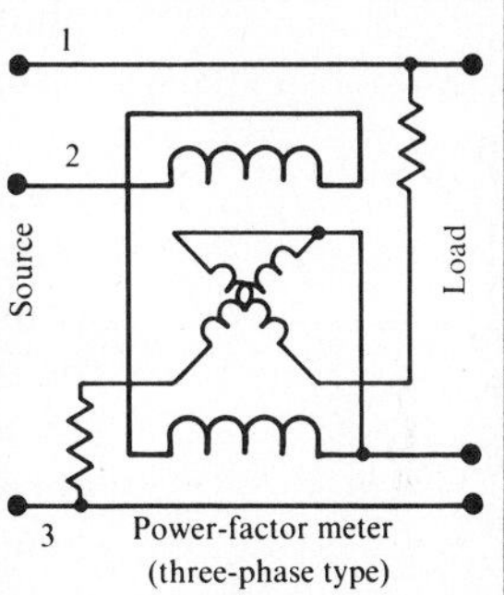

Power-factor meter
(three-phase type)

Crossed moving coils are connected to opposite legs of a three-phase system—fixed coils connected in series with line used as common for moving-coil connection. Moving system assumes a position depending on power factor of circuit. Correct for balanced loads only.

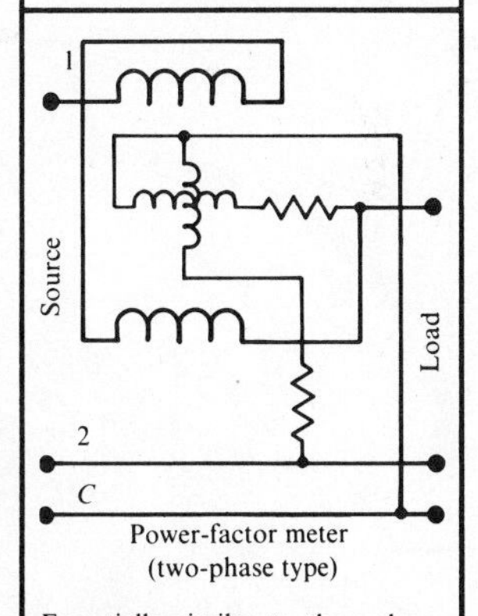

Power-factor meter
(two-phase type)

Essentially similar to three-phase except pointer is mounted in line with inner coil. Fixed coils not in the common line. For a single-phase power-factor meter the double moving coil is again used, with a reactance network to give the required phase difference in the currents in the moving system.

The fixed and moving coils are connected in series for voltmeter or ammeter applications; a voltmeter utilizes a multiplier resistor connected in series with the instrument circuit. Power measurements are made with an electrodynamic mechanism by connecting the fixed coils in series with the line, and connecting the moving coil through a multiplier resistor across the line. The instrument is often provided with switches so that the configuration can be changed quickly for measurement of current, voltage, or power values. A photograph of an electrodynamic mechanism is seen in Fig. 3.16.

If two complete field-coil systems are positioned one above the other, each including and acting upon its own moving coil, and if the two moving coils are mounted on a common shaft which carries the pointer, the instrument can be used to measure the total power in a polyphase ac system. Otherwise, two single-phase wattmeters must be used in a three-phase system. Since power for motors and industrial equipment is usually supplied by three-phase lines, it is important for us to understand how single-phase wattmeters are used to measure power in a three-phase system.

With reference to Fig. 3.17, two wattmeters are often used to measure three-phase power. The coil leads must be

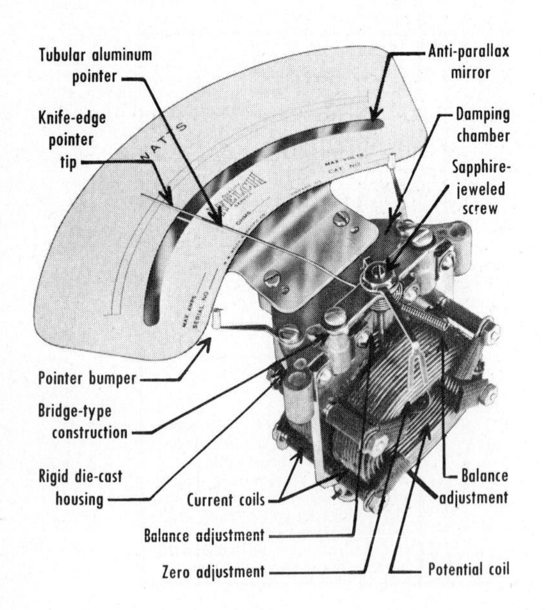

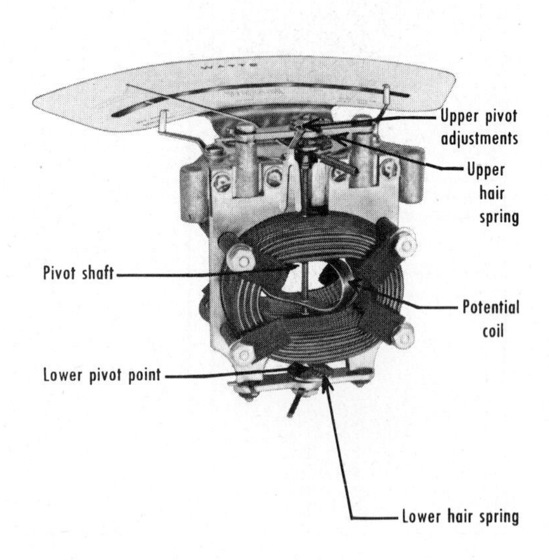

FIG. 3.16 *An electrodynamic mechanism. (Courtesy, Sargent-Welch Co.)*

connected as shown in the diagram to obtain proper indication. The sum of the two wattmeter readings is equal to the total three-phase power of the system. The student should note that a wattmeter indicates true power values; any reactive power that might be present is not indicated by a wattmeter. We can measure reactive power with a varmeter, a device somewhat similar to a wattmeter, but the current in the moving coil is rotated 90° in time phase, as depicted in Chart 3.1. Connections of crossed-coil electrodynamic instruments for measurement of power factor are depicted in Chart 3.1 also. The scale of a power-factor meter may be calibrated in phase-angle units, if desired.

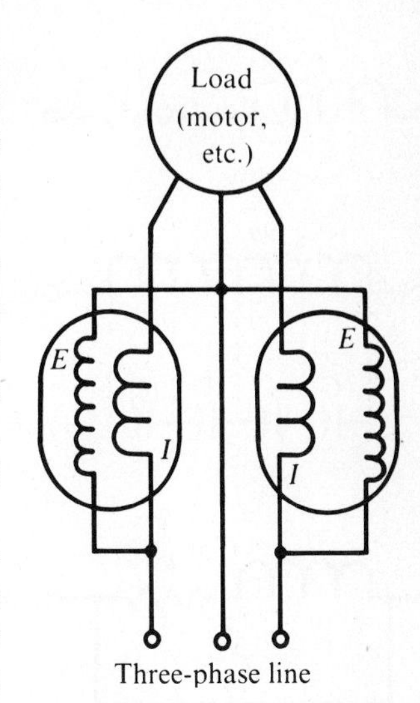

FIG. 3.17 *Basic two-wattmeter connections for measurement of three-phase power.*

Frequency measurements can be made with the electrodynamic instrument depicted in Fig. 3.18. Two tuned circuits are utilized; one tuned circuit resonates just below the low end of the scale, and the other resonates just above the high end of the scale. Thus, the center-scale frequency indication in this example is 60 Hz. Each tuned circuit is connected in series with half of the fixed-coil winding. At frequencies below 60 Hz, coil F_1 has the stronger effect on deflection and the pointer moves counterclockwise. At frequencies above 60 Hz, coil F_2 has the stronger effect and the pointer moves clockwise.

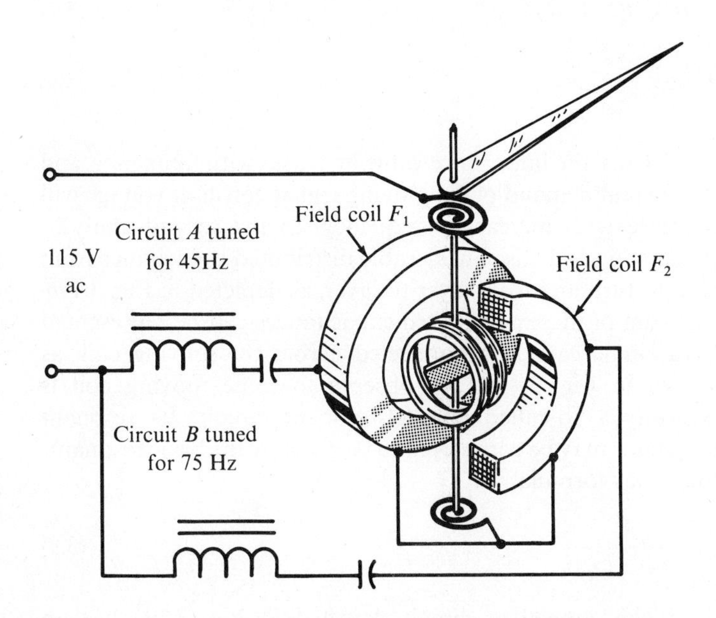

FIG. 3.18 *Schematic diagram of a resonant-circuit frequency meter. (Courtesy, General Electric Co.)*

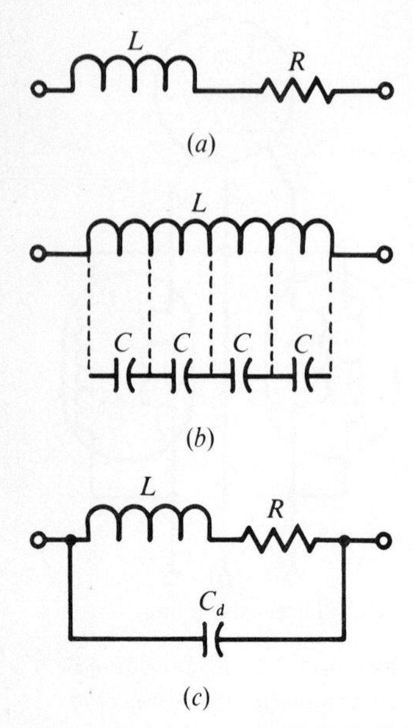

FIG. 3.19 *A moving coil (a) with inductance and resistance; (b) with distributed capacitance; and (c) as a resonant circuit.*

3.5 FREQUENCY CAPABILITIES OF BASIC AC INSTRUMENTS

The ac instruments that have been discussed have comparatively limited frequency ranges. Any ac meter is rated for a certain frequency range, and progressive indication error will occur if the meter is operated at higher frequencies than rated. For example, rectifier-type instruments are not simple resistive configurations. Instead, the equivalent circuit of a rectifier-type instrument is a network of inductance, capacitance, and resistance. The inductive and capacitive components become series or parallel resonant at certain frequencies. Capacitance that shunts resistance tends to bypass ac energy at progressively higher frequencies. Let us see how these reactive circuit actions impose a high-frequency limit in operation of a rectifier-type meter.

A D'Arsonval movement has inductance and resistance in the first analysis, as depicted in Fig. 3.19*a*. The impedance of the movement increases with frequency, and the inductive reactance components tend to reduce the ac current value. This circuit action is formulated as follows:

$$X_L = 2\pi f L \tag{3.6}$$

$$Z = \sqrt{R^2 + X_L{}^2} \tag{3.7}$$

$$I = \frac{E}{Z} \tag{3.8}$$

Thus, the impedance value increases with frequency, and the current demand of the moving coil at constant voltage will decrease with increasing frequency. In the second analysis, the moving coil has considerable distributed capacitance from turn to turn and from layer to layer, as depicted in Fig. 3.19*b*. The sum of these distributed capacitances can be represented by a single capacitor connected across the moving coil, as shown in Fig. 3.19*c*. We observe that the moving coil is basically a so-called parallel-resonant circuit. Its resonant frequency may be considered to be given by the basic resonant-frequency formula:

$$f = \frac{1}{2\pi\sqrt{LC}} \tag{3.9}$$

The equivalent circuit depicted in Fig. 3.19*c* has an

impedance at its resonant frequency that is stated by the approximate formula:

$$Z = \frac{L}{RC} \tag{3.10}$$

Since the resonant impedance of the moving coil is substantial, the rated frequency range of the instrument necessarily has a limit that is considerably below the resonant frequency. We will find that the equivalent circuit shown in Fig. 3.19*c* must be elaborated, to take the reactance of the instrument rectifier into account. This is approximately a parallel *RC* arrangement, as shown in Fig. 3.20*a*. The junction capacitance C_j tends to bypass ac energy around the unit as the operating frequency increases. In turn, less rectified current passes through the moving coil. The multiplier resistor also has an equivalent parallel *RC* circuit as a result of stray capacitance. The bypassing effect of the stray capacitance increases as the frequency increases, and more ac current is supplied to the rectifier. This effect is particularly evident when the value of the multiplier resistance is high.

Thus, the complete equivalent circuit for an ac voltmeter can be considered as a series-parallel *LCR* configuration, as depicted in Fig. 3.20*b*. Observe that conflicting effects on indication accuracy are present, as follows:

1. C_m tends to increase the scale indication as the frequency is increased.
2. C_j tends to reduce the scale indication as the frequency is increased.
3. C_m and C_j in series with L_{mc} tend to increase the scale indication as the frequency is increased; as the series-resonant frequency is approached, the internal impedance of the meter decreases, more ac current flows, and more pulsating dc passes through the moving coil.
4. C_{mc} tends to decrease the scale indication as the frequency is increased; as the so-called parallel-resonant frequency is

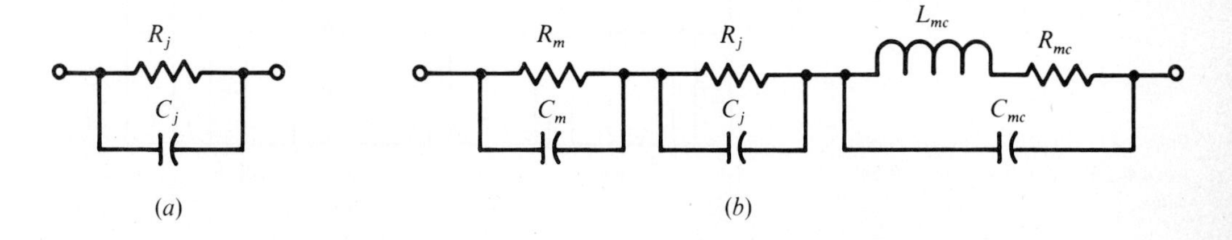

FIG. 3.20 *Equivalent circuit of (a) instrument rectifier, and (b) ac voltmeter.*

approached, the internal impedance of the meter increases, less ac current flows, and less pulsating dc passes through the moving coil.

Note that the series-resonant frequency and the parallel-resonant frequency of the equivalent circuit in Fig. 3.20*b* will not be the same in most cases. That is, the values of C_m, C_j, and C_{mc} vary greatly depending on the construction of individual ac voltmeters, and it would be a rare coincidence if the two resonant frequencies happened to be the same. Let us consider the indication error vs. frequency for a typical rectifier-type ac voltmeter when the multiplier resistor has a value of 50,000 Ω. In this example, the effect of C_m in Fig. 3.20*b* is dominant, and the scale indication increases rapidly at high frequencies, as depicted in Fig. 3.21. In other words, C_m has a reactance value that is less than the resistance value of R_m at frequencies higher than 100 kHz, and R_m is ineffective in controlling the current that passes through the rectifier and moving coil.

Next, let us consider the indication error vs. frequency when the multiplier resistor has a value of 10,000 Ω, and is used with the same rectifier and movement (Fig. 3.22). In this example, C_m has less bypassing action, and the increasing impedance of the moving coil is dominant. In turn, the scale indication decreases at high frequencies. It might be supposed that there is some value of R_m between 10,000 and 50,000 Ω that would result in zero frequency error. However, when this

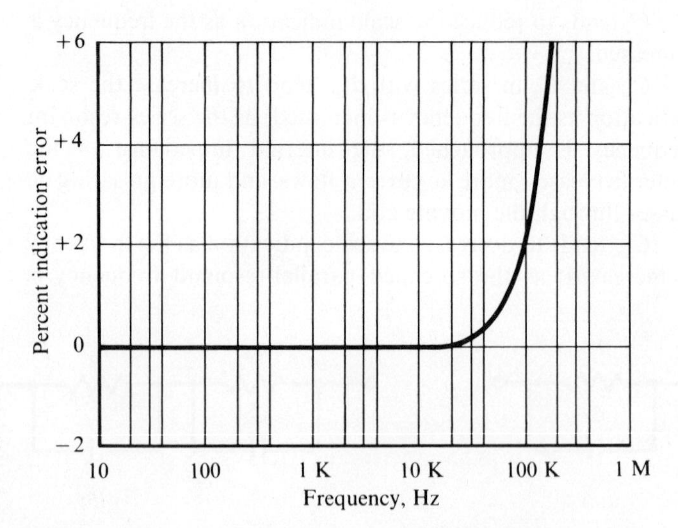

FIG. 3.21 *Indication vs. frequency when R_m has a comparatively high value.*

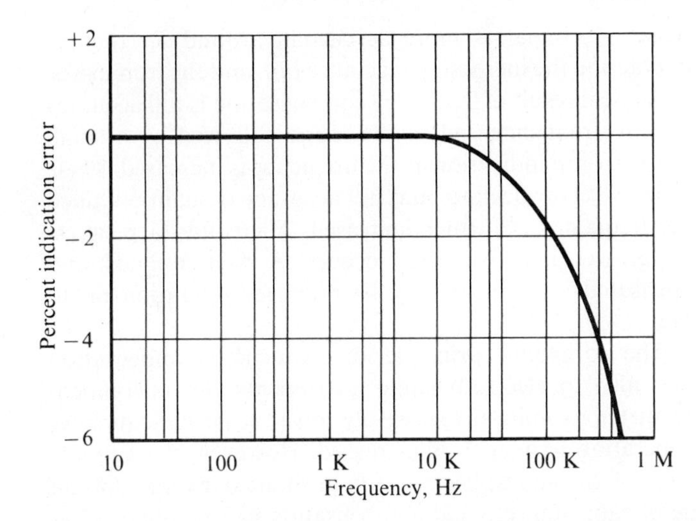

FIG. 3.22 *Indication vs. frequency when R_m has a comparatively low value.*

value of R_m is sought, we encounter successive positive and negative errors, terminating in a rapid negative error as the operating frequency is increased. This is the result of series and parallel resonance effects, which are terminated by increasing bypassing action of C_j in Fig. 3.20*b*.

Iron-vane instruments have a limited frequency range owing to increasing impedance of the winding at higher frequencies; that is, less current is drawn by the winding as the operating frequency is increased and, in addition, less magnetic flux is developed. Another limitation is imposed on frequency response by iron losses. In other words, eddy current and hysteresis losses are a function of frequency. Thus, the frequency range of an iron-vane instrument is limited to the lower audio frequencies. Most iron-vane meters are designed for application at power frequencies, and are rated typically for a frequency range from 25 to 133 Hz. Laboratory-type instruments are rated over ranges as great as 25 to 1000 Hz.

Wide-range iron-vane instruments provide maximum accuracy when they are designed with a *frequency-compensation network.* An uncompensated meter exhibits progressively lower scale indication as the operating frequency is increased. However, this characteristic can be compensated to a considerable extent by employment of a compensating capacitor, as depicted in Fig. 3.23. Note that the total multiplier resistance consists of R_m and R_c. The compensating capacitor C_c shunts the compensating resistance R_c. As the operating frequency

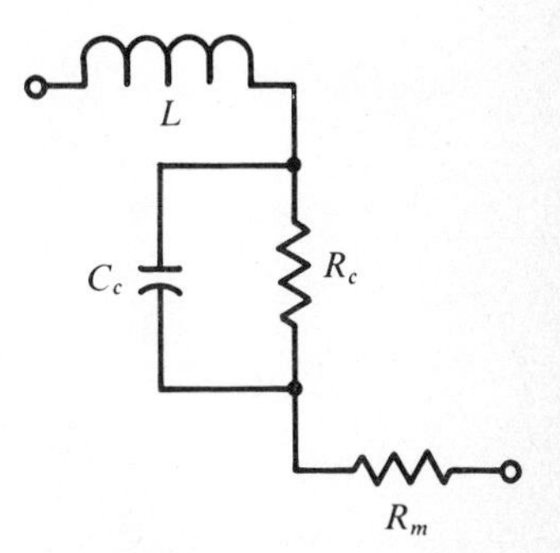

FIG. 3.23 *A frequency-compensating capacitor, C_c.*

increases, C_c bypasses more ac current around R_c, thereby counteracting the increasing reactance of L and the iron losses. The practical result of frequency compensation is full accuracy at the low-frequency end of the range, followed by a small positive error in indication as the frequency is increased. Then, the positive error declines and full accuracy is again obtained as the frequency is further increased. Thereafter, a progressively greater negative error occurs. A well-designed and compensated instrument may be rated for operation up to 2 kHz.

The foregoing principle of frequency compensation applies also to electrodynamic instruments. An uncompensated meter exhibits progressively lower scale indication as the operating frequency is increased. However, if a suitable portion of the multiplier resistor is shunted by the correct value of capacitance, good compensation can be obtained at frequencies up to 500 Hz. As would be anticipated, full accuracy is obtained at the low-frequency end of the range, followed by a small positive error. Full accuracy again occurs at approximately 250 Hz. At 500 Hz, the negative error reaches the rated accuracy limit of the instrument.

3.6 APPLICATIONS

In measuring ac voltage, current, or power, we are most often concerned with rms values. An rms value, though, is not as easy to measure as a peak value because its waveform may not be completely sinusoidal. To prevent error, rms values of complex waveforms should be measured with true-rms instruments. If the waveform is known to be truly sinusoidal, however, peak-responding instruments may be used to measure rms values correctly without incurring waveform error. In brief, rms values can be measured accurately (1) by a true-rms instrument regardless of waveform, and (2) by a peak-responding instrument only if the waveform is a good sine wave.

A peak-responding instrument (Chap. 6) will measure peak values accurately regardless of waveform. Basic nonsinusoidal waveforms are depicted in Fig. 3.24; these include square, pulse, sawtooth, half-rectified sine, and full-rectified sine waveforms. In each case, +V denotes the positive-peak voltage, and −V denotes the negative-peak voltage of the

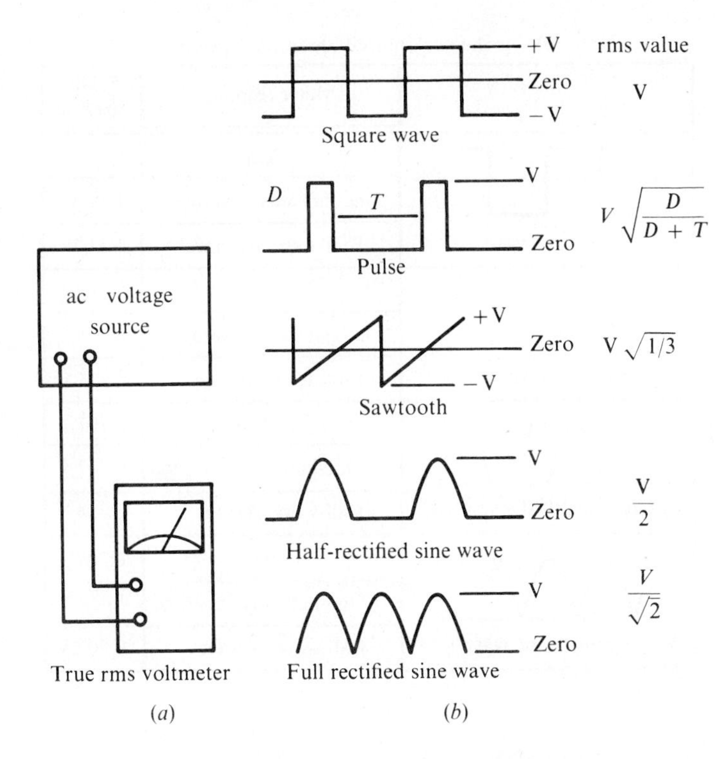

FIG. 3.24 *Using a true rms voltmeter. (a) Test setup; (b) rms values of basic waveforms.*

waveform. If we use a positive-peak responding voltmeter, the +V voltages can be measured accurately in each case. If we use a negative-peak responding voltmeter, the −V voltages can be measured accurately in each case. Finally, if we add the positive-peak reading to the negative-peak voltage reading, the peak-to-peak voltages can be determined accurately in each case. The rms voltage, expressed as a function of peak voltage, is noted for each waveform in Fig. 3.24. This is the rms value that will be indicated by a true-rms instrument for each of the nonsinusoidal waveforms.

Let us consider how much error will be incurred if we apply basic nonsinusoidal waveforms to a peak-responding meter, a half-wave average-responding meter, or a full-wave responding meter, as depicted in Fig. 3.25. Note that in each case the meter scale has been calibrated in terms of rms values for a sine wave. The waveform error that is incurred, in general, is very large (compare with the true-rms values in Fig. 3.24).

In summary, meters must be selected for the particular

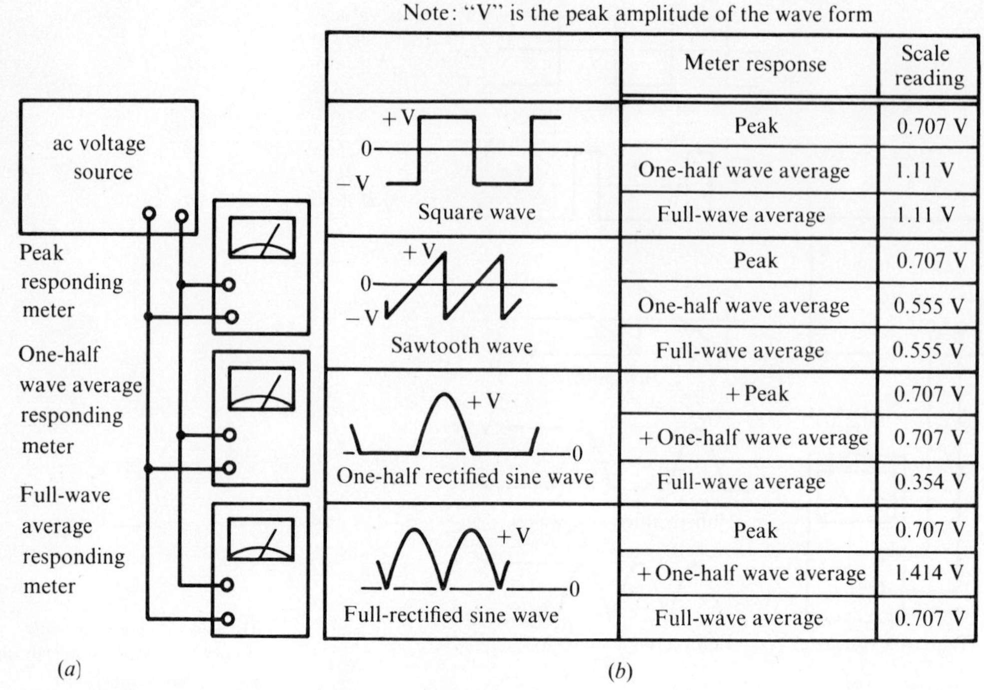

Waveform	Meter response	Scale reading
Square wave	Peak	0.707 V
	One-half wave average	1.11 V
	Full-wave average	1.11 V
Sawtooth wave	Peak	0.707 V
	One-half wave average	0.555 V
	Full-wave average	0.555 V
One-half rectified sine wave	+ Peak	0.707 V
	+ One-half wave average	0.707 V
	Full-wave average	0.354 V
Full-rectified sine wave	Peak	0.707 V
	+ One-half wave average	1.414 V
	Full-wave average	0.707 V

FIG. 3.25 *Comparing measurements on different types of ac voltmeters. (a) Three types connected in parallel; (b) comparative readings for nonsinusoidal waveforms.*

application, with due regard for waveform. When rms values are of interest, and it is not known whether the waveform may be strictly sinusoidal, it is good practice to employ a true-rms meter.

QUESTIONS AND PROBLEMS

1. What is the requirement for a voltmeter to indicate the true rms voltage of a complex waveform?
2. Name three types of voltmeters that will indicate either ac or dc volts.
3. How can a D'Arsonval meter be converted to measure ac voltage?
4. What value of voltage will a D'Arsonval-type voltmeter with a half-wave rectifier indicate when responding to the resulting pulsating dc voltage?
5. Why are ac voltmeter scales commonly calibrated in rms values?
6. Why is it impossible to obtain useful dc response on a D'Arsonval-type meter that is used for measuring ac voltage?

7. Explain the operation of diode D_2 in the meter circuit shown in Fig. 3.8.

8. What are the limitations of the tangent galvanometer shown in Fig. 3.9?

9. What is the purpose of the mirror in the galvanometer assembly shown in Fig. 3.10?

10. What are the indicating characteristics of the concentric vane instrument shown in Fig. 3.12?

11. How can the error introduced by residual magnetism be minimized in an iron-vane instrument?

12. Explain the operation of the radial vane mechanism shown in Fig. 3.15.

13. How is pointer damping accomplished in a radial vane meter mechanism?

14. Why is the electrodynamic mechanism more versatile than other meter types?

15. Discuss the connection of an electrodynamic mechanism as a voltmeter, varmeter, ammeter, wattmeter, and power factor meter.

16. Explain the operation of the two wattmeters in the circuit shown in Fig. 3.17.

17. Why is an ac voltmeter frequency sensitive?

18. Why are most iron-vane voltmeters limited to a frequency range of 25 to 135 Hz?

19. How can the frequency range of an iron-vane instrument be extended?

20. Discuss the graphs shown in Figs. 3.21 and 3.22.

21. At what frequency is the error greater than 2 percent in the instrument represented by the graph in Fig. 3.21?

22. At what frequency is the error greater than 3 percent in the instrument represented by the graph in Fig. 3.22?

23. What value will be indicated on a rectifier-type ac voltmeter if the average value of the measured voltage is 80 V?

24. What value will the foregoing voltmeter indicate when used to measure a rectified sine-wave voltage that has an average value of 96 V?

VOLT-OHM-MILLIAMMETERS

4.1 BASIC REQUIREMENTS

All volt-ohm-milliammeters (VOMs) have the basic plan depicted in Fig. 4.1. Note that three basic functions are provided: voltage measurement, current measurement, and resistance measurement. Switching facilities (or equivalent arrangements) are provided for selection of a chosen function. A volt-ohm-milliammeter always indicates electrical values by means of a D'Arsonval mechanism. Various ranges are available on each function; each range utilizes suitable multiplier or shunt resistors that are built into the instrument. Supplementary range functions for specialized applications are often provided by external multipliers, shunts, probe arrangements, instrument transformers, transducers, and other devices.

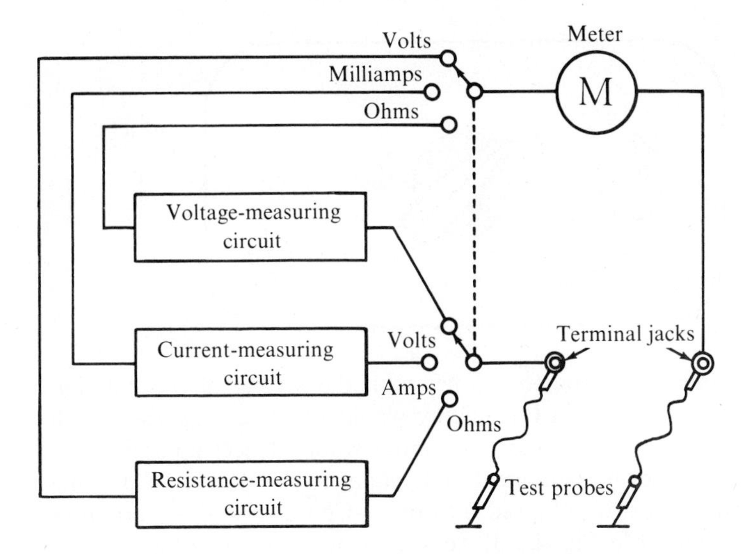

FIG. 4.1 *Three basic functions of volt-ohm-milliammeter.*

A VOM is designed as a utility instrument and provides moderate accuracy. An indication accuracy of ± 3 percent of full scale is typical. The sensitivity of a VOM on its dc voltage function is greater than that of most lab-type voltmeters designed for high indication accuracy. For example, a typical VOM has a rated sensitivity of 20,000 Ω/V. Thus, the sensitivity of a conventional VOM is greater than that of moderately accurate voltmeters used in power-distribution applications. On the other hand, the sensitivity of a VOM is much less than that of voltmeters designed to provide comparatively high input resistance or impedance. A VOM has the advantages of independence from a power source and of moderate cost.

4.2 RANGES AND SUBFUNCTIONS

Utility-type voltmeters require several ranges to meet the requirements of routine test situations. Thus, a typical VOM provides dc voltage ranges of 2.5, 10, 50, 250, 1000, and 5000 V, with a range switch as depicted in Fig. 4.2. In addition, a function switch is provided which has $-$dc and $+$dc positions, and is often called a *polarity-reversing switch.* Control of indication polarity is a convenience in practical test procedures because the test leads do not have to be reversed to obtain up-scale pointer deflection. That is, if the pointer happened to

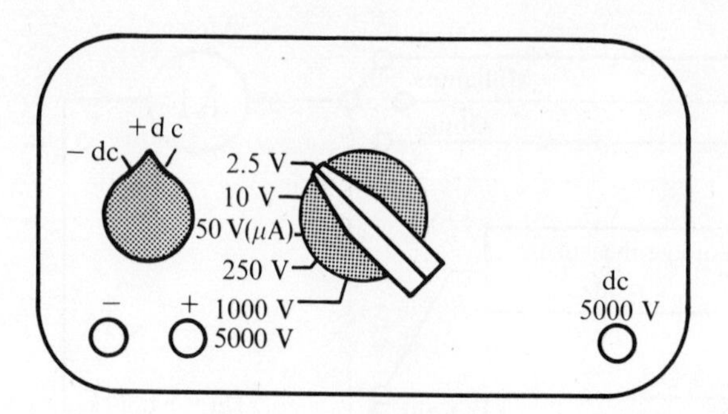

FIG. 4.2 *Range switch, function switch, and terminals of a typical VOM. (Courtesy, Simpson Electric Co.)*

deflect off-scale to the left with the setting of the function switch shown in Fig. 4.2, we would simply change the switch setting to +dc, and the pointer would deflect up-scale.

For measurement of dc voltage values up to 1000 V, the test leads are plugged into the −Common and + terminals depicted in Fig. 4.2. If we wish to measure dc voltage values between 1000 and 5000 V, the test leads are plugged into the −Common and 5000-V dc terminals. Note that the range switch will be set to the 1000–5000 V position. We observe that the 50-V position is also marked μA. To anticipate subsequent discussion, this switch position does double duty on the dc voltage and dc current functions. When the instrument is operating on its dc voltage function, we disregard the μA marking.

The scale plate for a VOM is provided with a number of scales, as seen in Fig. 4.3. When operating on the 2.5-V range, we utilize the 250-V scale and shift the decimal point two places to the left. When operating on the 10-V range, we read the 10-V scale directly. Similarly, the 25-V scale is read directly when operating on the 250-V range. To read values on the 1000-V range, we utilize the 10-V scale and shift the decimal point two places to the right. Finally, to read values on the 5000-V range, we employ the 50-V scale and shift the decimal point two places to the right.

In television service work, it is often necessary to measure dc voltage values up to 25,000 V. For this purpose, an external multiplier, called a *high-voltage dc probe*, is used (see Fig. 4.4). Three types of high-voltage probes are provided for the VOM in this example. The first probe extends the full-scale indication of the instrument to 10,000 V; the second probe extends the indication to 25,000 V; the third probe extends the indication to 50,000 V.

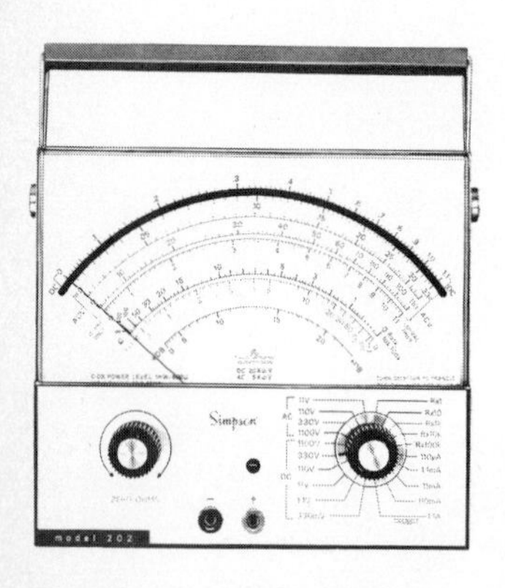

FIG. 4.3 *Scales of a typical VOM. (Courtesy, Simpson Electric Co.)*

The high-voltage probes in this example are designed for use on the 2.5-V dc range of a 20,000 Ω/V VOM. The particular scale to be used depends upon which probe is in use. Thus, when the 25,000-V probe is plugged into the instrument in place of the conventional test leads, we employ the 250-V scale in Fig. 4.3 and shift the decimal point two places to the right. If the 10,000-V probe is to be used, we read the 10-V scale and shift the decimal point three places to the right. If the 50,000-V probe is in use, we read the 50-V scale and shift the decimal point three places to the right.

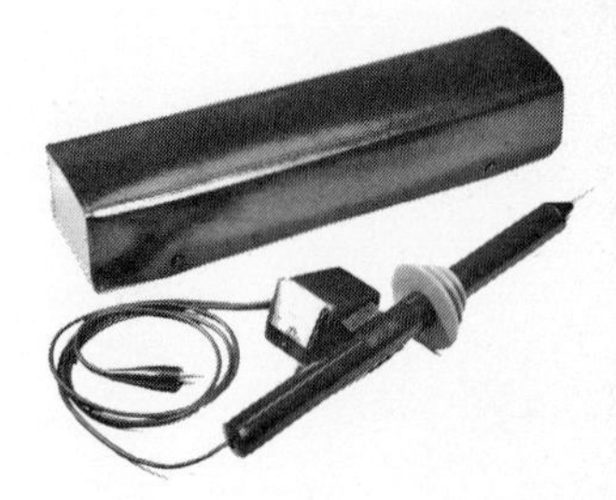

FIG. 4.4 *A high-voltage dc probe. (Courtesy, Precision Apparatus Co., division of B & K Dynascan Corp.)*

The student will note that most of the source voltage is dropped across the multiplier resistor in a high-voltage probe, thus necessitating that the multiplier resistor(s) have an adequate voltage rating. Moreover, since the power demand increases as the square of the applied voltage, the multiplier resistor may be required to dissipate a corresponding value of power. The physical design of a high-voltage probe entails precautions against arc-over that can endanger the operator. For example, most high-voltage probe have a rib or a series of ribbed sections to prevent any possibility of arc-over. This construction is supplemented by a grounded shield between the multiplier resistor(s) and the probe handle. Consequently, if excessive voltage is applied to the probe, or if the probe becomes damp, any arcing will be completed inside the probe housing, thus protecting the operator.

Next, let us consider the dc current ranges provided by the VOM in this example. Figure 4.5 shows how the range switch also serves as a function switch, with positions for 1 mA, 10 mA, 50 mA, 100 mA, 500 mA, and 10 A full-scale values. To measure dc current values from 1 mA to 500 mA, the test leads are plugged into the −Common and + terminals. If the 10-A range is employed, the test leads are

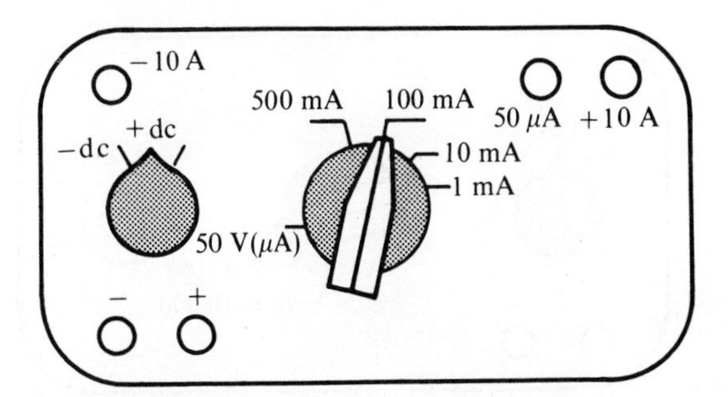

FIG. 4.5 *Dc current ranges and terminals. (Courtesy, Simpson Electric Co.)*

plugged into the -10 A and $+10$ A terminals. The reason for this terminal arrangement shall be clarified in the subsequent discussion of instrument circuitry.

We know that measurement of current requires that the circuit under test be opened and the meter connected in series with the circuit. Unless the polarity of connection is properly made, the pointer may deflect off-scale to the left. In such case, we merely change the setting of the polarity switch depicted in Fig. 4.5. Note, however, that the polarity switch is inoperative on the 10-A range; in this case, up-scale pointer deflection can be obtained only by observing the required polarity when the test leads are connected into the circuit under test.

A subfunction provided by the 50-μA range in this example is an auxiliary 0.25-V range. In other words, the input resistance of the VOM on its 50-μA range is 5000 Ω. With reference to Fig. 4.2, the next highest dc voltage range is 2.5 V. The availability of a 0.25-V range is very useful in checking bias voltages in transistor circuits. The student may show that the sensitivity of the VOM on its 0.25-V range is 20,000 Ω/V. Note that dc current values are indicated on the same scales that are used for dc voltage measurements in Fig. 4.3.

Let us now consider the resistance ranges and terminals provided by the VOM in this example. As seen in Fig. 4.6, three resistance ranges are available: $R \times 1$, $R \times 100$, and $R \times 10{,}000$. These are basically multiplying factors for the ohms scale depicted in Fig. 4.3. For example, we read the scale directly when the VOM is set to its $R \times 1$ position; if the instrument is operated on its $R \times 100$ range, we multiply the scale reading by 100; and if the $R \times 10{,}000$ range is in use, we multiply the scale reading by 10,000. The test leads are

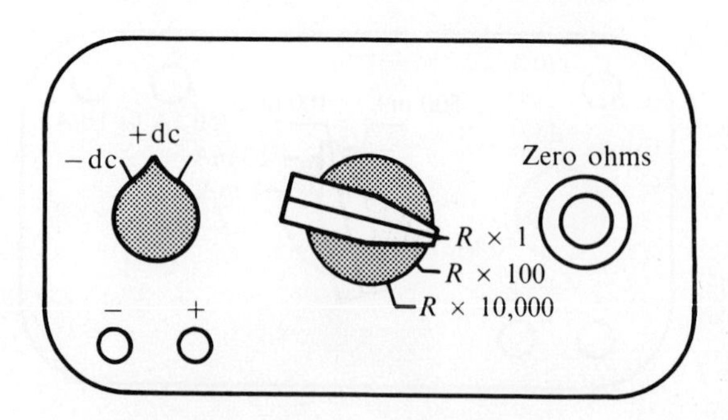

FIG. 4.6 *Resistance ranges and terminals. (Courtesy, Simpson Electric Co.)*

plugged into the −Common and + terminal when the ohmmeter function is employed.

Before making a resistance measurement, the test leads must be connected together (short-circuited), and the Zero ohms control adjusted to bring the pointer exactly to zero on the ohms scale. Otherwise, the resistance measurement is likely to be incorrect. This control compensates for aging of the internal ohmmeter battery, as explained subsequently.

Ohmmeter accuracy is calculated on the basis of dc voltage indication accuracy. In other words, if a VOM has an accuracy of ± 3 percent on its dc voltage function, this range of uncertainty corresponds to an *arc of error* on the ohms scale. The ohmmeter might be rated for an accuracy of ± 3 degrees of arc. Since an ohmmeter scale is nonlinear, it is apparent that the best measurement accuracy will be realized if a range that provides indication on the right-hand half of the scale is used. The polarity switch in Fig. 4.6 can be used to reverse the polarity of the test voltage applied across the component whose resistance is to be measured. This is a useful feature of the instrument when measuring the front-to-back resistance ratio of a semiconductor diode or transistor junction.

Since an ohmmeter scale is nonlinear, as seen in Fig. 4.3, best accuracy is obtained when a range is chosen which provides an indication on the right-hand portion of the ohms scale. Note that the center-scale resistance values in this example are 12, 1200, and 120,000 Ω, and that the ohmmeter indicates dc resistance values. That is, if we measure the resistance of an inductor, the ohmmeter indicates the dc winding resistance of the coil. In general, this dc resistance value is less, and might be very much less, than the ac resistance of the coil. A VOM cannot measure ac resistance values.

Next, let us consider the ac voltage ranges provided by the VOM in this example. Figure 4.7 shows how ac voltages are measured when the function switch is set to its ac position. The ac voltage ranges are the same as provided on the dc voltage function. Test leads are plugged into the −Common and + terminals for operation on all ranges except the 5000-V ac range. When voltages are to be measured in the range from 1000 to 5000 V, the range switch is set to its 1000–5000 V position, and the test leads are plugged into the −Common and 5000-V ac terminals. Polarity markings have no significance on the ac voltage function of the VOM. Note that the 50-V μA position of the range switch provides only a 50-V

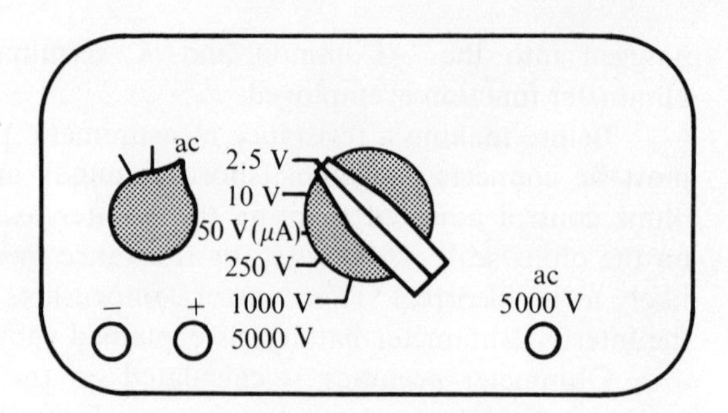

FIG. 4.7 *Ac voltage ranges and terminals. (Courtesy, Simpson Electric Co.)*

ac range; the μA marking has no significance on the ac voltage function.

With reference to Fig. 4.3, we read the 2.5-V ac only scale when the VOM is operated on its 2.5-V ac range. This scale is employed because nonlinear indication occurs at values less than 2 V. Higher ac voltages are read on the same scales as dc voltages. A VOM indicates rms values of sine-wave ac voltages. Unless the input waveform is sinusoidal, the scale indication will be incorrect. Note also that a VOM has a somewhat limited frequency range. At frequencies above 10 kHz in this example, the scale indication will be too high or too low, depending upon the range that is in use. Figure 4.8 shows these frequency characteristics.

Note in Fig. 4.3 that a decibel scale is also provided on the scale plate. This scale is used to measure dB values on the ac voltage ranges of the VOM. The scale is direct-reading when

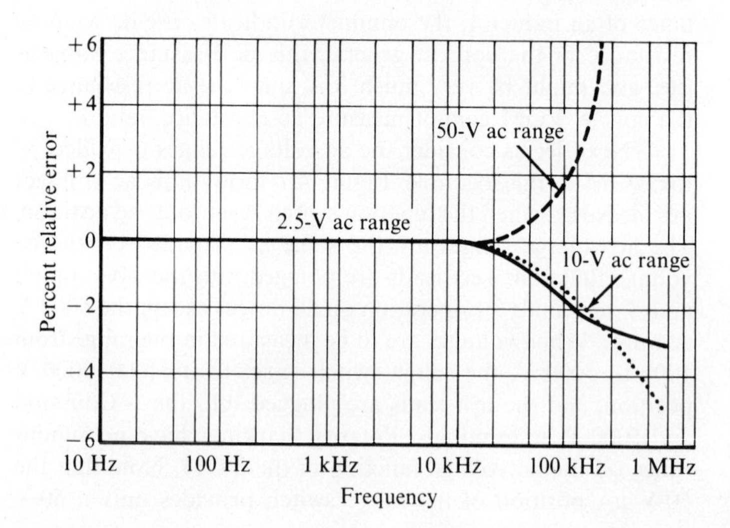

FIG. 4.8 *Frequency characteristics for a VOM on its ac voltage ranges. (Courtesy, Simpson Electric Co.)*

operating on the 2.5-V ac range, provided that the test leads are connected across a 600-Ω resistive load. On higher ac voltage ranges, a scale factor must be added to the reading; thus, we add 12 dB to the reading when operating on the 10-V ac range, and so on. These scale factors are indicated on the scale plate. Table 4.1 depicts a decibel chart that tabulates the voltage and power ratios corresponding to various dB values.

Since dB values correspond to power ratios, the decibel scale of a VOM reads correctly only when ac voltages are

TABLE 4.1

NEG (−)			POS (+)		NEG (−)			POS (+)	
Voltage ratio	*Power ratio*	*−dB+*	*Voltage ratio*	*Power ratio*	*Voltage ratio*	*Power ratio*	*−dB+*	*Voltage ratio*	*Power ratio*
1.0000	1.0000	**0**	1.000	1.000	0.3162	.1000	**10.0**	3.162	10.00
0.9772	0.9550	0.2	1.023	1.047	0.3090	0.09550	10.2	3.236	10.47
0.9550	0.9120	0.4	1.047	1.096	0.3020	0.09120	10.4	3.311	10.96
0.9333	0.8710	0.6	1.072	1.148	0.2951	0.08710	10.6	3.388	11.48
0.9120	0.8310	0.8	1.096	1.202	0.2884	0.08318	10.8	3.467	12.02
0.8913	0.7943	**1.0**	1.122	1.259	0.2812	0.07943	**11.0**	3.548	12.59
0.8710	0.7586	1.2	1.148	1.318	0.2754	0.07586	11.2	3.631	13.18
0.8511	0.7244	1.4	1.175	1.380	0.2692	0.07244	11.4	3.715	13.80
0.8318	0.6918	1.6	1.202	1.445	0.2630	0.06918	11.6	3.802	14.45
0.8128	0.6607	1.8	1.230	1.514	0.2570	0.06607	11.8	3.890	15.14
0.7943	0.6310	**2.0**	1.259	1.585	0.2512	0.06310	**12.0**	3.981	15.85
0.7762	0.6026	2.2	1.288	1.660	0.2455	0.06026	12.2	4.074	16.60
0.7586	0.5754	2.4	1.318	1.738	0.2399	0.05754	12.4	4.169	17.38
0.7413	0.5495	2.6	1.349	1.820	0.2344	0.05495	12.6	4.266	18.20
0.7244	0.5248	2.8	1.380	1.905	0.2291	0.05248	12.8	4.365	19.05
0.7079	0.5012	**3.0**	1.413	1.995	0.2239	0.05012	**13.0**	4.467	19.95
0.6918	0.4786	3.2	1.445	2.089	0.2188	0.04786	13.2	4.571	20.89
0.6761	0.4571	3.4	1.479	2.188	0.2138	0.04571	13.4	4.677	21.88
0.6607	0.4365	3.6	1.514	2.291	0.2089	0.04365	13.6	4.786	22.91
0.6457	0.4169	3.8	1.549	2.399	0.2042	0.04169	13.8	4.898	23.99
0.6310	0.3981	**4.0**	1.585	2.512	0.1995	0.03981	**14.0**	5.012	25.12
0.6166	0.3802	4.2	1.622	2.630	0.1950	0.03802	14.2	5.129	26.30
0.6026	0.3631	4.4	1.660	2.754	0.1905	0.03631	14.4	5.248	27.54
0.5888	0.3467	4.6	1.698	2.884	0.1862	0.03467	14.6	5.370	28.84
0.5754	0.3311	4.8	1.738	3.020	0.1820	0.03311	14.8	5.495	30.20
0.5623	0.3162	**5.0**	1.778	3.162	0.1778	0.03162	**15.0**	5.623	31.62
0.5495	0.3020	5.2	1.820	3.311	0.1738	0.0320	15.2	5.754	33.11
0.5370	0.2884	5.4	1.862	3.467	0.1698	0.02884	15.4	5.888	34.67
0.5248	0.2754	5.6	1.905	3.631	0.1660	0.02754	15.6	6.026	36.31
0.5129	0.2630	5.8	1.950	3.802	0.1622	0.02630	15.8	6.166	38.02

measured across the rated reference value of resistance; in this example, this resistance value is 600 Ω. The basic bel and decibel formulas are as follows:

$$\text{Bels} = \log \frac{P_2}{P_1} \tag{4.1}$$

$$\text{Decibels} = 10 \log \frac{P_2}{P_1} \tag{4.2}$$

where P_1 and P_2 denote power values, with P_2 conventionally denoting the largest power value; in turn, a power loss will be indicated by placing a minus sign before the dB value.

It is helpful in audio work to compare power levels in terms of dB values because the decibel unit is proportional to ear response. On the other hand, power, voltage, or current levels are not proportional to ear response. A gain or loss of 1 dB is just perceptible. As a decibel level is increased from 1 dB to 2 dB, then from 2 dB to 3 dB, and is followed by a gain of 3 dB in a system, the system gain is 5 dB. If a loss of 30 dB is followed by a gain of 5 dB, the system loss is 25 dB.

Since a VOM responds to ac voltage values and its decibel scale is calibrated in dB values, let us calculate the electrical relations that are involved. If we make a pair of ac voltage measurements across the same value of resistance in both cases, we can employ the power law and write:

$$P = \frac{E^2}{R} \tag{4.3}$$

$$\text{Decibels} = 10 \log \frac{{E_2}^2}{{E_1}^2} \tag{4.4}$$

$$\text{Decibels} = 20 \log \frac{E_2}{E_1} \tag{4.5}$$

However, if we measure the values of E_2 and E_1 across *different* values of resistance, the dB values that we read on the VOM decibel scale must be corrected. With reference to Fig. 4.3, we observe that the dB scale indication might be either positive or negative. It follows from the foregoing formulas that the algebraic sign of a dB value depends simply upon the reference power level which has been chosen for 0 dB indication by the instrument designer. Thus, 0 dB in this example corresponds to 2.45 V across 600 Ω. If less than

2.45 V are dropped across 600 Ω, we read a negative dB value; if more than 2.45 V are dropped across 600 Ω, we read a positive dB value.

If we measure dB values across resistances other than 600 Ω in this example, two guiding points must be kept in mind. If the resistance values are different from the reference value of 600 Ω, but are *equal* in value, we can subtract a pair of dB readings to calculate the true gain or loss in decibels. For example, we could make a pair of ac voltage measurements across 250 Ω at the input and output of a system, and the difference between the dB scale readings will represent the true loss or gain of the system. However, if a pair of measurements is made across unequal values of resistance, a correction factor must be applied to find the true dB gain or loss. It is helpful to use the chart shown in Fig. 4.9 for this purpose. For example, if the first measurement is made across 1200 Ω, and the second across 60 Ω, the resistance ratio is 20. The chart in Fig. 4.9 also shows that we must add or subtract 13 dB to the apparent gain or loss in decibels as read on the meter scale. If a chart is not available, the true dB value can be calculated as follows:

$$\text{Decibels} = 10 \log \frac{(E_o)^2/R_o}{E_{\text{in}}{}^2/R_i}$$

$$= 10 \log \frac{(E_o)^2}{(E_{\text{in}})^2} \frac{R_i}{R_o}$$

$$= 20 \log \frac{E_o}{E_{\text{in}}} + 10 \log \frac{R_i}{R_o} \tag{4.6}$$

$$\text{Decibels} = 10 \log \frac{I_o{}^2 R_o}{I_{\text{in}}{}^2 R_i}$$

$$= 10 \log \frac{(I_o)^2}{(I_{\text{in}})^2} \frac{R_o}{R_i}$$

$$= 20 \log \frac{I_o}{I_{\text{in}}} + 10 \log \frac{R_o}{R_i} \tag{4.7}$$

To continue the foregoing example, the ratio of R_i/R_o is equal to 20, and the log of 20 is equal to 1.3. Therefore, we must add 13 dB to the number of decibels gain indicated by the difference between the two scale readings; or we must subtract 13 dB from the number of decibel loss indicated by

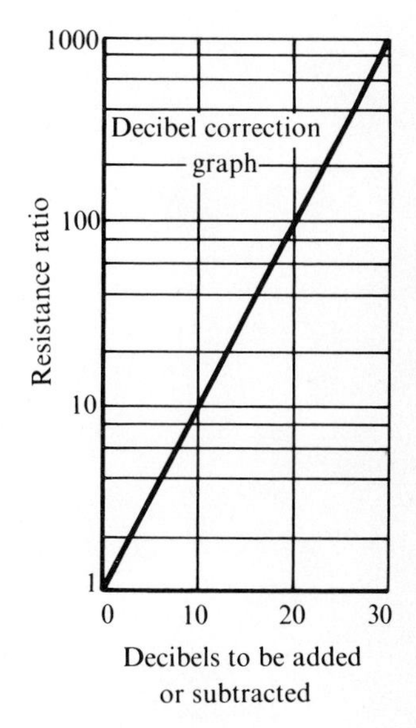

FIG. 4.9 *Decibel correction chart. (Courtesy, Simpson Electric Co.)*

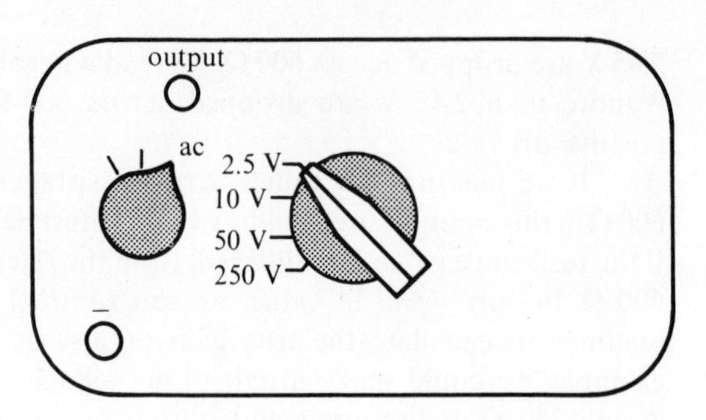

FIG. 4.10 *Output function ranges and terminals. (Courtesy, Simpson Electric Co.)*

the scale readings. Note that the same principle applies to ac current ratios, as stated by Formula (4.7). However, a VOM cannot measure ac current values conveniently.

Next, let us consider a subfunction of the ac voltage function provided by the VOM in this example. The ac voltage function indicates correct values only when no dc component is present in the sine-wave input. If we are measuring ac voltage at the collector of a transistor, for example, a dc component is present and we must then use the output function. When the output function is employed, a blocking capacitor is connected in series with the ac voltage-measuring circuit. Thus, the instrument responds to ac voltage but does not respond to a dc voltage component.

Figure 4.10 shows the control settings and terminals that are used on the output function of the VOM in this example. Either the ac voltage scales or the decibel scale may be used. Because a series blocking capacitor is inserted in series with the instrument circuit, the frequency characteristics of the instrument are affected to some extent. The most prominent effect is reduced low-frequency response, particularly on the 2.5-V range. Figure 4.11 depicts the frequency characteristics for the VOM in this example when operated on its output function. Circuit details are discussed subsequently.

4.3 BASIC TYPES OF VOLT-OHM-MILLIAMMETERS

Figure 4.12 illustrates the external appearance of the VOM under discussion. This is a comparatively accurate instrument of the utility type. It is provided with a mirrored scale to minimize parallax error, and is rated for an accuracy of ±1.5 percent of full scale on all dc voltage ranges up to

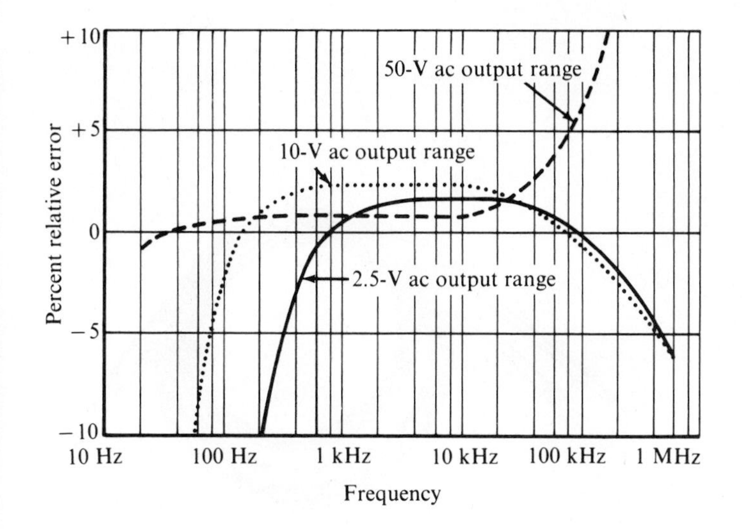

FIG. 4.11 *Frequency characteristics of a typical VOM on its output function. (Courtesy, Simpson Electric Co.)*

1000 V. Its accuracy rating on the 5000-V dc range is ±2.5 percent of full scale. On its ac voltage ranges up to 1000 V, the instrument is rated for an accuracy of ±3 percent of full scale; its accuracy rating on the 5000-V scale is ±4 percent of full scale. Sensitivity is 20,000 Ω/V on dc voltage ranges, and 5000 Ω/V on the ac voltage ranges.

Note that the output function has a rated input limit of 250 V, which is imposed by the rated working voltage of the series blocking capacitor. Rated accuracy on the dc current function is ±1 percent of full scale on the 50-μA range, and ±1.5 percent of full scale on the higher dc current ranges. The ohmmeter function is rated for an accuracy of ±2° of arc on the $R \times 1$ range, and ±1.5° of arc on the higher ranges. Note that the ohmmeter accuracy rating becomes less as the internal batteries weaken. The rated accuracies are specified for operation of the VOM in a horizontal plane. If operated in a vertical plane, or at various angles to a horizontal plane, the rated accuracy may be somewhat less, depending upon the residual unbalance in the meter mechanism.

Another basic type of VOM is illustrated in Fig. 4.13. This instrument is comparatively compact, and provides multiple terminals instead of a range switch. Its functions provide for measurement of dc and ac voltage, and resistance, but not dc current. Thus, the instrument is classified as a volt-ohmmeter. The decibel scale is omitted; however, dB measurements can be made with the aid of Table 4.1 and

FIG. 4.12 *Simpson volt-ohm-milliammeter Model 261. (Courtesy, Simpson Electric Co.)*

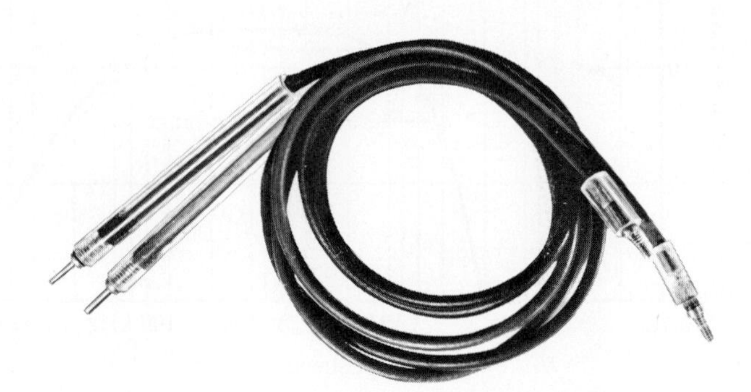

FIG. 4.13 *A VOM that provides multiple terminals instead of a range switch.* (*Courtesy, Simpson Electric Co.*)

Fig. 4.9. This type of VOM can be used as an output meter on its ac voltage function if a suitable blocking capacitor is connected in series with one of the test leads. The rated sensitivity of the instrument is 10,000 Ω/V on both dc and ac voltage functions.

Figure 4.14 illustrates a VOM that has a comparatively high sensitivity rating of 100,000 Ω/V on its dc voltage function. On its ac voltage function, the instrument is rated for a sensitivity of 5000 Ω/V. Rated accuracies are ±3 percent of full scale on dc voltage ranges, and ±5 percent of full scale on ac voltage ranges. An external multiplier probe is provided for operation on the 4000-V dc range. The chief advantage of a high-sensitivity VOM is its reduced current demand on voltage-measuring functions, which reduces measurement error due to circuit loading in high-resistance or high-impedance circuits.

Since a multiplicity of scales presents possible confusion to inexperienced operators, the design illustrated in Fig. 4.15 may be preferred for use on production lines or in school laboratories. Note that a corresponding scale is rotated into view on each position of the range switch. A mirror is provided to minimize parallax error. Rated sensitivity of the VOM is 20,000 Ω/V. The range switch does double duty as a function switch, except that a separate polarity-reversing switch is provided. As in the case of the VOMs discussed previously, the dc voltage function can be extended for measurement of

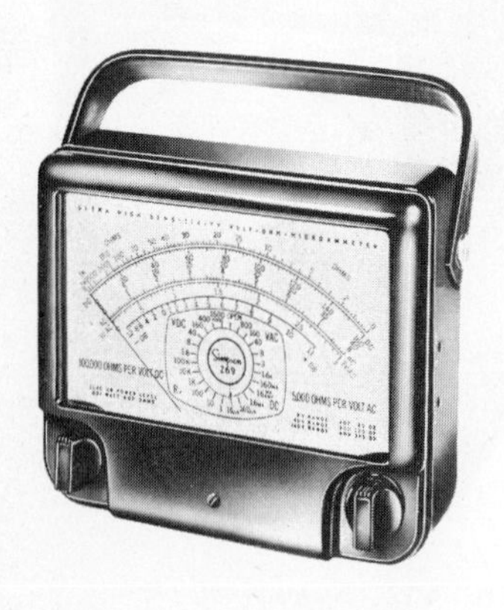

FIG. 4.14 *A 100,000-Ω/V VOM.* (*Courtesy, Simpson Electric Co.*)

voltages up to 25,000 V by using a supplementary high-voltage dc probe.

FIG. 4.15 *A VOM that provides individual scales for each range. (Courtesy, B & K Manufacturing Co.)*

Economy-type VOMs have a sensitivity of 1000 Ω/V on both dc and ac voltage functions. A typical instrument in this category is illustrated in Fig. 4.16. It is a complete VOM, although a decibel scale is omitted. Because of its comparatively low input resistance, this type of VOM is used chiefly for testing in circuits that have moderate or low internal resistance (or impedance). It can be used to measure dc voltages up to 25,000 V by means of a supplementary high-voltage dc probe. As explained subsequently, a probe used with a 1000 Ω/V VOM has a lower resistance than a probe used with a 20,000 Ω/V VOM.

A few VOMs feature extremely high sensitivity, such as the 1,000,000 Ω/V instrument illustrated in Fig. 4.17. The dc voltage ranges are from 0.6 to 6000 V. Thus, the input resistance of the instrument is 600,000 Ω on its 0.6-V range, and 6000 MΩ on its 6000-V range. Although this type of VOM has very high sensitivity, it is classified as a utility instrument because its accuracy is considerably less than that of laboratory-type instruments. For example, the VOM in this example has a rated accuracy of ± 1.5 percent of full scale on its dc voltage function. On the other hand, a moderately accurate laboratory voltmeter has a rated accuracy of ± 0.5 percent of full scale, and a highly accurate instrument is rated for an accuracy of ± 0.1 percent of the scale reading.

The VOM illustrated in Fig. 4.17 has a rated sensitivity of 20,000 Ω/V on its ac voltage ranges, and has uniform frequency response up to 100 kHz. The ohmmeter function indicates resistance values up to 100 M. Center-scale indication on the $R \times 1$ range is 4 Ω. The first dc current range is from 0 to 120 μA. Although a 0 to 1 μA dc current range could be provided, extremely small current values are not measured in the fields of application for which the instrument is designed. A suspension meter movement is used in this type of VOM to provide its unusually high sensitivity.

FIG. 4.16 *A 1000-Ω/V VOM. (Courtesy, Triplett Elec. Inst. Co.)*

4.4 APPLICATIONS

A VOM can be used to check a coupling capacitor for leakage (in a radio receiver, for example) as shown in Fig. 4.18. In the first method, it is good practice to remove (pull) the tube that is driven by the coupling capacitor in order to eliminate possible

FIG. 4.17 *A 1,000,000-Ω/V VOM. (Courtesy, Triplett Elec. Inst. Co.)*

confusion from grid contact potential. If the grid resistor R is not grounded, the avc[1] (or agc[1]) line should be short-circuited to ground. The VOM is connected across the grid resistor, and is operated on its dc-voltage function. In most cases, the lowest dc-voltage range will be employed. Any dc-voltage reading indicates that the coupling capacitor is leaky and should be replaced. It is evident that a VOM with a high ohms-per-volt rating will provide a more sensitive test than a VOM with a low ohms-per-volt rating.

When a VOM with a high ohms-per-volt rating is used, maximum sensitivity in a coupling-capacitor leakage check is provided by a disconnect test as depicted in Fig. 4.18*b*. This method provides maximum sensitivity because the grid resistor does not provide a shunt path for current flow in parallel with the VOM. It is advisable to make initial tests with the VOM set to a range that is greater than the $B+$ value. This precaution ensures that the meter will not be burned in the event that the capacitor happens to be short-circuited. If the pointer does not deflect visibly, the operating range can then be reduced as required. If no pointer deflection is visible on the lowest VOM range, the coupling capacitor is judged to be in good condition within the limits of the test method.

A VOM can be used to measure avc or agc voltage values in a radio or television receiver. However, if the circuit has a high internal resistance, a loading error may need to be taken into account. The loading error can be minimized by various expedients, one of which will be described below. With reference to Fig. 4.19, the agc circuit illustrated has a comparatively high internal resistance. When the value of the control voltage is measured with the VOM, the indicated voltage depends upon the input resistance of the meter. As we have learned, the input resistance is different for each range of the instrument. The effect of range switching is exemplified in Fig. 4.20.

Let us consider how loading error can be minimized by means of an expedient called the range-switching correction method. We connect the VOM test leads to the circuit under test (such as an agc line) and observe the meter readings on two of the dc-voltage ranges. If the circuit has substantial internal resistance, the two readings will differ. We calculate the actual voltage value from the formula:

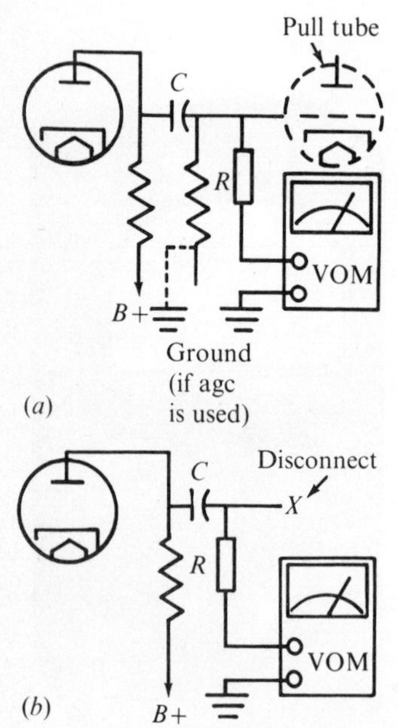

FIG. 4.18 *Testing capacitors with a VOM. (a) In-circuit test for capacitor leakage; (b) disconnect test for capacitor leakage.*

[1] Abbreviations for automatic volume control and automatic gain control.

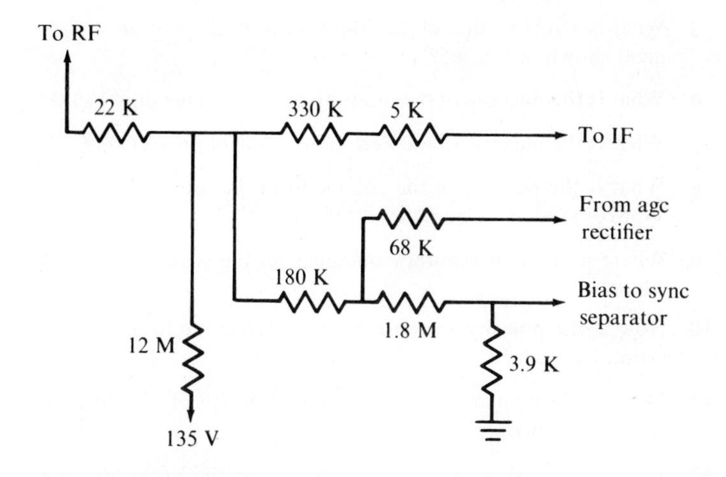

FIG. 4.19 *Ac circuit with substantial internal resistance for point X.*

$$E = \frac{E_1 E_2 (R_2 - R_1)}{E_1 R_2 - E_2 R_1} \tag{4.8}$$

where E = true agc voltage
E_1 = meter reading of first range setting
E_2 = meter reading of second range setting
R_1 = input resistance of VOM for first range setting
R_2 = input resistance of VOM for second range setting

The range-switching correction method is strictly valid only for linear systems. However, a conventional avc or agc system approximates a linear system, so that the nonlinearity error involved in the measurement and calculation is practically negligible.

Range, V	Indication, V	Actual error, %
2.5	−1.7	78
10.0	−4.1	47
50.0	−6.6	15

FIG. 4.20 *Typical ac voltage indications with a 20,000-Ω/V VOM. True agc voltage is −7.8 V.*

QUESTIONS AND PROBLEMS

1. What is the ohms-per-volt rating of a typical VOM?
2. What are two important advantages of a VOM?
3. What is the function of a high-voltage probe and how does it serve that function?
4. What is the condition of the polarity switch of the VOM shown in Fig. 4.5 when the 10-A range is selected?

5. What is a subfunction of the 50-μA current range of the instrument shown in Fig. 4.5?
6. What is the purpose of the multiplying scale factor on a VOM?
7. What is the purpose of the Zero ohms control on a VOM?
8. What is the polarity of the voltage from the test leads when a VOM is used as an ohmmeter?
9. Where is the best accuracy obtained on the scale of an ohmmeter?
10. How is the polarity switch used to measure the front-to-back ratio of a diode?
11. Do the polarity markings on the input terminals of a VOM have any significance in the ac position? Why?
12. Why is the 2.5-V ac only scale employed on the VOM featured in Fig. 4.3?
13. In reference to Fig. 4.8, compare the frequency characteristics for the ac scales.
14. What is the purpose of the decibel chart shown in Table 4.1?
15. What is the purpose of the output function on a VOM?
16. Why does a VOM have greater low-frequency distortion when used on the output function position?
17. Why is the output function of a VOM limited to a lower voltage reading than the ac function?
18. Does the position in which an ohmmeter is placed when reading resistance have any effect on the accuracy? Explain your answer.
19. What effect does the weakening of the internal battery have on the accuracy of an ohmmeter?
20. Explain how you would make a dB measurement with the VOM shown in Fig. 4.13.
21. Explain how you would measure an ac voltage that has a dc component, using the voltmeter shown in Fig. 4.13.
22. What is the advantage of the multimeter shown in Fig. 4.15 over other types?
23. Can the high-voltage probe that is used with a 20,000-Ω/V meter be used with a 1000-Ω/V instrument?
24. Why isn't the VOM shown in Fig. 4.12 classified as a laboratory instrument?

25. Suppose the chart in Fig. 4.9 is valid for a VOM used to measure an input voltage of 10 mV across 600 Ω and an output voltage across the 100-Ω output of 10 V; what is the dB gain of the amplifier?

26. The input resistance of an amplifier is 2 K and the output resistance is 100 Ω; what is the dB gain if a VOM represented by the chart in Fig. 4.9 measures an input of 1 V and an output of 1 V?

27. The following values are read for an amplifier with a VOM represented by the chart in Fig. 4.9; V_{in} = 100 V across 1 K; V_{out} = 200 mV across 3 K. What is the dB gain of the amplifier?

VOLT-OHM-MILLIAMMETER CIRCUITRY

5.1 BASIC VOM

Circuit configurations used in volt-ohm-milliammeters are based on the principles explained in Chap. 2. VOM circuitry is comparatively complex because of the various functions and multiplicity of ranges that are provided. Numerous multipliers and shunts are utilized in combination with the necessary switching circuits. Since a VOM is essentially a utility instrument, it is often applied by inexperienced personnel. In turn, the design of instrument circuitry includes various overload protection features, such as fast-acting relays, fuses, and silicon diodes. Ring-shunt configurations are always used in the current-measuring section.

The meter mechanism is seldom fused, for reasons that were explained in Chap. 2. To review briefly, an instrument fuse has a nonlinear resistance characteristic that calls for specially calibrated scales. Such a requirement would not be objectionable if instrument fuses could be manufactured to close tolerances. Their wide production tolerances, however, create replacement difficulties, and thus the modern design trend usually avoids this method of overload protection. Large fuses are often utilized in the ohmmeter section of a VOM, for they can be operated at comparatively low temperature, thereby minimizing their nonlinear resistance characteristic.

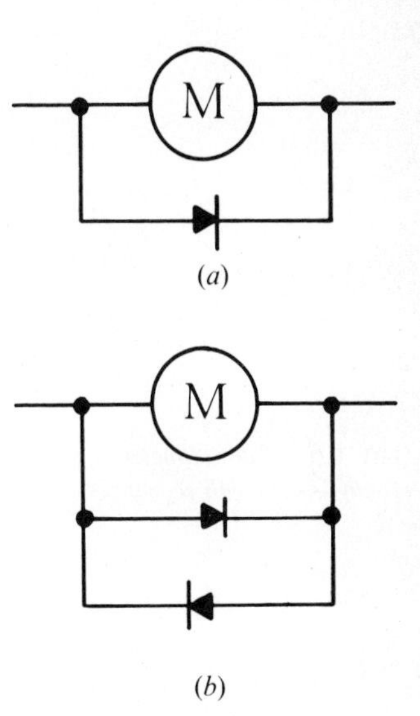

FIG. 5.1 *Meter mechanism protection. (a) Single-diode protection against overload in one direction; (b) two-diode protection for both directions of current.*

Silicon diodes have a practical advantage over instrument fuses for overload protection of meter mechanisms. A silicon diode is connected directly across the mechanism, as depicted in Fig. 5.1. Utilizing a pair of diodes in reverse polarity protects the mechanism against overload regardless of the direction of the current. That is, if the operator applies the test leads in either polarity to a voltage source, the arrangement in Fig. 5.1*b* will provide protection against overload damage to the meter movement. As shown in Fig. 5.2, approximately 0.6 V must be applied in the forward direction before a silicon diode conducts appreciably. Thereafter, the internal resistance of the diode falls to a very low value.

The voltage drop across the meter terminals at full-scale indication is less than 0.6 V. If the pointer is driven off-scale, the protective diode then conducts, and excess current is shunted around the meter terminals. This feature prevents accidental burnout of the D'Arsonval coil, but it does not protect resistors in the instrument circuitry against overload damage. Therefore, fuse protection is also provided in many instruments, particularly in the ohmmeter circuit. This fuse will blow if the ohmmeter test leads are accidentally applied to a voltage source. Additional VOM protection features are explained subsequently.

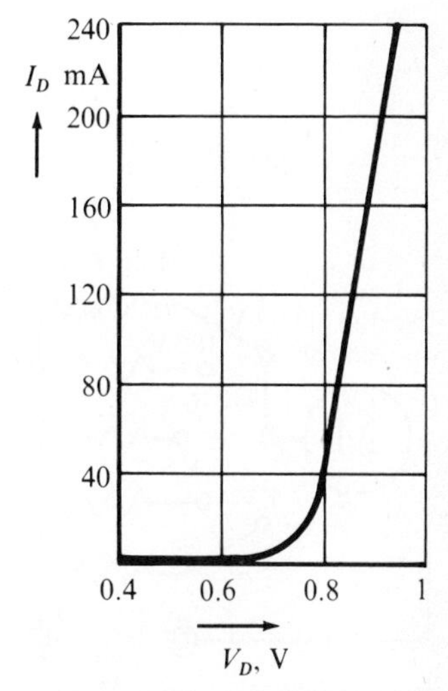

FIG. 5.2 *Silicon diode that starts to conduct at approximately 0.6 V in the forward direction.*

5.2 DC VOLTAGE-MEASURING CIRCUITRY

A typical VOM provides six dc voltage ranges, as shown in Fig. 5.3. In this example, the instrument is switched to its 2.5-V range. The test leads are plugged into the Pos and Neg terminals. Since the meter movement has an internal resistance of 5000 Ω, the total resistance of the instrument circuit is

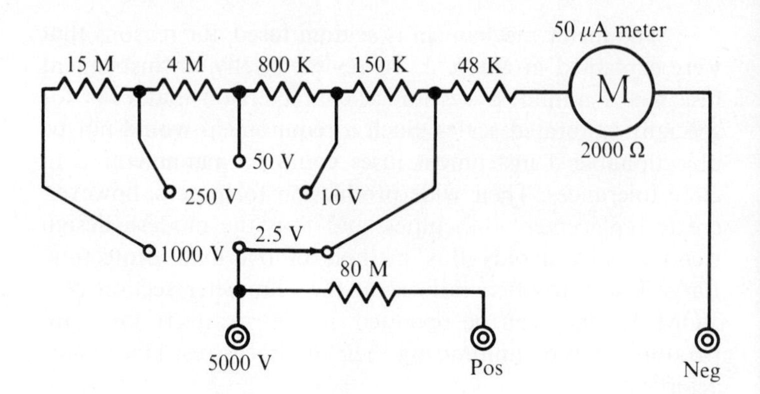

FIG. 5.3 *A dc voltmeter configuration with six ranges.*

50,000 Ω on its 2.5-V range. In turn, the voltmeter circuit draws 50 μA at full-scale indication:

$$I = \frac{2.5}{50,000} = 50 \times 10^{-6} \text{ A} \tag{5.1}$$

The student may show that the instrument circuit draws 50 μA at full-scale indication on each of the remaining five ranges. Note that the test leads are plugged into the 5000-V dc and Neg terminals when operating on the 5000-V range. A separate terminal is required on this range, because a rotary switch cannot withstand potentials in excess of 1000 V. It is not necessary to employ a series multiplier circuit, as exemplified in Fig. 5.3. Separate multiplier resistors are sometimes provided on each range, as shown in Fig. 5.4. The only disadvantage of the arrangement in Fig. 5.4 is that high-value multiplier resistors must dissipate somewhat more power than is required in a series configuration.

With reference to Fig. 5.3, the student may show that the sensitivity of the voltmeter circuit is 20,000 Ω/V on each range. If the meter movement is employed directly to obtain a low-voltage range, the full-scale indication will be 100 mV:

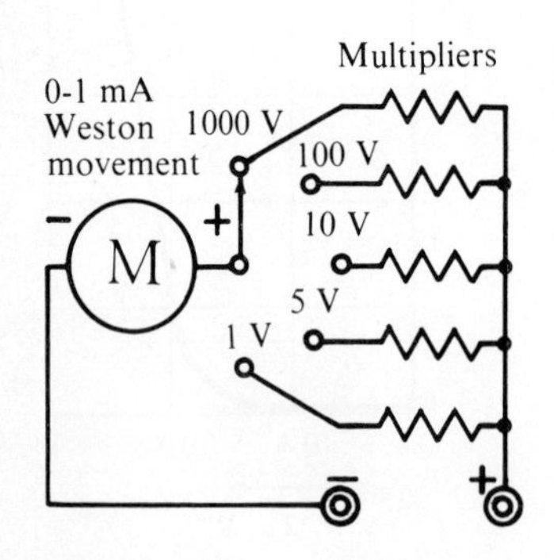

FIG. 5.4 *Another multiplier configuration.*

$$E = IR = 50 \times 10^{-6} \times 2000 = 0.1 \text{ V} \tag{5.2}$$

It is evident that this low-voltage range also provides a sensitivity of 20,000 Ω/V:

$$\text{Ohm/V} = 2000/0.1 = 20,000 \ \Omega/\text{V} \tag{5.3}$$

Note that the voltmeter sensitivity on any range is also given by the reciprocal of the full-scale current demand:

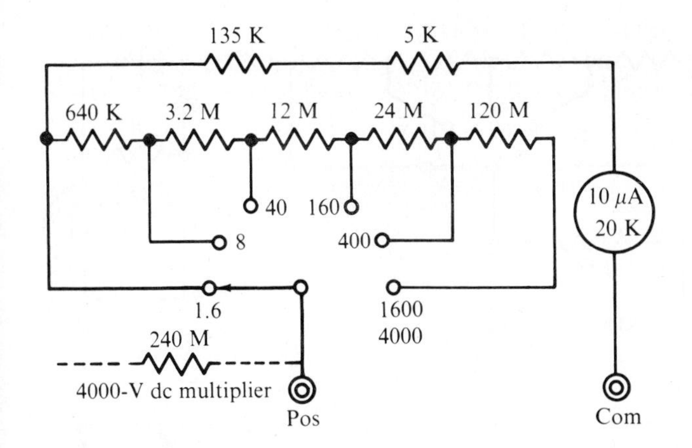

FIG. 5.5 *Configuration of a 100,000-Ω/V dc voltmeter.*

$$\text{Ohm/V} = \frac{1}{50 \times 10^{-6}} = 20{,}000\ \Omega/\text{V} \tag{5.4}$$

When a 10-μA D'Arsonval movement is utilized, as depicted in Fig. 5.5, the sensitivity of the dc voltmeter is 100,000 Ω/V. In this example, the internal resistance of the meter movement is 20,000 Ω, and the voltage drop across its terminals at full-scale indication is 200 mV. The chief advantage of high sensitivity is reduced circuit loading. For example, let us consider the test circuit depicted in Fig. 5.6. We observe that the instrument resistance is 1 M, in accordance with Ohm's law. Therefore, the effective load resistance is reduced to 0.5 M by the voltmeter circuit, and the scale indication is 25 V instead of the true value of 37.5 V. On the other hand, if we use a 100,000-Ω/V instrument, the scale indication will be 34.1 V, approximately.

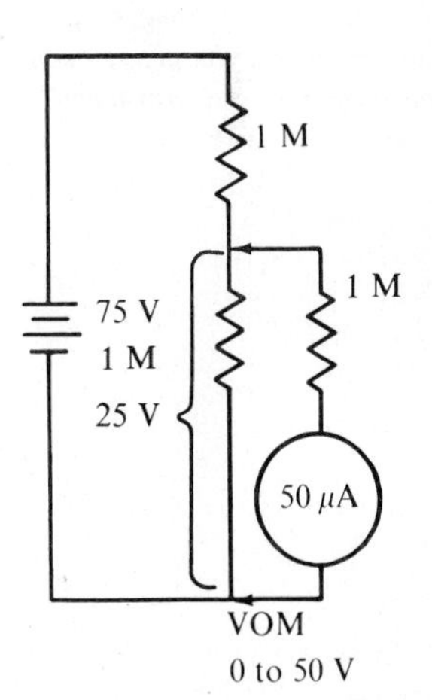

FIG. 5.6 *Voltage measurement in a high-resistance circuit.*

5.3 AC VOLTAGE-MEASURING CIRCUITRY

A typical VOM provides six ac voltage ranges, as depicted in Fig. 5.7*a*. The ac multiplier configuration is basically similar to the dc arrangement in Fig. 5.3, except that the input resistance on each of the ac voltage ranges is one-twentieth of the input resistance on a corresponding dc voltage range. In other words, the sensitivity of the ac voltmeter configuration in this example is 1000 Ω/V. There are two basic reasons for this comparatively low sensitivity on the ac volt function of the instrument:

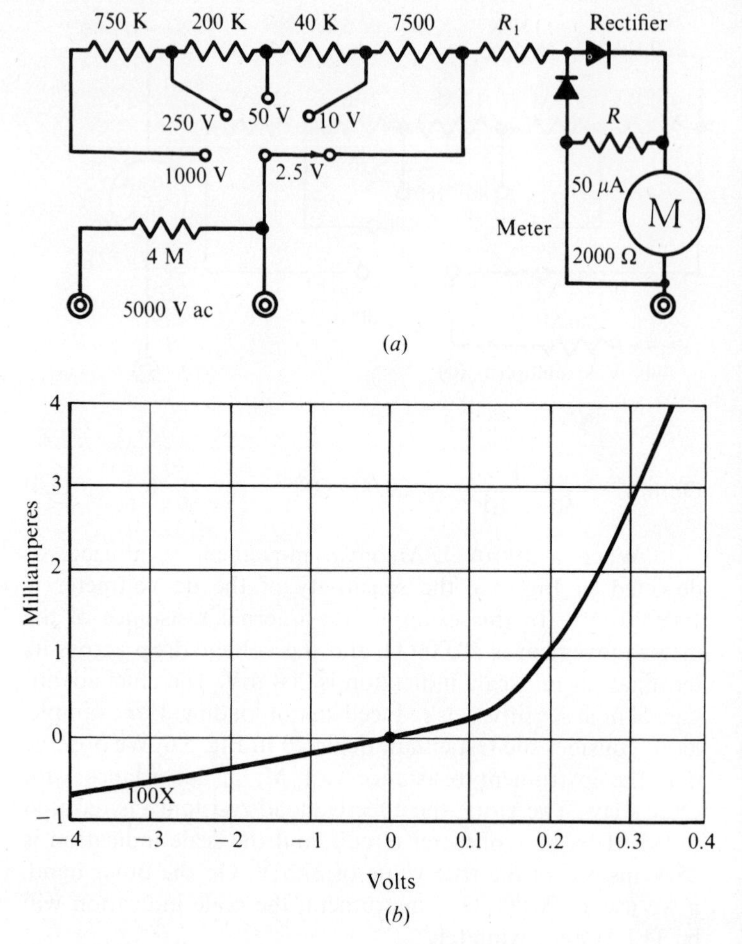

FIG. 5.7 *(a) An ac voltmeter configuration with six ranges; (b) voltage-current characteristic of a typical copper-oxide instrument rectifier.*

1. The applied ac voltage is necessarily rectified, and the meter movement responds to 0.318 of peak value in this example. In turn, the sensitivity of the ac voltmeter circuit is thereby reduced to 31.8 percent of its dc value, or 6360 Ω/V.
2. Rectification occurs on a highly nonlinear characteristic, as seen in Fig. 5.7*b*. In turn, a swamping resistor *R* is employed to minimize "bunching" at the lower end of the ac scale. The result is a further reduction in ac voltmeter sensitivity to 1000 Ω/V.

We know that bunching at the low end of the ac scale is caused by the curvature in the forward-current characteristic of the instrument rectifier. In other words, a semiconductor diode has a comparatively high internal resistance when a small forward voltage is applied, and a comparatively low internal resistance when a substantial forward voltage is applied. Next, we recall that a fixed resistor, such as a wire-wound resistor, has the same resistance value regardless of the applied voltage. Therefore, if we connect a fixed resistor in series or in shunt with a semiconductor diode, the resulting circuit characteristic will be more linear than the diode alone, but will be less linear than the resistor alone. Accordingly, the design engineer makes a tradeoff between improved linearity and reduced sensitivity when he utilizes a swamping resistor in combination with an instrument rectifier.

Note in Fig. 5.7*a* that the current demand of the meter movement is supplied by the series rectifier. The shunt rectifier is provided only to *bypass* each alternate half cycle of ac voltage around the series rectifier and meter movement. This is called a *three-terminal half-wave* arrangement. Inclusion of the shunt rectifier minimizes the reverse voltage that must be withstood by the series rectifier. Accordingly, the series rectifier is protected against breakdown, particularly on the higher ac voltage ranges. Similarly, the series rectifier protects the shunt rectifier against reverse-voltage breakdown.

Since the rated sensitivity of the ac voltmeter depicted in Fig. 5.7*a* is 1000 Ω/V, its input resistance on the 2.5-V range is approximately 2500 Ω. This input resistance is somewhat less when the shunt rectifier is conducting. When the series rectifier conducts, the total input resistance comprises the sum of the multiplier (calibrating) resistance R_1, the forward resistance of the series rectifier, and the parallel resistance of the meter movement and its swamping resistance R. This total input resistance tends to increase at low input voltages, owing to increase in the forward resistance of the rectifier, as seen in Fig. 5.7*b*.

To review briefly, the meter movement in Fig. 5.7*a* responds to the average value of a half-rectified sine wave, or to 0.318 of peak value. On the other hand, the ac voltage scales are calibrated to indicate rms values of sine waves, or 0.707 of peak value. Therefore, if we apply a dc voltage in the forward direction to the ac voltmeter configuration, the pointer will indicate 2.22 times the value of the dc voltage:

$$E_{\text{ind}} = \frac{\text{rms voltage}}{\text{average voltage}} = \frac{0.707}{0.318} = 2.22 + \qquad (5.5)$$

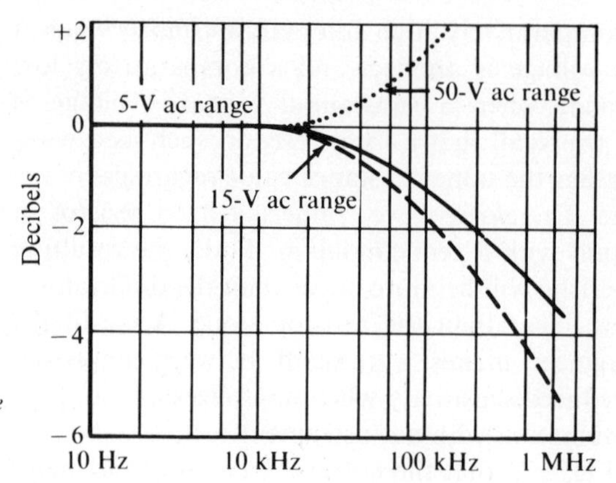

FIG. 5.8 *Rated frequency response of a VOM. (Courtesy, Simpson Electric Co.)*

If we apply a dc voltage in the reverse direction to the ac voltmeter configuration, the pointer remains practically at zero, since the shunt rectifier bypasses nearly all the current around the indicating circuit. Note that the exact values for R_1 and R are selected in production to compensate for tolerances on instrument rectifiers, so that maximum indication accuracy is obtained. Rectifier characteristics tend to change with age; the forward resistance gradually increases, and the reverse resistance gradually decreases. Accordingly, periodic calibration checks are advisable.

It is evident that the ac voltmeter circuit depicted in Fig. 5.7*a* is not purely resistive—the moving coil has inductance and distributed capacitance, the rectifiers have junction capacitance, and the multiplier system has stray capacitance. In turn, the frequency response of the instrument circuit is limited to approximately 10 kHz, and high-frequency errors are different on each range, as seen in Fig. 5.8. The practical low-frequency limit is approximately 10 Hz, inasmuch as pointer vibration becomes objectionable at lower frequencies.

5.4 AC VOLTMETER FUNCTION AT 5000 Ω/V

The configuration for a 100,000-Ω/V dc voltmeter was shown in Fig. 5.5. Since a 10-μA meter movement is utilized, the VOM operates at a sensitivity of 5000 Ω/V on its ac voltage function. Figure 5.9 depicts the corresponding ac voltmeter circuit. As in the foregoing example, a three-terminal half-wave rectifier configuration energizes the meter movement; however, germanium diodes are used instead of copper-oxide instrument rectifiers. The lower junction capacitance of a germanium diode provides extended frequency response, particularly on the low ac voltage ranges. For example, the ac voltmeter is rated for full accuracy up to 200 kHz on its 3-V ac range.

Since the forward resistance of a germanium diode is very small in comparison with the resistance of the instrument circuit in Fig. 5.9, we can neglect this forward resistance value in a preliminary analysis. When the range switch is set to its 3-V position, the input resistance is 15,000 Ω with respect to forward conduction of the series diode. Note that the meter movement is connected in series with an 85-K resistor and a 5-K resistor; in turn, these components are shunted by a 13-K resistor and a 2-K resistor connected in series. Thus, the series diode works into a resistance of 13,200 Ω:

$$R = \frac{(85\text{ K} + 5\text{ K} + 20\text{ K})(13\text{ K} + 2\text{ K})}{85\text{ K} + 5\text{ K} + 20\text{ K} + 13\text{ K} + 2\text{ K}} = 13.2\text{ K} \qquad (5.6)$$

Next, we observe that this value of 13.2 K is connected in series with an 1800-Ω resistor during forward conduction of the series diode. In turn, the total input resistance is 15,000 Ω on the 3-V ac range. Since half-wave rectification is utilized, full-scale deflection will be obtained if $3 \div 2.22 = 1.35$ dc volts are applied in the forward direction. The student may show that the *nominal* meter current is then slightly in excess of 10 μA. In actual operation, this nominal value is reduced to

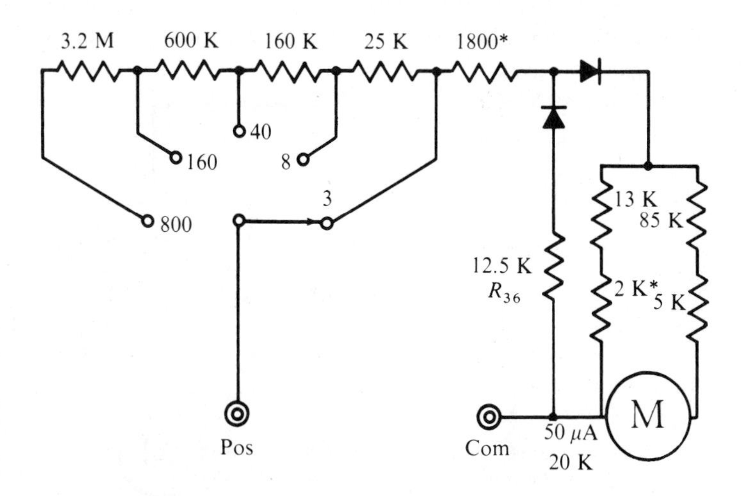

FIG. 5.9 *A 5000-Ω/V ac voltmeter configuration. (Courtesy, Simpson Electric Co.)*

10 μA due to finite forward resistance of the series diode. The student may also show that the input resistance of the circuit is approximately 14,300 Ω during conduction of the shunt diode.

Full-wave instrument-rectifier circuits are used in some VOMs. Either a full-wave bridge or a full-wave half bridge, as depicted in Fig. 5.10, may be employed. A four-diode bridge is more efficient, but presents greater production difficulties because of tolerances on the diodes. A two-diode bridge requires less time in selection of suitable diode pairs from production stock. Full-wave rectification provides twice the sensitivity of half-wave rectification, because the average value of a full-rectified sine wave is 0.636 of peak value.

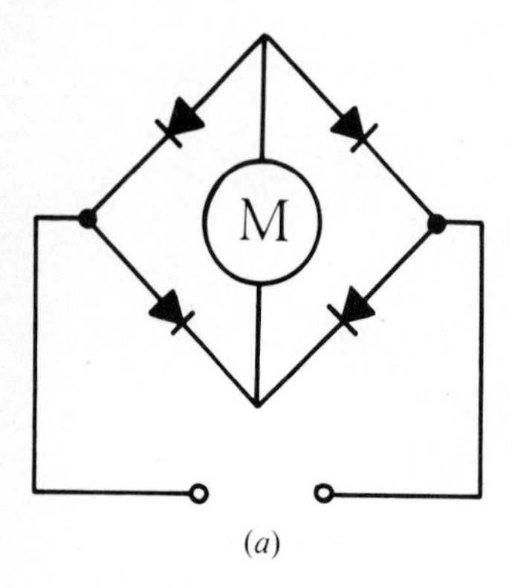

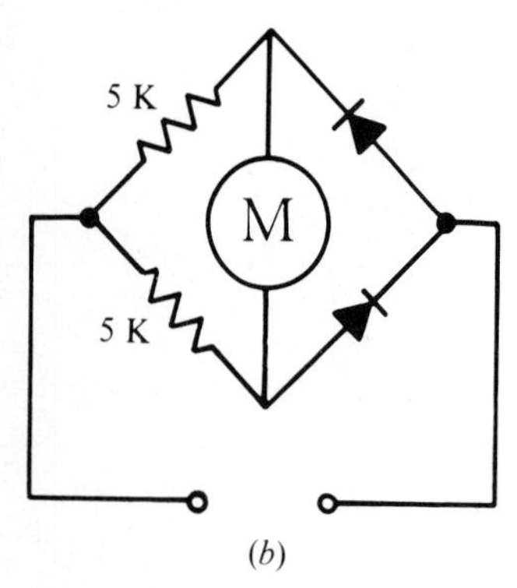

FIG. 5.10 *Bridge circuit. (a) A full-wave instrument-rectifier bridge; (b) a full-wave half-bridge configuration.*

5.5 OUTPUT-METER FUNCTION

As noted previously, the output-meter function of a VOM is similar to its ac voltage function except that a blocking capacitor is connected in series with the instrument circuit, as shown in Fig. 5.11. This capacitor permits passage of ac but blocks passage of dc. In turn, a VOM may be applied on its output-meter function to measure the ac voltage at, say, the collector of a transistor where the voltage is a pulsating dc, as depicted in Fig. 5.12. Unless the output-meter function is used, the dc component of the waveform would cause the ac voltage indication to be in serious error.

From an application viewpoint, the chief limitation of this output-meter function is encountered in low-frequency

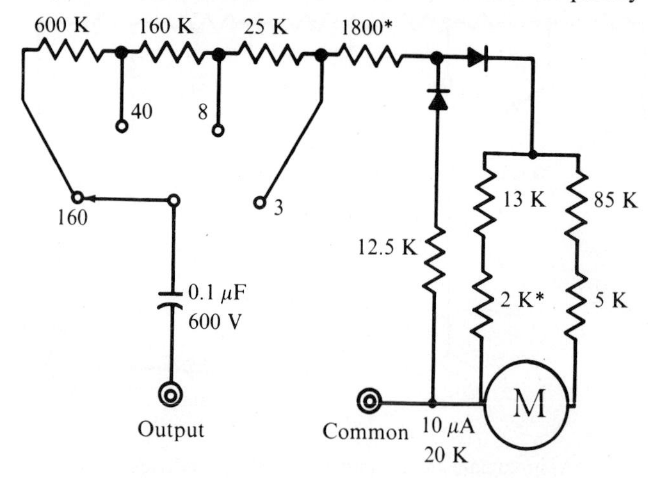

FIG. 5.11 *Typical output-meter configuration. (Courtesy, Simpson Electric Co.)*

tests. For example, the reactance of the 0.1-μF capacitor in Fig. 5.11 becomes appreciable at low frequencies, and indication accuracy is progressively impaired at frequencies below 1 kHz on the 3-V range. On the 8-V range, accuracy is maintained down to 250 Hz owing to the higher input resistance on this range. A blocking capacitor also reduces the indication accuracy at high frequencies because of resonances within the instrument circuit. For example, on the 40-V range, the rated upper frequency limit is 4.5 kHz.

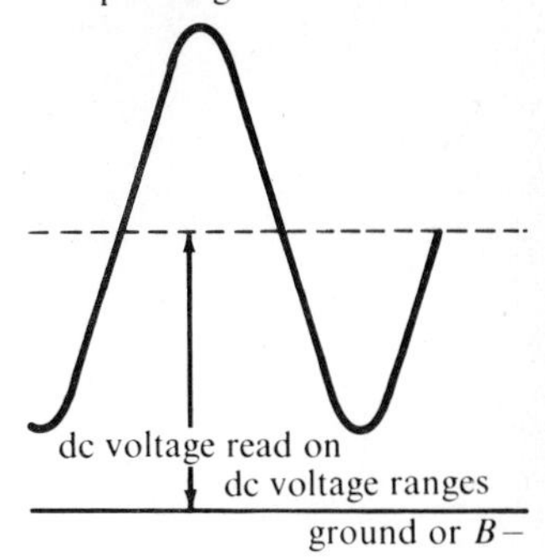

FIG. 5.12 *An example of a pulsating dc voltage.*

5.6 DC CURRENT FUNCTION

We have learned that a milliammeter consists basically of a meter movement connected across a resistive shunt. A VOM provides several dc current ranges by means of a *ring-shunt* configuration, as seen in Fig. 5.13*a*. The chief advantage of a ring shunt is that the meter movement always remains shunted when the range switch is turned from one position to another. The movement is protected against accidental burnout when switching ranges with the test leads connected into a "live" circuit. Note that the 10-A range is brought out to a separate terminal. This feature provides maximum indication accuracy, inasmuch as switch-contact resistance is thereby eliminated on the high-current range.

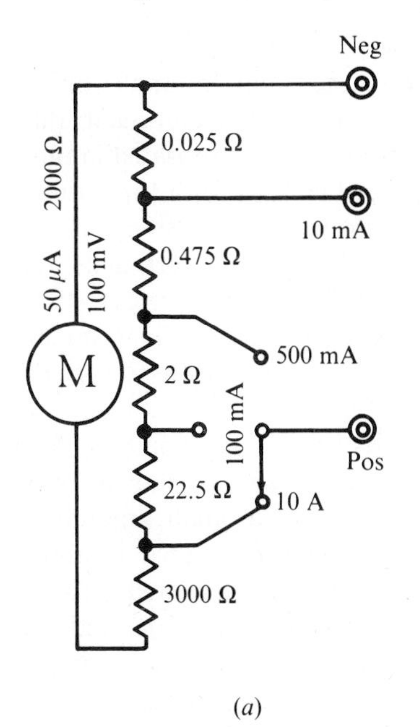

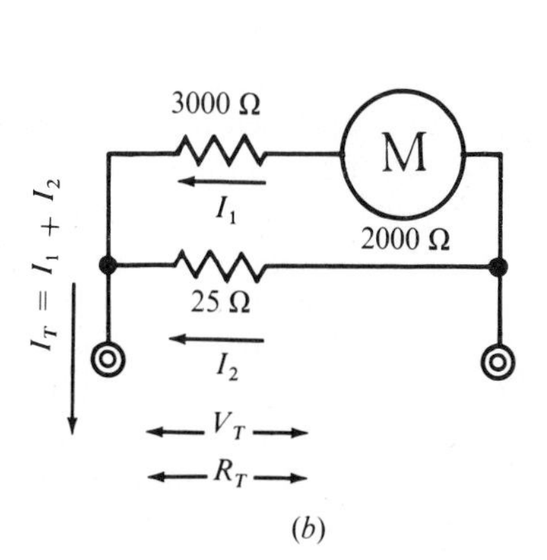

FIG. 5.13 *Basic ammeter circuit. (a) Ring-shunt configuration; (b) equivalent circuit.*

Let us consider the operation of the milliammeter on its 10-mA range. The equivalent circuit is depicted in Fig. 5.13*b*. When the test leads are connected in series with a 10-mA current source, the following relations are established in accordance with Ohm's and Kirchhoff's laws:

$$R_T \simeq 25\ \Omega \tag{5.7}$$

Note that the input resistance of the circuit is 25 Ω from a practical viewpoint, since 5000 Ω is 200 times as great as 25 Ω. In turn, a 10-mA current produces a terminal voltage drop that is formulated practically:

$$V_T = 25 \times 0.01 \simeq 0.25\ \text{V} \tag{5.8}$$

Since the drop across the meter movement is given by Kirchhoff's voltage law, we write

$$V_M = \frac{2000 \times 0.25}{5000} \simeq 100\ \text{mV} \tag{5.9}$$

Thus, a current of 10 mA produces full-scale pointer deflection from a practical point of view. It is left as an exercise for the student to calculate the error that results from the assumption that $R_T = 25\ \Omega$, and to compare this error with rated accuracy of the milliammeter (± 3 percent of full scale).

5.7 OHMMETER FUNCTION

We know that an ohmmeter employs an internal voltage source and can be regarded as an analog computer that indicates solutions to Ohm's law on a scale calibrated in ohm values. Figure 5.14 shows simplified circuits for a three-range ohmmeter. Let us consider instrument operation on the $R \times 1$ range. If we short-circuit the ohmmeter terminals, the zero-adjust control can be set to bring the pointer exactly to the zero calibration on the ohms scale. Then, if the terminals are open-circuited, the pointer will fall back to the ∞ calibration on the scale.

Observe that the input resistance is nominally 11.5 Ω in Fig. 5.14*b*. However, the 1.5-V cell has a small internal resistance, and the test leads also have a small amount of resistance. Therefore, the actual input resistance is about 12 Ω. This means that if the test leads are connected across a 12-Ω resistor, the pointer will deflect to half scale, and the resistance indication will be 12 Ω. Or, if the test leads are

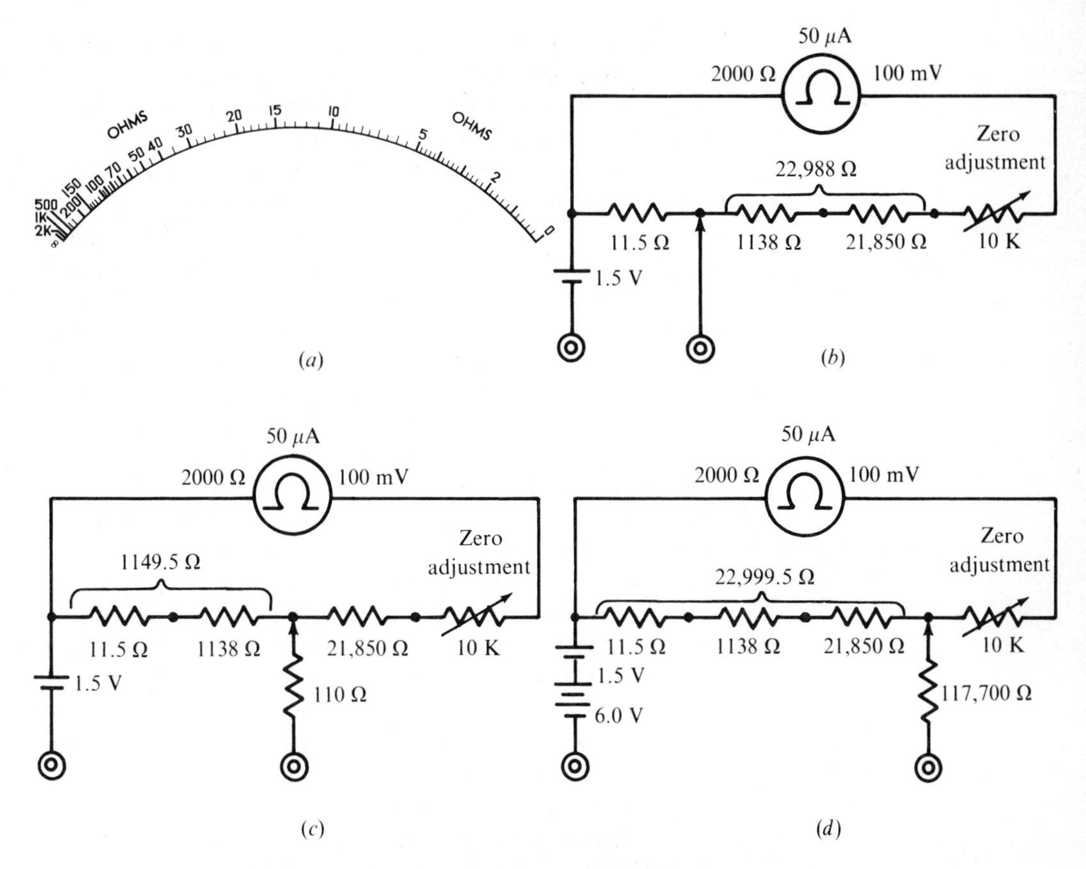

FIG. 5.14 *Basic ohmmeter. (a) Ohmmeter scale; (b) nominal circuit on R × 1 range including resistance of test leads (not shown); (c) a 110-Ω resistor switched in on the R × 100 range to compensate for resistance of test leads; (d) a 117,700-Ω resistor switched in on the R × 10,000 range to compensate for resistance of test leads.*

connected across a 6-Ω resistor, the pointer will deflect to the 6-Ω point on the scale. It is left as an exercise for the student to show why the ohmmeter scale is nonlinear.

It is desirable to employ the same scale on all ranges of the ohmmeter. This requires that the circuit proportions remain fixed on each range. For example, let us consider ohmmeter operation on the $R \times 100$ range in Fig. 5.14*c*. To maintain a proportional resistance in the test-lead and cell circuit, a 110-Ω resistor is automatically switched in series with one of the test leads. As before, if the ohmmeter terminals are short-circuited, the pointer will indicate zero on the ohms scale (the zero-set control may have to be reset to compensate for component tolerances). If the ohmmeter terminals are open-circuited, the pointer falls back to ∞ on the scale; if the test leads are connected across a 120-Ω resistor, the pointer deflects to half scale and indicates 120 Ω.

Next, let us consider ohmmeter operation on the $R \times 10{,}000$ range, as depicted in Fig. 5.14*d*. Because additional battery voltage is required on this high-resistance range, a 6-V battery is automatically switched in series with the 1.5-V cell. Note that a 117.7-K resistor must now be included in series with one of the test leads to maintain the basic circuit proportions. If the test leads are connected across a 120-K resistor, the student may show that half-scale deflection results. Note that ohmmeter accuracy becomes impaired as the internal battery ages, because the zero-adjust control must then be set to a subnormal value. This tends to upset the proportions of the nominal ohmmeter circuit.

5.8 COMPLETE VOM CONFIGURATION

The complete configuration for a simple VOM is shown in Fig. 5.15. This instrument provides dc voltage, ac voltage, ohmmeter, dc current, decibel, and output-meter functions. Half-wave instrument rectification is used, and ac voltages are calibrated in rms values for sine waves. Both dc and ac voltages to 5000 V can be measured; dc currents to 10 A are provided for. A maximum resistance value of 20 M can be indicated. The instrument sensitivity is 20,000 Ω/V on the dc function, and 1000 Ω/V on the ac function. No protective diodes or fuses are included in this configuration.

A slightly more elaborate VOM configuration is depicted in Fig. 5.16. In this arrangement, a silicon diode is provided across the meter movement for overload protection. Note that a 1-A fuse is also included in series with the common lead to prevent burnout of the ohmmeter shunt resistors in the event the test leads are accidentally applied in a live circuit on the ohmmeter function. Germanium diodes are used in a half-bridge full-wave rectifier circuit on the ac voltage function. Adjustable series and shunt resistors are provided in the meter-movement circuit as a maintenance feature for compensating magnet aging in the meter mechanism. As the permanent magnet becomes weaker with age, the associated resistors can be readjusted as required to restore calibration accuracy.

5.9 METER APPLICATION NOTES

Difficulties in voltage measurement due to circuit loading have been previously explained. A useful quick check to

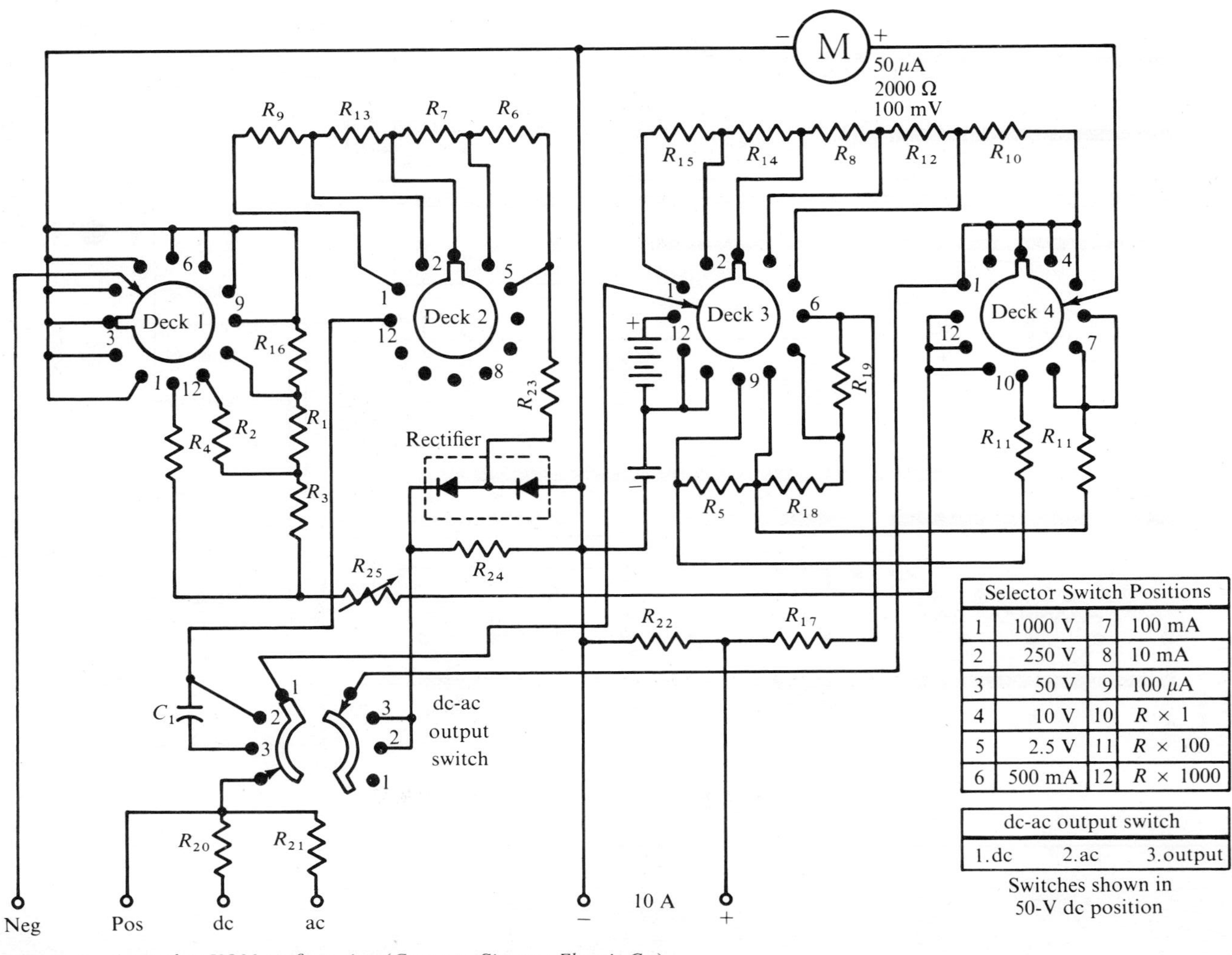

FIG. 5.15 *A complete VOM configuration. (Courtesy, Simpson Electric Co.)*

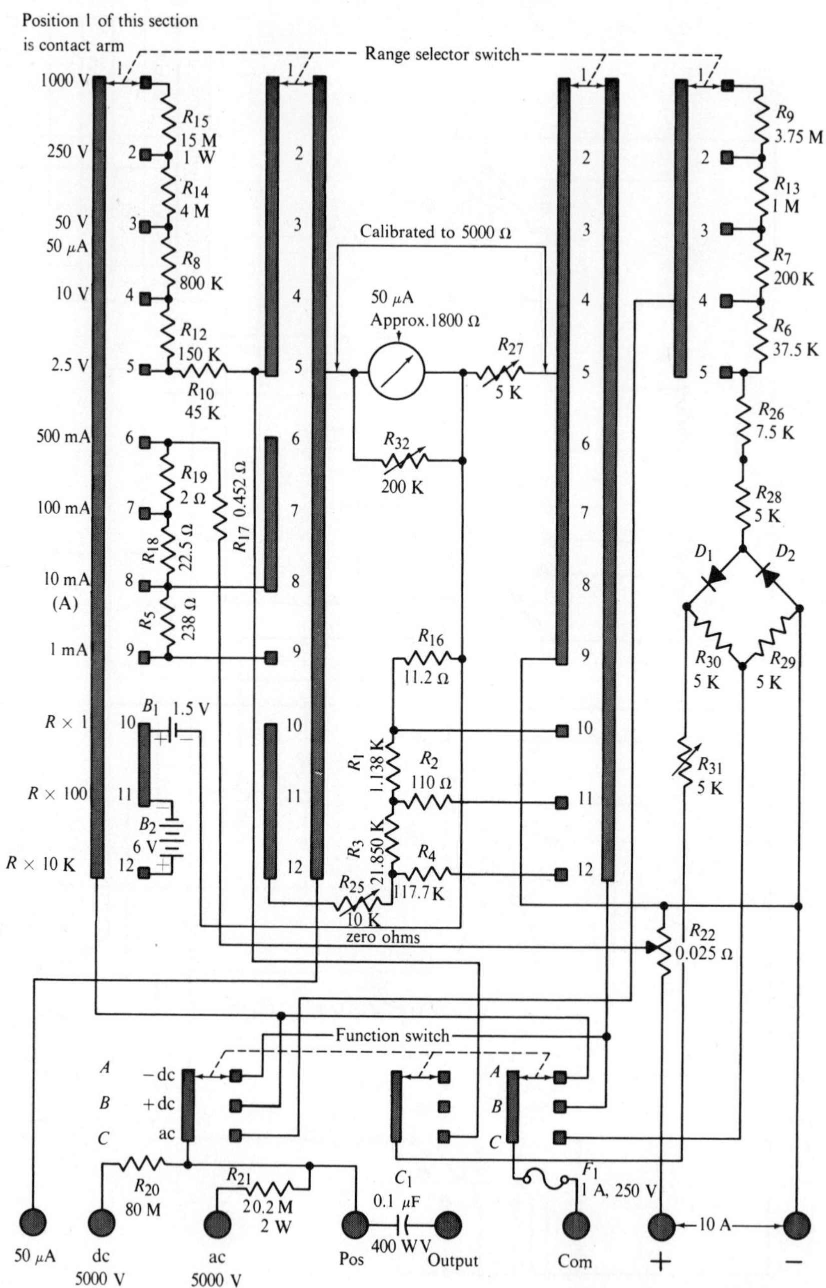

FIG. 5.16 *Another type of VOM configuration.* (*Courtesy, Simpson Electric Co.*)

determine whether objectionable circuit loading is present can be made by checking the voltage indication on an adjacent range (usually the next higher range). Since a higher range provides a higher value of input resistance, substantially different readings on the two ranges disclose that objectionable circuit loading is present. In such case, a VOM with a higher ohms per volt rating must be used to obtain an accurate measurement. For example, if a voltmeter rated at 20,000 Ω/V is unsuitable, we may use a 100,000 Ω/V VOM. Some VOMs have a rating of 1,000,000 Ω/V, and are well suited to measurement of low-voltage values in high-resistance circuits.

Comparatively few difficulties are ordinarily encountered in resistance measurements. However, some transistors can be damaged if a forward-resistance measurement is made on the $R \times 1$ range of typical VOMs. It is good practice to use the $R \times 100$ range when testing transistors in order to reduce the test current to a low value. The forward resistance of any junction device will exhibit different resistance values on various ohmmeter ranges because the junction resistance is nonlinear and varies with the test voltage that is applied. Note also that the filament resistance of small light bulbs or vacuum tubes will also exhibit different resistance values on various ohmmeter ranges owing to the positive temperature coefficient of the filament material.

Aside from circuit loading, one of the most common sources of error in ac voltage measurements is the waveform error that occurs when the ac waveform is not sinusoidal. For example, if the output from an audio amplifier is not a true sine wave, a voltage measurement made with a VOM will be more or less incorrect. Decibel measurements will also be incorrect if the waveform is not sinusoidal. Even if the waveform is sinusoidal, an error will occur if dB measurements are made across a load impedance instead of a load resistance. In other words, dB values are defined with respect to a resistive load. A reactive load develops both real power and reactive power; accordingly, a dB measurement is deceptively high when made across a load impedance.

5.10 VOM MAINTENANCE

A VOM requires little maintenance in normal use. If the pointer does not rest exactly at zero, the zero-set screw below the scale plate should be turned slightly as required.

Sometimes the zero-set adjustment must be changed when the instrument is moved from a vertical to a horizontal position. This is due to a small unbalance of the pointer assembly. The ohmmeter battery or batteries must be replaced when they become noticeably weak.

Ohmmeter calibration is easily checked by measuring the resistance of a 1-percent precision resistor. Voltage calibration can be checked within practical limits by measuring the terminal voltages of mercury cells or batteries. Comparison checks can be made by connecting more than one voltmeter across the same voltage source.

Highly accurate voltage checks must be made by use of standard Weston cells and precision potentiometer arrangements. These may be available in school laboratories. Otherwise, a voltmeter must be sent to the manufacturer or to a meter repair depot for high-accuracy voltage calibration. If a voltmeter does not indicate correctly, the permanent magnet in the meter mechanism may have lost some of its normal field strength, or multiplier resistors may be off-value. As noted previously, multiplier resistors can be damaged by substantial overload.

Alternating current voltage calibration can be checked within practical limits by measuring the terminal voltages of mercury cells or batteries. Remember that the ac voltage indication will read 2.22 times the dc source voltage if a half-wave instrument rectifier is present in the instrument. The ac voltage indication will read 1.11 times the dc source voltage if a full-wave instrument rectifier is present. Most ac voltage indication errors are caused by defective instrument rectifiers. Unless the operator is experienced, instrument rectifiers should be replaced by a meter repair depot, and the ac calibration resistor(s) readjusted if necessary.

5.11 VOM ACCESSORIES

For measurement of ac current, an adapter is generally used with a VOM. One type of adapter is illustrated in Fig. 5.17*a*. The adapter is designed to plug into the terminals on the front panel of the VOM. Full-scale ac current ranges from 250 mA to 25 A are provided by binding posts on the front panel of the adapter. A current transformer is used to energize the VOM, as depicted in Fig. 5.17*b*; the transformer is designed for measurement purposes, with its primary winding con-

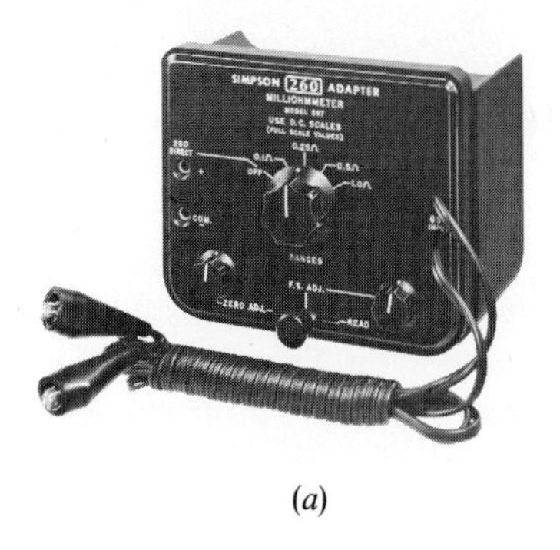

(a)

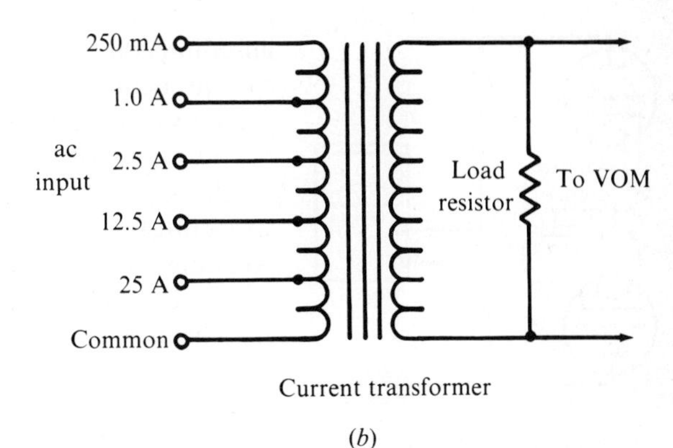

(b)

FIG. 5.17 *Photograph (a) and circuit (b) of current adapter for a VOM. (Courtesy, Simpson Electric Co.)*

nected in series with the circuit carrying the current that is to be measured. All current transformers employ a precision resistor connected in shunt to the secondary winding.

The adapter illustrated in Fig. 5.17 is operated with the VOM set to its 2.5-V ac range. The 2.5-V scale reads directly in amperes when the 2.5-V input terminal of the adapter is connected. When other adapter terminals are connected into the circuit under test, the 2.5-V scale of the VOM indicates values that must be multiplied by a suitable scale factor, as noted on the front panel of the adapter. The rated frequency range is from 50 to 3000 Hz. As in the case of ac voltage measurements, correct indication is obtained only when the current has a true sine waveform.

Since the frequency range of a VOM is comparatively limited, as was depicted in Fig. 5.8, an external rectifier probe must be used to make tests in high-frequency circuits. For example, radio and television technicians often employ the signal-tracing probe shown in Fig. 5.18. This probe is used in conjunction with the dc voltage ranges of a VOM. The probe is not suitable for precise measurements of high-frequency voltages, but has useful application to determine the presence or absence of signal. It also provides a rough indication of comparative signal levels in low-impedance circuits. Since the

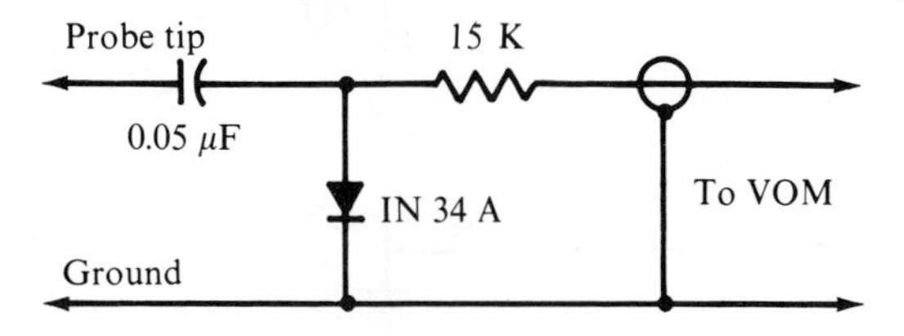

FIG. 5.18 *Signal-tracing probe for a VOM.*

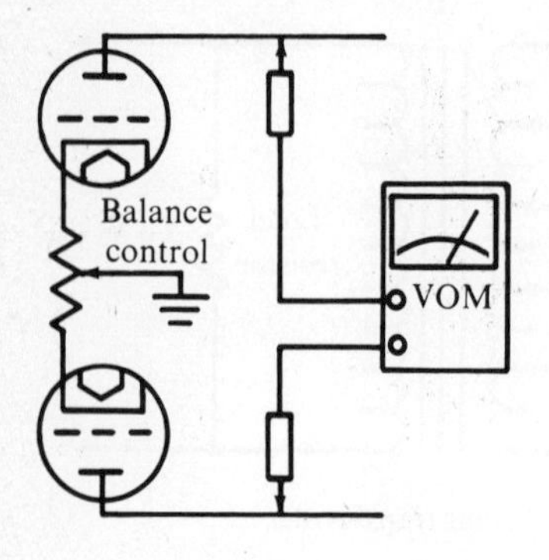

FIG. 5.19 *Checking a push-pull amplifier for a dc balance.*

input impedance of the probe is low (as a result of substantial current demand by the VOM), it is impractical to make precise voltage measurements in most cases.

Various other accessories are often used with a VOM to check transistors and batteries, to measure power and temperature, to establish precise voltage values in the microvolt range, to measure capacitance, and so on. These accessories will be discussed subsequently under the pertinent instrument classification.

5.12 APPLICATIONS

Some applications for a VOM entail two equal input voltages, as depicted in Fig. 5.19. In this situation, the VOM operates basically as a galvanometer, and the zero indication is sought. When the balance control for the push-pull amplifier is correctly adjusted, the VOM reads zero and the amplifier is in dc balance. It is advisable to start the test on a comparatively high range of the VOM to protect the meter against an overload in case the amplifier is considerably out of balance. As the balance condition is approached, the range switch of the VOM may be set to lower positions to obtain more critical readings.

In other applications we measure the dc component in a pulsating-dc waveform, as exemplified in Fig. 5.20. This test is often made to determine whether the preceding stages are operative, with a zero reading on the meter indicating that there is no output signal from the video detector. This test is also employed to determine whether lack of receiver operation

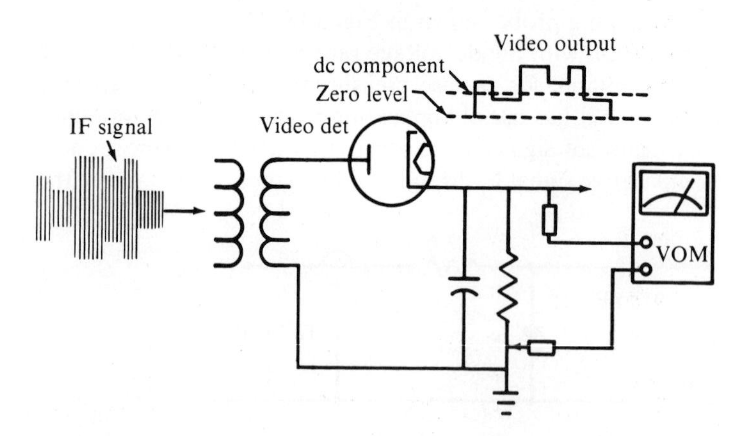

FIG. 5.20 *Measurement of dc component in a pulsating dc waveform.*

may be caused by self-oscillation in the if section. That is, if there is spurious oscillation in the if section, the dc voltage output from the video detector will be many times the normal value. For example, the dc-voltage output might be 0.5 V in normal operation, whereas a reading of 10 to 15 V is typical of oscillation in the if amplifier.

The ac function of a VOM is useful for checking power-line voltages, transformer output voltages, heater voltages in radio or tv receivers, and for more specialized tests such as depicted in Fig. 5.21. With the test setup shown, the VOM serves to indicate whether the insulation resistance of the electrical appliance is adequate. The VOM is operated on its ac voltage function. With the range switch set to a high value, the plug is inserted into the 117-V outlet, and the meter reading (if any) is observed. If no pointer deflection is visible, the range switch is set to lower positions for a more sensitive check. If any voltage reading occurs on the lowest range, the insulation resistance of the electrical appliance is too low and should be corrected. If repair is impractical, the appliance should be discarded. Poor insulation resistance is hazardous, because the user can receive a shock if standing on a damp floor or touching other equipment.

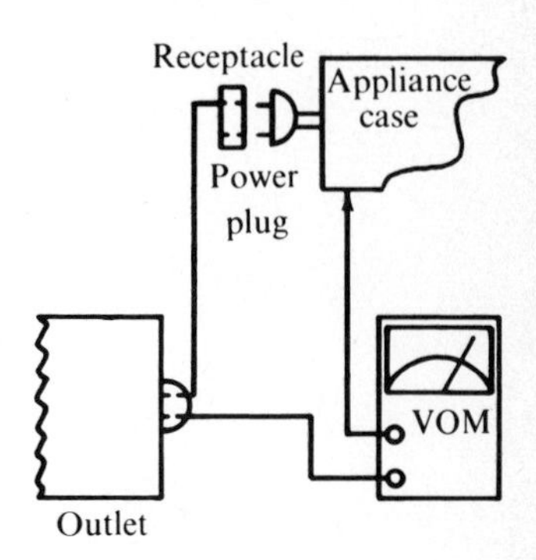

FIG. 5.21 *Checking on appliance for insulation resistance at line voltage.*

QUESTIONS AND PROBLEMS

1. What four safety features are sometimes built into a VOM?
2. Why is the meter mechanism in a VOM seldom fused?
3. What is the purpose of connecting a silicon diode across the meter movement in a VOM?
4. Why is only the ohmmeter section of a VOM fused?
5. Why can't the 5000-V range of a VOM be connected through a rotary switch?
6. Prove that the full-scale current on the 250-V range of the meter in Fig. 5.3 is 50 μA.
7. What is the disadvantage of the voltmeter configuration shown in Fig. 5.4?
8. What is the ohms-per-volt rating of the voltmeter shown in Fig. 5.5?
9. How does a voltmeter load a high resistance circuit?
10. Why is the ohms-per-volt rating on ac ranges of a VOM less than that of the dc ranges?

11. What is the purpose of the swamping resistor R in the meter circuit shown in Fig. 5.7*a*?

12. Explain the action of the three-terminal half-wave rectifier as it functions in an ac voltmeter.

13. Explain why the input resistance of the ac voltmeter circuit, shown in Fig. 5.7*a*, is not constant for alternate half cycles.

14. An ac voltmeter set to the 100-V range is connected across a 50-V dc source; what value of voltage does it indicate?

15. Why isn't the ac voltmeter circuit shown in Fig. 5.7 purely resistive?

16. Why are germanium diodes used in the voltmeter shown in Fig. 5.7, instead of copper-oxide instrument rectifiers?

17. Prove that the input resistance of the instrument shown in Fig. 5.9 is 14.3-K when the shunt diode is conducting.

18. What is the disadvantage of the full-wave bridge circuit shown in Fig. 5.10*a*?

19. What is the advantage of a full-wave rectifier in an ac meter over a half-wave rectifier?

20. What is the chief limitation of the output meter function on a VOM?

21. Why does a blocking capacitor reduce the accuracy of a VOM at high frequencies?

22. What is the primary advantage of a ring-shunt meter configuration?

23. Why is the 10-A current range brought out to a separate terminal in the ammeter shown in Fig. 5.13?

24. Calculate the error (assuming that $R_T = 25\ \Omega$) that is incurred on the 10-mA range of the ammeter shown in Fig. 5.13.

25. Prove that the ohmmeter scale of the instrument shown in Fig. 5.14 is nonlinear.

26. Why does the accuracy of an ohmmeter become impaired as the internal battery ages?

27. With reference to the meter circuit shown in Fig. 5.16, what is the purpose of resistors R_{27} and R_{32}?

28. Why will the junction resistance of a diode vary on different ohmmeter ranges?

29. Give four sources of possible error when a VOM is used to measure the dB gain of an amplifier.

30. Discuss the routine maintenance that may be necessary for a VOM.

31. Explain the operation of the ac current adapter shown in Fig. 5.17.

32. Explain the operation of the signal-tracing probe for a VOM shown in Fig. 5.18.

VACUUM-TUBE VOLTMETERS

6.1 ADVANTAGES AND DISADVANTAGES OF VTVM

A vacuum-tube voltmeter is defined as an instrument that utilizes the characteristics of one or more vacuum tubes for measuring voltage values. The chief advantage of a vacuum-tube voltmeter (VTVM) is its high input resistance, particularly on low-voltage ranges. For example, a utility-type VTVM provides an input resistance of 11 MΩ on all dc voltage ranges. The first range might have a full-scale value of 1 V, corresponding to a sensitivity of 11 MΩ/V. On the other hand, a typical VOM provides a sensitivity of 20,000 Ω/V and a high-quality VOM has a sensitivity of 1 MΩ/V, which is only one-eleventh the sensitivity of an ordinary VTVM. Some service-type VTVMs have an input resistance of 22 MΩ on all ranges, which provides twice the sensitivity noted above.

Laboratory-type VTVMs provide much higher values of input resistance, a typical instrument having an input resistance of 100 MΩ on all dc ranges. Another lab-type VTVM provides open-grid input on dc ranges, as explained subsequently. This design results in an extremely high value of input resistance, since a negatively biased grid draws practically no current.

Another advantage of a VTVM as compared with a VOM is its extremely high value of input impedance on all voltage ranges. For example, a conventional VTVM has an input capacitance of 1 pF, whereas a VOM has an input capacitance of 50 pF, approximately. We will find that low input capacitance affords much higher accuracy when measuring dc voltages in high-frequency tuned circuits. In other words, low input capacitance minimizes detuning of the circuit under test; the dc voltage distribution of some high-frequency circuits is changed considerably when test conditions cause substantial detuning.

A VTVM also has certain disadvantages compared with a VOM. For example, most VTVMs are line operated and cannot be used in locations where electric outlets are not available. (However, a few VTVMs are battery operated.) The production cost of a VTVM is substantially greater than that of a VOM with equal rated accuracy. A VTVM tends to drift slowly during operation, and the zero-adjustment control requires occasional resetting; by way of comparison, a VOM seldom requires zero-set adjustment, except on ohmmeter ranges. Most VTVMs are somewhat bulkier and heavier than VOMs that have the same ranges and functions. Because of power requirements of a VTVM, a few instruments are designed as vacuum-tube volt-ohmmeters (VTVOMs) or combination VTVM-VOMs. The chief advantage of a VTVOM is that it meets general service requirements in any location.

6.2 THE VACUUM-TUBE BRIDGE

Early VTVMs were designed with a single vacuum tube, and we still find this basic design in adapters that convert a VOM into a VTVM. As seen in Fig. 6.1, the tube operates as an electronic impedance transformer; it employs a high-resistance grid circuit that controls the current in a low-impedance cathode circuit. That is, the grid circuit draws only a very

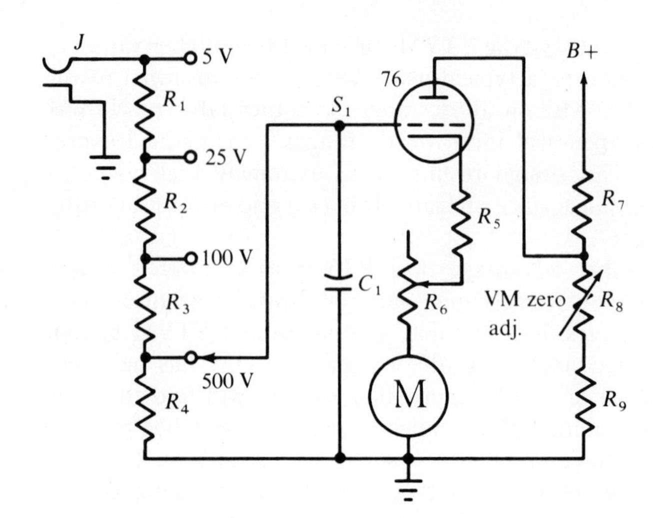

FIG. 6.1 *Basic one-tube VTVM configuration.*

slight current although it controls a substantial meter current in the cathode circuit. The input resistance of this instrument is 10 M on all ranges. Since the tube operates in Class A to obtain optimum linearity of operation, a quiescent current is present in the cathode circuit. Therefore, a zero-center meter movement is used. One advantage of this arrangement is that either positive or negative dc voltages can be measured without switching the instrument circuitry.

To zero-set the pointer in Fig. 6.1, a zero-set control is provided; this is R_8 and it serves to adjust the operating point of the tube. R_8 is an operating control which requires occasional attention. In order to obtain an accurate full-scale indication on the meter, the gain of the dc amplifier circuit must be suitably adjusted. This is the function of R_6, which varies the amount of negative feedback. R_6 is a maintenance control, and is readjusted only when the tube is changed, or if the VTVM is to be recalibrated after an extended period of service. R_6 is supplemented by R_5, which contributes to the total amount of negative feedback for stability of operation. The rated accuracy of this configuration is ± 5 percent of full scale.

It is essential that only dc voltage be applied to the grid of the tube in Fig. 6.1. Any incidental ac voltage will tend to drive the tube out of its linear operating region and cause inaccurate indication. To prevent this effect, a grid-bypass capacitor C_1 is provided. We observe that the meter operates in a single-ended circuit in Fig. 6.1; in other words, this is not a vacuum-tube bridge circuit. Although a large amount of negative feedback is provided, the instrument is not as stable as could be desired. The zero-set adjustment requires comparatively frequent attention. This inherent drift in the measuring circuit is chiefly due to slight grid-current flow. Traces of gas, grid emission, and contact potential, aggravated by thermoelectric effects, contribute to the generation of small varying grid currents. Therefore, the better-designed VTVMs use a pair of tubes in a bridge circuit to provide differential dc amplification.

The basic vacuum-tube bridge circuit is depicted in Fig. 6.2. This configuration differs from that shown in Fig. 6.1 in that the meter movement is connected into a bridge circuit. Tubes V_1 and V_2 comprise two of the bridge arms, and resistors R_1 and R_2 provide the other two bridge arms. Note that the plate resistances of V_1 and V_2 are the varying bridge components. Although the values of R_1 and R_2 are fixed, the plate resistances of V_1 and V_2 will vary when an input voltage is applied to the grid of V_1. For example, if a positive input

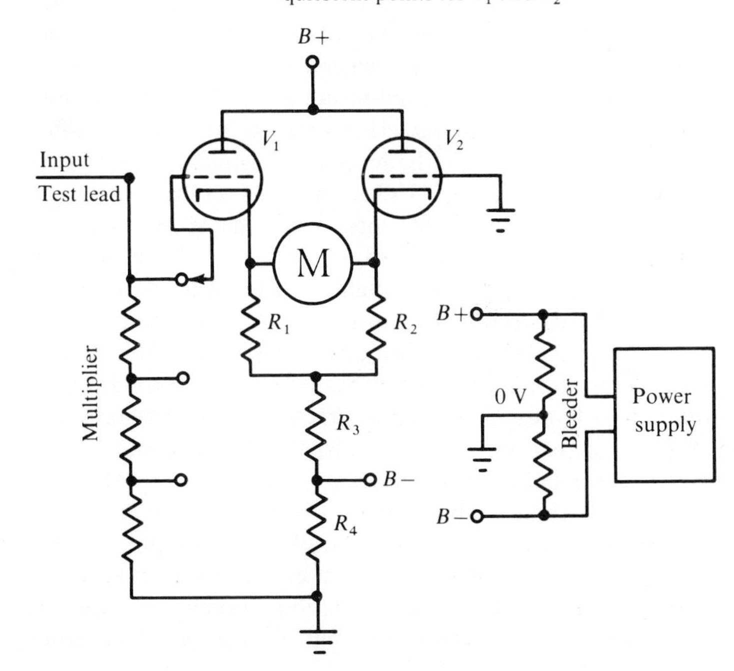

FIG. 6.2 *Basic VTVM bridge circuit.*

voltage is applied to the grid of V_1, its plate resistance decreases: simultaneously, the increased cathode-current demand biases the cathode of V_2 more positive, and its plate resistance increases. Thus, the bridge circuit is unbalanced and the meter draws current.

Let us analyze the bridge arrangement in Fig. 6.2 to see why it is much more stable than the simple configuration of Fig. 6.1. One cause of instability is line-voltage fluctuation, which causes the $B+$ voltage to change in value. In Fig. 6.1, it is evident that an increase in $B+$ voltage will cause the meter reading to increase. On the other hand, an increase in $B+$ voltage in Fig. 6.2 causes a plate-current increase in both V_1 and V_2. Although both cathode voltages increase, their difference remains practically constant. Therefore, the VTVM bridge circuit provides stable operation over a rather wide range of line-voltage fluctuation.

Although a regulated power supply could be used in the configuration of Fig. 6.1, there are sources of instability other than $B+$ voltage variation to be contended with; we will find that these can be minimized by use of a vacuum-tube bridge circuit. For example, most commercial resistors have a positive temperature coefficient of resistance. In other words, a resistor tends to increase in value as it becomes warm. This is just another way of saying that the values of R_5, R_6, R_7, R_8, and R_9 in Fig. 6.1 increase slightly when the instrument is switched on. Even after thermal equilibrium is established, the nonlinearity of the tube characteristics results in a slightly shifted operating point. In turn, the zero-set adjustment tends to vary during operation of the VTVM. This zero-set drift, caused by resistor temperature coefficients, is minimized by the bridge circuit in Fig. 6.2, because a slight increase in the value of R_1 is largely balanced out by a slight increase in the value of R_2.

Furthermore, the plate resistance of a vacuum tube tends to decrease slightly as the tube becomes hotter. This drift in plate resistance is not compensated in Fig. 6.1. However, since both V_1 and V_2 operate at practically the same temperature in Fig. 6.2, this source of drift is largely compensated. Still another stabilization feature provided by a VTVM bridge circuit is its ability to balance out *contact potential* in the grid circuits. Contact potential is basically due to slight voltages generated by dissimilar metals in a circuit. Contact potential is usually considered to include residual potentials produced by traces of gas, grid emission, and the velocity of electrons emitted from the cathodes. Contact potential also varies with the temperature of the tubes. It is evident that any contact potential developed by V_1 in Fig. 6.2 will be largely balanced out by a similar contact potential developed by V_2; that is, the *difference* in contact potential between the grids of V_1 and V_2 will remain practically zero.

When we analyze the effects of contact potential in somewhat greater detail, we shall recognize that the simple configuration of Fig. 6.2 cannot provide complete compensation. That is, contact potential is associated with a slight grid-current flow, even when tubes are carefully selected. Since the multiplier in Fig. 6.2 has a very high resistance, such as 10 M, grid-current flow in V_1 causes an *IR* drop across the multiplier. On the other hand, grid-current flow in V_2 produces no *IR* drop because the grid of V_2 is grounded. Therefore, the more elaborate VTVMs employ a dummy multiplier in the grid circuit of V_2 as depicted in Fig. 6.3. We observe that the dummy multiplier provides the same resistance value from grid to ground for each tube on any setting of the range switch. In turn, if the bridge tubes are properly selected, they will have practically the same value of grid current, and the zero setting will not shift appreciably when the range switch is rotated.

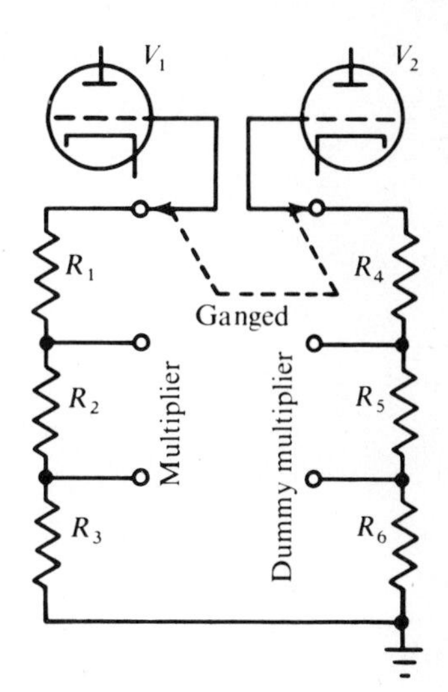

FIG. 6.3 *Plan of a dummy multiplier.*

In older types of VTVMs, separate triodes were generally used in the bridge circuit. However, it was subsequently determined that twin triodes tend to be well balanced in characteristics, and most modern VTVMs now employ them. Tube selection is still necessary, though, because some twin triodes are found to be better balanced than others. After a twin triode has been selected that provides minimum zero shift when the range switch is rotated, it is also common practice to "age" the tubes by leaving the instrument turned on for 48 hr before it is placed in use. If there is practically no zero shift on various settings of the range switch after the aging period, the twin triode is deemed satisfactory and the instrument is placed in service, preferably after a calibration check. Calibrating techniques are explained subsequently.

Older types of VTVMs employed full rated values of $B+$ voltage and heater voltage. Subsequently, it was found that improved stability could be obtained by operating the triodes at somewhat reduced values of plate and heater voltage. In other words, tube characteristics drift to a lesser extent when the plate currents are moderate and when the cathode temperature is reduced somewhat. In addition to improved stability, limited cathode emission and low plate-current saturation provide improved meter protection. If excessive test voltage is applied to the grid of a bridge tube, the tube saturates soon after the pointer is driven off-scale on the meter, and damage to the meter mechanism is prevented. Figure 6.4 visualizes the control of saturation limiting by choice of operating temperature.

A typical VTVM operates a 12AU7 tube with a plate-cathode voltage of 72 V and a heater voltage of 5.8 V instead of the full rated values of 330 V and 6.3 V, respectively. Some VTVMs employ a full rated 6.3 V for heater operation, but plate-cathode voltages greater than 72 V are seldom utilized. Meter movements may have full-scale values as low as 100 μA or as high as 400 μA. Sensitive meter movements provide somewhat greater accuracy, because the bridge tubes are then operated over a smaller interval of their dynamic range. A typical VTVM is illustrated in Fig. 6.5. Note that the dc voltage scales are practically linear; scale linearity results from operating the bridge tubes in Class A. There is a slight residual nonlinearity in the dc voltage scales, however, because the response of an unbalanced bridge is not strictly linear.

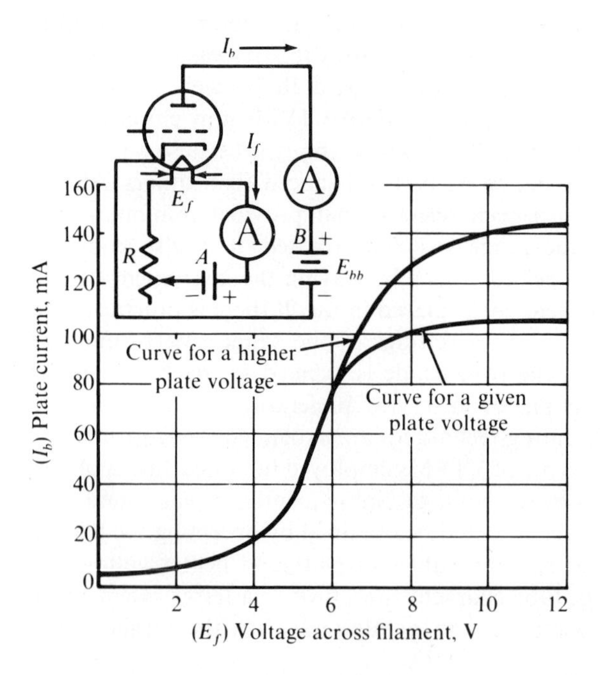

FIG. 6.4 *Visualization of saturation limiting process.*

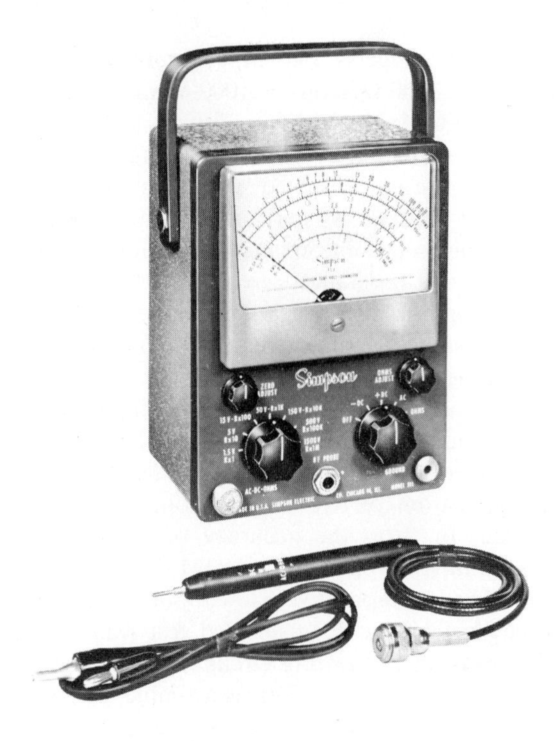

FIG. 6.5 *Appearance of a typical VTVM. (Courtesy, Simpson Electric Co.)*

6.3 MULTIPLIERS AND BRIDGE CONFIGURATIONS

A VTVM differs from a VOM in that the former provides a constant input resistance on all ranges. As an example, we observe in Fig. 6.6 that the input resistance is 22 M on all ranges. Seven dc voltage ranges are provided, with full-scale values from 1.5 V to 1000 V. The 7-M resistor is called an isolating resistor, and is contained in the tip of the dc test probe (see Fig. 6.5). Its function is to effectively separate the capacitance of the shielded input cable from the circuit under test. That is, the input cable might have a capacitance of 100 pF; however, the isolating resistor places 7 M between this capacitance and the circuit in which the dc voltage measurement is to be made. Thereby, the input cable capacitance has practically no loading effect on the circuit under test. It is left as an exercise for the reader to verify that the values of the multiplier resistors in Fig. 6.6 provide the full-scale values indicated.

Since the input resistance of a VTVM is constant on all ranges, it follows that its ohms per volt sensitivity is different on each range. For example, in Fig. 6.6, the ohms-per-volt

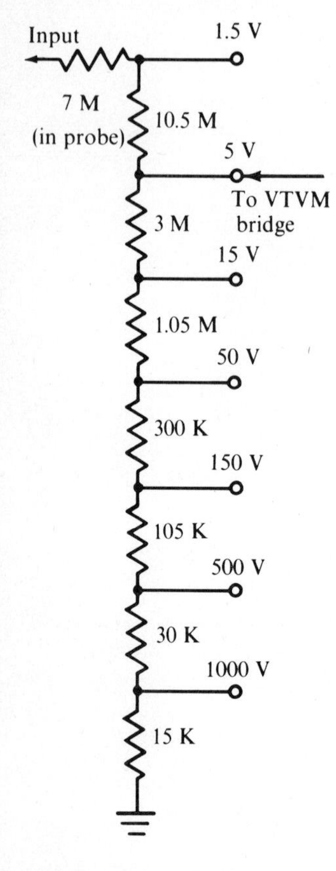

FIG. 6.6 *Schematic of a multiplier.*

sensitivity on the 1.5-V range is 14.6 M/V, approximately, while the sensitivity is 22,000 Ω/V on the 1000-V range. Thus, in comparison to a VOM, the chief sensitivity advantage provided by a VTVM is realized on its low-voltage ranges. Conventional VTVMs are not designed with an input resistance greater than 22 M. The reason for this limitation is that vacuum tubes tend to become erratic in operation when extremely high grid-circuit resistances are utilized. Moreover, under conditions of high humidity, it is difficult to maintain ample insulation resistance if an extremely high value of grid-circuit resistance is employed.

It is necessary to use a shielded input cable with a VTVM because of its high input resistance and electronic bridge circuit. Open test leads would pick up strong stray fields which would result in application of overload ac voltages to the vacuum-tube bridge. In turn, the accuracy of dc voltage indication would be impaired. Because a VOM does not use vacuum tubes, and has a comparatively low input resistance, it can be used satisfactorily with unshielded test leads. An isolating resistor forms an integrating circuit with the cable capacitance, as depicted in Fig. 6.7*a*. This is a simple lowpass filter which permits passage of dc but bypasses ac to ground. When measuring the dc voltage at the plate of an oscillator tube or at the collector of an oscillator transistor, an ac voltage is superimposed on the dc level, as visualized in Fig. 6.7*b*. When this combined dc and ac voltage is applied to a lowpass filter, only the dc voltage appears at the filter output. Thereby, the ac voltage is prevented from gaining entry to the bridge circuit in the VTVM.

To obtain optimum lowpass filter action, the multiplier is usually followed by a shunt capacitor (see the 0.01-μF capacitor in Fig. 6.8). This arrangement increases the time

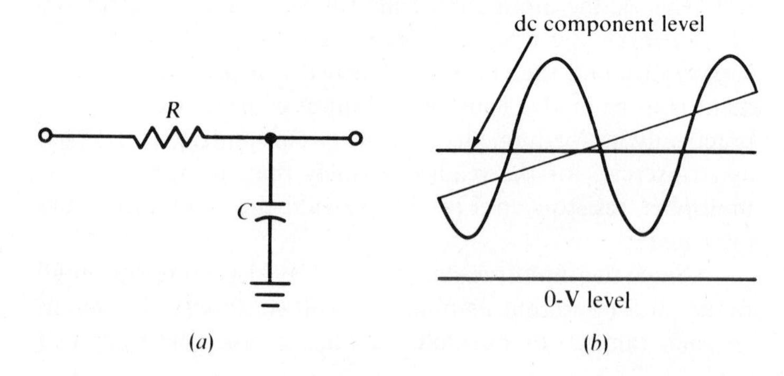

FIG. 6.7 (*a*) *Equivalent circuit of isolating resistor and cable capacitance;* (*b*) *sine-wave ac voltage with a dc component.*

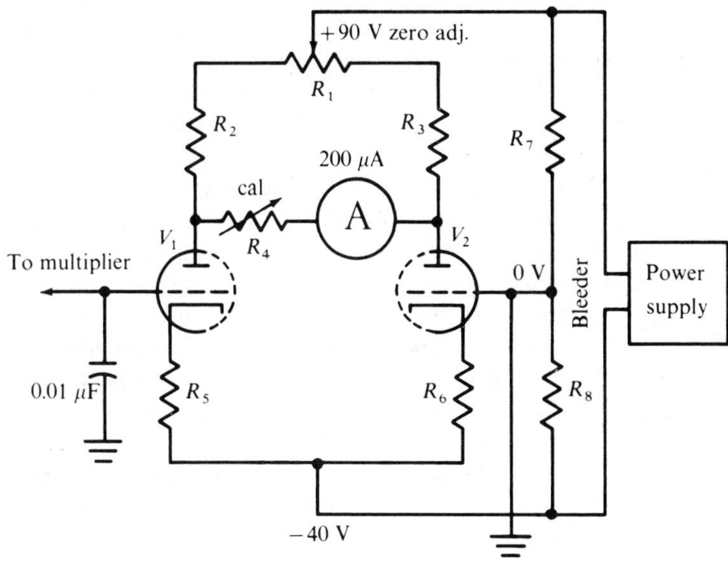

FIG. 6.8 *Another type of vacuum-tube bridge circuit.*

constant of the integrator and ensures that low-frequency ac voltages will be adequately bypassed to ground. This schematic diagram also shows another commonly used bridge circuit; the meter movement is connected between the plates of the triodes. Thus, V_1, V_2, R_2, and R_3 form the arms of the bridge configuration. This bridge operates basically in the same manner as the bridge depicted in Fig. 6.2. Note the zero-adjust control; R_1 is a potentiometer that permits equalization of the plate voltages by changing the ratio of the plate-load resistances. R_4 is a calibrating potentiometer that provides adjustment of the meter sensitivity.

The accuracy of a VTVM in dc voltage measurements depends upon the tolerances of the multiplier resistors and tubes, the accuracy rating of the meter movement, and the stability of the bridge circuit. Each resistor in Fig. 6.6 has a typical tolerance of ± 1 percent. In turn, the tolerance of a complete multiplier is ± 1 percent. The meter movement has a typical tolerance of ± 2 percent of full-scale indication. This is an additive tolerance; therefore, the tolerance on the multiplier and movement combination will be ± 3 percent of full-scale indication. This is the nominal accuracy rating of the VTVM. Note, however, that as the tubes weaken, the scale indication becomes inaccurate as a result of increasing nonlinearity of the bridge circuit. Although the calibrating control can be adjusted for correct full-scale indication, the accuracy at half-scale indication might then be ±4 percent of full scale, instead of ±3 percent of full scale. In summary, it is good practice to replace the bridge tubes when they show evidence of weakening, instead of recalibrating the instrument.

A typical lab-type VTVM employs a four-tube bridge, as depicted in Fig. 6.9. When carefully matched tubes are utilized, this arrangement provides maximum operating stability. That is, each tube tends to drift by the same amount and in the same direction, resulting in maintenance of bridge balance. The four-tube bridge also provides a considerably greater dynamic range than a two-tube bridge. Thus, the VTVM in Fig. 6.9 provides ranges up to 150 V without switching the input multiplier. Instead, six ranges are provided by switching six values of resistance in series with the meter movement.

Note also that two input circuits are provided in Fig. 6.9. When the two upper terminals are connected together by a link, the input resistance is 10 M, as in a conventional VTVM. On the other hand, when the two upper terminals

FIG. 6.9 *Four-tube bridge circuit used in a lab-type VTVM. (Courtesy, General Radio Corp.)*

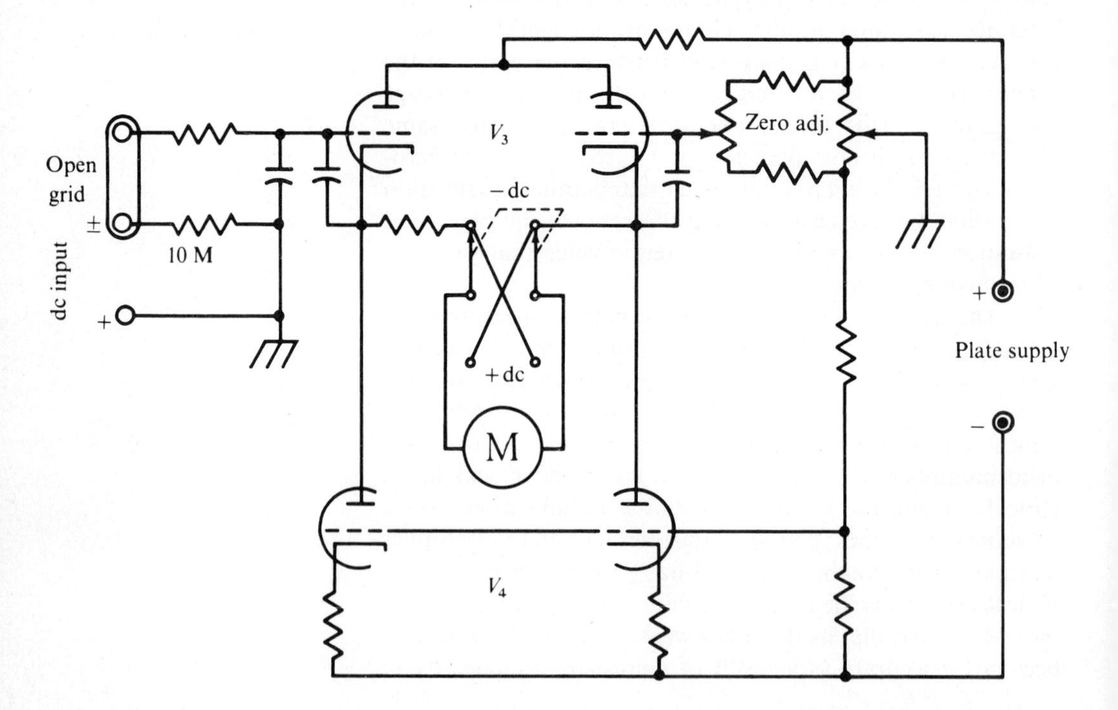

are not short-circuited, the grid-input circuit is open (open-grid operation) and the input resistance is practically infinite. In this mode of operation, the input voltage is applied between the upper and lower terminals. The advantage of open-grid operation is that circuit loading is minimized in case the circuit under test has extremely high internal resistance. This instrument is rated for an accuracy of ± 2 percent of full-scale indication, compared with ± 3 percent for a conventional utility-type VTVM.

6.4 OHMMETER CONFIGURATIONS AND RESISTANCE MEASUREMENT

A basic VTVM ohmmeter input circuit is depicted in Fig. 6.10*a*. The meter movement is arranged to provide full-scale indication when 1.5 V are applied to the bridge circuit. Therefore, the meter reads full scale when the test leads are open-circuited, and reads zero when the test leads are short-circuited. If the unknown resistance to be measured (R_X) happens to have the same resistance as R_1, the meter will evidently read half scale; that is, R_X and R_1 then operate as a voltage divider, and 0.75 V is applied to the bridge circuit. A VTVM ohms scale is quite nonlinear, as seen in Fig. 6.11; *R* is the ohms scale, which may be compared with the dc volts scales A and B. Note that the resistance values increase in an opposite direction on a VTVM ohms scale, compared with resistance values on a VOM ohms scale.

Figure 6.10*b* depicts an ohmmeter multiplier configuration for a VTVM. Since the range switch is set to the 10-K multiplier resistor, the meter will read half scale if R_X has a value of 10 K. Again, if the range switch were set to the 10-M multiplier resistor, the meter would read half scale if R_X has a value of 10 M. When the range switch is set to the 9.75-Ω multiplier resistor, the meter reads 10 Ω at half scale if R_X has a value of 10 Ω. This multiplier resistor does not have a value of 10 Ω because the 1.5-V cell has internal resistance and the test leads have a small amount of resistance. This total residual resistance has an average value of 0.25 Ω. As the cell weakens, the accuracy of ohmmeter indication becomes impaired through increasing internal resistance of the cell.

It is necessary to provide seven resistance ranges in the foregoing example, because the ohmmeter scale indications

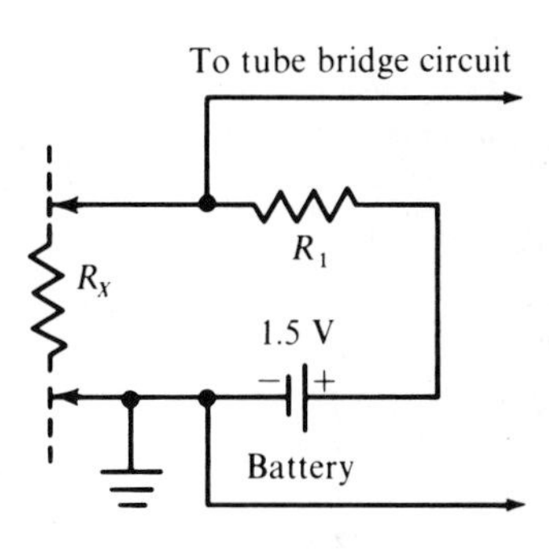

(*a*)

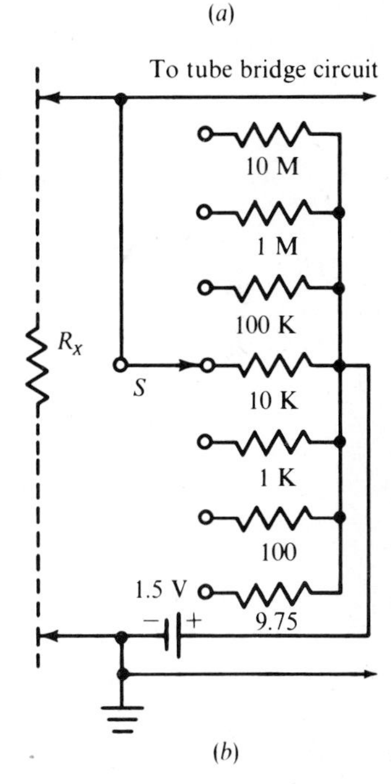

(*b*)

FIG. 6.10 *Basic ohmmeter. (a) Circuit diagram; (b) ohmmeter multiplier configuration.*

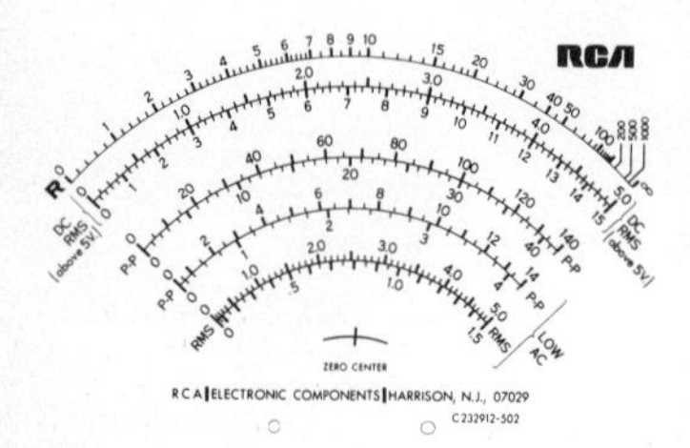

FIG. 6.11 *Scale plate for a VTVM.* (*Courtesy, Radio Corporation of America.*)

are excessively crowded at the right-hand end, as seen in Fig. 6.11. That is, it is impossible to read high-resistance values accurately when the ohmmeter is set to a low-resistance range. Conversely, it is impossible to read low-resistance values accurately when the ohmmeter is set to a high-resistance range. In general, we choose an ohmmeter range that provides indication in the center-scale region, when this is possible. Of course, very low resistance values must necessarily be measured at the low end of the scale, and very high values must be measured at the high end of the scale.

Ohmmeter multipliers commonly employ ± 1 percent resistors, and the meter movement has a typical tolerance of ± 2 percent of full-scale indication. In turn, the ohmmeter has a nominal accuracy of ± 3 percent of full scale. Since the ohmmeter scale is nonlinear, this full-scale accuracy rating is applied as an arc of error in resistance readings. (The meaning of an arc of error was previously explained in Chap. 4.) The nominal accuracy of the ohmmeter function in a VTVM is realized only when the ohmmeter cell is in good condition and the bridge tubes are normal. Note in Fig. 6.10*b* that a maximum test voltage of 1.5 V is applied across R_X under any conditions. Therefore, a VTVM has an advantage over a VOM in making resistance measurements of delicate components such as certain semiconductor devices.

Note in Fig. 6.5 that a VTVM has both a zero-adjust control and an ohms-adjust control. The zero-adjust control must be set for zero indication on the scale when the test leads are short-circuited on the ohmmeter function; the ohms-adjust control must be set for full-scale indication when the test leads are open-circuited. Thus, a VTVM ohmmeter is not quite as simple to operate as a VOM ohmmeter. In an ideal design, the settings of the zero-adjust and ohms-adjust controls would be the same on any range of the ohmmeter. However, practical tolerances in the instrument configuration usually require that the settings of both controls be checked when the setting of the ohmmeter range switch is changed. This requirement becomes aggravated as the ohmmeter cell weakens, and the cell should then be replaced.

A few VTVMs employ a special power supply instead of a cell to energize the ohmmeter circuit. This is a maintenance advantage, inasmuch as no cell replacement is required. A low-voltage power supply for an ohmmeter circuit must have a very low internal resistance, and be capable of supplying adequate current on the $R \times 1$ range. An ohmmeter power supply must also be well regulated, so that its output voltage remains constant in spite of line-voltage fluctuations. Zener diode regulation is suitable for ohmmeter power-supply circuits.

6.5 AC VOLTAGE MEASUREMENT

As seen in Fig. 6.11, a VTVM commonly provides for both rms and peak-to-peak ac voltage measurements (with proper scale). We will find that the instrument rectifier circuitry is designed in a manner that permits rms voltage measurement of sine waveforms only (see Fig. 6.12). In other words, rms voltages of nonsinusoidal waveforms cannot be measured. This restriction was previously noted for ac voltage measurements with a VOM. On the other hand, it will be explained how the peak-to-peak voltage scales of a VTVM indicate peak-to-peak voltages of either sine waves or complex (nonsinusoidal) waves.

Figure 6.13 shows the basic configuration of a VTVM

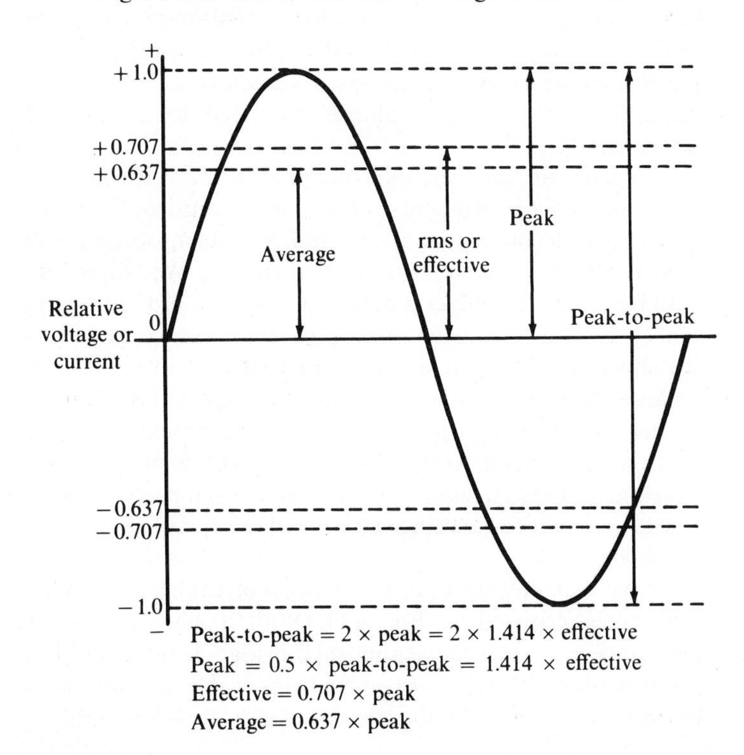

FIG. 6.12 *Relationship between peak-to-peak, effective, and average values of voltage or current.*

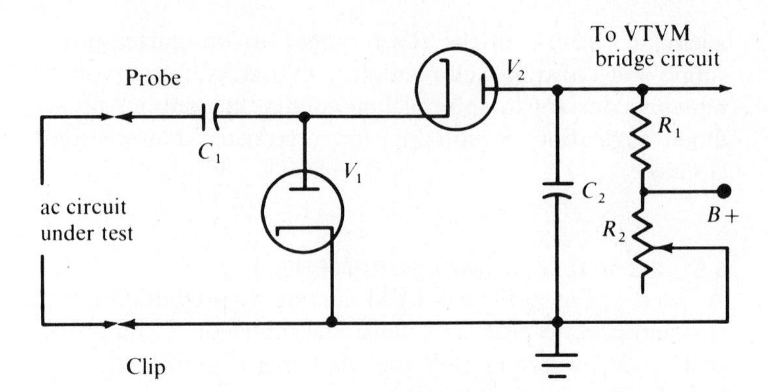

FIG. 6.13 *Peak-to-peak instrument rectifier circuit.*

peak-to-peak instrument-rectifier circuit. This is sometimes called a voltage doubler arrangement. Its sequence of operation is as follows: when the positive half cycle of an ac waveform is applied to the probe, it is coupled to diode V_1 via C_1. In turn, V_1 conducts and charges C_1 to the positive-peak value of the applied voltage. Next, on the negative half cycle of the input waveform, the combined negative peak voltage and the charge potential of C_1 are coupled to the cathode of V_2. In turn, V_2 conducts and charges C_2 to the peak-to-peak voltage of the input waveform. That is, C_2 is charged to the sum of the positive-peak and negative-peak voltages of the waveform. Thus, this peak-to-peak voltage is applied to the VTVM bridge circuit.

To obtain accurate ac voltage indication, the contact potential of the instrument-rectifier circuit must be balanced out. This is accomplished by R_1 and R_2, which operate as a voltage divider for the $B+$ power supply. We adjust R_2 until the negative contact potential at the plate of V_2 is exactly canceled; then the meter reads zero when the ac test leads are short-circuited. An open-circuit check is not advisable, because the probe in Fig. 6.13 is simply a direct-through connection, and the open leads might pick up appreciable stray fields. R_2 is not an operating control, but a maintenance adjustment. Once R_2 is correctly set for a given pair of diodes, it will not require resetting until the diodes weaken and are replaced.

We can now recognize the basis of calibration of the rms scales depicted in Fig. 6.11. With a sine-wave input voltage (Fig. 6.12), the vacuum-tube bridge in the VTVM is energized by 2 V dc. However, the rms scales are calibrated to read 0.707 V. That is, the rms scales indicate 0.358 of the dc voltage actually applied to the bridge. In a more exact analysis, calibration of the ac voltage scales must also take the nonlinearity of the instrument-rectifier diodes into account. A vacuum diode has two regions of opposite curvature, as seen in Fig. 6.14. The lower nonlinear portion tends to produce scale cramping; that is, the internal resistance of the diode decreases as we progress into its linear region. The upper nonlinear portion corresponds to approach of emission saturation, and the internal resistance of the diode changes at a different rate in this region.

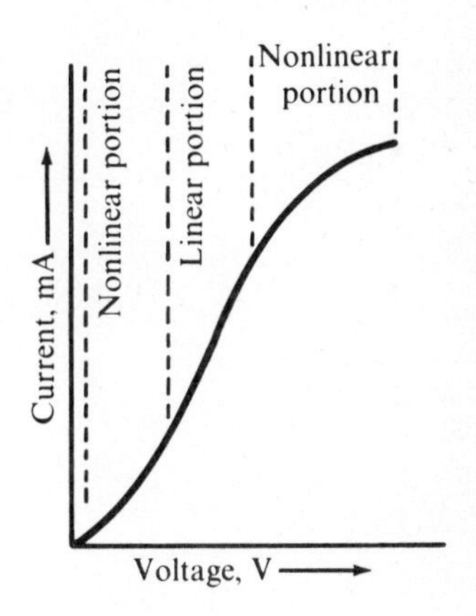

FIG. 6.14 *A voltage-current characteristic for a vacuum diode.*

Because of diode nonlinearities, separate ac voltage scales are provided in Fig. 6.11 for the various ranges. Thus, if we are operating on the 1.5-V rms ac range, we read scale G. On the other hand, if we are operating on the 1500-V rms ac range, we read scale B. That is, the low ac scales are used on the initial ac voltage ranges, and the upper scales are used on the high ac voltage ranges. The ac voltage function of a VTVM is usually rated to the same nominal accuracy as the dc voltage function. However, this accuracy cannot be realized unless the instrument-rectifier diodes are in good operating condition. Therefore, the ac calibration should be checked occasionally, and the diodes replaced if necessary.

Next, let us consider peak-to-peak voltage measurements. It is evident that the instrument-rectifier configuration depicted in Fig. 6.13 permits peak-to-peak voltages of either sine waves or complex waves to be indicated directly on the peak-to-peak voltage scales of the VTVM. For example, the rms voltage of the pulse waveform shown in Fig. 6.15 cannot be measured with a VTVM. However, its peak-to-peak voltage is indicated directly. The fact that its positive-peak and

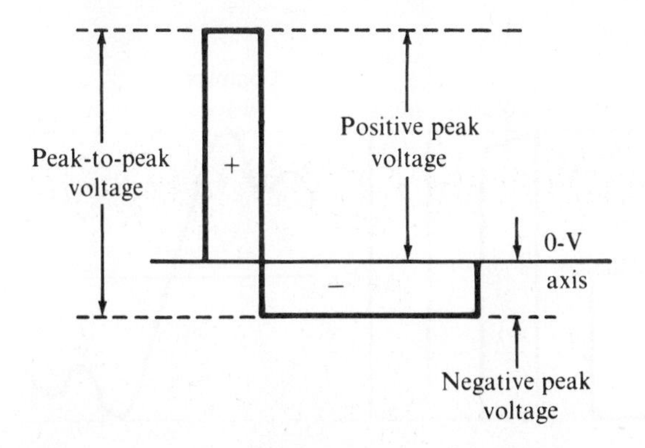

FIG. 6.15 *Voltage relations in a pulse waveform.*

negative-peak voltages are unequal is inconsequential. In Fig. 6.13, the instrument rectifiers simply add the peak voltages and indicate on the peak-to-peak scales. Since a VOM cannot indicate peak-to-peak voltage values, a VTVM has an important advantage in this respect.

In summary, a peak-to-peak voltage value is independent of the ac waveform. Figure 6.16 illustrates five waveforms that have the same peak-to-peak voltage. The beginning student should note that many shapes of complex waveforms are encountered in electronic equipment; Fig. 6.17 shows four encountered in a television receiver. The peak-to-peak voltages of these waveforms also vary over a wide range; they are specified for the ac waveform alone, and any dc component is disregarded. That is, a VTVM rejects any dc component that might be present with an ac waveform. This is just another way of saying that C_1 in Fig. 6.13 blocks the flow of dc.

Figure 6.18 shows the complete circuit diagram for a VTVM. Although the configuration appears complicated, we can trace the dc voltage, ac voltage, and resistance functions. As previously explained for a VOM, a VTVM has an upper frequency limit, due to stray capacitances associated with its ac multiplier network. It also has a lower frequency limit, due to the reactances of C_1 and C_2 in the instrument-rectifier circuit. Thus, the ac voltage function of the VTVM in Fig. 6.18 is rated for the frequency range from 30 Hz to 3 MHz. The input impedance on the ac voltage function is much lower than on the dc voltage function. For example, the input resistance and capacitance on the first three ac voltage ranges are 0.83 M shunted by 70 pF. On the 500-V ac range, the input resistance is 1.3 M shunted by 60 pF.

Note in Fig. 6.11 that a zero-center scale mark is pro-

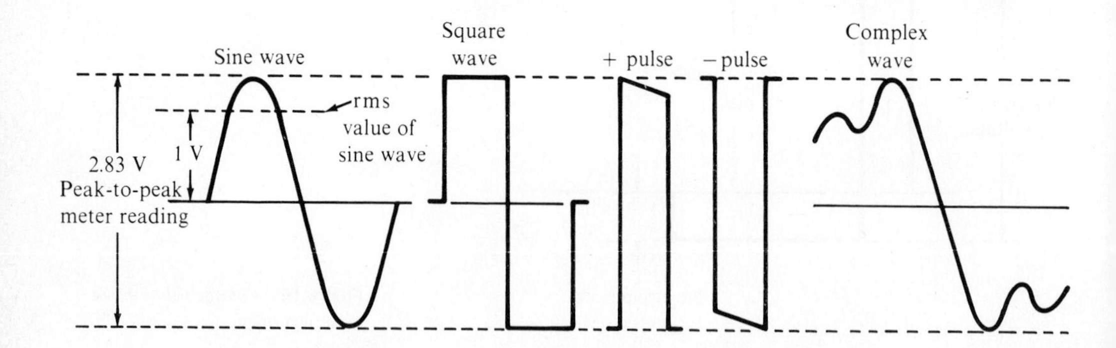

FIG. 6.16 *Various waveforms with the same peak-to-peak voltage.*

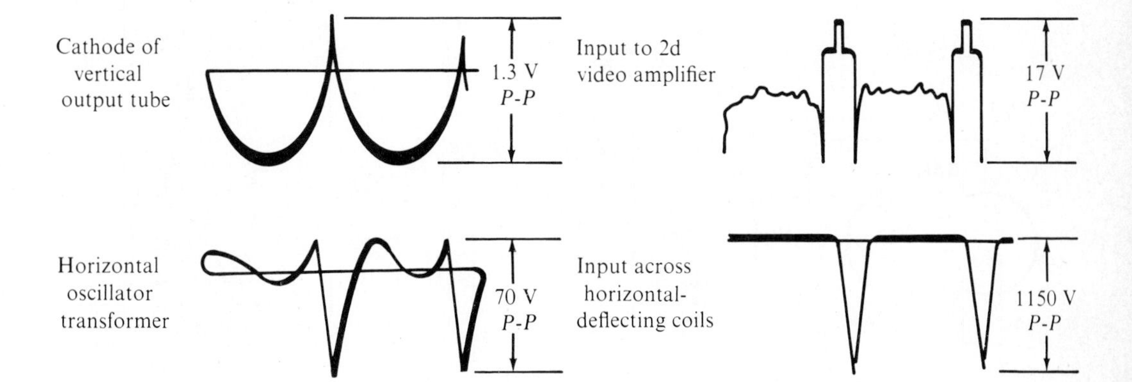

FIG. 6.17 *Television receiver waveforms.*

vided. To operate the VTVM with zero-center scale indication, the zero-adjust control (R_{22} in Fig. 6.18) is turned accordingly. This feature is used on the dc voltage ranges when aligning the fm discriminator in a radio or television receiver. It provides convenience, because the discriminator output reverses in polarity as the alignment slug or trimmer is tuned through the center frequency of the circuit. Note that −DC and +DC positions are provided on the function switch for polarity reversal when operating with the pointer in the normal zero-left position. With reference to Fig. 6.18, polarity reversal is accomplished by reversing the meter connections to the plates of the bridge tubes.

Although polarity reversal can be obtained with a VOM by reversing the test-lead connections to the circuit under test, this is not practical with a VTVM, since the braid of the coaxial input cable is grounded to the case of the instrument. Therefore, VTVM operation is possible only with the cable braid grounded to the chassis of the equipment under test. In turn, all VTVMs have polarity-reversing switches, but the simpler types of VOMs do not have polarity-reversing switches. Unless the ground system of a VTVM is connected to the ground system of the equipment under test, the indication accuracy is likely to be very poor.

6.6 COMBINATION VOM-VTVM INSTRUMENTS

As was mentioned earlier, a few instruments are designed as combination VOM-VTVM arrangements. This type of instrument is often called a VTVOM. A typical configuration is shown in Fig. 6.19. The chief advantage of a combination

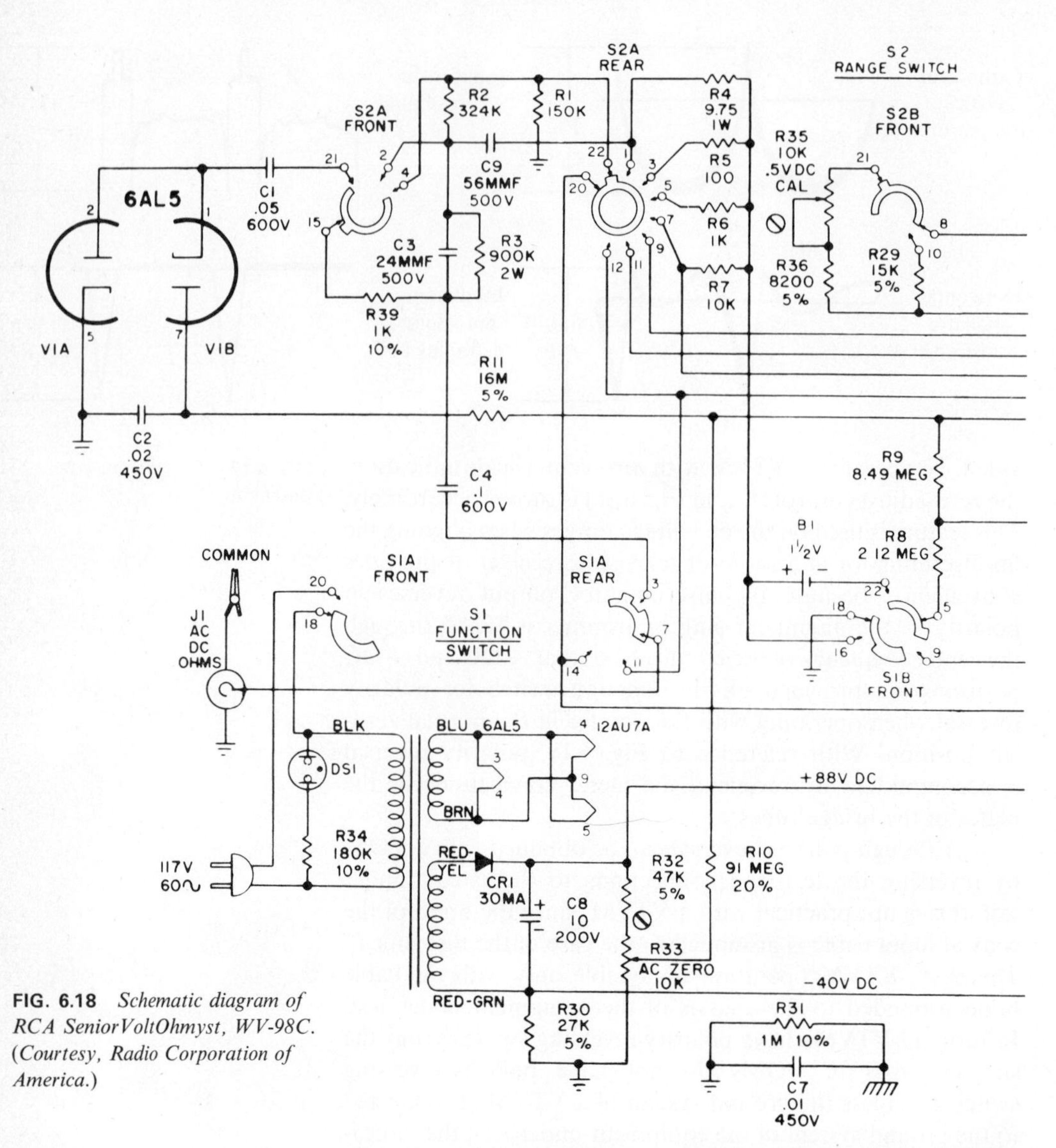

FIG. 6.18 *Schematic diagram of RCA SeniorVoltOhmyst, WV-98C. (Courtesy, Radio Corporation of America.)*

instrument is that it can be operated as a VOM in case a power outlet is not available to energize the VTVM section. Another advantage is provision of a dc current-measuring function; conventional VTVMs do not measure current values. Figure 6.20 illustrates the appearance of a VTVOM. The same number of controls are provided as in a conventional VTVM; however, the function switch has more positions in order to accommodate both VOM and VTVM operation.

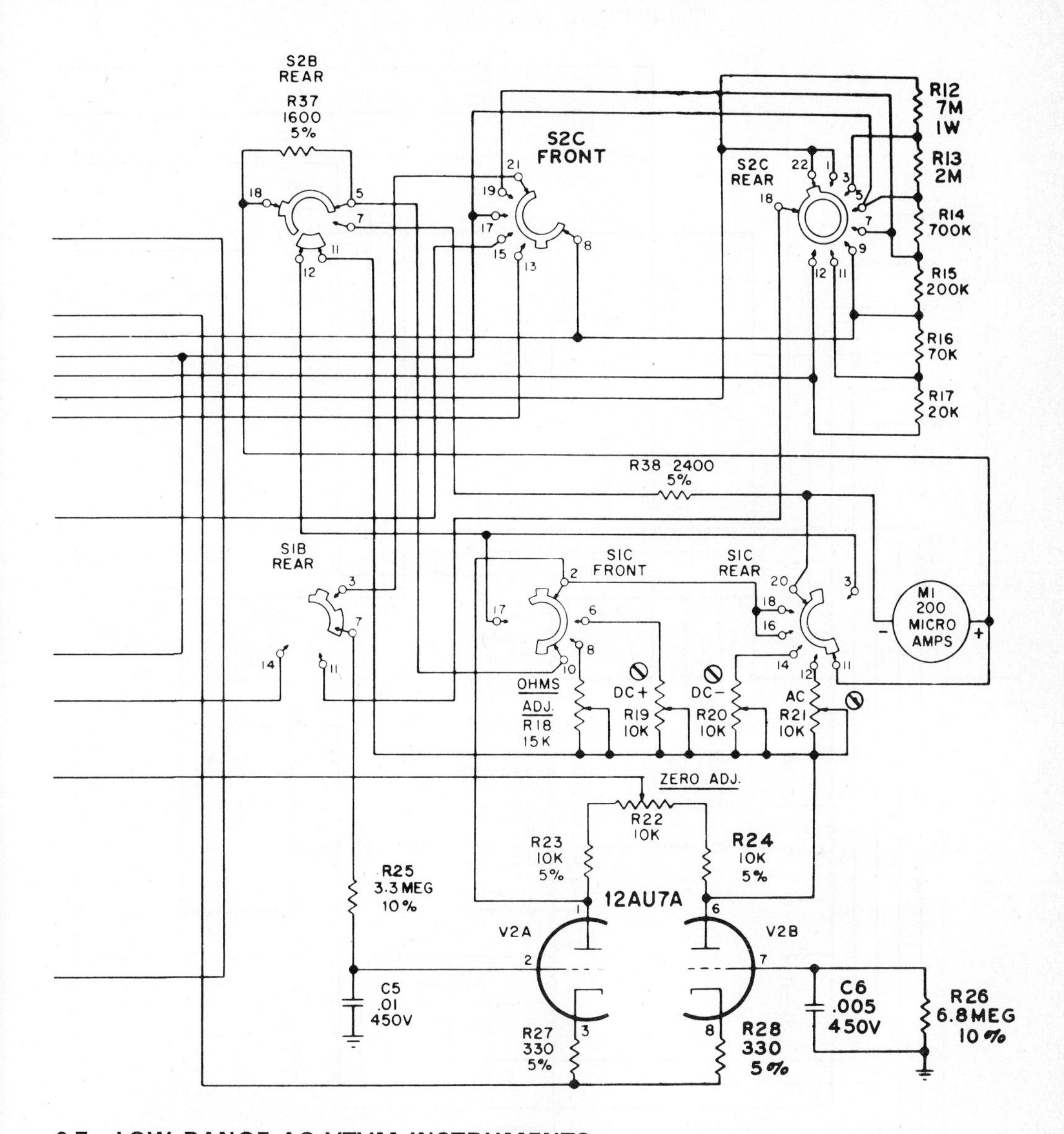

6.7 LOW-RANGE AC VTVM INSTRUMENTS

The VTVMs that have been discussed are not designed to measure small ac voltage values. Tests of audio-frequency equipment often require measurement of voltages as low as 10 mV rms. Special types of VTVMs used for this purpose are sometimes called audio VTVMs. A typical instrument is illustrated in Fig. 6.21. Note that only ac voltages are accommodated; no dc voltage or resistance functions are provided.

FIG. 6.19 *Schematic of a Sencore SM 112 servicemaster combination VOM-VTVM. (Courtesy, Sencore, Inc.)*

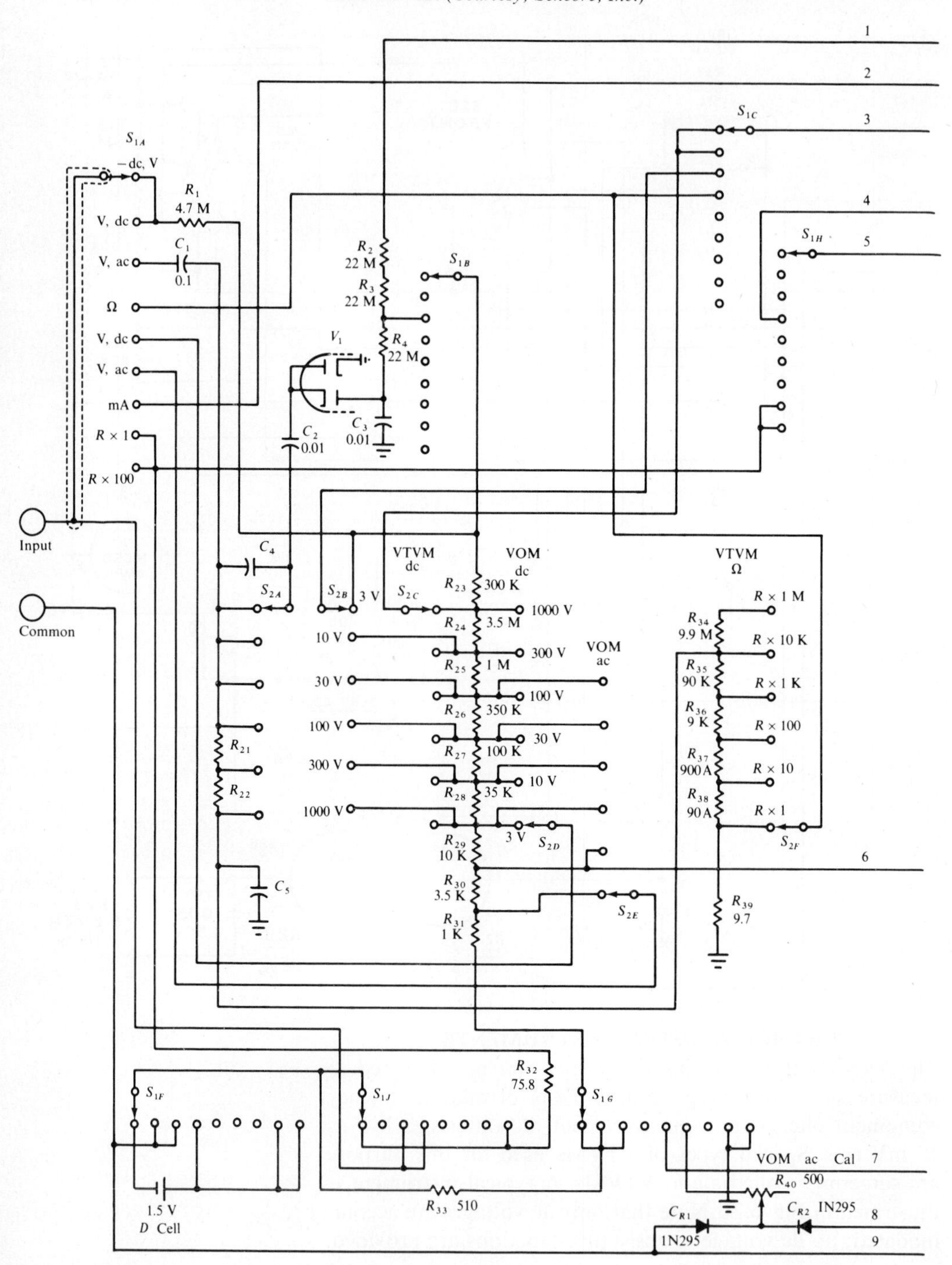

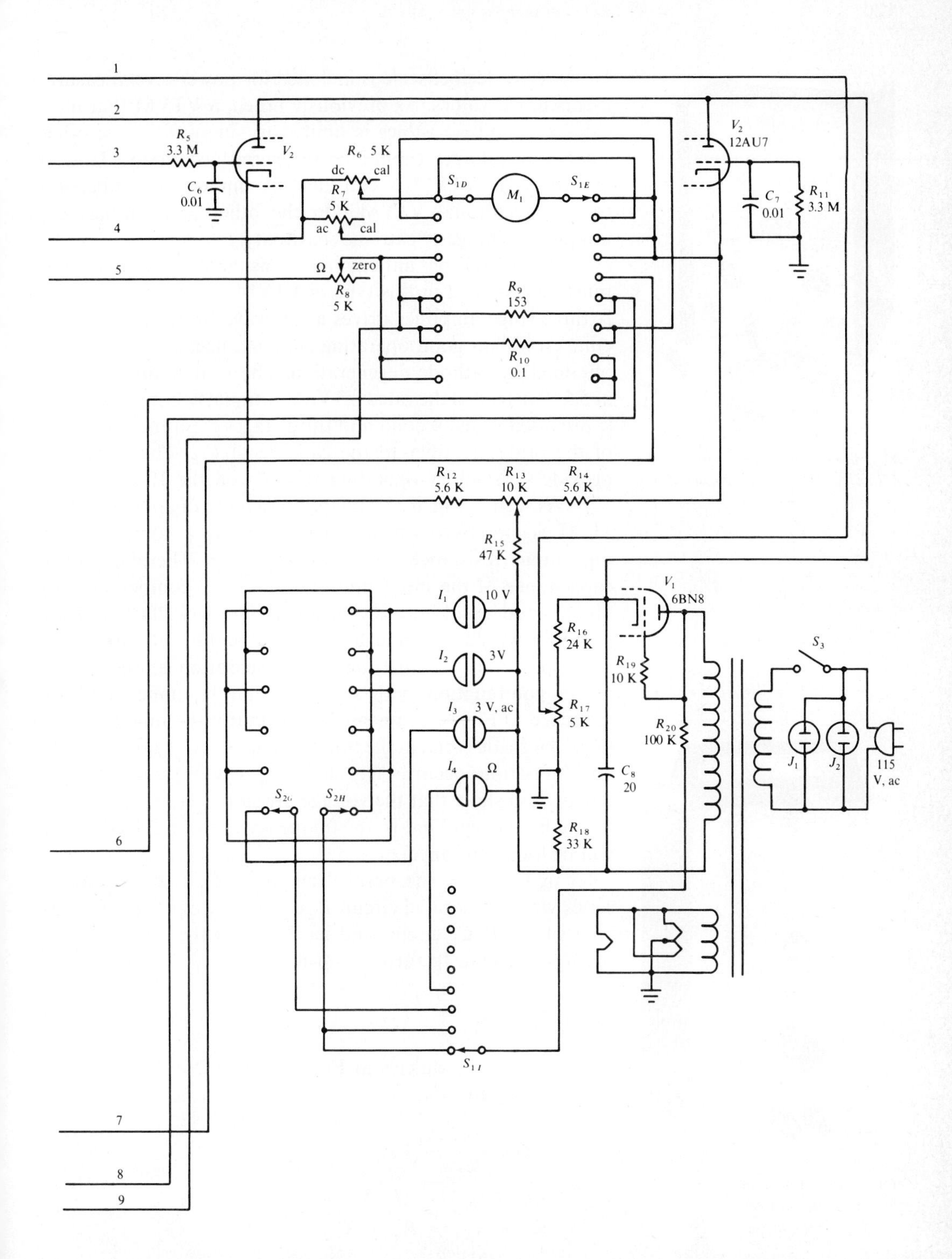

1
2
3
4
5
6
7
8
9
R_5
3.3 M
C_6
0.01
V_2
R_6 5 K
dc
cal
R_7
5 K
ac
cal
Ω
zero
R_8
5 K
S_{1D}
M_1
S_{1E}
R_9
153
R_{10}
0.1
V_2
12AU7
C_7
0.01
R_{11}
3.3 M
R_{12}
5.6 K
R_{13}
10 K
R_{14}
5.6 K
R_{15}
47 K
I_1
10 V
I_2
3V
I_3
3 V, ac
I_4
Ω
S_{2G}
S_{2H}
S_{1I}
V_1
6BN8
R_{16}
24 K
R_{19}
10 K
R_{17}
5 K
R_{20}
100 K
C_8
20
R_{18}
33 K
S_3
J_1
J_2
115
V, ac

However, a decibel scale is included for power-level measurements in dB values. As previously noted, a VTVM that indicates rms voltage values is limited to sine-wave tests; other waveform voltages cannot be measured accurately. Thus, a low-range ac VTVM is considerably limited in comparison to a conventional VTVM. On the other hand, it has the unique advantage of extreme sensitivity.

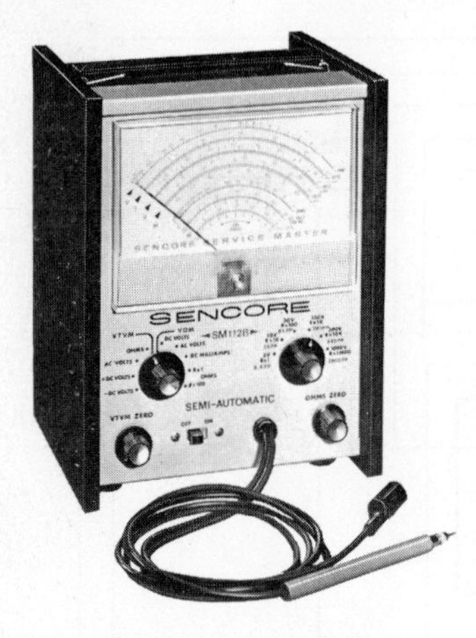

FIG. 6.20 *Appearance of a VTVOM. (Courtesy, Sencore, Inc.)*

As would be anticipated, considerable amplification must be provided in this type of VTVM, as seen in Fig. 6.22. A three-stage amplifier drives a full-wave bridge-rectifier circuit. To obtain good operating stability, negative feedback is obtained by cathode degeneration. The heaters are operated at 5.5 V instead of the rated 6.3 V. Additional negative feedback is provided in the second and third stages from the lower end of the bridge rectifier to the cathode of V_{1B}. These features provide satisfactory operating stability on the 10-mV range.

As seen in Fig. 6.22, an input resistance of approximately 11 M is employed on all ranges. This is a comparatively high input resistance for an ac voltmeter. Therefore, stray capacitance of the input multiplier must be compensated to obtain good high-frequency response; otherwise, the frequency response would be quite limited on the 10-, 30-, 100-, and 300-V positions. High-frequency compensation is provided by C_2 in combination with C_3. This can be understood by reference to Fig. 6.23; we shall show that if the time constant (R_1C_1) is made equal to the time constant (R_2C_2), no frequency discrimination occurs. We shall apply Ohm's law to the circuit, and show that the voltage division is constant regardless of frequency. That is, the frequency terms in the equation will be found to cancel out.

In Fig. 6.23, the parallel circuit R_1C_1 is connected in series with the parallel circuit R_2C_2. Let Z_1 equal the impedance of the R_1C_1 circuit and let Z_2 equal the impedance of the R_2C_2 circuit. In turn, we write

$$\frac{e_{out}}{e_{in}} = \frac{Z_2}{Z_1 + Z_2} \times \frac{i}{i} \tag{6.1}$$

Note that i cancels in Formula (6.1). Next, we expand the terms Z_1 and Z_2:

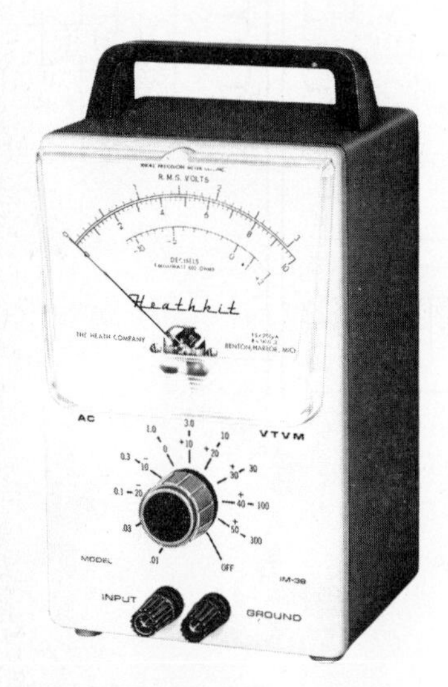

FIG. 6.21 *Appearance of a low-range ac VTVM. (Courtesy, Heath Co.)*

$$\frac{e_{out}}{e_{in}} = \frac{\dfrac{R_2X_{C2}}{R_2 + X_{C2}}}{\dfrac{R_1X_{C1}}{R_1 + X_{C1}} + \dfrac{R_2X_{C2}}{R_2 + X_{C2}}} \tag{6.2}$$

To show that the frequency terms cancel out, we next expand the reactance terms:

$$\frac{e_{out}}{e_{in}} = \frac{\dfrac{R_2/j\omega C_2}{R_2 + 1/j\omega C_2}}{\dfrac{R_1/j\omega C_1}{R_1 + 1/j\omega C_1} + \dfrac{R_2/j\omega C_2}{R_2 + 1/j\omega C_2}} \tag{6.3}$$

To start our cancellation of the frequency terms, we rearrange Formula (6.3) as follows:

$$\frac{e_{out}}{e_{in}} = \frac{\dfrac{R_2/j\omega C_2}{(j\omega C_2 R_2 + 1)/j\omega C_2}}{\dfrac{R_1/j\omega C_1}{(j\omega C_1 R_1 + 1)/j\omega C_1} + \dfrac{R_2/j\omega C_2}{(j\omega C_2 R_2 + 1)/j\omega C_2}} \tag{6.4}$$

It is evident that the denominators cancel out, and we obtain

$$\frac{e_{out}}{e_{in}} = \frac{\dfrac{R_2}{j\omega C_2 R_2 + 1}}{\dfrac{R_1}{j\omega C_1 R_1 + 1} + \dfrac{R_2}{j\omega C_2 R_2 + 1}} \tag{6.5}$$

If we stipulate that $R_1 C_1 = R_2 C_2$, the remaining denominators cancel out, and we write

$$\frac{e_{out}}{e_{in}} = \frac{R_2}{R_1 + R_2} \tag{6.6}$$

Therefore, if the time constants of the two sections are made equal in Fig. 6.23, the output voltage is independent of frequency. When C_2 is properly adjusted in Fig. 6.22, the multiplier provides accurate voltage division at any frequency. Note, however, that an upper frequency limit of 1 MHz is imposed by the subsequent circuitry. To maintain the high impedance provided by the input multiplier, V_{1A} is operated as a cathode follower. This stage has a low output impedance suitable for energizing the second multiplier. Because of its comparatively low resistance, the second multiplier does not require frequency compensation. The cathode-follower stage operates basically as an electronic impedance transformer; its voltage gain is slightly less than unity.

Voltage amplification is provided by V_{1B} and V_2 in a conventional RC-coupled amplifier configuration. Thereby, an input signal level of 10 mV rms produces full-scale meter deflection. Since a low-frequency phase shift is imposed by

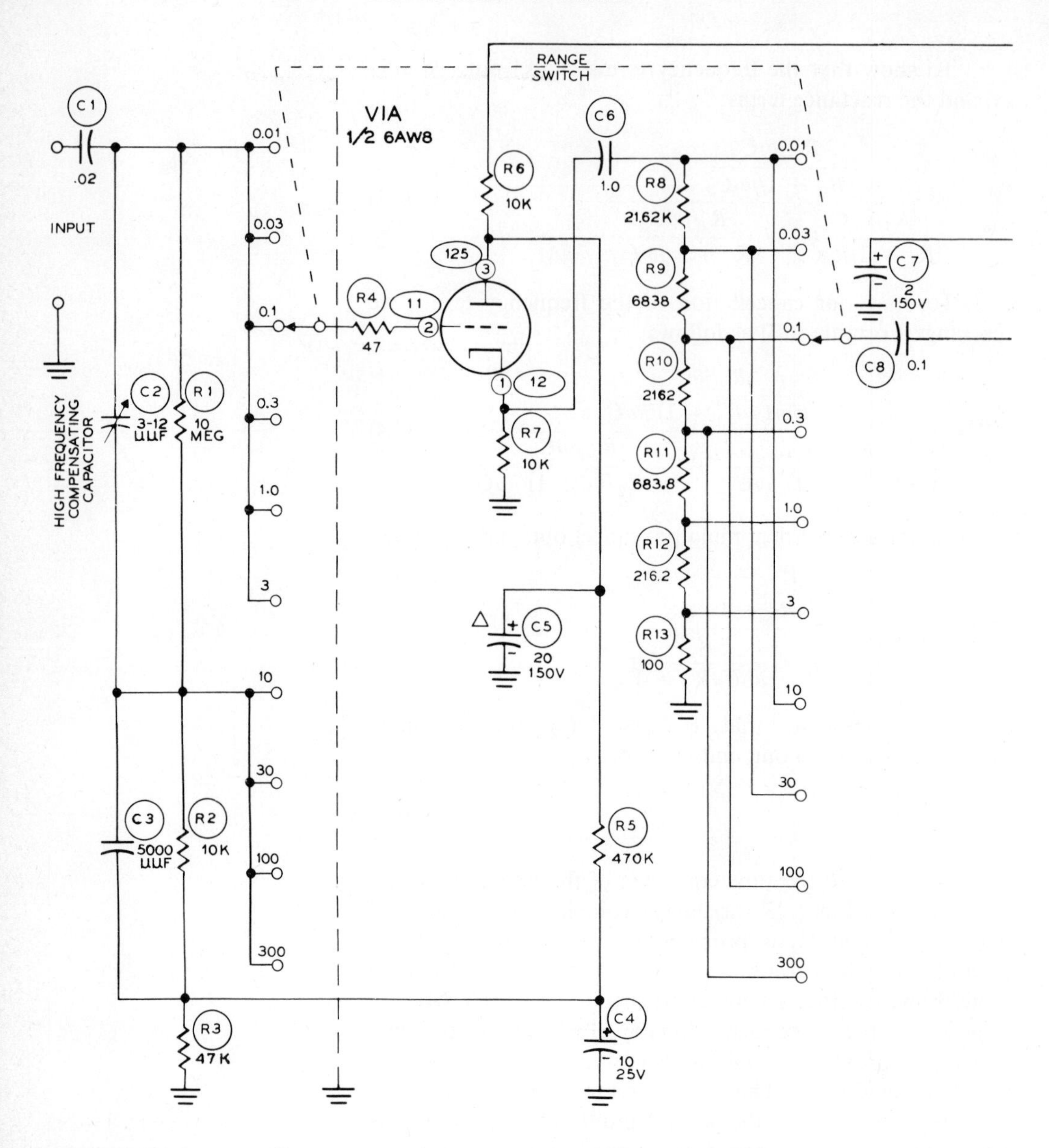

FIG. 6.22 *Configuration of low-range ac VTVM. (Courtesy, Heath Co.)*

C_{13} and R_{25}, a phase-compensating branch consisting of C_9 and R_{19} is provided. This phase shift maintains negative feedback to the cathode of V_2 over the entire frequency range. Since there is stray capacitance associated with plate-load resistors R_{16} and R_{24}, the frequency response tends to decrease at high frequencies. Accordingly, high-frequency response is

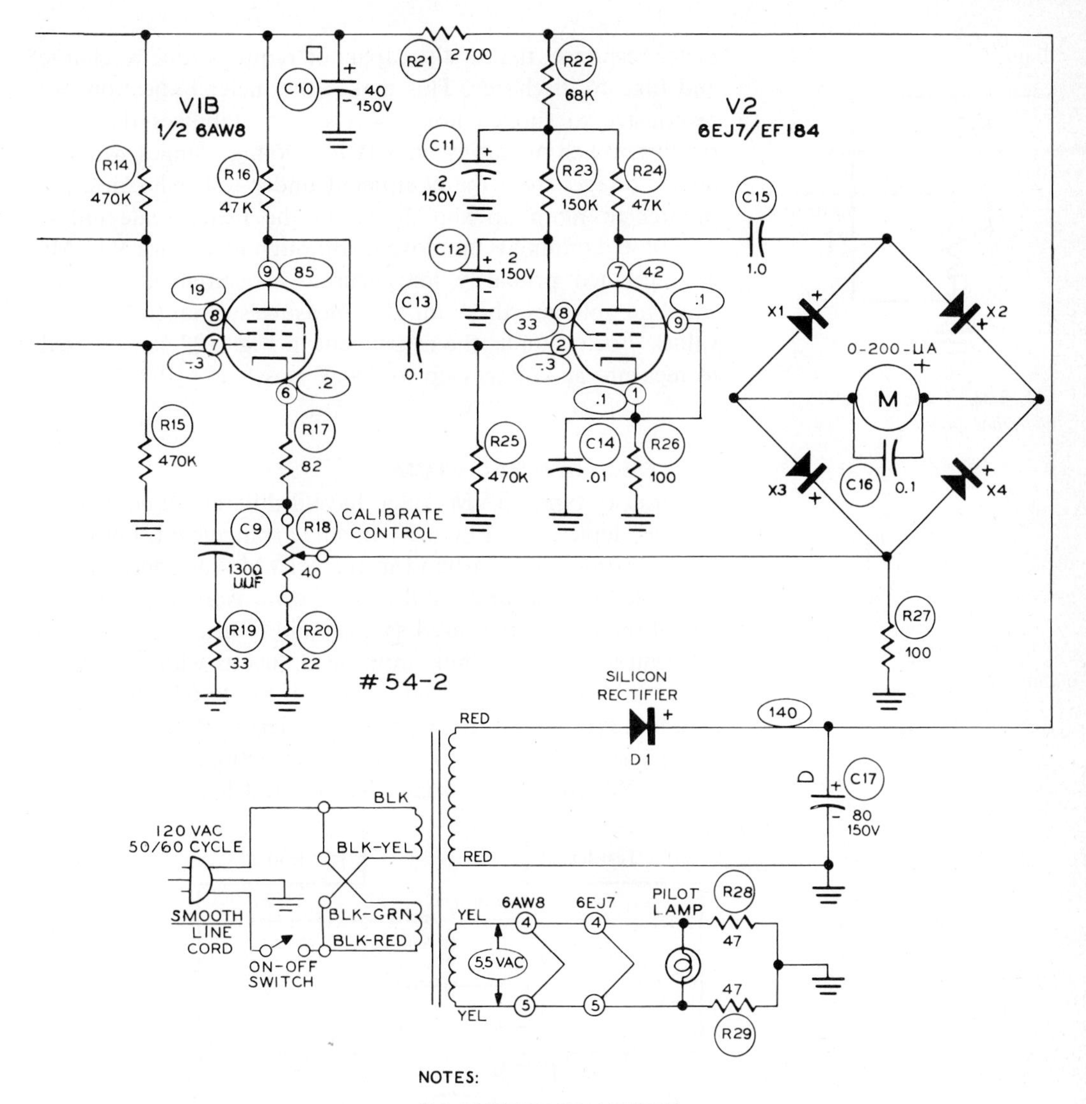

maintained by partial bypassing of R_{26}; at low frequencies, C_{14} has little bypassing action and the stage is degenerative—however, at high frequencies, C_{14} bypasses the cathode signal to ground and the stage is not degenerative.

Note capacitor C_{16} connected across the meter terminals. This capacitor provides electrical damping of the

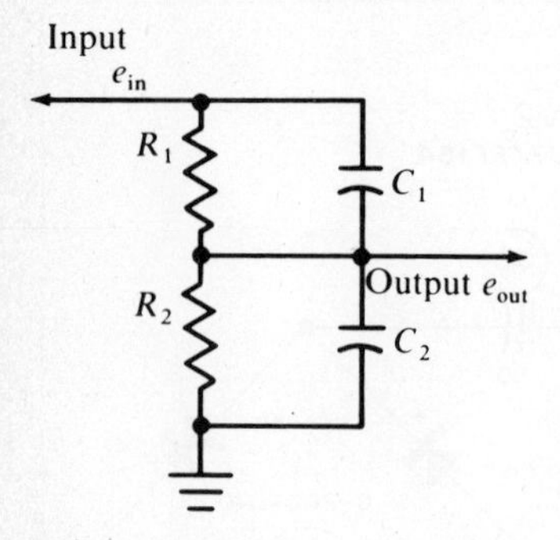

FIG. 6.23 *Compensated RC multiplier circuit.*

meter response; that is, the capacitor requires time to charge and time to discharge. This makes the meter indication unresponsive to sudden noise bursts, and stabilizes the scale reading. When operating on very low voltage ranges, random noise voltages from the equipment under test will cause the pointer to jump up and down on the meter scale unless adequate damping is employed. Although low-range VTVMs are designed primarily for accurate measurement of very small ac voltages, they can also measure substantial voltage values. For example, the instrument in Fig. 6.22 can be used to measure ac voltages up to 300 V rms.

6.8 DIFFERENTIAL VTVM

Laboratory-type VTVMs often have a differential configuration, as depicted in Fig. 6.24. In this arrangement, both test leads operate above ground in the VTVM; each test lead is connected to a separate multiplier. To measure a dc voltage, one of the test leads is used as a ground return and the other test lead is used as a "hot" input lead, just as with a conventional VTVM. Note that either one of the test leads can be used as a ground return; reversal of the test leads to the circuit under test merely reverses the meter deflection, as in the case of a VOM. In other words, when one test lead is connected

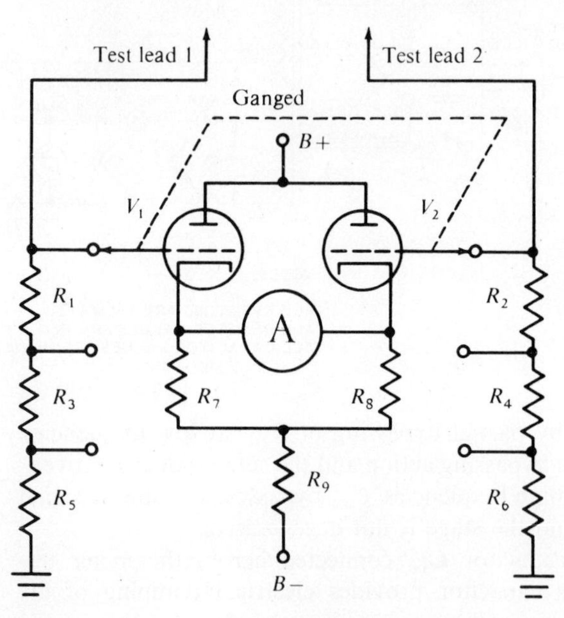

FIG. 6.24 *Basic differential VTVM circuit.*

to a ground point, the grid of the associated bridge tube operates at zero volts; the voltage to be measured is applied to the grid of the other bridge tube.

Next, we will find that a differential VTVM input circuit has an advantage over a conventional VTVM in that both of its test leads can be connected to circuit points operating above ground. For example, suppose that we wish to check a push-pull amplifier stage for dc balance. To make this test with a differential VTVM, we connect one test lead to the collector of one push-pull transistor, and connect the other test lead to the collector of the other push-pull transistor. If the push-pull stage is balanced, the differential VTVM reads zero volts. On the other hand, if the stage is not balanced, the differential VTVM reads the *difference* between the voltages applied to the two test leads. Note that if a conventional VTVM were used in this application, a pair of measurements would have to be made to obtain an accurate balance check.

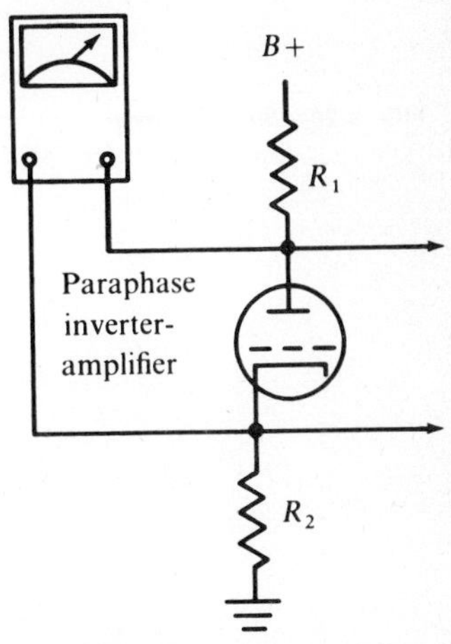

FIG. 6.25 *Measurement of plate-cathode voltage with a differential VTVM.*

Another example of differential VTVM application is shown in Fig. 6.25. The plate-cathode voltage of the paraphase inverter-amplifier tube can be measured directly with a differential VTVM. On the other hand, if a conventional VTVM were used, a pair of measurements would have to be made, and the cathode voltage subtracted from the plate voltage to find the plate-cathode voltage. To repeat an important point, a conventional VTVM is very likely to indicate incorrectly if applied as shown in Fig. 6.25. As we continue with our study of electronic test instruments, we shall find that differential amplifiers provide additional advantages, from the standpoint of both application flexibility and operating stability.

Instruments that employ differential input circuits are sometimes said to have balanced input, push-pull input, or double-ended input. These are equivalent terms. By way of comparison, instruments that operate with one of the test leads grounded to the case of the instrument are said to have single-ended input. An instrument with differential input can be used in any application for which an instrument with single-ended input is suitable, plus various applications for which single-ended instruments are unsuitable.

6.9 SENSITIVE AC VTVM

Since a conventional VTVM cannot measure low ac voltage values, the arrangement shown in Fig. 6.26 is used to measure values in the millivolt range. The chief distinction in this type

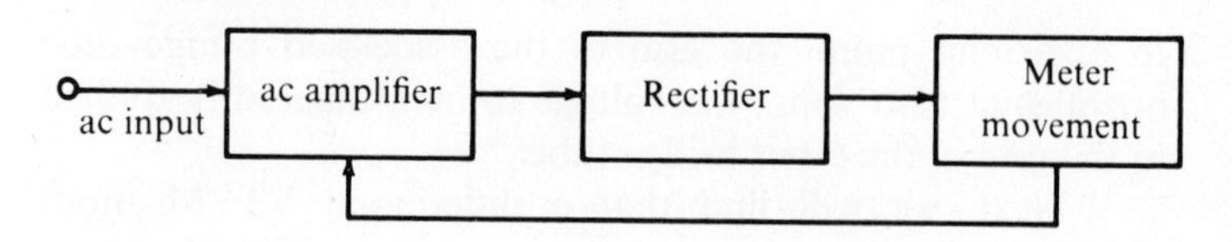

FIG. 6.26 *Block diagram of a sensitive VTVM.*

of instrument is the provision of high-gain amplifiers in a stabilized configuration. A large amount of negative feedback is employed to stabilize the amplifier characteristics. With reference to Fig. 6.27, negative feedback is obtained via R_1, R_2, and R_3. Calibration is made by adjusting the value of R_3. A full-wave semiconductor bridge rectifier is utilized, and the meter indicates rms values of sine waves.

A commercial configuration for a sensitive ac VTVM used in audio work is seen in Fig. 6.28. The multiplier (not shown) has nine steps to provide ranges of 0.01, 0.03, 0.1, 0.3, 1, 10, 30, 100, and 300 rms. The first amplifier stage is a *cascade* type; that is, the plate of the first triode drives the cathode of the second triode in a series arrangement. This configuration provides substantial negative feedback. Note that the grid of the second triode is grounded for ac. The plate of the second triode drives a third triode, which in turn energizes a half-bridge instrument rectifier. Negative feedback

FIG. 6.27 *Basic configuration for a sensitive ac VTVM.*

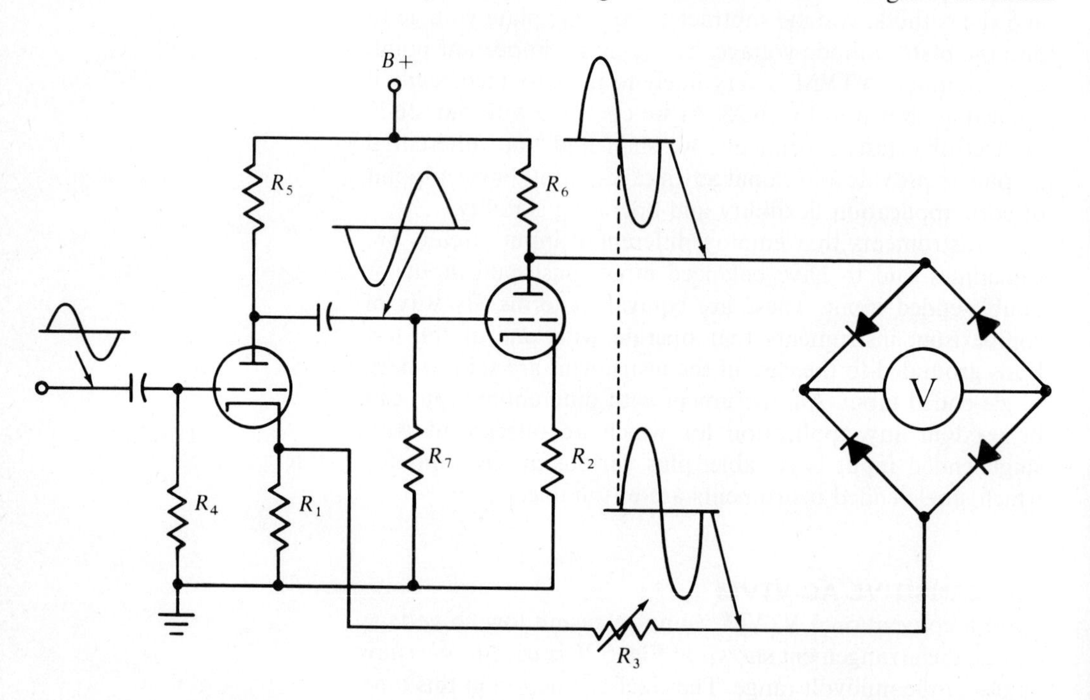

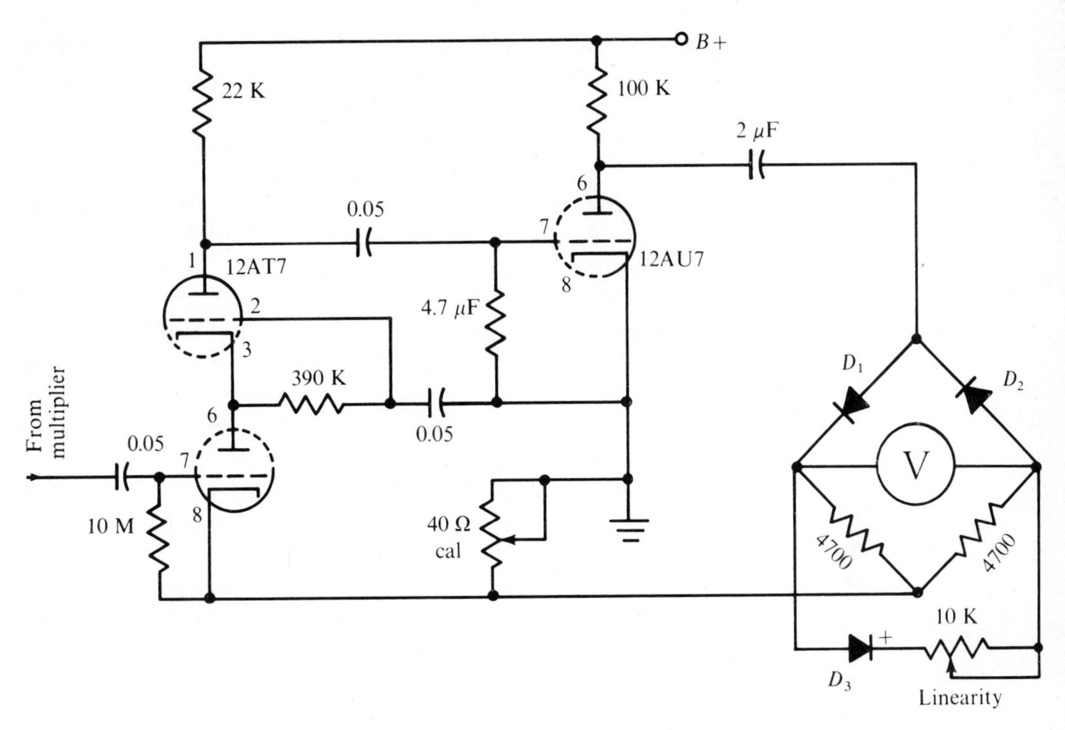

is provided by the 40-Ω calibration control and by a feedback loop from the instrument bridge to the cathode of the first triode.

FIG. 6.28 *A commercial configuration of a sensitive ac VTVM. (Courtesy, Heath Co.)*

Calibration control is also provided by a linearity circuit connected across the meter movement. This circuit comprises a diode and a 10-K potentiometer. It permits adjustment of center-scale indication independently of full-scale indication. Thus, the linearity control improves indication accuracy. Amplifier gain is adjusted to obtain full-scale indication with a 10-mV input on the first range. The VTVM is rated for an accuracy of ± 5 percent of full scale. Triode tubes are used in this type of instrument to minimize noise voltages. However, the residual noise level is sufficiently high that the lower end of the scale is not useful on the 10-mV range. That is, the residual noise causes the pointer to "bounce" continuously over the first several scale divisions.

6.10 ELECTROMETER-TYPE VTVM

An electrometer is a highly sensitive electrical instrument for measurement of voltages or currents. Most present-day electrometers employ electrometer tubes in VTVM or hybrid configurations. An electrometer tube has unusually low grid current, such as 10^{-12} A. By way of comparison, a tube used in a conventional VTVM may have a grid current 10^4 times as great as an electrometer tube. To assist in minimizing residual grid current, a plate voltage of approximately 10 V is utilized. A typical hybrid electrometer-type VTVM has the configuration shown in Fig. 6.29.

This electrometer is used for measurement of ac and both positive and negative dc voltages, up to 1500 V. It is basically a very stable dc voltage amplifier with a vacuum-tube input stage. A probe containing a vacuum diode rectifies ac signals for ac voltage measurements. For resistance measurements, a regulated voltage source, in series with one of four range-determining resistors, is connected to the input terminals so that the unknown resistance can be determined by measuring the voltage drop across it.

Note that the meter section functions as a pair of cathode followers, driving the meter movement from cathode to cathode. Each cathode follower is made up of two vacuum tubes and one transistor, connected to approximate ideal cathode-follower operation. The first tube is connected as a simple cathode follower and is operated at reduced plate current and reduced heater voltage, to minimize grid current. The transistor and second tube make up a circuit with an extremely high emitter input impedance. Since the gain of the first tube is very nearly $\mu/(\mu + 1)$, the circuit operation is practically independent of tube aging.

Since the gain of the transistor-tube circuit is highly stabilized by feedback, the voltage amplification of the entire circuit is minimized with decrease in grid control of tubes due to aging. The output impedance of the circuit is less than 1 Ω, because of the high feedback loop gain. In turn, the output impedance remains practically constant as the tubes age. Independence from line-voltage fluctuation is provided by a highly regulated power supply.

6.11 APPLICATIONS

One of the basic advantages of the VTVM is its low input capacitance, compared with that of a VOM. In turn, a VTVM imposes minimum loading on high-frequency oscillator circuits, such as depicted in Fig. 6.30. We commonly measure the signal-developed bias voltage at the grid of an oscillator tube

(a)

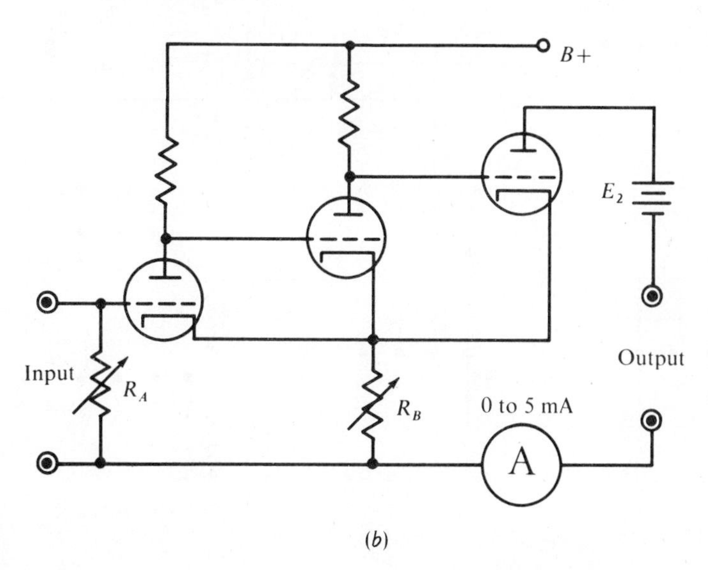

(b)

FIG. 6.29 *An electrometer-type VTVM.* (*a*) *Appearance,* (*Courtesy, General Radio Corp.*)*;* (*b*) *basic circuit;* (*c*) *complete circuit* (*pages* 130 *and* 131). (*Courtesy, Sencore, Inc.*)

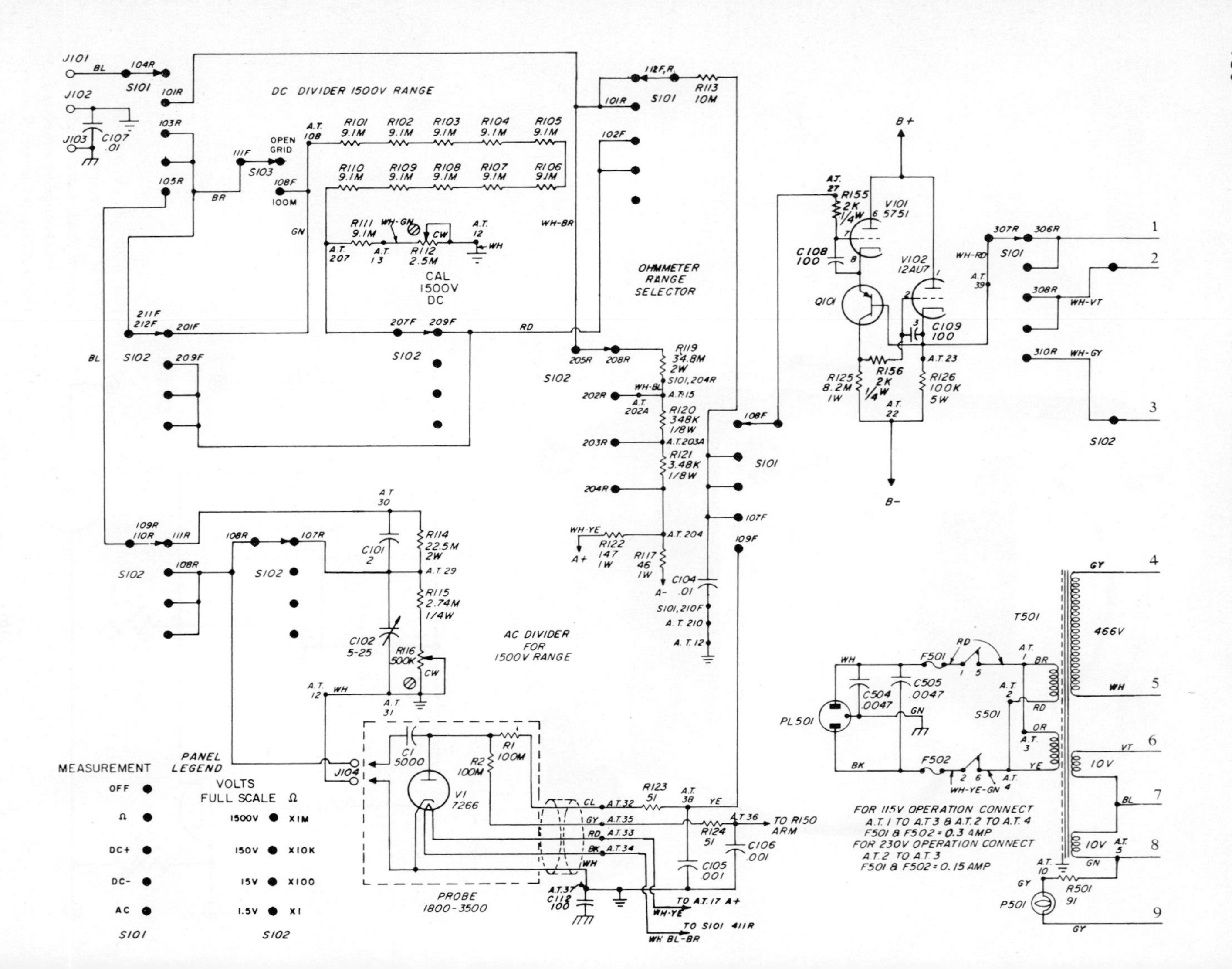
DC DIVIDER 1500V RANGE
OHMMETER RANGE SELECTOR
CAL 1500V DC
AC DIVIDER FOR 1500V RANGE
PROBE 1800-3500
FOR 115V OPERATION CONNECT A.T.1 TO A.T.3 & A.T.2 TO A.T.4 F501 & F502 = 0.3 AMP
FOR 230V OPERATION CONNECT A.T.2 TO A.T.3 F501 & F502 = 0.15 AMP
MEASUREMENT
OFF
Ω
DC+
DC-
AC
S101
PANEL LEGEND
VOLTS FULL SCALE Ω
1500V X1M
150V X10K
15V X100
1.5V X1
S102

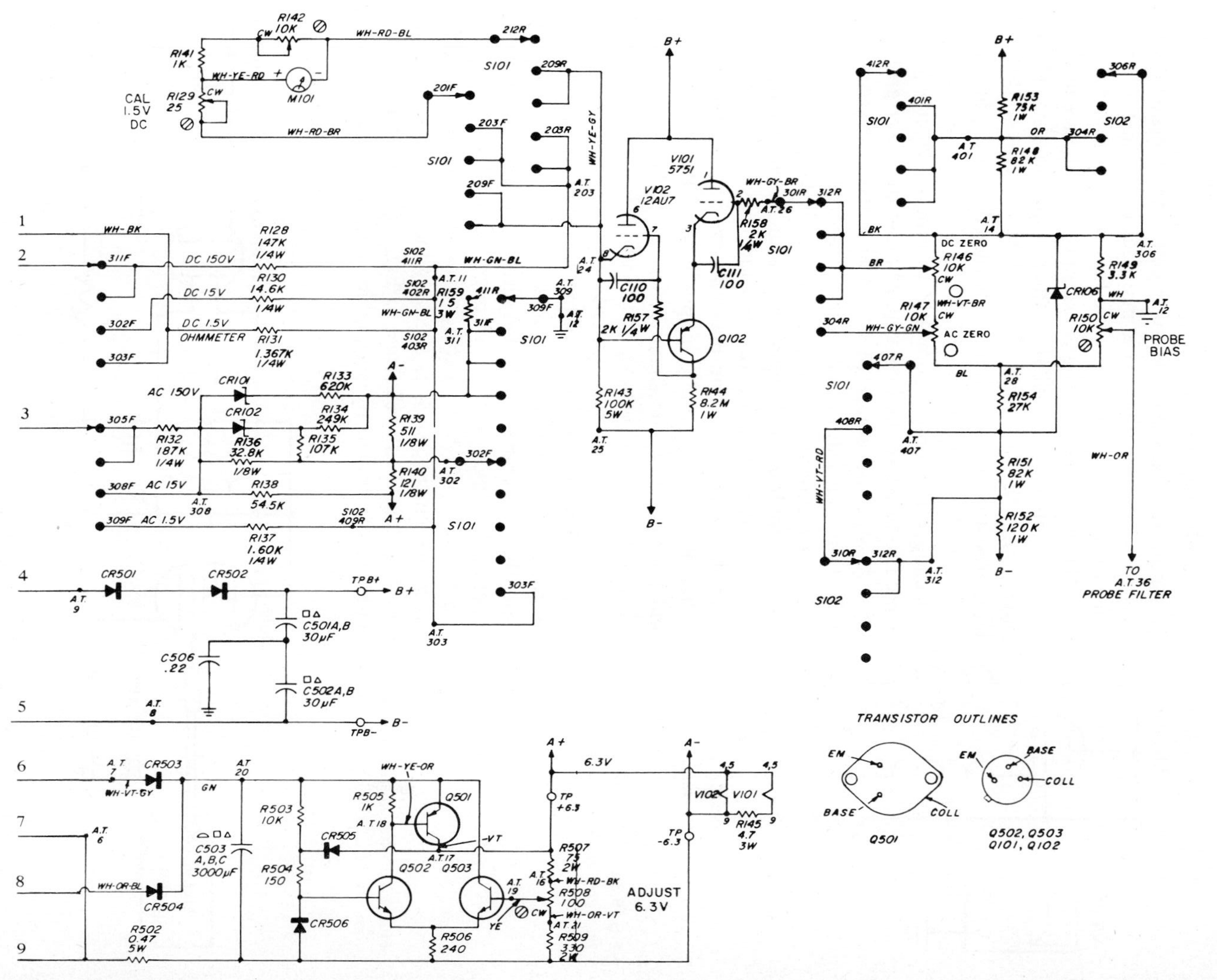

FIG. 6.29c (continued)

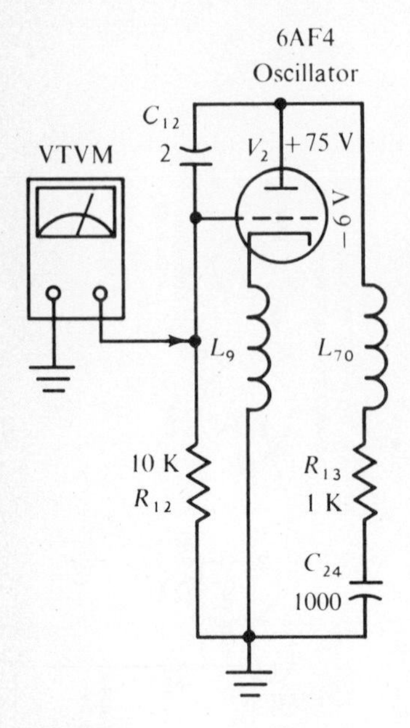

FIG. 6.30 *A VTVM or TVM measures oscillator and grid bias with minimum loading.*

FIG. 6.31 *Alignment of FM discriminator transformer secondary.*

to determine whether the circuit is functioning properly. Zero bias indicates that the oscillator circuit is not operating; weak bias indicates that there is a marginal defect in the circuit. If a VOM is used to measure the bias voltage at an oscillator grid, the capacitive loading is excessive with the result that a normally operating oscillator often stops working. On the other hand, the dc probe of a VTVM contains a 1-M isolating resistor that greatly reduces the effective input capacitance of the instrument. In turn, a minimum value of capacitance is shunted from grid to ground when a VTVM is used to measure the bias voltage at an oscillator grid.

Another advantage of a VTVM is its center-zero scale indication. This feature is commonly utilized in the alignment of fm discriminator circuits, as depicted in Fig. 6.31. The test signal is applied by a signal generator between the grid of the limiter tube and ground. A blocking capacitor is employed to avoid drain-off of grid-bias voltage. The VTVM is connected between the audio-output lead of the discriminator and ground. With the signal generator tuned to 10.7 MHz (standard fm-if frequency), the VTVM is operated on its dc-voltage function and adjusted for center-zero scale indication. To align the secondary of the discriminator transformer, its tuning slug is adjusted for zero-volts meter indication. Note that the pointer will swing from one side of zero to the other as the slug is adjusted through 10.7 MHz.

Since a VTVM has a decibel scale, it is useful for making dB measurements, such as checking the separation value in a stereo system. With reference to Fig. 6.32, the stereo system under test is driven by a suitable generator signal, and the

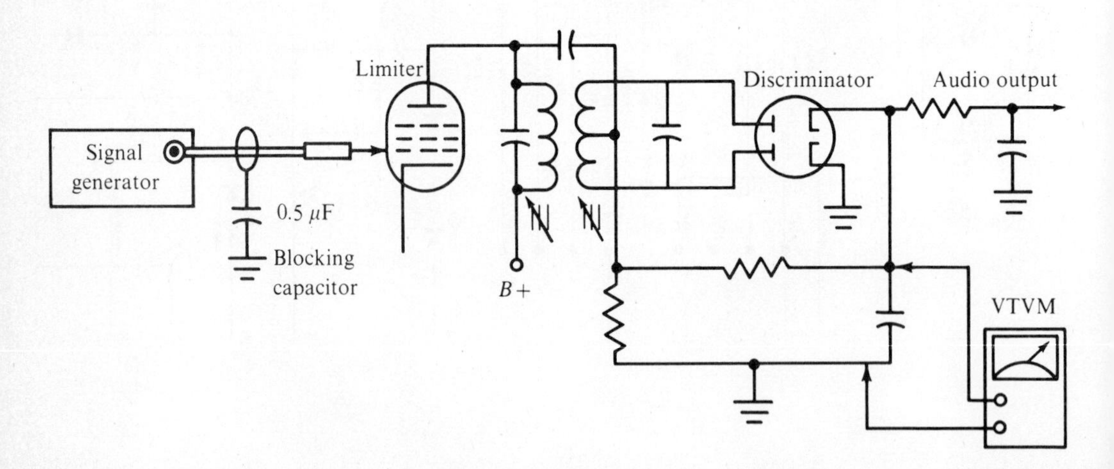

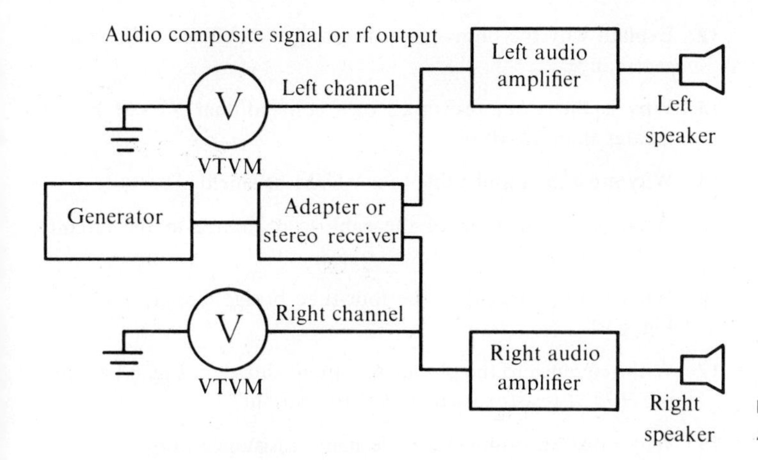

FIG. 6.32 *Setup for testing separation in a stereo system.*

output levels in the left and right channels are measured with a pair of VTVMs (or a single VTVM connected first to one channel and then to the other channel). In a typical test, we might apply a left signal to the system, and measure a left-channel output of 25 dB. In turn, the right-channel output might measure −10 dB. The conclusion of these tests states that the separation value is equal to 35 dB.

QUESTIONS AND PROBLEMS

1. What are the chief advantages of a VTVM?
2. What is a typical value of input resistance for a VTVM?
3. State three disadvantages of VTVMs and VOMs.
4. What is a VTVOM?
5. Explain the operation of the basic VTVM bridge circuit shown in Fig. 6.2.
6. What are two advantages of a VTVM bridge circuit over a single tube configuration?
7. Explain *contact potential* as the term is applied to a VTVM.
8. Why are tubes aged before being installed in a VTVM?
9. Why are modern VTVMs operated with reduced values of plate and heater voltage?
10. Why do sensitive meter movements provide greater accuracy in a VTVM?
11. What is the function of the resistor in the dc test probe of a VTVM?

12. Explain why the ohms-per-volt rating of a VTVM differs on each range.
13. Why is the input resistance of a conventional VTVM never greater than 22 MΩ?
14. Why must the input cable to a VTVM be shielded?
15. What is the function of the cable capacitance in the circuit shown in Fig. 6.7?
16. What is the purpose of the four-tube bridge circuit shown in Fig. 6.9?
17. With reference to the ohmmeter circuit shown in Fig. 6.10, why is a 9.75-Ω resistor used in the 10-Ω circuit?
18. Why must an ohmmeter have many resistance ranges?
19. What is the nominal accuracy of an ohmmeter which uses ± 1 percent resistors and a ± 2 percent meter?
20. What advantages does a VTVM have over a VOM in measuring semiconductor-device parameters?
21. Why do some VTVMs use an internal supply to energize the ohmmeter? What are the requirements of such a supply?
22. Why does a VTVM or VOM indicate an inaccurate rms value for a complex-waveform voltage?
23. How is the contact potential of the instrument-rectifier circuit in a VTVM canceled out?
24. What is the purpose of potentiometer R_2 in Fig. 6.13, and under what conditions is it adjusted?
25. A VTVM set for ac operation is connected across a 10-V dc battery. What is the indicated value of voltage?
26. Why are separate ac voltage scales provided on a VTVM for various ranges?
27. Explain how a VTVM can indicate the peak-to-peak value of a waveform such as shown in Fig. 6.15.
28. Is the peak-to-peak voltage value of an ac waveform independent of the dc voltage?
29. How is a VTVM adjusted to read a zero-center scale indication?
30. What is the purpose of the zero-center scale indication of a VTVM?
31. Why do VTVMs have polarity-reversing switches?

32. Why does the VTVOM have more positions on its function switch than a VTVM?

33. What is the purpose of a low-range ac VTVM?

34. Discuss the stability features of the VTVM shown in Fig. 6.22.

35. What is the purpose of the *RC* multiplier circuit in Fig. 6.23?

36. What is the configuration of the input amplifier (V_{1A}) in the VTVM shown in Fig. 6.22?

37. What is the purpose of capacitor C_{16} in the schematic of the VTVM in Fig. 6.22?

38. What is an advantage of the differential VTVM over a conventional VTVM?

39. What are three names for an instrument which uses a differential input amplifier?

40. What is the purpose of the negative feedback in the VTVM circuit shown in Fig. 6.28?

41. What is the advantage of an electrometer-type VTVM over conventional types? How is this advantage developed in an electrometer-type VTVM?

TRANSISTOR VOLTMETERS

7.1 ADVANTAGES

Transistor voltmeters (TVMs) have various advantages over the conventional VTVM; a TVM requires no warm-up time, is comparatively compact and lightweight, and may be designed as a battery-operated instrument for use in locations where power outlets are not available. A hybrid TVM employs both transistors and vacuum tubes; for example, a vacuum tube input stage provides a much higher input impedance than can be obtained with a conventional transistor. However, modern field-effect transistors have practically as high input impedance as vacuum tubes, and hybrid TVM's are becoming obsolescent. It is instructive to start our analysis of the TVM with the simplest configuration employing a conventional junction transistor. This basic TVM arrange-

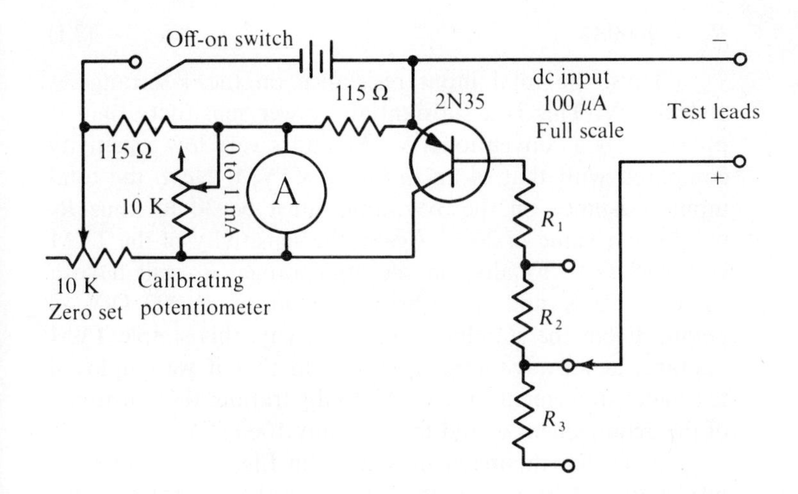

FIG. 7.1 *Basic TVM configuration.*

ment is analogous to the basic VTVM circuit discussed in the preceding chapter.

Figure 7.1 depicts a simple TVM circuit; although it appears similar to a one-tube VTVM configuration, there is an important distinction to be observed. The junction transistor is a current-operated device; accordingly, the input circuit does not employ a voltage multiplier—R_1, R_2, and R_3 have values related to the base-current demand of the transistor. In this example, a base-current input of 100 μA produces full-scale deflection of the 1-mA meter movement. The collector current is equal to beta times the base current. Since the base-emitter junction of the transistor has a comparatively low resistance (in the order of 1000 Ω), most of the applied input voltage drops across R_2 and R_1. Note that the input resistance changes on each position of the range switch, in the same manner as discussed previously for a VOM.

We shall now calculate the values of R_1, R_2, and R_3 that are required in Fig. 7.1 to provide dc voltage ranges of 1, 3, and 10 V. On the 1-V range, the resistance of R_1 plus the base input resistance must produce a current of 100 μA when 1 V is applied to the test leads. Assuming that the base input resistance is 1000 Ω, we write

$$100 \times 10^{-6} = \frac{1}{R_1 + 1000}$$

or

$$R_1 = 9000\ \Omega \tag{7.1}$$

Thus, the total input resistance on the 1-V range is 10,000 Ω/V. This is a substantially lower sensitivity than is provided by a conventional VOM, and a very low sensitivity compared with that of an ordinary VTVM. Next, the total input resistance on the 3-V range must be 30 K; thus R_2 will have a value of 20 K. Again, the sensitivity of the TVM is 10,000 Ω/V. Finally, on the 10-V range, R_3 will have a value of 70 K and the sensitivity value is 10,000 Ω/V as before. From the standpoint of sensitivity, this simple TVM operates at an advantage, however, in that if we employed the meter movement in a VOM configuration, the sensitivity of the arrangement would then be only 1000 Ω/V.

Note that the meter movement in Fig. 7.1 is connected into a bridge circuit, so that the saturation current of the transistor can be balanced out. If the bridge circuit were omitted, this saturation current would cause a pointer offset from zero. The amount of offset would vary with the ambient temperature; however, with provision of the zero-set control, the operator can bring the pointer to zero scale indication regardless of the transistor temperature. The calibrating control permits adjustment of the meter sensitivity to match the beta value of the particular transistor. This simple arrangement does not provide high indication accuracy, because the beta value is a function of temperature and also because the nonlinear characteristic of the transistor changes with temperature.

7.2 CASCADED TRANSISTORS

To increase the sensitivity of a TVM, a pair of transistors can be cascaded, as shown in Fig. 7.2. Note that the input stage employs a *PNP* transistor, while the output stage uses an *NPN* transistor. This complementary configuration minimizes the number of circuit components. The sensitivity of this arrangement is 200,000 Ω/V, which is an order of magnitude higher than a conventional VOM, although an ordinary VTVM provides much greater sensitivity on its low-voltage ranges. Excessive drift is the chief difficulty encountered in operation, and acceptable accuracy can be obtained only at a specific ambient temperature.

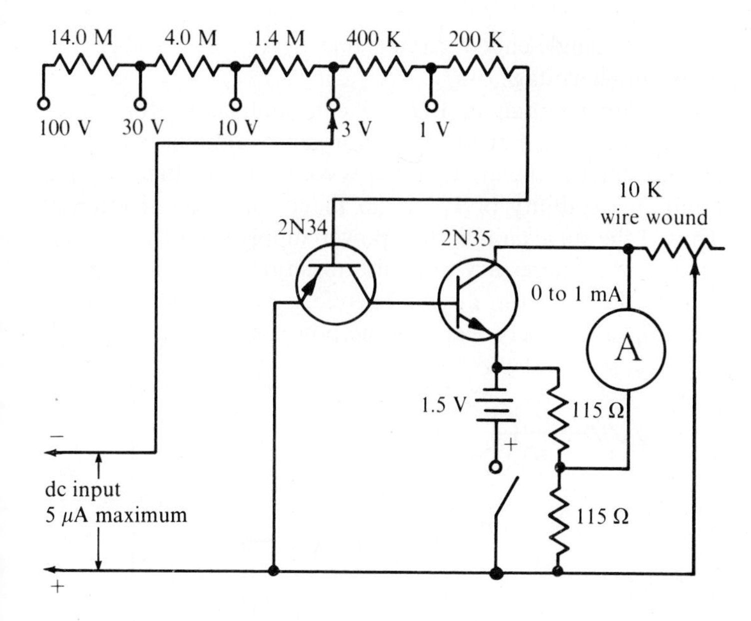

FIG. 7.2 *TVM circuit with two-stage amplifier.*

7.3 COMPENSATED CONFIGURATIONS

We shall find that the major portion of the drift voltage in a TVM is due to variation of the collector current. If a pair of transistors is chosen at random, their temperature-dependent characteristics will usually track within 10 percent. By selection of matching pairs, it is practical to utilize transistors in production that track within 1 percent. In the following discussion, we shall consider units that are matched within 4 percent. Figure 7.3 shows two basic compensated dc amplifier configurations. These circuits provide reduction of emitter voltage variation to $\pm 100\ \mu V/°C$.

In Fig. 7.3*a*, a diode connected in series with the emitter of the transistor provides compensating action. Resistor R_3 has a high value in order to simulate a constant-current source from V_{EE}. Potentiometer R_2 provides adjustment for optimum compensation. As the temperature increases, the emitter junction of Q_1 decreases in resistance, and the junction of D_1 also decreases in resistance. Since D_1 then draws a greater proportion of the supply current, the voltage across the emitter junction of Q_1 remains essentially constant. We observe that the configuration in Fig. 7.3*b* has similar compensating action, except that Q_2 is utilized instead of a diode.

This is a single-ended circuit that serves to stabilize the base-emitter voltage of Q_1.

Neither circuit in Fig. 7.3 can compensate for power supply variations nor for transistor gain variations. Although the drift level of the input circuit is controlled within practical limits, the stability of V_{CC} is an independent consideration. Thus, if the stage gain is 10, a power-supply stability of better than 1 mV is necessary to avoid objectionable fluctuation or drift in the collector circuit. Therefore, battery bias must be utilized, or a closely regulated ac power supply.

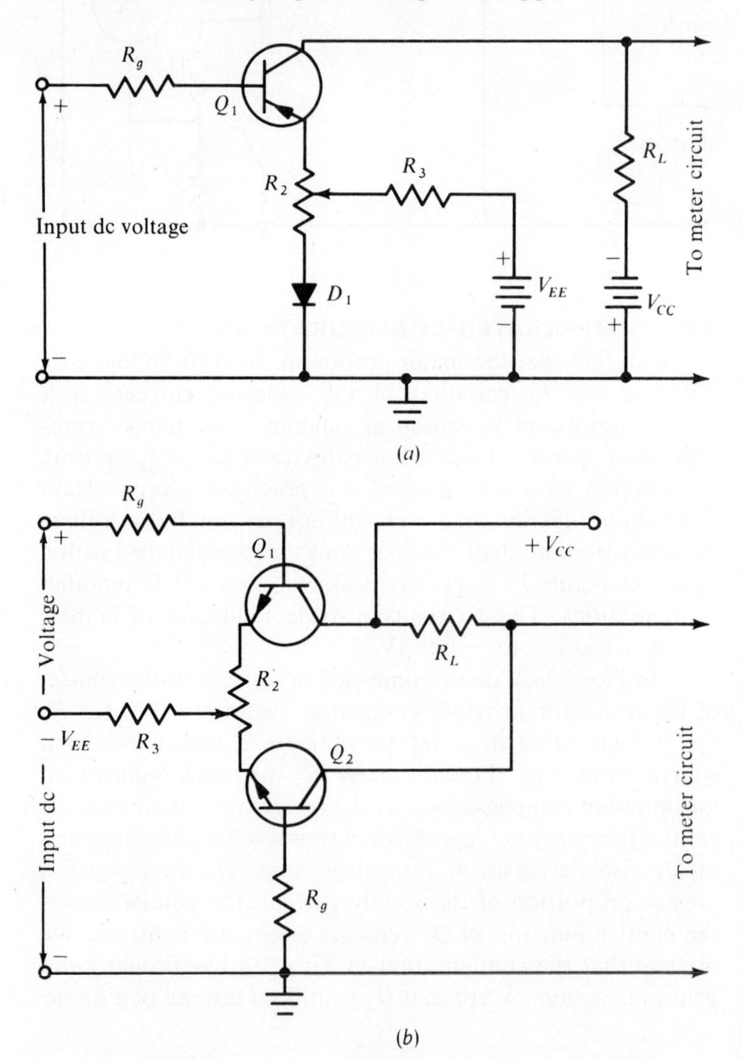

FIG. 7.3 *Dc amplifier with (a) compensating diode, and (b) compensating transistor.*

7.4 TRANSISTOR BRIDGE CIRCUITS

Transistor bridge circuits provide stability in TVM operation in much the same way that vacuum-tube bridge circuits provide stability in VTVM operation. A bridge circuit with substantial negative feedback minimizes drift due to variations in beta, I_{co}, and power-supply fluctuations. A basic configuration is shown in Fig. 7.4. The bridge arms consist of R_{L1}, R_{L2}, and the internal collector resistances of transistors Q_1 and Q_2. If a positive voltage is applied to the base of Q_1, increased emitter current flows. In turn, Q_2 draws decreased emitter current due to the increased voltage drop across R_1. Thus, the bridge is unbalanced and current flows through the meter movement.

If the ambient temperature increases, the beta value of both Q_1 and Q_2 increases (Fig. 7.4). Assuming that the transistors are perfectly matched, the current increase through R_{L1} is the same as the current increase through R_{L2}, and the bridge remains balanced. The saturation current I_{co} through each transistor will also increase by the same amount as the temperature increases, and the bridge remains balanced. Power-supply fluctuations result in variation of V_{CC} and V_{EE} values; and since the bias change is the same on both transistors, the bridge remains balanced. However, maintenance of calibration accuracy requires the action of a substantial amount of negative feedback, in addition to a stable balance condition, as will be explained.

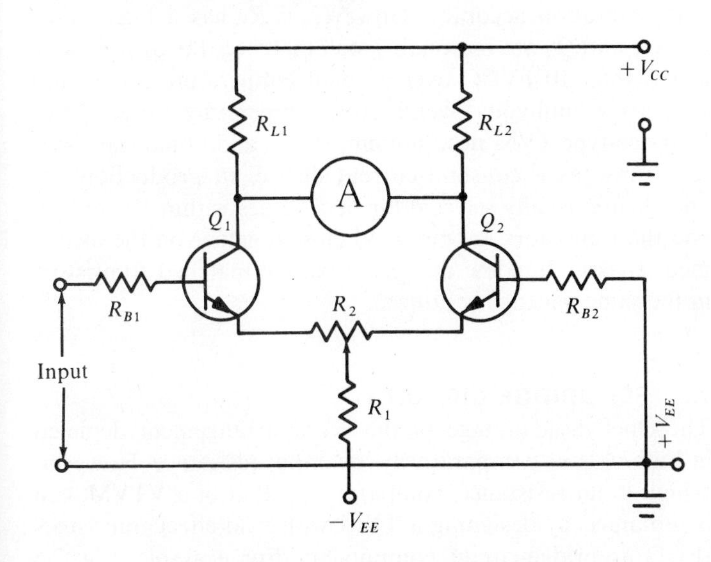

FIG. 7.4 *A basic TVM bridge configuration.*

First let us suppose that R_{g1}, R_{g2}, R_2, and R_1 are replaced by short circuits. The TVM will remain operative, provided the value of V_{EE} is suitably chosen. Now, if the ambient temperature increases, the beta value of Q_1 and Q_2 increases. In turn, when a positive voltage is applied to the base of Q_1, its collector current is greater than at the original temperature, and the meter reads high. Note that Q_2 is effective merely to maintain quiescent balance in this example, because R_1 was replaced by a short circuit. Thus, it is evident that some compensating action, such as negative feedback, must be utilized to maintain calibration accuracy.

Note that the zero-set control R_2 in Fig. 7.4 also provides resistance in series with the emitters of Q_1 and Q_2. If the ambient temperature increases, Q_1 and Q_2 draw more emitter current, but the increase is limited to some extent by the increased voltage drop across the emitter resistance. This negative feedback is supplemented by the action of R_1, which usually has a very high value. We observe that the TVM operates effectively from a constant-current source if R_1 is sufficiently high in value, and that the drop across R_1 provides negative feedback for Q_1 and Q_2. Accordingly, calibration accuracy is maintained over a comparatively wide temperature range.

In a more elaborate arrangement, R_1 in Fig. 7.4 may be replaced by a transistor employed as a constant-current device. This refinement provides maximum operating stability and calibration accuracy. However, if R_1 has a high value, and Q_1 and Q_2 are reasonably well matched, the drift voltage is less than 10 μV/°C over a small temperature range, and only a few millivolts over a 50°C temperature range. Thus, a service-type TVM need not employ an additional transistor to operate as a constant-current device. In production, Q_1 and Q_2 are usually selected for beta values within 10 percent, and the transistors are mounted close together on the instrument chassis. In some designs, a pair of matched transistors in the same package is utilized.

7.5 FET BRIDGE CIRCUITS

The chief disadvantage of the TVM arrangement depicted in Fig. 7.4 is its comparatively low input resistance. However, a high input resistance, comparable to that of a VTVM, can be obtained by designing a TVM with field-effect transistors (FET) as bridge-circuit components. For example, Fig. 7.5

FIG. 7.5 *(a) Basic FET TVM configuration; (b) analogous FET and vacuum-tube elements; (c) FET drain characteristics.*

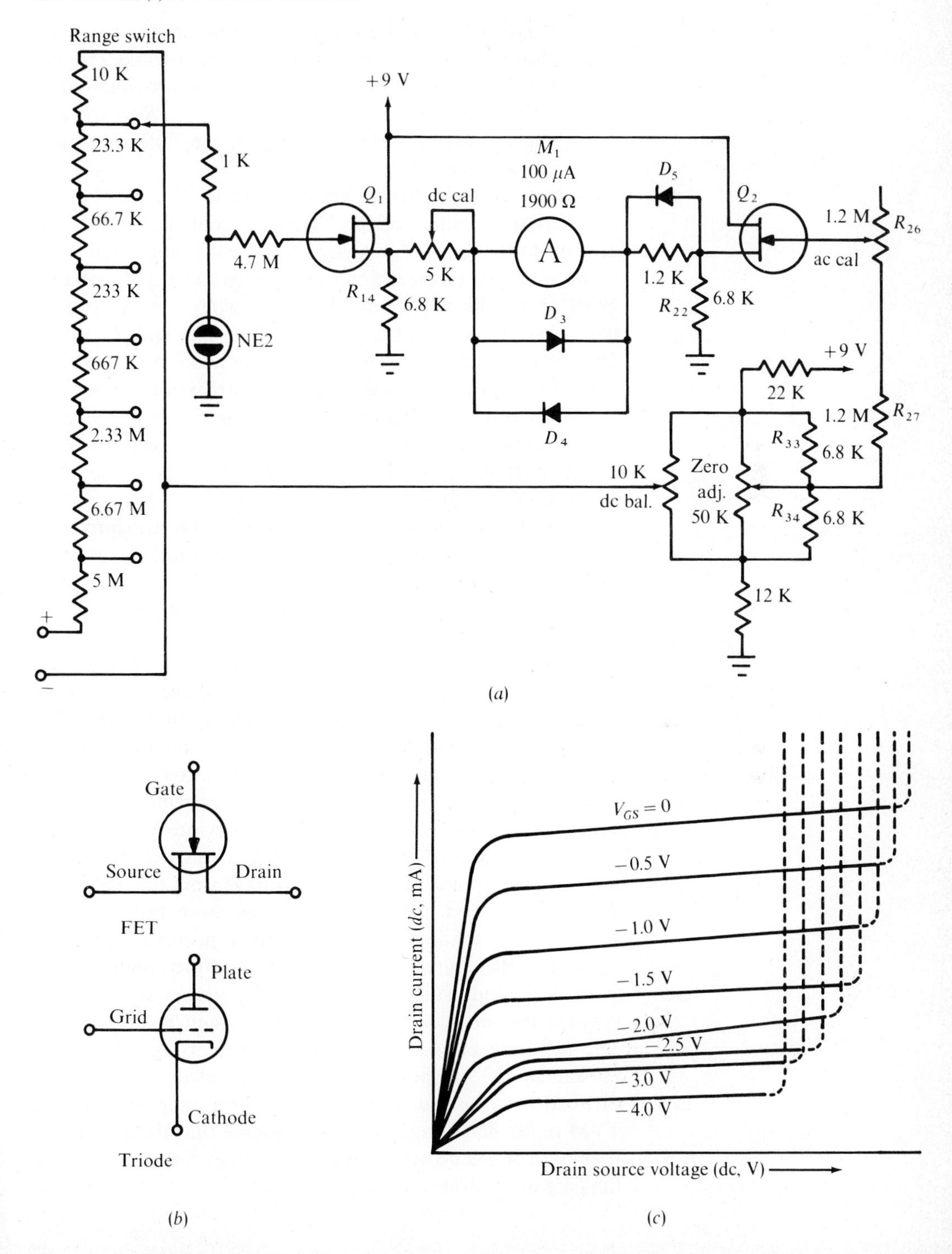

shows a basic FET TVM arrangement for measuring dc voltage values. Since the gate of an FET is comparable to the grid of a vacuum tube, an input resistance of 15 M is provided on all ranges. The bridge circuit is made up of Q_1, Q_2, R_{14}, and R_{22}. Its operating features are described in the following paragraphs.

Note that the bridge arrangement in Fig. 7.5 is also a differential amplifier circuit. That is, control voltages are applied to the gates of both Q_1 and Q_2; the meter indication is proportional to the difference between these gate voltages. With zero volts applied to the input of Q_1, the zero-adjust control is set to make the voltages across R_{14} and R_{22} equal. Differential amplifier action results in zero meter current under this condition. The dc balance control is not an operating control, but a maintenance control. It functions in the same basic manner as the zero-adjust control, except that its design purpose is to compensate for production tolerances and component aging.

Next, when a dc voltage value is to be measured, the voltage is applied to the input of Q_1. In turn, the balance between Q_1 and Q_2 is upset. The unbalanced current passes through the meter, and pointer deflection is proportional to the value of the voltage drop. A neon bulb, NE2, is provided for protection of Q_1 against accidental application of excessive input voltages. If an input voltage exceeding the range setting is applied, the neon bulb fires and thereby limits the input voltage to Q_1. Diode D_5 is a temperature-compensating component for maintenance of calibration accuracy over a wide temperature range. D_3 and D_4 provide protection against meter overload, as explained previously.

Note that in Fig. 7.5 Q_2 conducts less if Q_1 conducts more, due to application of a gate voltage from the multiplier. That is, the drop across R_{14} becomes more positive if Q_1 increases its conduction. In turn, this positive voltage is applied to the source electrode of Q_2, and the conduction of Q_2 is decreased. Thus, the voltage drop across R_{22} tends to become less positive, and the bridge circuit is unbalanced. The resulting unbalanced current flows through the meter movement. A 5-K potentiometer is provided in series with the lead between the source electrodes for calibration of the TVM on its dc voltage function. Observe that the ac calibration control has no effect on the dc voltage function, because the gate of Q_2 draws negligible current.

Next, let us consider the ac voltage function of the TVM in this example. The essential circuitry is shown in Fig. 7.6. A compensated multiplier is employed to provide accurate indication at high frequencies. The ac voltage to be measured is applied to Q_1, which operates as an electronic impedance transformer to drive the peak-to-peak type detector comprising D_1, D_2, C_{10}, and C_{11}. In turn, the dc output voltage from the detector is applied to the gate of Q_2 through a voltage divider consisting of R_{25}, R_{26}, and R_{27}. Diode D_6 provides temperature compensation for this circuit.

To summarize briefly, Q_2 is driven on the ac voltage function of the TVM in Fig. 7.6, whereas Q_1 is driven on the dc voltage function (Fig. 7.5). The reason for this change in driving point is that the detector has a substantial current

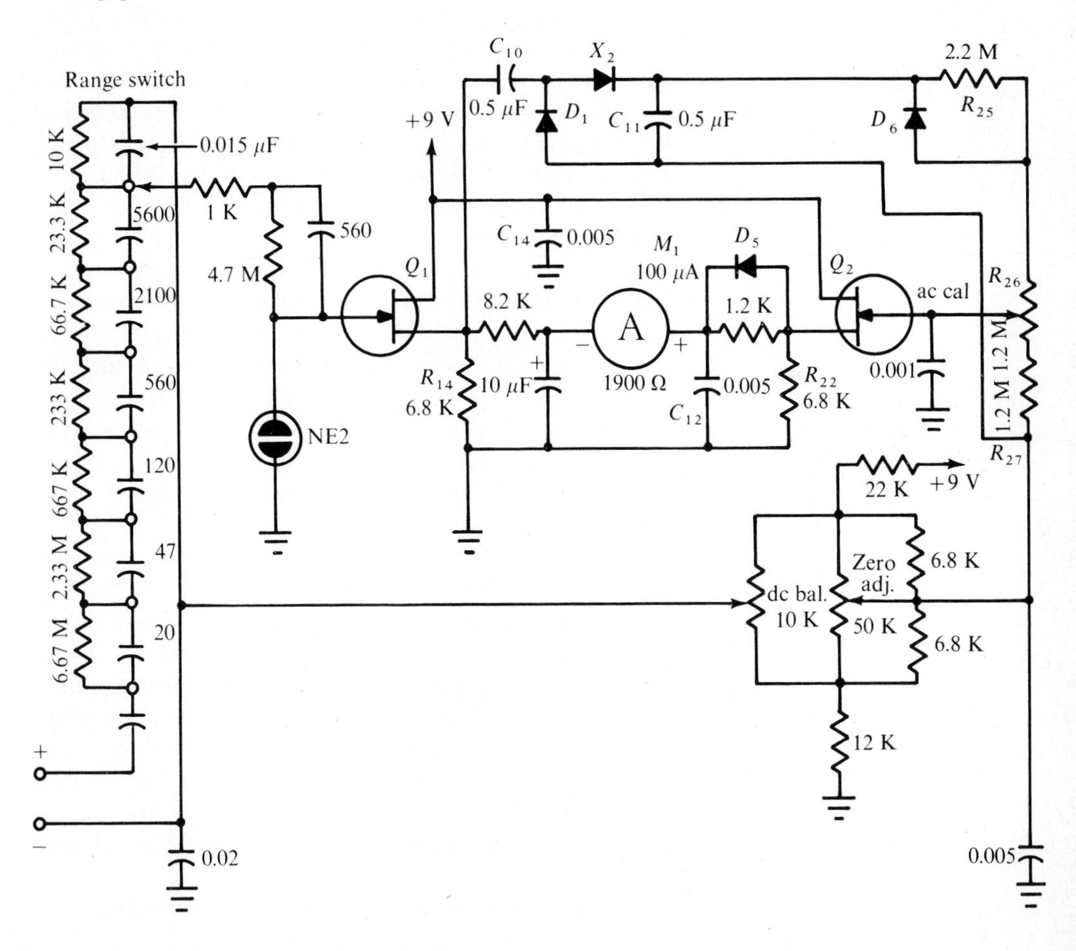

FIG. 7.6 *Ac voltage-measuring configuration for a TVM.*

demand. In other words, a high input impedance could not be maintained on the ac voltage function unless one FET is used as an impedance transformer. From the standpoint of practical circuit design, it is most economical to employ Q_1 as an impedance transformer. Note that since Q_2 is driven in Fig. 7.6, the ac calibration control affects the meter indication, and we adjust R_{26} for maximum accuracy of ac voltage indication.

The input impedance of the TVM in Fig. 7.6 comprises 10 M shunted by 29 pF. Frequency response is practically flat from 25 Hz to 1 MHz. Rated accuracy on the ac voltage function is ± 5 percent of full scale, compared with ± 3 percent of full scale on the dc voltage function. As explained previously for other rectifier-type instruments, the ac scale is calibrated in rms values for sine waves, whereas the rectifier circuit develops the peak-to-peak value of the applied ac waveform. Therefore, the rms scale indicates correct values only when the input waveform is sinusoidal. However, peak-to-peak scales are also provided for the TVM, and the peak-to-peak voltage indication is accurate, regardless of waveform.

Let us consider the operation of the FET bridge in Fig. 7.6. Although the input ac voltage is applied to the gate of Q_1, the resulting source output produces no indication on the meter scale, because the average value of any ac waveform is zero. However, the positive rectified output voltage from the detector causes Q_2 to increase its conduction. In turn, the source electrode of Q_2 becomes more positive, and this positive voltage is applied to the source of Q_1, which causes Q_1 to decrease its conduction. Thus the bridge becomes unbalanced, and the unbalanced current flows through the meter movement.

The external appearance of the TVM in this example is seen in Fig. 7.7, and the complete configuration is shown in Fig. 7.8. The ohmmeter function is similar to that of a VTVM. That is, a battery voltage is applied across a voltage divider comprising the unknown resistance and a known resistance. For example, R_9 is the known resistance on the $R \times 1$ range, R_{10} is the known resistance on the $R \times 10$ range, and so on. Thus, ohmmeter operation is conventional. It follows from the value of R_9 that the center-scale indication is 10 Ω. Indication accuracy on the resistance scale is rated to ± 3 degrees of arc.

Note that a transistor volt-ohmmeter is designated by

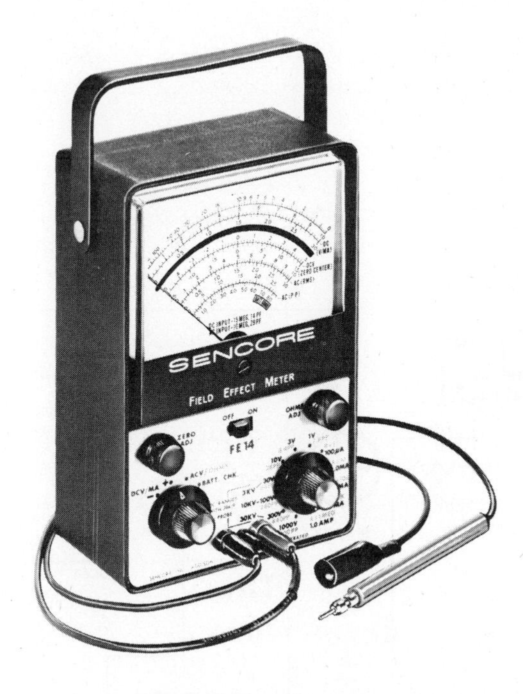

FIG. 7.7 *Sencore Model FE14 FET meter. (Courtesy, Sencore, Inc.)*

the abbreviation TVOM. This abbreviation is also applied to a transistor volt-ohm-milliammeter.

Dc current measurements are also provided by the TVOM in this example. Note that the FET bridge is not used during current measurements. That is, the milliammeter function is the same as explained previously for a VOM. Indication accuracy on the dc current function is rated to ± 3 percent of full scale. Since the FET bridge is not used during current measurements, the power may be switched off if desired. Rated accuracy on other functions cannot be realized unless the 9-V battery has a terminal voltage above a certain minimum value. Accordingly, a "battery-check" position is provided on the function switch. If the pointer does not deflect into the battery-check sector on the scale when this test is made, the battery should be replaced.

7.6 FET INPUT WITH STABILIZER STAGE

The TVM configuration depicted in Fig. 7.5 utilizes an FET to provide high input resistance; stabilization is obtained by means of another FET connected into a bridge circuit, and supplemented by a semiconductor diode in the meter branch.

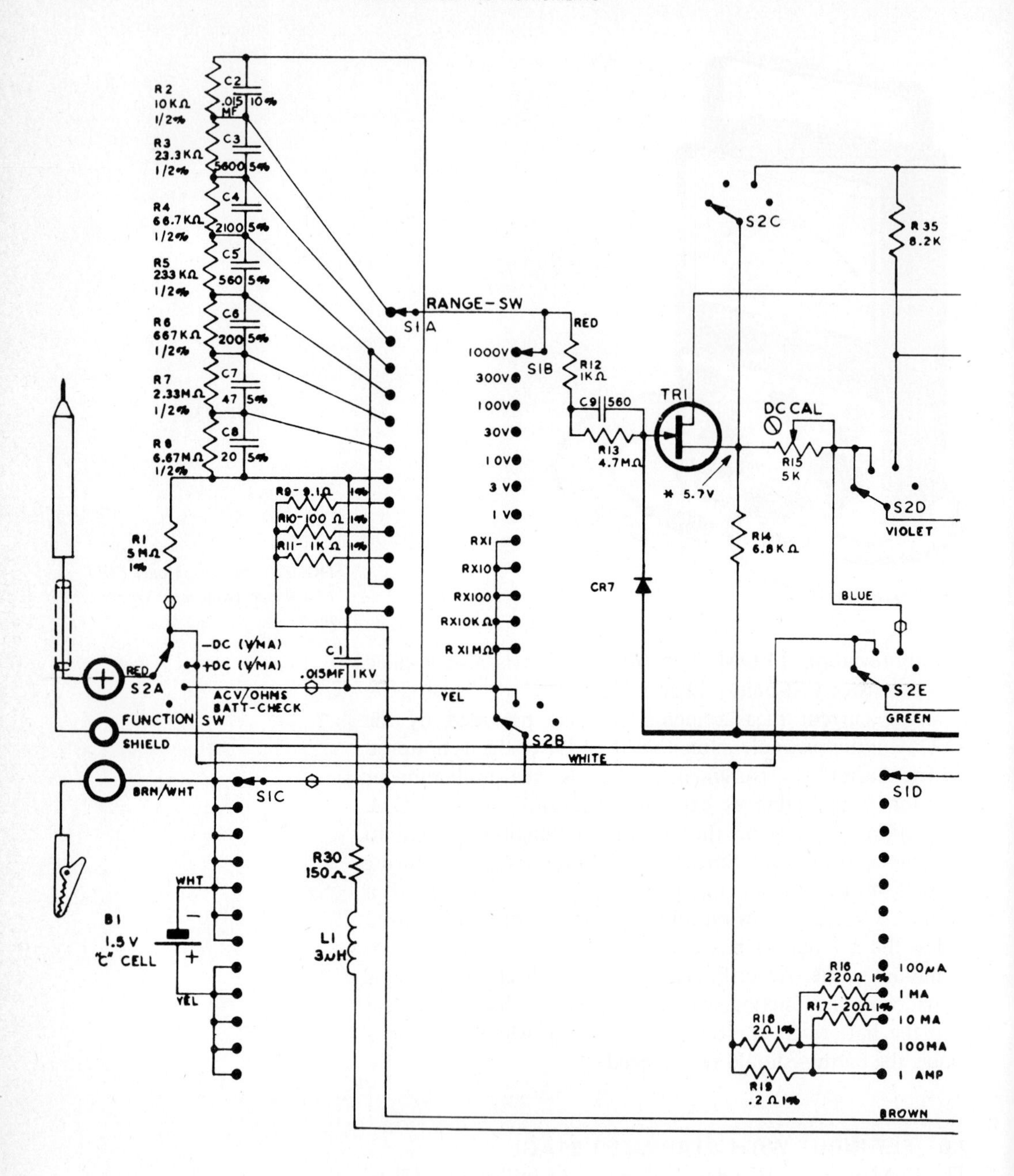

FIG. 7.8 *Configuration of TVM illustrated in Fig. 7.7. (Courtesy, Sencore, Inc.)*

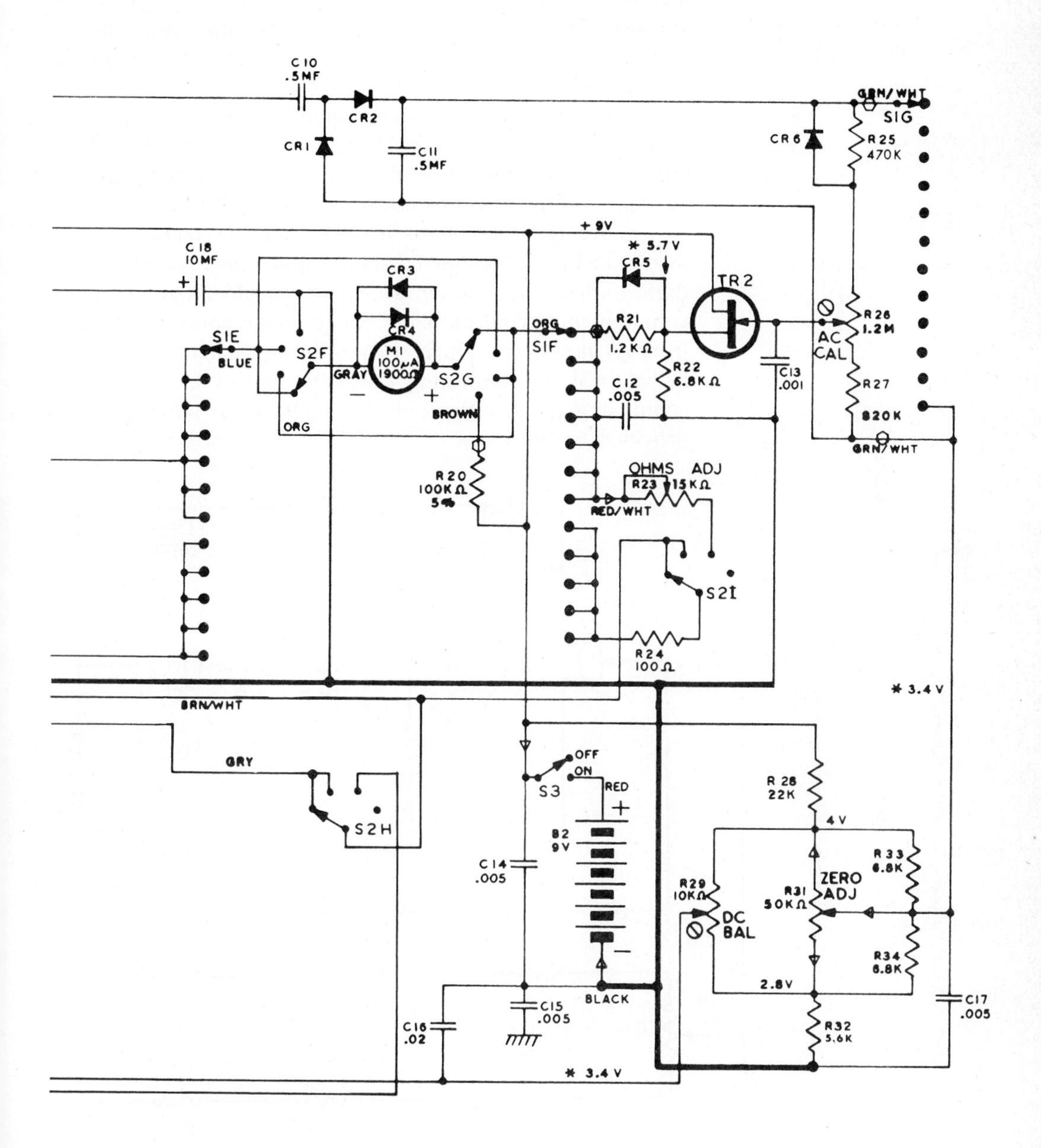
C10
.5MF
CR2
CR1
C11
.5MF
GRN/WHT
SIG
CR6
R25
470K
+9V
* 5.7V
C18
10MF
CR3
CR5
TR2
CR4
R21
1.2KΩ
R26
1.2M
S1E
BLUE
S2F
GRAY
M1
100μA
1900Ω
S2G
ORG
S1F
R22
6.8KΩ
AC
CAL
C13
.001
R27
820K
C12
.005
BROWN
ORG
GRN/WHT
R20
100KΩ
5%
OHMS ADJ
R23 15KΩ
RED/WHT
S2I
R24
100Ω
* 3.4V
BRN/WHT
GRY
OFF
ON
S3
RED
R28
22K
4V
S2H
B2
9V
C14
.005
R33
6.8K
R29
10KΩ
R31
50KΩ
ZERO
ADJ
DC
BAL
R34
6.8K
2.8V
C15
.005
BLACK
C16
.02
C17
.005
R32
5.6K
* 3.4V

Let us compare this arrangement with the FET and junction-transistor circuitry shown in Fig. 7.9. High input resistance is provided by Q_1, since its gate has practically no current demand. Instead of a neon bulb, as in Fig. 7.5, semiconductor diode D_1 operates in the Zener mode in Fig. 7.9 to protect Q_1 against overload damage. The input resistance of the TVM (10.6 M) is determined by the multiplier resistance plus the resistance of the dc probe not shown in Fig. 7.9).

Since Q_1 operates as an electronic impedance transformer in Fig. 7.9, ample current output is available from its drain to drive the junction transistor Q_2, which operates as a common-emitter dc amplifier. In turn, the collector current from Q_2 flows through the meter. Zero-left meter indication is obtained by connecting the meter into a bridge circuit comprising E_1, E_2, Q_2, and R_2. Precise zero adjustment is provided by means of the zero-adjust control which bleeds

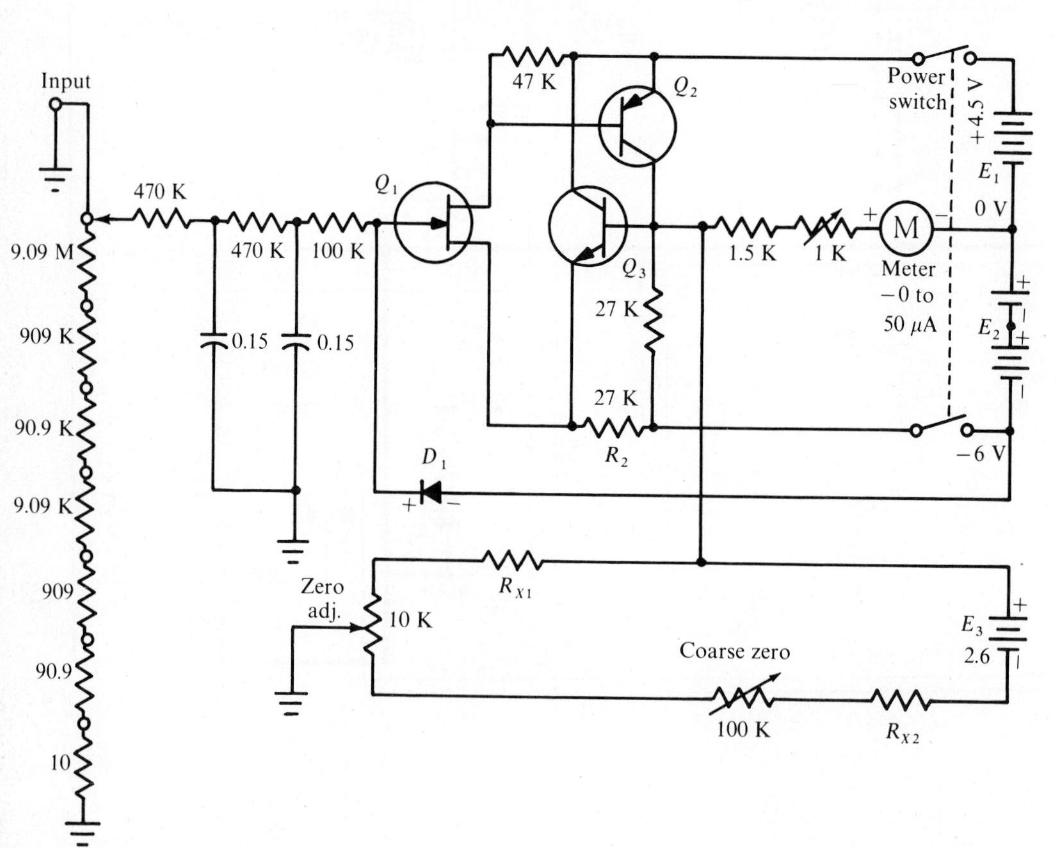

FIG. 7.9 *Simplified schematic of dc voltmeter.*

a small amount of current into the meter branch from E_3. Transistor Q_3 operates as a stabilizing component by supplying negative feedback to minimize the effect of temperature variations, battery aging, and component tolerances. Let us consider the operation of this negative-feedback branch.

Suppose that the ambient temperature increases, with a resulting tendency for the dc amplifier to increase its gain in Fig. 7.9. In turn, the operating point is shifted, and more current flows into the base of Q_3. Accordingly, more current is drawn by the collector of Q_3, and less current flows through Q_2 from emitter to collector. The end result is to minimize the change in system gain and to minimize the shift in operating point. Although perfect compensation cannot be obtained by negative-feedback action, the residual error is very small and can be neglected in practice.

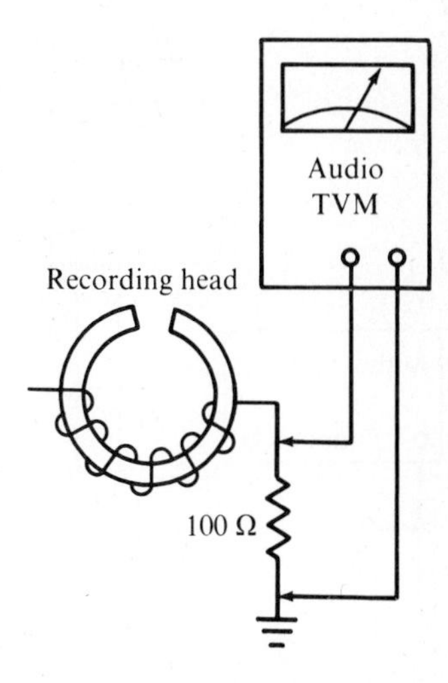

FIG. 7.10 *Measurement of bias current.*

7.7 APPLICATIONS

Although a TVM does not necessarily have ac current ranges, it can be used to measure the bias current in the recording head of a tape recorder, for example. With reference to Fig. 7.10, the return lead from the recording head is opened, and a precision (± 1 percent) 100-Ω resistor is inserted. The TVM is connected across the resistor. We operate the TVM on its ac-voltage function, switch the recorder to its recording function, and note the meter reading. The value of the bias current can then be calculated from Ohm's law. Each 0.1 V dropped across the 100-Ω resistor corresponds to 1 mA of current. Normal bias-current values range from 1 mA to several, depending upon the head design.

A good-quality audio oscillator in normal operating condition has a practically constant output voltage over its entire frequency range. Figure 7.11 shows how an audio oscillator can be checked for output uniformity. The TVM is connected across the output terminals of the oscillator. We operate the TVM on its ac-voltage function and tune the audio oscillator through its complete frequency range. A lab-type audio oscillator in good operating condition has a typical output variation of less than ± 1 dB over its entire frequency range. A constant output voltage facilitates test work, because the attenuator of the oscillator does not have to be reset when the frequency is changed in order to maintain a constant signal level.

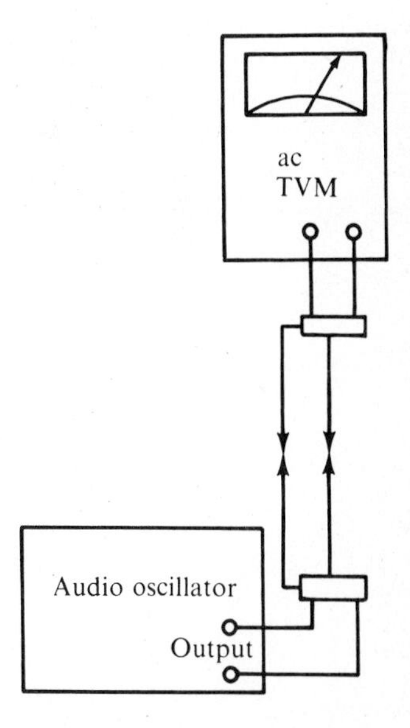

FIG. 7.11 *Measurement of output voltage from audio oscillator.*

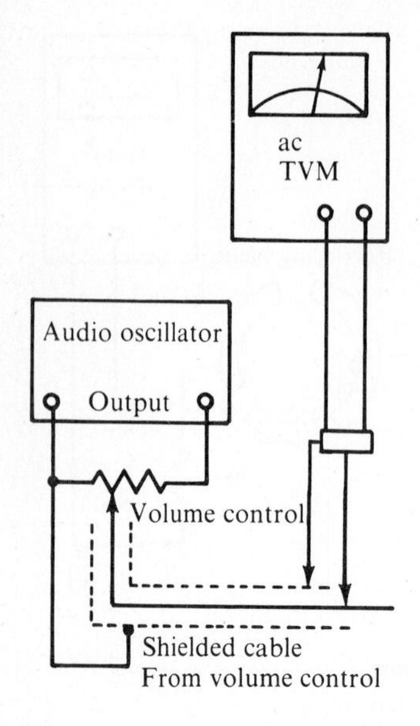

FIG. 7.12 *Measurement of volume-control frequency response.*

As an example of typical test work, Fig. 7.12 shows how the frequency response of a volume control is measured. An audio oscillator is used as the signal source, and a TVM is connected at the output terminals of the volume control (often a shielded cable). The TVM is operated on its ac function, and any change in output level is noted as the audio oscillator is tuned from 20 Hz to 20 kHz. We usually observe a fall-off in high-frequency response, because of the shunt capacitances in the volume control arrangement. This variation in frequency response changes with the setting of the volume control; best frequency response is usually found when the control is set for maximum output, and vice versa.

QUESTIONS AND PROBLEMS

1. State three advantages of transistor voltmeters over vacuum-tube voltmeters.
2. What are the two main disadvantages of the basic TVM circuit shown in Fig. 7.2?
3. How does the addition of the diode D_1 in Fig. 7.3 increase the stability of the basic TVM?
4. How does the addition of transistor Q_2 in Fig. 7.3 increase the stability of the basic TVM?
5. What circuit variations are we concerned with that cannot be compensated for with either of the circuits shown in Fig. 7.3?
6. What are the advantages of the bridge circuit in Fig. 7.4 over the basic TVM circuits in Fig. 7.3?
7. Explain the operation of the bridge circuit in Fig. 7.4 when a positive voltage is applied to the input terminals.
8. What is the purpose of resistor R_2 in the TVM bridge circuit in Fig. 7.4?
9. How does the TVM in Fig. 7.5 function during an ac voltage measurement?
10. Does the ac calibration control in Fig. 7.5 have any effect on the dc voltage function?
11. How is a high-input impedance maintained in the TVM circuit in Fig. 7.5?
12. What two factors determine the high-frequency cutoff of the TVM in Fig. 7.5?

13. Why is the accuracy on the ac function of the TVM in Fig. 7.5 less than on the dc function?

14. How does the TVM in Fig. 7.8 measure current?

15. Explain the operation of negative feedback to compensate for drift due to a temperature change in the TVM in Fig. 7.9.

BRIDGE-TYPE INSTRUMENTS

RESISTANCE BRIDGES

8.1 BRIDGE BALANCE REQUIREMENTS

Previous discussion has introduced us to the concept of a bridge circuit. For example, we observed a *resistance bridge* configuration in Chap. 6. We also learned the condition of *bridge balance*, wherein no unbalanced current flows across the bridge. We are now in a good position to consider bridge-type instruments in greater detail, and to analyze the operation of high-precision bridges. It is important that we discuss two general classes of measuring instruments:

1. Indicating instruments, wherein measurement is effected by means of pointer deflection along a calibrated scale
2. Comparison instruments, wherein measurement is effected by means of a balance condition between the unknown parameter and a reference or standard parameter which is physically present in the comparison instrument

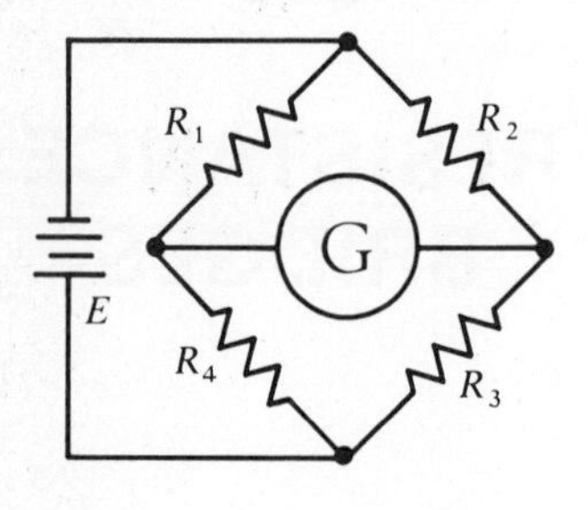

FIG. 8.1 *Basic resistance bridge.*

All high-precision bridge instruments are comparison instruments. Their basic operational feature is that they operate on the null principle. In other words, the instrument bridge is adjusted to *minimize* the unbalanced current insofar as possible. This adjustment can be made with high precision, because the zero indication of a galvanometer is more stable and more accurate than any off-zero indication. For example, if the permanent magnet in a galvanometer should weaken, the accuracy of zero indication is unaffected. On the other hand, the accuracy of any off-zero indication would be affected.

8.2 RESISTANCE BRIDGES

The basic resistance bridge, depicted in Fig. 8.1, was invented by Prof. James Christie. However, it was generally ignored until Sir Charles Wheatstone pointed out the advantages of the resistance bridge. Thereafter, the arrangement was called the Wheatstone bridge, and came into wide use. It follows from considerations of symmetry that if $R_1 = R_2 = R_3 = R_4$, then the galvanometer will be connected between equipotential points and no pointer deflection will occur. We recognize also that the bridge will be balanced if $R_1 = R_2$ and $R_3 = R_4$. It can be shown that balance is maintained if $R_1/R_4 = R_2/R_3$. From the standpoint of strict terminology, note that the resistors are called the *legs* or the *arms*, and the galvanometer branch is called the *bridge*.

Next, let us consider the relations in an unbalanced bridge circuit. Figure 8.2 depicts an unbalanced configuration in which each leg has a different value. The two parallel legs are $R_1 - R_4$ and $R_3 - R_5$. Thus R_2 is the bridge. We shall now determine the current through and the voltage drop across each resistor. Observe that the current of 0.1 A flows into junction *a* and divides into two parts. The part that flows through R_1 is denoted I_1, and the part that flows through R_3 is equal to $0.1 - I_1$. Similarly, at junction *b*, I_1 divides, with one part flowing through R_2 and the remaining part flowing through R_4. The current through R_2 is designated I_2, and the current through I_4 is equal to $I_1 - I_2$. We assume a direction of current flow through R_2 arbitrarily.

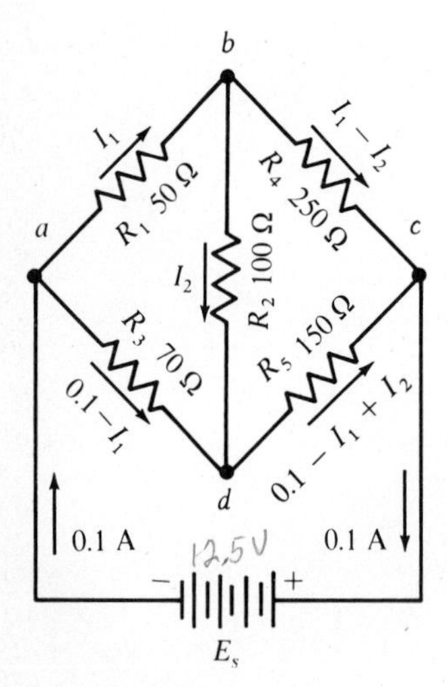

FIG. 8.2 *Unbalanced resistance bridge.*

We shall find that if our solution is obtained with a positive value for I_2, the assumed direction of flow was correct. On the other hand, a negative solution denotes that an incorrect direction of flow was assumed; the numerical value of the solution will still be correct. We observe that the current through R_4 is equal to $I_1 - I_2$. At junction d, the currents may be analyzed in a similar manner. Current I_2 through R_2 joins current $0.1 - I_1$ from R_3; furthermore, the current through R_5 is equal to $0.1 - I_1 + I_2$.

Next, the unknown currents I_1 and I_2 in Fig. 8.2 may be determined by writing two voltage equations in which both currents appear. We shall solve these equations for I_1 and I_2 in terms of the given values of current and resistance. The first voltage equation is written by tracing clockwise around the closed circuit containing resistors R_1, R_2, and R_3. We start at junction a, and then proceed to b, to d, and back to a. The algebraic sum of the voltages around this circuit is equal to zero, in accordance with Kirchhoff's voltage law. We express the voltages in terms of resistance and current. Going from a to b, the voltage drop is in the direction of the arrow, and is equal to $-50I_1$; the drop across R_2, going from b to d, is $-100I_2$; and the voltage from d to a (in the opposite direction to the arrow) is $70(0.1 - I_1)$. In turn, we write

$$-50I_1 - 100I_2 + 70(0.1 - I_1) = 0 \tag{8.1}$$

Multiplying both sides by -1, we obtain

$$50I_1 + 100I_2 - 70(0.1 - I_1) = 0 \tag{8.2}$$

Transposing and simplifying, we write

$$120I_1 + 100I_2 = 7 \tag{8.3}$$

We proceed to establish our second voltage equation by tracing clockwise around the circuit, which includes resistors R_4, R_5, and R_2. Starting at junction b, we proceed to c, to d, and then return to b. The voltage across R_4, from b to c, is $-250(I_1 - I_2)$; the voltage across R_5, from c to d, is $+150(0.1 - I_1 + I_2)$; and the voltage across R_2, from d to b, is $+100I_2$. In turn, we write

$$-250(I_1 - I_2) + 150(0.1 - I_1 + I_2) = 100I_2 = 0 \tag{8.4}$$

which yields

$$400I_1 - 500I_2 = 15 \tag{8.5}$$

Equations (8.3) and (8.5) can now be solved simultaneously; we multiply Eq. (8.3) by 5, and then add the equations to eliminate I_2, as follows:

$$400I_1 - 500I_2 = 15$$

$$600I_1 + 500I_2 = 35$$

$$1000I_1 = 50 \quad \text{or} \quad I_1 = 0.05 \text{ A} \tag{8.6}$$

Substituting the value of 0.05 for I_1 in Eq. (8.3), we solve for I_2:

$$120(0.05) + 100I_2 = 7 \qquad 100I_2 = 1$$

$$I_2 = 0.01 \text{ A} \tag{8.7}$$

Thus, we have found that the current in R_1 is $I_1 = 0.05$ A; the current in R_2 is $I_2 = 0.01$ A; the current in R_3 is $0.1 - I_1 = 0.05$ A; the current in R_4 is $I_1 - I_2 = 0.04$ A; and the current in R_5 is $0.1 - I_1 + I_2 = 0.06$ A. Next, we find the voltages E_1, E_2, E_3, E_4, and E_5, as follows:

$$E_1 \text{ across } R_1 \text{ is } I_1R_1 = 0.05 \times 50 = 2.5 \text{ V} \tag{8.8}$$

$$E_2 \text{ across } R_2 \text{ is } I_2R_2 = 0.01 \times 100 = 1.0 \text{ V} \tag{8.9}$$

$$E_3 \text{ across } R_3 \text{ is } (0.1 - I_1)R_3 = 0.05 \times 70 = 3.5 \text{ V} \tag{8.10}$$

$$E_4 \text{ across } R_4 \text{ is } (I_1 - I_2)R_4 = 0.04 \times 250 = 10 \text{ V} \tag{8.11}$$

$$E_5 \text{ across } R_5 \text{ is } (0.1 - I_1 + I_2)R_5 = 0.06 \times 150 = 9.0 \text{ V} \tag{8.12}$$

The source voltage E_s is equal to the sum of the voltages across R_3 and R_5, or across R_1 and R_4. In other words, we calculate E_s as follows:

$$E_s = E_1 + E_4 = 2.5 + 10 = 12.5 \text{ V} \tag{8.13}$$

or

$$E_s = E_3 + E_5 = 3.5 + 9 = 12.5 \; V \tag{8.14}$$

Note that the voltage across R_2 is the difference in the voltages across R_3 and R_1. This voltage is also the difference in the voltages across R_4 and R_5.

8.3 RESISTANCE MEASUREMENT WITH THE WHEATSTONE BRIDGE

Figure 8.3*a* depicts the circuit diagram for a basic Wheatstone bridge used to measure resistance values. R_1, R_2, and R_3 are precision adjustable resistors; these resistors are often adjusted in convenient steps. R_X is the resistor that is to be

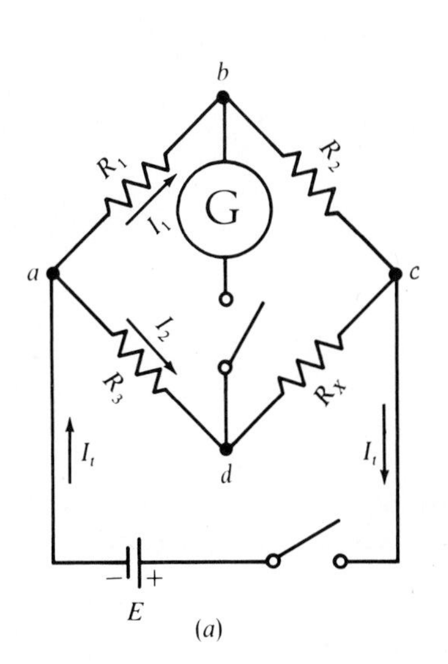

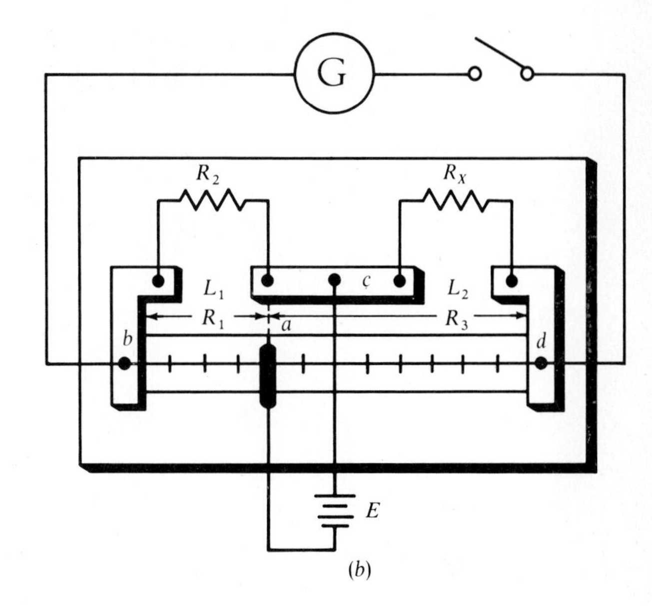

FIG. 8.3 *Wheatstone bridge circuit. (a) Schematic circuit; (b) slide-wire bridge.*

measured for its resistance value. After the bridge has been balanced, the unknown resistance value may be calculated by means of a simple formula. As explained previously, a galvanometer G is connected between terminals b and d to indicate the condition of balance. This occurs when the pointer deflection is zero. The galvanometer circuit contains a switch, which is opened when R_X is not connected; this precaution prevents damage to the galvanometer.

Operation of the bridge may be summarized as follows. A switch in the battery circuit is provided to prevent depletion of the battery when the bridge is not in use. If the switch is closed, electrons flow to point a and divide through R_1 and R_2 and through R_3 and R_X. Currents I_1 and I_2 unite at point c and return to the positive terminal of the battery. We observe that the value of I_1 depends on the sum of R_3 and R_X. In accordance with Ohm's law, the current value is inversely proportional to the resistance through which it flows.

We adjust the values of R_1, R_2, and R_3 so that no deflection of the pointer occurs on the galvanometer scale when the bridge switch is closed. There is, under this condition, no potential difference between points b and d. It follows that the voltage drop E_1 across R_1, between points a and b, is the same as the voltage drop E_3 across R_3, between points a and *d*. Similarly, the voltage drops across R_2 and R_X—that is, E_2 and E_X—are also equal. Thus, we write

$$E_1 = E_3 \tag{8.15}$$

or,

$$I_1 R_1 = I_2 R_3 \tag{8.16}$$

and

$$E_2 = E_X \tag{8.17}$$

or

$$I_1 R_2 = I_2 R_X \tag{8.18}$$

Now, if we divide the voltage drops across R_1 and R_3 by the respective voltage drops across R_2 and R_X, we obtain

$$\frac{I_1 R_1}{I_1 R_2} = \frac{I_2 R_3}{I_2 R_X} \tag{8.19}$$

or,

$$\frac{R_1}{R_2} = \frac{R_3}{R_X} \tag{8.20}$$

whence,

$$R_X = \frac{R_2 R_3}{R_1} \tag{8.21}$$

The resistance values of R_1, R_2, and R_3 may be determined, for example, from the switch positions on a resistance box, such as illustrated in Fig. 8.4. This is an example of a utility-type resistance box. Figure 8.5 illustrates a high-precision decade resistance box. A decade arrangement is so-called because its steps are decimally related. Another design, often used in school laboratories, employs a slide-wire arrangement as depicted in Fig. 8.3*b*. The slide wire (*b* to *d*) corresponds to R_1 and R_3 in Fig. 8.3*a*. The wire is commonly an alloy such as German silver or Nichrome, with a precise and uniform cross section; its total resistance is typically 100 Ω. Point *a* is established where the slider contacts the wire, and the bridge is balanced by moving the slider to a suitable point along the wire.

FIG. 8.4 *A utility-type resistance box.* (*Courtesy, Heath Co.*)

Our equation to solve for R_X in the slide-wire bridge is similar to the equation that was previously derived. However,

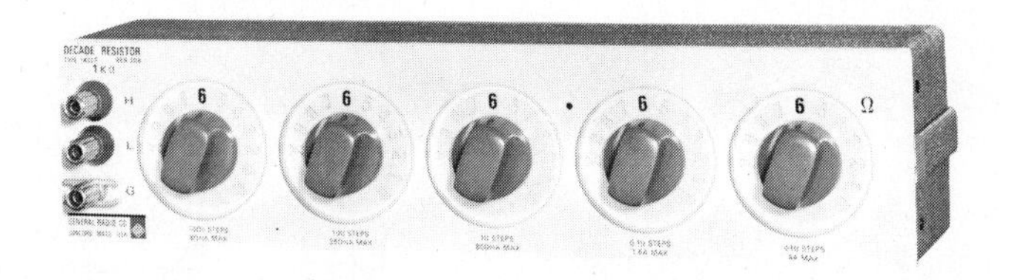

FIG. 8.5 *Precision-type decade resistance box. (Courtesy, General Radio Corp.)*

in a slide-wire bridge, the length L_1 corresponds to resistance R_1, and the length of L_2 corresponds to resistance R_3 in Fig. 8.3. Thus, L_1 and L_2 may be substituted for R_1 and R_3 in the bridge equation. The resistance of L_1 and L_2 varies uniformly with the slider movement, because of the uniform cross section utilized in fabrication of the slider wire. In turn, we write

$$R_X = \frac{L_2 R_2}{L_1} \tag{8.22}$$

A meter rule is usually mounted underneath a slide wire, so that the lengths of L_1 and L_2 can be easily observed. A refinement used in high-precision bridges that employ sensitive galvanometers is a *meter shunt*, which is normally connected across the galvanometer, but which can be open-circuited by pressing a button. The shunt is a protective device that prevents overload damage to the galvanometer while balance is being approached. Final balance is obtained with the shunt button depressed. Since the galvanometer sensitivity can be increased in any bridge by utilizing a higher battery voltage, this might seem to be an economical way of obtaining high precision with a low-sensitivity galvanometer. However, we must limit the battery voltage to the maximum power dissipation ratings of the resistors in the circuit.

8.4 MEASUREMENT OF SMALL RESISTANCE VALUES

Resistance values less than 0.1 Ω are difficult to measure precisely with a simple bridge, because of contact resistances in connections of the unknown to the bridge terminals. Therefore, a slightly elaborated arrangement called the *Kelvin double bridge* is used to measure low-resistance values. A typical configuration is shown in Fig. 8.6. We note in part *a* that two additional resistors, r_a and r_b, have been connected in series, and shunted across R; we regard R as being connected in series with the standard resistance and the unknown

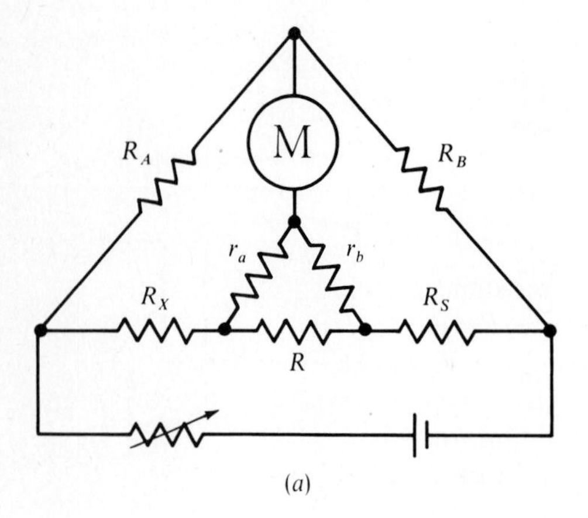

(*a*)

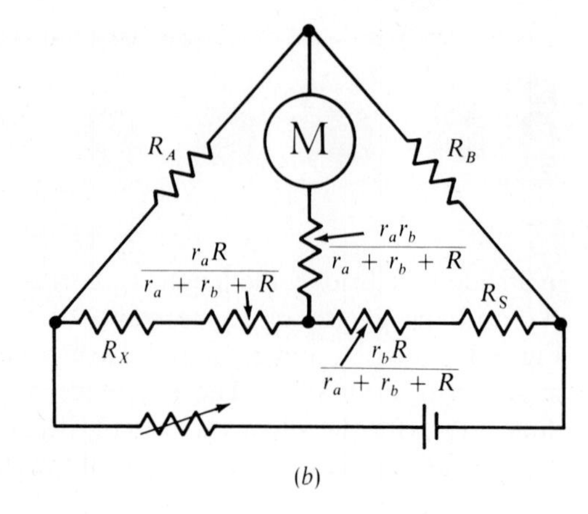

(*b*)

FIG. 8.6 *Kelvin bridge and equivalent circuit. (a) Kelvin bridge; (b) equivalent Wheatstone bridge.*

resistance. The galvanometer connects to the junction of r_a and r_b. Note that in performing the balance adjustments, we must set r_a/r_b equal to R_A/R_B. When this is done, we can then compute the value of the unknown resistance from the equation:

$$R_X = R_S(R_A/R_B) \tag{8.23}$$

where R_X = low-valued resistance to be measured
R_S = value of standard resistance
R_A = value of resistance in arm A
R_B = value of resistance in arm B

The advantage of the double bridge is found in the fact that the value of R is eliminated from consideration, as seen from Eq. (8.23). In Fig. 8.6*b*, an equivalent Wheatstone bridge circuit is shown for the Kelvin bridge. This equivalent circuit is found by means of the wye-delta relations that you studied in your basic-electricity course. The student may show that the elimination of R and the form of Eq. (8.23) result from calculation of the equivalent circuit. Although the effect of contact resistance is minimized in the Kelvin arrangement, it is nevertheless good practice to use firm low-resistance contacts throughout the circuit.

8.5 BRIDGE-TYPE INDICATING INSTRUMENTS

We have learned that bridge configurations are used primarily in comparison instruments, in which the value of an unknown is compared with the value of a standard. Bridge arrangements are also used to a considerable extent in various types of indicating instruments. One prominent example has been discussed under the topics of vacuum-tube voltmeters and transistor voltmeters. A bridge that is used in an indicating instrument necessarily operates as an unbalanced bridge, and a D'Arsonval meter indicates the magnitude of the unbalanced current. It can be shown that the magnitude of the unbalanced current is nearly proportional to the amount of resistive unbalance in a bridge arm, provided that the percentage of resistive unbalance is comparatively small with respect to the total resistance of the arm.

Another example of an unbalanced bridge instrument is shown in Fig. 8.7. It is a temperature-measuring instrument that employs a bridge circuit in which one arm of the bridge is utilized as an external thermistor probe. A thermistor is a type of resistor which exploits the large coefficient of temperature exhibited by certain metallic oxides. Most thermistors have a negative temperature coefficient; in other words, their hot resistance is less than their cold resistance. The temperature meter illustrated in Fig. 8.7 has its scale calibrated from -50° to $+70^\circ$F. Indication accuracy is rated within ± 1 percent at center scale, and within $\pm 2^\circ$ at either end of the scale.

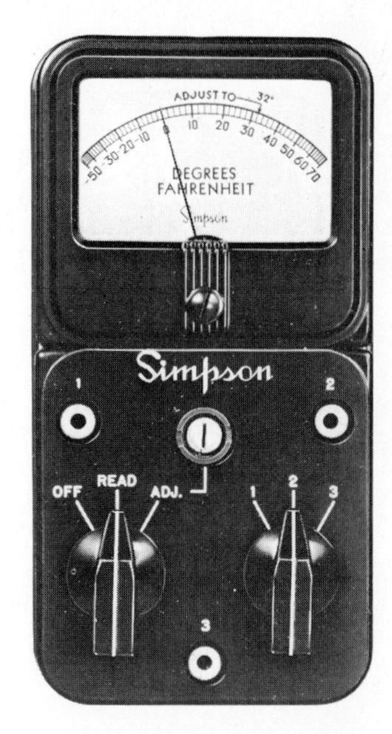

FIG. 8.7 *A temperature-measuring instrument, Model 385. (Courtesy, Simpson Electric Co.)*

Figure 8.8 shows the configuration of this temperature meter. It is basically a Wheatstone bridge circuit; however, the bridge is balanced only at a temperature that provides a thermistor probe resistance of 8900 Ω. At lower temperatures, the pointer will deflect below its resting position, and vice versa. Since the meter is energized by unbalanced current, the bridge cannot be driven by an arbitrary voltage value, as in the case of a true Wheatstone bridge. Instead, the battery voltage must be effectively adjusted to a critical value prior to a temperature measurement. Let us consider this aspect of bridge operation.

The switch in Fig. 8.8 has three positions. An Off position is provided to disconnect the battery when the instrument is not in use. In the Adjust position of the switch, the thermistor probe is not in the circuit; instead, the probe is replaced by a 3100-Ω resistor connected in series with a 400-Ω recalibrator potentiometer. A reference mark is observed on the meter scale, and the 20,000-Ω potentiometer is adjusted to bring the pointer to this mark, thereby compensating the effect of battery aging. Finally, the switch is thrown to its Read position to connect the thermistor probe into the bridge circuit. In turn, the meter indicates the temperature of the probe by means of the unbalanced current that flows.

Because a thermistor has a small tolerance, and because

Degrees, F	Resistance	Microamps movement with shunt	Microamps movement only, no shunt
−50°	51.00 K	35.9 μA	30.4 μA
−40°	35.50 K		26.8 μA
−30°	24.50 K		21.8 μA
−20°	17.40 K		15.6 μA
−10°	12.20 K		8.2 μA
0°	8.90 K	0 μA	0 μA
10°	6.40 K		9.3 μA
20°	4.65 K		19.2 μA
30°	3.50 K		28.4 μA
32°	3.28 K		30.5 μA
40°	2.60 K		37.5 μA
50°	1.99 K		45.6 μA
60°	1.54 K		52.4 μA
70°	1.20 K	68 μA	58.1 μA

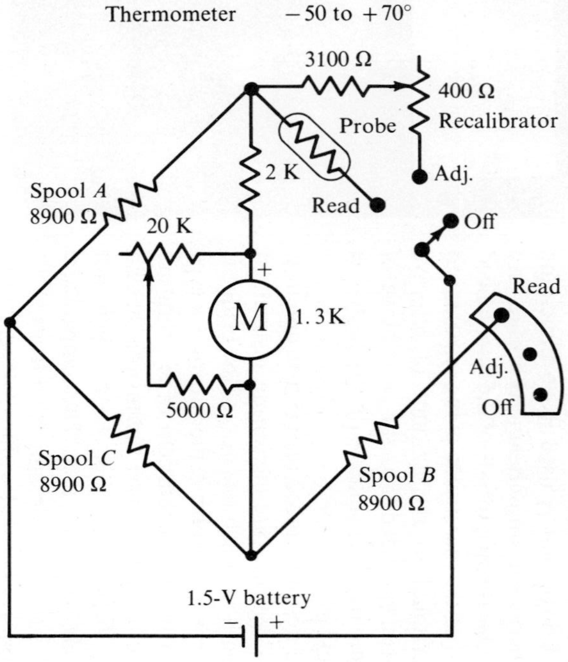

FIG. 8.8 *Temperature-meter configuration.* (*Courtesy, Simpson Electric Co.*)

its characteristics may change slightly with age, a recalibrator potentiometer is provided in the Adjust branch of the circuit in Fig. 8.8. This is not an operating control, but a maintenance control. To check the temperature calibration of the instrument, the probe is placed in a vessel with melting ice. This establishes a reference temperature of 32°F. The recalibrator potentiometer is then adjusted as required to obtain a scale reading of 32°F. Note that when the battery ages sufficiently such that the pointer cannot be brought to the reference mark on the scale by adjustment of the 20,000-Ω potentiometer, the battery must then be replaced.

8.6 STRAIN-GAGE BRIDGE CONFIGURATION

When mechanical stresses or strains are to be measured, it is often convenient to change the mechanical forces into proportional electrical voltages or currents, and to measure the forces in terms of unbalanced bridge currents. Strain gages such as depicted in Fig. 8.9 are commonly used as transducers. This type of strain gage employs a small-diameter resistance wire, typically fabricated from constantan, which has a very low temperature coefficient of resistivity. Thus, the gage resistance remains essentially constant at various ambient temperatures.

The resistance wire in a strain gage is often cemented to a thin insulating sheet called the carrier sheet. In turn, the engineer cements the strain gage assembly to the mechanical surface that is being analyzed. When extensive or compressive forces are applied to the strain gage, its resistance changes

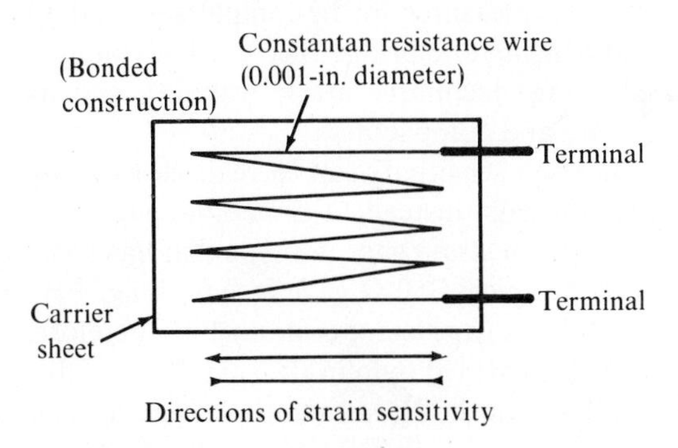

FIG. 8.9 *A strain-gage arrangement.*

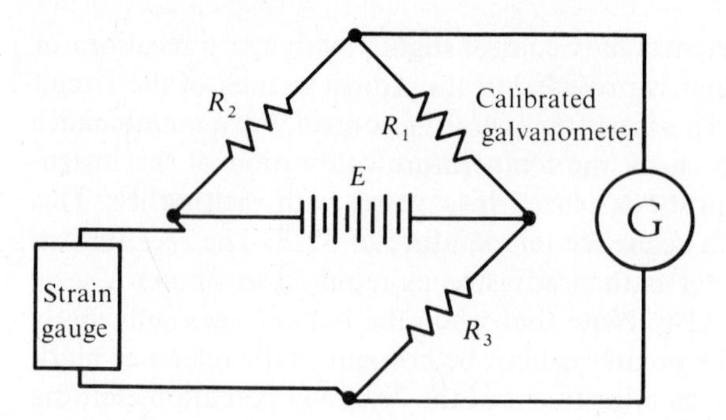

FIG. 8.10 *Basic strain-gage bridge configuration.*

more or less. Extension produces a resistance increase, and vice versa. The basic bridge arrangement is depicted in Fig. 8.10. As in the case of a temperature meter, the unbalanced voltage energizes a calibrated galvanometer.

When the mechanical surface is strained, the wire in the gage is also strained. Its percentage change in resistance is formulated:

$$\frac{\Delta R}{R} = K \frac{\Delta L}{L} \tag{8.24}$$

where R = change in resistance
R = original resistance value
K = a proportionality constant called the *gage factor*

If a strain gage is fabricated from constantan, the value of K is approximately 2. Thus, a strain of 1 percent will produce a resistance change of 2 percent in this example. The nominal resistance of a strain gage is typically 120 Ω. When high precision is desired, the strain gage is not used in an unbalanced bridge configuration. Instead, the gage is treated as the unknown and its value is measured by the comparison method with a conventional Wheatstone bridge. The chief advantages of the unbalanced bridge technique are its rapidity and its comparative simplicity and compactness.

Improved indication sensitivity can be realized by using a semiconductor strain gage instead of a metallic gage. For example, the flexible silicon strain gage is a type that has come into fairly wide use. A typical 350-Ω gage is 1 in. long, $\frac{1}{2}$ in. wide, and 0.005 in. thick. Its gage factor is about 130; therefore, far greater sensitivity is provided than in the case of a metallic gage. However, since silicon has a substantial negative temperature coefficient of resistance, the bridge circuit must include a temperature-compensating circuit.

8.7 APPLICATIONS

A resistance bridge provides a convenient means of measuring the forward resistance of a semiconductor diode, as depicted in Fig. 8.11. The bridge is set up in the usual manner, except that the bridge driving voltage is variable, thus permitting the voltage across the diode (or the current through the diode) to be varied. As would be anticipated, the value of R_x changes when the amount of diode current is changed. The value of diode current (or voltage) at balance is easily calculated from the bridge control settings, or a voltmeter can be connected across R_x and a milliammeter can be connected in series with R_x for direct reading of voltage or current values. The advantage of a resistance bridge over an ohmmeter in this application is that the test voltage of an ohmmeter is fixed and is not under the control of the operator.

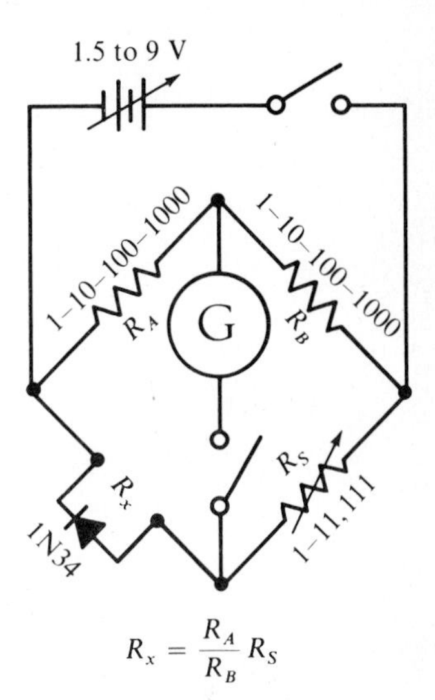

FIG. 8.11 *Bridge measurement of forward resistance of a germanium diode.*

Another application for a resistance bridge is in the measurement of resistivity of various metal-wire samples. For example, in quality-control processes, it is sometimes desirable to measure the resistivity of production samples at intervals. For this purpose, the Kelvin bridge is commonly utilized, as depicted in Fig. 8.12. The same bridge arrangement is used on resistance-thermometer detectors and on point-resistance thermometers. These devices are based on the temperature coefficient of a metal such as copper, often wound into a small coil. Resistance thermometers are useful to measure small changes in temperature. Kelvin bridges are also used to measure the temperature of field coils in large generators by calculation from their measured resistance value.

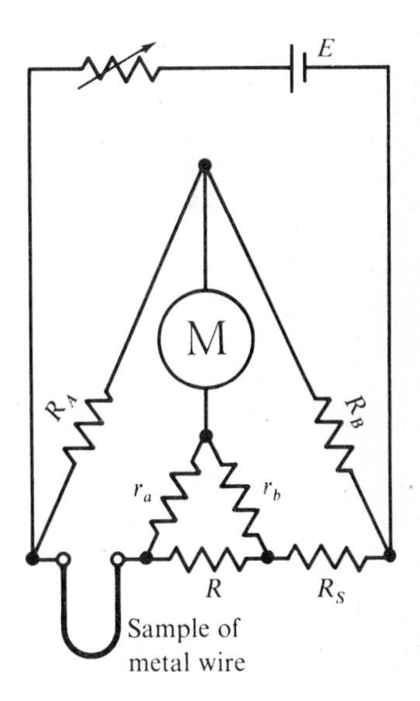

FIG. 8.12 *Measurement of resistivity of a metal-wire sample.*

QUESTIONS AND PROBLEMS

1. What are two general classes of measuring instruments that utilize bridge circuits?
2. Which of the above types is used in all high-precision bridge instruments?
3. What is the advantage of a comparison type of bridge circuit?
4. Who was the inventor of the Wheatstone resistance bridge?

5. What are the conditions for balance in the bridge circuit shown in Fig. 8.1?
6. Suppose the values of the resistors in the bridge circuit shown in Fig. 8.1 are: $R_1 = 1$ K, $R_2 = 5$ K, and $R_3 = 6$ K. What is the value of resistor R_4 that will balance the bridge?
7. Explain the operation of the bridge circuit shown in Fig. 8.3 as it is used to measure an unknown value of resistance.
8. Why are resistances with values less than 0.1 Ω difficult to measure with a simple bridge?
9. What is the purpose of the Kelvin double-bridge circuit?
10. Explain how a bridge circuit may be modified to measure temperature.
11. Explain the operation of the temperature meter circuit shown in Fig. 8.8.
12. What is the purpose of the recalibration potentiometer in the circuit shown in Fig. 8.8 and why is it necessary?
13. How is the instrument shown in Fig. 8.8 calibrated?
14. When should the battery be replaced in the instrument shown in Fig. 8.9?
15. Explain the operation of the strain gage shown in Fig. 8.9.
16. Explain how stress or strain on a strain gage is measured with a Wheatstone bridge.
17. Why must temperature compensation be included in an instrument used with a semiconductor strain gage?

CAPACITANCE BRIDGES

9.1 BRIDGE VOLTAGE REQUIREMENTS

Capacitance values are commonly measured with bridge-type instruments. A capacitance bridge must be driven by ac voltage, because the null indication is based on capacitive-reactance values. Either 60-Hz or 1-kHz oscillators are ordinarily provided. We shall find that the bridge-driving frequency is not particularly critical, although optimum accuracy is provided by an operating frequency which is neither extremely low nor extremely high. That is, the reactance of a small capacitor becomes very high at very low frequencies; at very high frequencies, bridge resistors "look like" impedances. Either of these extremes will impair indication accuracy. Laboratory-type capacitance bridges usually operate at 1 kHz; service-type bridges usually operate at 60 Hz. Low-cost bridges often utilize a 60-Hz voltage from the power line.

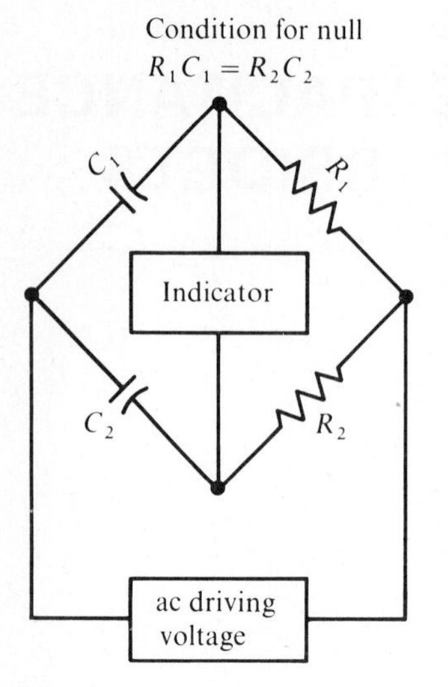

FIG. 9.1 *Circuit diagram of basic capacitance bridge.*

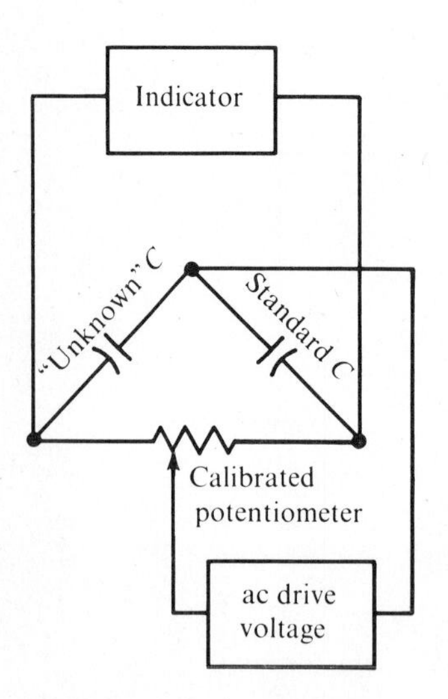

FIG. 9.2 *A simple service-type capacitor bridge.*

9.2 BASIC CAPACITANCE BRIDGE

Small mica, ceramic, and paper capacitors that are not defective can be regarded as pure capacitances in many practical situations. For example, the capacitance value of this type of capacitor can be measured with the simple capacitance bridge arrangement shown in Fig. 9.1. The bridge will be balanced when the reactance ratio is equal to the resistance ratio:

$$\frac{X_{C_1}}{X_{C_2}} = \frac{R_1}{R_2} = \frac{C_2}{C_1} \tag{9.1}$$

$$R_1 C_1 = R_2 C_2 \tag{9.2}$$

or

$$C_1 = \frac{R_2 C_2}{R_1} \tag{9.3}$$

Electron-ray tubes are often used as null indicators in economy-type capacitance bridges. The grid of the tube is operated at cathode potential (zero bias) so that grid current is drawn. Thus, the tube operates as its own rectifier, and shadow-angle response is obtained as if dc voltage were being applied from an auxiliary rectifier. The configuration of Fig. 9.1 is suitable for routine measurement of capacitance values, provided that the capacitor under test has negligible leakage resistance and practically zero power factor. In other words, this type of bridge has no provision for balancing any series or shunt resistance that might be associated with the capacitor.

The range of a basic capacitance bridge is typically from 10 pF to 0.5 μF. Rated accuracies of low-cost bridges are usually from ± 10 to ± 20 percent. A calibrated potentiometer is commonly employed in the resistance-arm section, as depicted in Fig. 9.2. Various ranges are obtained by switching different values of capacitance into the standard-capacitance arm of the bridge. For example, a standard capacitance of 200 pF may be switched into the bridge arm, and the potentiometer may be calibrated from 10 to 5000 pF on this range. Again, if a standard capacitance of 0.02 μF is switched into the bridge arm, a range of 0.001 to 0.5 μF will be provided.

When measuring small values of capacitance, the stray capacitance of a pair of test leads can cause serious experimental error. Therefore, small capacitors must be connected directly to the terminals of the bridge. Another source of experimental error is leakage resistance in the capacitor under test. In other words, if the "unknown" capacitor in Fig. 9.2 is leaky, it will be effectively shunted by resistance. Consequently, the currents in the two capacitive arms will be out of phase. The result of out-of-phase arm currents is to cause an incorrect setting (false null) of the calibrated potentiometer; another result is an incomplete null. An incomplete, or shallow null occurs when there is residual unbalanced current that cannot be eliminated by bridge adjustment.

Leakage is a common capacitor defect. Therefore, most capacitance bridges provide a leakage-resistance test. It is good practice to make a leakage-resistance test before checking the capacitance value, because there is no point in attempting to measure capacitance if the capacitor is leaky. Leakage resistance is usually measured or checked with a Wheatstone bridge arrangement and a dc source, as shown in Fig. 9.3. Since various capacitors may be rated for widely different working voltages, a control is usually provided for adjusting the dc bridge-driving voltage. For example, a typical service bridge provides a range of 0 to 500 V for leakage tests. It is preferable to apply rated working voltage to a capacitor under test, because leakage resistance might be overlooked at comparatively low applied voltage.

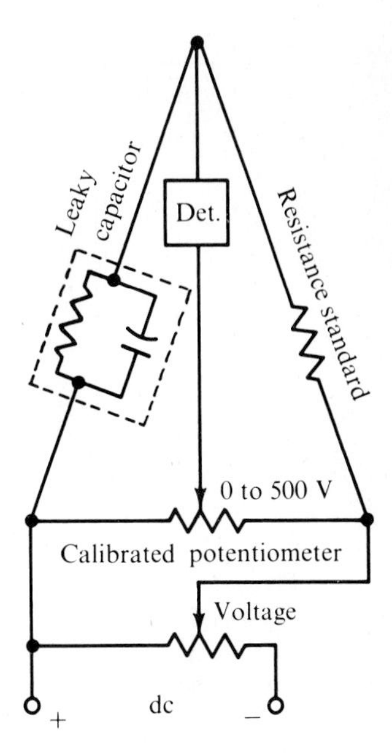

FIG. 9.3 *Circuit diagram of a leakage-resistance bridge.*

If a bridge provides scales for resistance measurement, leakage resistance can be measured in ohms, kilohms, or megohms. However, the simpler types of capacitance bridges usually have go/no go indication provided by an electron-ray tube. If any leakage is observed in terms of a change in the shadow angle, the capacitor is generally discarded without further test. It should be stressed that these considerations apply to paper, mica, and ceramic capacitors only. For example, electrolytic capacitors such as utilized in power-supply filters normally have some small amount of leakage current, and this type of capacitor is checked with a more elaborate bridge arrangement.

9.3 POWER-FACTOR MEASUREMENT

Next, let us consider a configuration for a capacitor bridge that measures both capacitance values and power factor. Electrolytic capacitors have inherent resistance in many cases which is effectively in series with the capacitance, as depicted in Fig. 9.4. Evidently, this effective series resistance cannot be

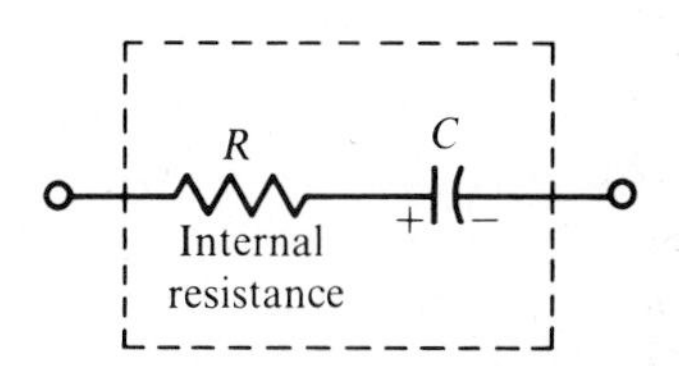

FIG. 9.4 *Capacitor showing internal resistance in series with capacitance.*

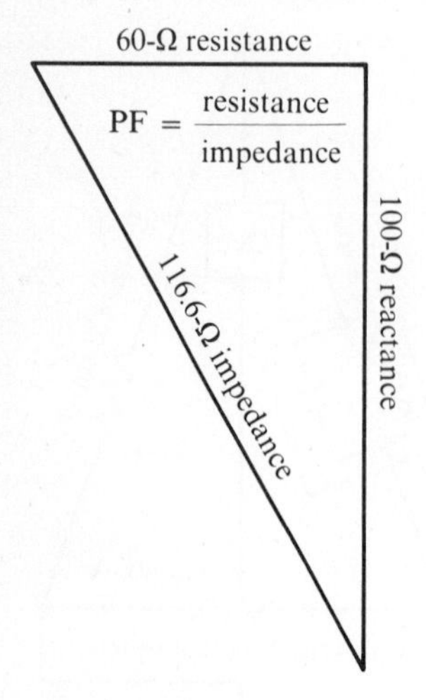

FIG. 9.5 *Impedance triangle showing effect of series resistance on power factor.*

measured with either a Wheatstone bridge or with a simple capacitance bridge. It is helpful to briefly review the definition of power factor for a capacitor. With reference to Fig. 9.5, the resistance and reactance vectors have been drawn at right angles to each other. In turn, the impedance of the capacitor is denoted by the hypotenuse of the triangle. The power factor is defined as the ratio of resistance to impedance. In most cases, we express a power factor as a percentage.

Figure 9.6 shows the configuration for a basic capacitance bridge with a power-factor control. Rheostat R_{PF} is ordinarily calibrated from zero to 80 percent. The calibrated potentiometer indicates capacitance values, as explained previously. Note that when the power-factor control is misadjusted, a complete null cannot be obtained by adjustment of the potentiometer. On the other hand, when both controls are correctly adjusted, the bridge detector indicates a complete null. The value of the unknown capacitance is read on the potentiometer scale, and the power factor is read on the rheostat scale. Thus, we do not measure the value of R_c directly; however, its effective value may be calculated, if desired, from the known capacitance, power factor, and frequency values.

It is left as an exercise for the student to analyze the bridge operation at 1 kHz in Fig. 9.6 to determine whether one or both of the control settings may be frequency responsive. Note the current-limiting resistance in series with the bridge-driving source voltage. This resistance protects the potentiometer and rheostat against accidental damage in case the unknown terminals should be short-circuited. The value of the bridge-driving voltage must not exceed the rated working voltage of the standard capacitor or of the capacitor under test. This arrangement, which employs a 50-V ac source, is sometimes used to measure the capacitance and power factor of electrolytic capacitors rated for operation from 150 to 400 V. However, better accuracy can be obtained if a dc polarizing voltage is also provided, as explained subsequently.

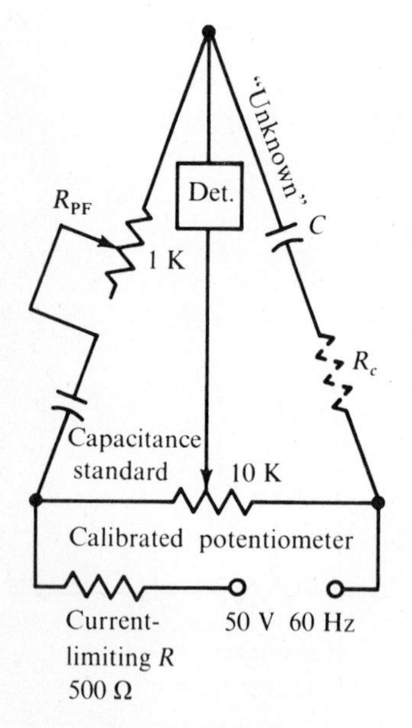

FIG. 9.6 *Bridge for measuring power factor that includes a series rheostat.*

Note that if a leaky capacitor is tested in the bridge configuration of Fig. 9.6, incorrect readings of both capacitance value and power factor will result. The reason for this inaccuracy is seen from the familiar series-parallel impedance triangle in Fig. 9.7. That is, the bridge in Fig. 9.6 is designed to balance series resistance in one arm against series resistance in the other arm. If the capacitor under test is leaky, the test situation becomes as shown in Fig. 9.8. We observe in Fig. 9.7 that the bridge will respond as if the parallel resistance R_p were a series resistance R_S, and as if the parallel reactance X_{CP} were a series reactance X_{CS}. That is, the readings will be in very substantial error in this example.

Since this possibility of error exists for the bridge configuration of Fig. 9.6, it is good practice to make a leakage test of an electrolytic capacitor before attempting to measure its capacitance and power factor. A Wheatstone bridge arrangement is commonly used to measure or indicate leakage, as was shown in Fig. 9.3. However, it is usual to employ an indicator such as an electron-ray tube in a go/no go electrolytic leakage test; less indicator sensitivity is used because an electrolytic capacitor is not discarded unless it has substantial leakage. Note that a satisfactorily high value of leakage resistance will not upset the bridge operation objectionably in Fig. 9.8. On the other hand, if the electrolytic capacitor under test has substantial leakage resistance, it should be rejected without an attempt to measure its capacitance value or power factor.

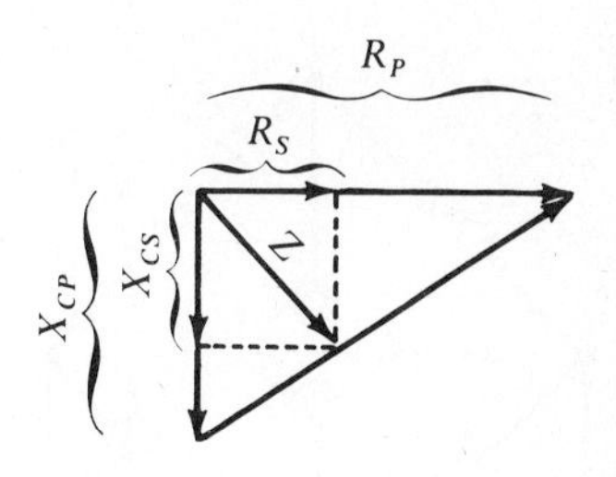

FIG. 9.7 *Series and parallel components of a capacitive impedance.*

9.4 CAPACITANCE BRIDGE CLASSIFICATIONS

An example of a *series-resistance* capacitive bridge was shown in Fig. 9.6. Another version of this basic arrangement is seen in Fig. 9.9. Headphones are used as a null indicator, as is common practice in laboratories. The ratio arms consist of R_a and R_b, with R_a adjustable so that the R_a/R_b ratio can be varied. In the standard arm, a calibrated variable capacitor C_s is connected in series with an adjustable resistor R_s. Capacitor C_x is the unknown capacitance, and R_{cx} represents the effective series resistance of the capacitor. When the bridge is balanced, points P and P' are at the same ac potential, and no current flows through the headphones. A high-quality laboratory bridge has a typical rated accuracy of ± 1 percent.

Next, let us consider the *Schering* type of capacitance bridge; a basic configuration is shown in Fig. 9.10. In this arrangement, the effective series resistance R_{cx} of the unknown capacitor is balanced by adjustment of C_a. In other words, the resistive component of the unknown capacitor is *not* balanced by resistance variation, as in Fig. 9.9, but is balanced instead by capacitance variation. This is a fundamental distinction between the Schering bridge and the series-resistance capaci-

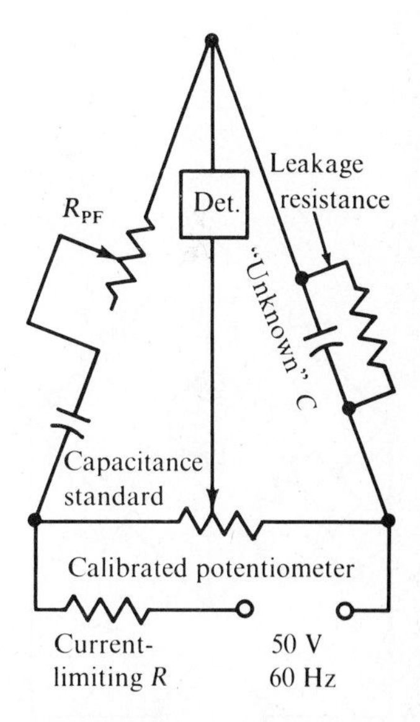

FIG. 9.8 *Situation in which incorrect bridge indications are obtained.*

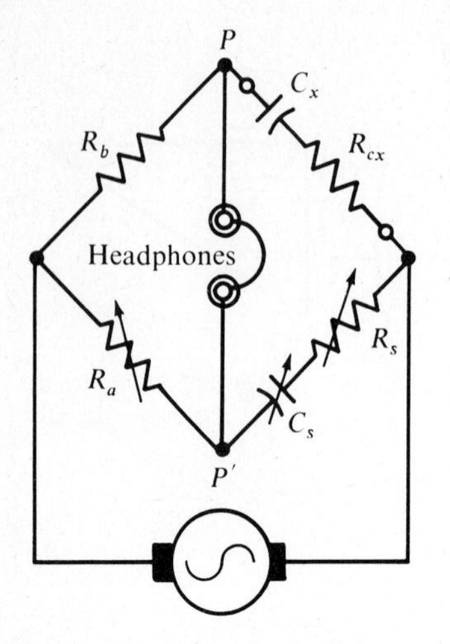

FIG. 9.9 *Another example of a series-resistance capacitive bridge.*

tive bridge. An adjustable ratio arm is provided by R_b in Fig. 9.10. For a fixed ratio of R_a/R_b, the value of the standard calibrated capacitor C_s is similarly proportional to the value of the unknown capacitor C_x.

Now, let us consider the equations for balance of a series-resistance capacitive bridge and a Schering bridge. With reference to Fig. 9.9, the value of the unknown capacitance is formulated:

$$C_x = C_s \frac{R_a}{R_b} \tag{9.4}$$

$$R_{cx} = R_s \frac{R_b}{R_a} \tag{9.5}$$

Engineers and technicians are chiefly concerned with the power factor of an electrolytic capacitor, instead of the value of R_{cx}. Therefore, the scale of rheostat R_s (Fig. 9.9) is ordinarily calibrated in percentage power factor, as noted previously for the bridge depicted in Fig. 9.6. Similarly, engineers are chiefly concerned with the dissipation factor of a paper, mica, or ceramic capacitor. The dissipation factor D is related to the power-factor (PF) value and to the Q value as follows:

$$D = \frac{R}{X_c} \tag{9.6}$$

$$\text{PF} = \frac{R}{Z} \tag{9.7}$$

$$Q = \frac{X_c}{R} \tag{9.8}$$

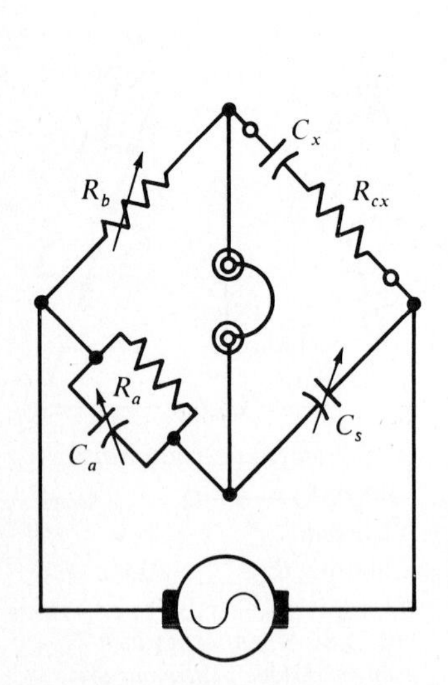

FIG. 9.10 *Basic Schering bridge.*

It is evident from Fig. 9.11 that the reactance and the impedance of a capacitor have practically the same value when the power factor is very small. Therefore, if R_s (Fig. 9.9) is calibrated in power-factor values, the scale also reads dissipation values from a practical viewpoint when the power factor is very small. Of course, the scale of R_s can be calibrated in D values, if desired. Although the scale could be calibrated in Q values, this is customarily avoided because very large numbers would be involved.

The alert student will perceive that the design of a capacitance bridge becomes progressively more difficult as lower values of D are contended with. Accuracy becomes a problem, because standard capacitors cannot be ideal. In turn, intensive effort is required on the part of the design engineer to fabricate standard capacitors that have a considerably lower D value than the capacitor to be tested. The design of laboratory standard capacitors is an extensive and complex subject, and the interested student is referred to specialized engineering textbooks.

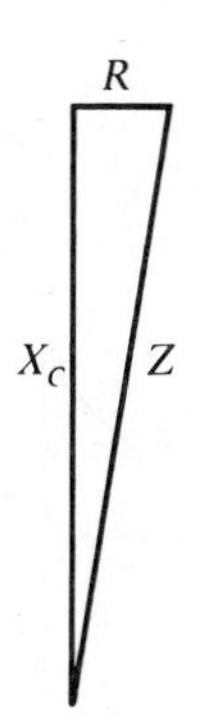

FIG. 9.11 *Nearly identical reactance and impedance values with very small power factor.*

Next, let us consider the impedances of the Schering bridge arms in Fig. 9.10, and the balance equations. The impedance of the unknown arm is written:

$$Z_x = R_{cx} - \frac{j}{\omega C_x} \tag{9.9}$$

Proceeding clockwise around the bridge, we write

$$Z_s = X_s = \frac{-j}{\omega C_s} \tag{9.10}$$

$$Z_a = \frac{1}{\dfrac{1}{R_a} + j\omega C_a} \tag{9.11}$$

$$Z_b = R_b \tag{9.12}$$

The general balance equation is written:

$$\frac{-jR_b}{\omega C_s}(1 + j\omega C_a R_a) = R_a\left(R_{cx} - \frac{j}{\omega C_x}\right) \tag{9.13}$$

In turn, the useful relations for C_x and R_{cx} are obtained:

$$C_x = C_s \frac{R_a}{R_b} \tag{9.14}$$

$$R_{cx} = \frac{C_a}{C_s} R_b \tag{9.15}$$

As noted previously, the power factor or D value is of more interest than R_{cx} in most engineering work. Therefore, the Schering bridge is commonly provided with scales calibrated in terms of D and C_x. It follows from Formula (9.6) that the dissipation factor D is equal to the tangent of the phase-defect angle δ in Fig. 9.12. Similarly, it follows from Formula (9.7) that the power factor PF is equal to the cosine

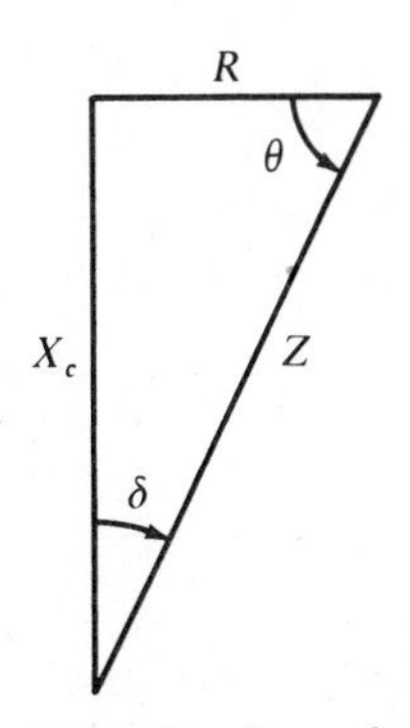

FIG. 9.12 *Power-factor angle (θ) and phase-defect angle (δ).*

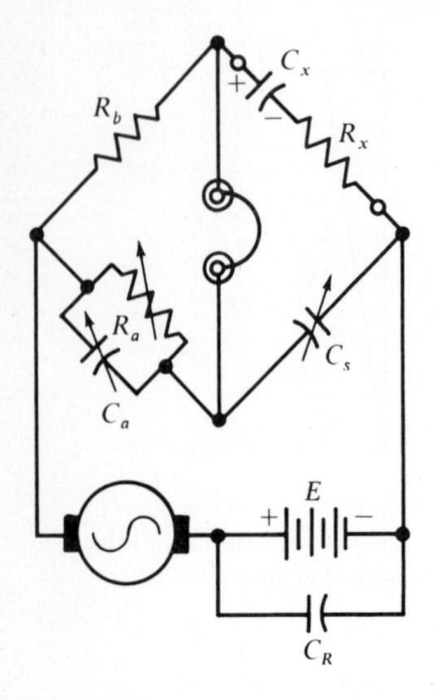

FIG. 9.13 *Schering bridge with dc polarizing voltage included.*

of the phase angle θ in Fig. 9.12. These angles are related by the equation:

$$\theta + \delta = 90^\circ \tag{9.16}$$

To measure capacitance values of electrolytic capacitors with high accuracy, a polarizing voltage is employed as depicted in Fig. 9.13. Capacitor C_R is merely a large bypass capacitor to minimize the ac voltage drop across the battery. Note that C_x must be connected to the bridge terminals in correct polarity. Inclusion of a polarizing voltage improves the accuracy of measurement because the capacitor is tested under normal working conditions. The ac and dc source voltages may be adjusted to the values that are utilized in the intended application of the electrolytic capacitor. Both the capacitance and the effective resistance of an electrolytic capacitor depend significantly upon the working voltage. It is good practice to check the scale readings at 20-min intervals until the readings level off; this procedure ensures maximum measurement accuracy by giving the capacitor an opportunity to "form."

We will find that some ac bridges are frequency responsive, while others indicate independently of the operating frequency. For example, the series-resistance capacitive bridge depicted in Fig. 9.9 is basically frequency independent. In other words, the null settings of C_s and R_s will remain the same, whether the operating frequency is increased or decreased. Of course, there are practical limits to be noted, since the bridge will become very insensitive at a very low operating frequency. At such a frequency, the reactances of C_x and C_s become very high, and the currents in these arms become very small. In turn, the unbalanced current becomes very small, and it is difficult to determine when the null has been reached.

Next, let us consider the practical high-frequency operating limit in Fig. 9.9. Although not shown explicitly, stray capacitances are implicit in the diagram. In other words, each component has a small amount of stray capacitance with respect to every other component. Thus, there is stray coupling (electrostatic coupling) among the bridge components and to nearby objects such as the operator's body. These stray couplings permit spurious ac current paths which can be neglected in practice when the operating frequency is moderate. On the other hand, at a high operating frequency these spurious paths conduct sufficient current between various components that the indication accuracy becomes poor.

Now, we shall observe an important example of a frequency-responsive bridge measurement. With reference to Fig. 9.10, it follows from Formulas (9.14) and (9.15) that measurements of C_x and R_x are independent of frequency (an omega term does not appear in the balance equations). On the other hand, suppose that this bridge is provided with a Q-value scale. It can be shown that the Q value of the unknown capacitor is formulated:

$$Q_x = \frac{1}{\omega C_a R_a} \tag{9.17}$$

Since this balance equation contains an omega term, it is obvious that any shift in operating frequency will cause an error in Q-value indication. As noted previously, a Schering bridge is usually calibrated to indicate D values. Thus, the balance equation in this case is written:

$$D_x = \omega C_a R_a \tag{9.18}$$

Since the foregoing expression for D_x contains an omega term, it is evident that an experimental error will result unless the bridge is operated at its exact design frequency. The majority of bridges in this classification operate at 1 kHz using either external or built-in ac generators. Considerations of frequency stability in generators are presented subsequently.

9.5 SUBSTITUTION METHOD OF CAPACITANCE MEASUREMENT

When a small value of capacitance is to be measured accurately, a substitution method may be used. The resistances of the two ratio arms are made equal, and a capacitor with a small known value is connected across the unknown terminals. This capacitor should have a value in the same general order of magnitude as the unknown capacitor. The bridge is balanced and the capacitance scale reading is noted. Then, the unknown capacitor is connected in parallel with the known capacitor, and the bridge is balanced again. The difference between the second reading and the first reading gives the value of the unknown capacitor with optimum accuracy.

One of the advantages of the substitution method is that

the experimental error due to stray capacitances is minimized. Another advantage is that the pair of readings are made farther away from the end of the scale; some bridges have least accuracy at the extreme low end of the scale. Still another advantage is that the scale reading in the first measurement can be compared with the known small capacitance value; this serves as a check on the experimental error that may be anticipated due to stray capacitances. Although the substitution method cannot eliminate stray-capacitance error, this source of experimental error is effectively minimized.

FIG. 9.14 *Basic bridge indications in which standard values correspond to (a) unknown series values, (b) unknown equivalent parallel values, (c) unknown equivalent series values, (d) unknown parallel values, and in which a graphical solution is given, (e) of series and parallel equivalent values.*

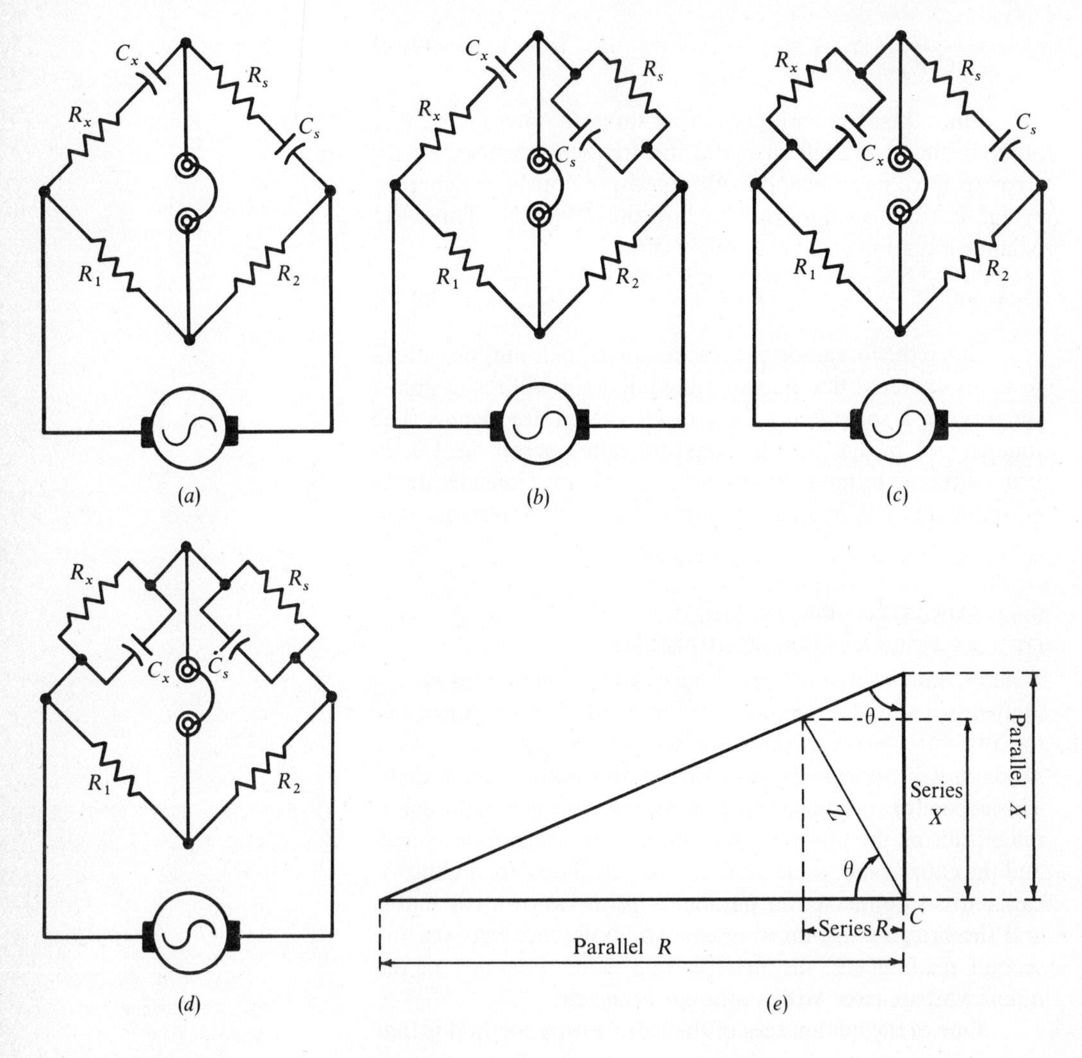

9.6 SERIES OR PARALLEL COMPONENT INDICATION

In general, when an unknown capacitor is connected to a bridge, we do not know whether it may have leakage resistance, or series resistance, or both. Whether balance is obtained on the basis of series R and C, or on the basis of parallel R and C depends on how the standards are connected in the bridge circuit. The fundamental considerations are shown in Fig. 9.14.

We observe in Fig. 9.14*a* that the standard and the unknown arms both have series configurations. In turn, if the ratio arms are equal, $C_s = C_x$ and $R_s = R_x$ at balance. On the other hand, in Fig. 9.14*b*, the standard arm has a parallel configuration, whereas the unknown arm has a series configuration. Consequently, balance is obtained when C_s has a value equal to the equivalent parallel capacitance of C_x and R_x, and when R_s has a value equal to the equivalent parallel resistance of C_x and R_x. These equivalents are seen in Fig. 9.14*e*.

Next, we observe in Fig. 9.14*c* that the standard arm has a series configuration, whereas the unknown arm has a parallel configuration. Accordingly, balance is obtained when C_s has a value equal to the equivalent series capacitance of C_x and R_x. These relations are seen in Fig. 9.14*e*. Finally, we observe in Fig. 9.14*d* that the standard and unknown arms have parallel configurations. Therefore, balance is obtained when $C_s = C_x$ and $R_s = R_x$. In summary, we need not know whether the unknown includes capacitance and resistance in series or in parallel; in any case, the bridge will indicate series components if the standards are connected in series, or the bridge will indicate parallel components if the standards are connected in parallel.

It is left as an exercise for the student to derive the equivalent series components for an electrolytic capacitor which has both series resistance and leakage (parallel) resistance. It is also instructive to derive the equivalent parallel components for the same situation.

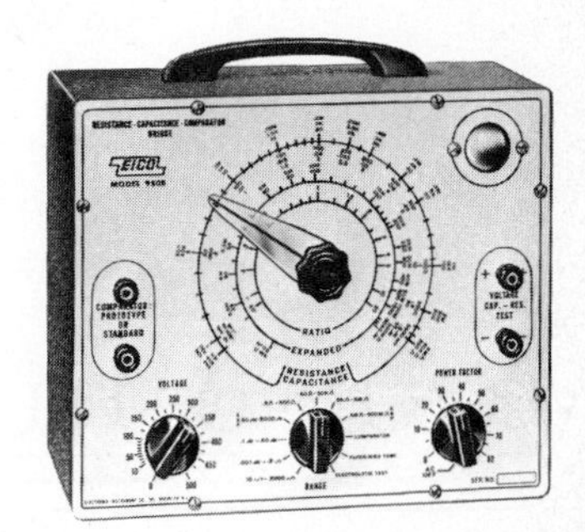

FIG. 9.15 *Appearance of a simple RC bridge. (Courtesy, Electronic Instrument Co., Inc.)*

9.7 COMMERCIAL CAPACITANCE BRIDGES

A service-type RC bridge is illustrated in Fig. 9.15. The circuit for the instrument is shown in Fig. 9.16. When set to its

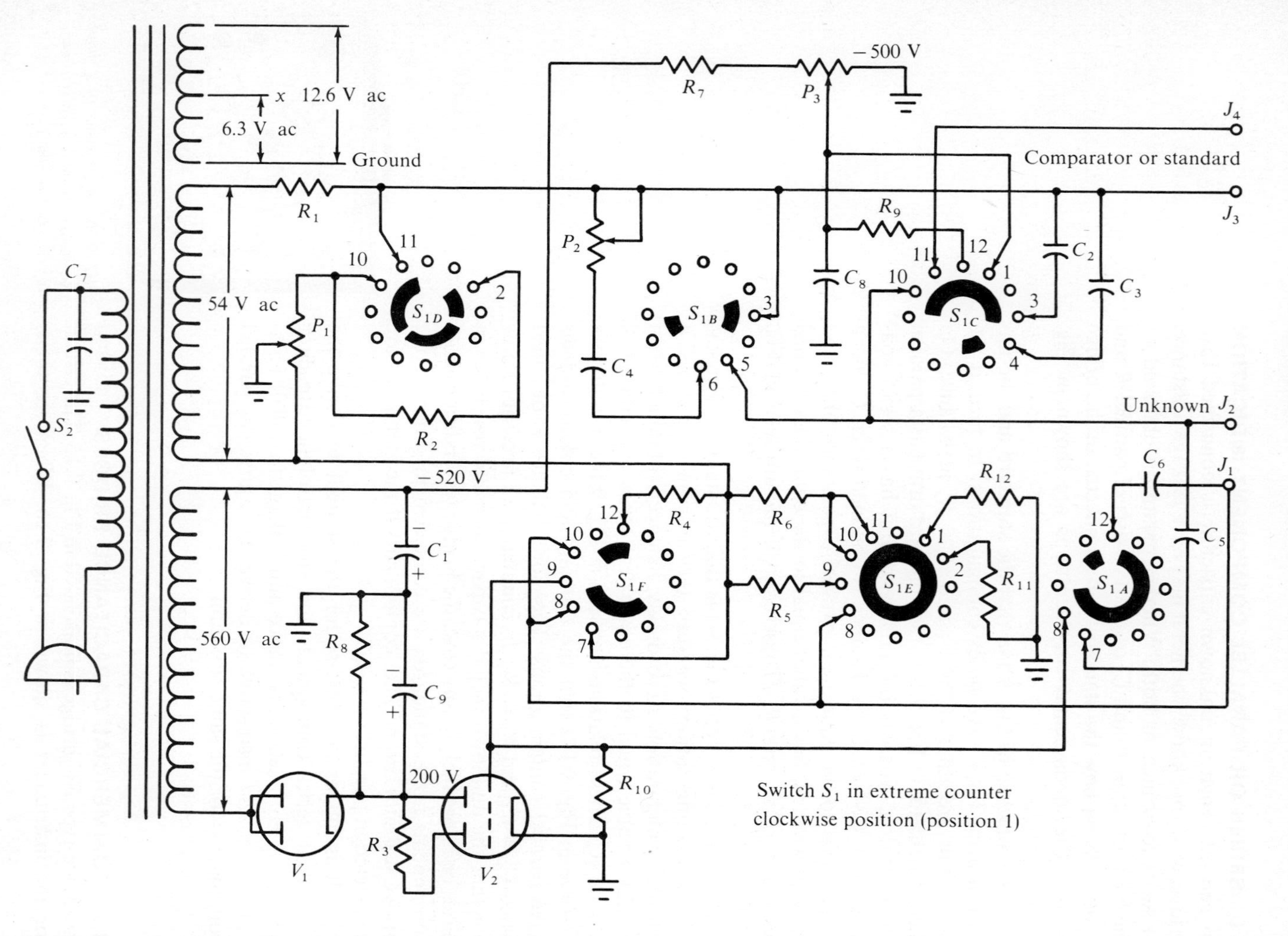

FIG. 9.16 *Circuit of the Electronic Instrument Co., Inc., Model 950B resistance-capacitance-comparator bridge.*

resistance function, a Wheatstone configuration is utilized to measure resistance values from 0.5 to 500 Ω, 50 to 50,000 Ω, 5000 Ω to 5 M, and 5 to 500 M. An electron-ray tube is employed as a null indicator. Potentiometer P_1 is the calibrated ratio-arm component. The Wheatstone arrangement is powered by dc voltage. Terminals J_1 and J_2 are connected to the unknown resistor. A current-limiting resistor is inserted between the power supply and the bridge to avoid damage to the components in case the unknown terminals are short-circuited.

On the first three capacitance ranges, this bridge employs the configuration shown in Fig. 9.2. These ranges are from 1 to 5000 pF, 0.001 to 0.5 μF, and 0.1 to 50 μF. The simple capacitance bridge circuit is supplemented by the leakage detector arrangement depicted in Fig. 9.3. A fourth capacitance range from 50 to 5000 μF is provided for checking electrolytic capacitors; this range utilizes the series-resistance capacitive, bridge configuration shown in Fig. 9.6. As noted in the diagram, the 1-K rheostat is calibrated in terms of percentage power factor; values from zero to 80 percent are indicated. A leakage detector facility is provided as shown in Fig. 9.3.

Note the terminals J_3 and J_4 in Fig. 9.15. An external standard capacitor can be connected to these terminals, if desired. However, technicians generally use terminals J_3 and J_4 in comparator application to measure the turns ratio of a transformer as shown in Fig. 9.17. Note that if a 1:1 transformer is under test, the bridge will balance when the ratio-arm potentiometer is set to its midpoint. The potentiometer is provided with a scale calibrated in ratios from 0.05 to 20. Let us suppose that a 4:1 transformer is being checked; a null will be obtained at 4 on the ratio scale. If the primary and secondary connections are reversed, a null will be obtained at 0.25 on the ratio scale (1:4 ratio).

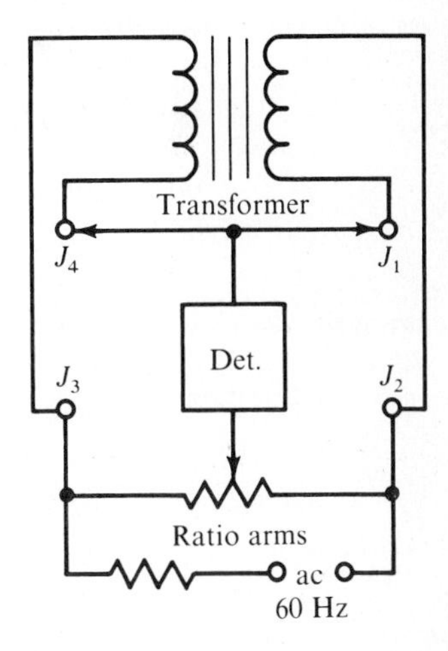

FIG. 9.17 *Comparator arrangement for measuring turns ratio of a transformer.*

9.8 APPLICATIONS

In addition to the conventional applications for a capacitance bridge, such as measuring the capacitance values of fixed or variable capacitors, the bridge can be used to measure the length of a coaxial cable, as depicted in Fig. 9.18. Its length is measured in terms of capacitance per foot. For example, if the particular type of cable has a capacitance of 30 pF/ft, and we measure a capacitance value of 0.009 μF, it follows that

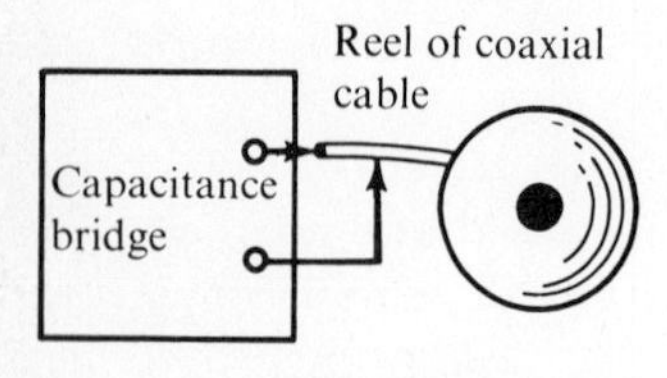

FIG. 9.18 *Measurement of length of a coaxial cable.*

the length of the cable is 300 ft. The same method can be used to find the distance to a break in a cable, even if it is buried underground; with the distance to the break known, considerable time can be saved in the repair procedure. Note that this method applies only when the far end of the cable is open-circuited. If it is short-circuited, other types of instruments must be employed.

A capacitance bridge can also be used to measure the winding capacitance of the primary or secondary of a transformer to core (ground), as shown in Fig. 9.19. This measurement is occasionally helpful in selecting a replacement transformer in pulse equipment. If the winding inductance is comparatively small, a 1-kHz capacitance bridge will provide an accurate measurement; if the winding inductance is substantial, a 60-Hz capacitance bridge should be utilized. In other words, an accurate capacitance measurement requires that the bridge frequency be sufficiently low that inductive-reactance and resonance effects can be neglected.

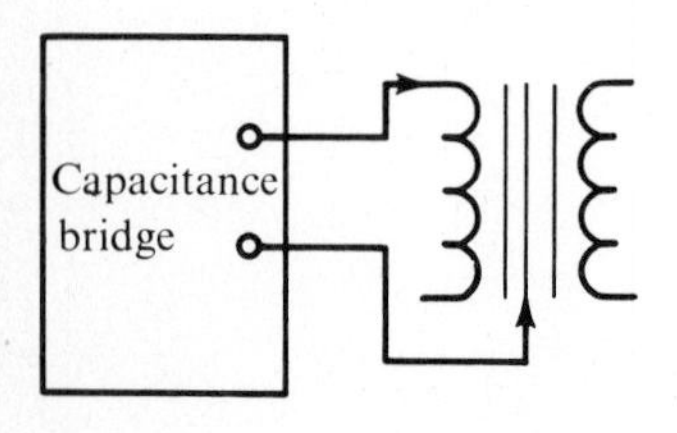

FIG. 9.19 *Measurement of winding capacitance to core (ground).*

If a ratio or comparator function is provided by a capacitance bridge, the relation of the sections of a ganged variable capacitor can be determined as shown in Fig. 9.20. Such capacitors are often designed with different total-capacitance values, and with a different capacitance-vs.-rotation relation for each section. With the test setup shown, the capacitance ratio at any setting of the variable capacitor can be quickly measured.

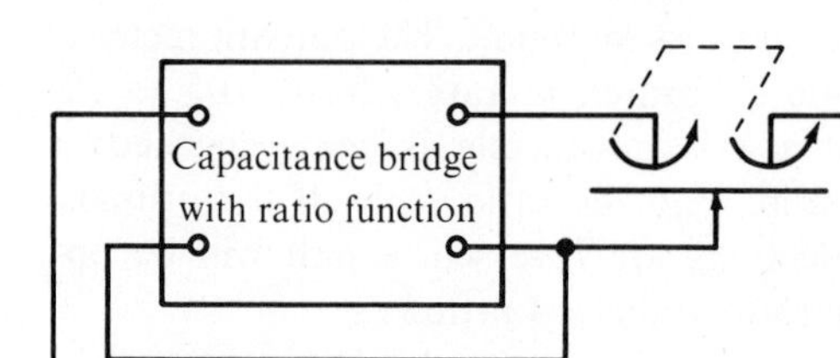

FIG. 9.20 *Measurement of capacitance ratio of a two-section ganged variable capacitor.*

QUESTIONS AND PROBLEMS

1. Why is the accuracy of a capacitance bridge somewhat dependent upon the bridge-driving frequency?
2. Why is an electron-ray tube used in a capacitance bridge?
3. What is the usual operating frequency of laboratory-type capacitance bridges?

4. How are the various ranges switched in the capacitance bridge shown in Fig. 9.2?

5. What precautions must be taken when measuring small values of capacitance?

6. What may be the cause of a false or incomplete null of a capacitance bridge when making a capacitance measurement?

7. Why should a capacitor be given a leakage-resistance test before measuring its capacitance?

8. Why does the testing of an electrolytic capacitor require a more elaborate bridge arrangement than the testing of paper, mica, and ceramic capacitors?

9. What precautions should be taken when measuring the leakage current of a capacitor?

10. What is the significance of the power-factor measurement of a capacitor?

11. How is the power factor of a capacitor expressed?

12. What is the purpose of the 500-Ω resistor in the bridge-circuit diagram in Fig. 9.6?

13. What is the typical rating of a high-quality laboratory capacitance bridge?

14. What is the fundamental distinction between the Schering bridge and the series-resistance capacitance bridge?

15. Calculate the values of the components in the bridge circuit in Fig. 9.10 when the bridge is balanced and $R_a = 1000\ \Omega$, $R_b = 5000\ \Omega$, and $C_s = 100$ pF. What is the value of the unknown capacitor?

16. What factors limit the practical high-frequency operating limit of the bridge shown in Fig. 9.9?

17. Why must the Schering bridge be operated at its exact design frequency?

18. Explain the substitution method for measuring small values of capacitance with optimum accuracy.

19. List three advantages of using the substitution method for measuring small values of capacitance.

20. Explain why it is not necessary to know whether the internal resistance of a capacitor is in series or parallel when an unknown capacitor is analyzed on a capacitance bridge (p. 181).

21. What are the null points on the potentiometer of the capacitance bridge, shown in Fig. 9.15, when a 3:2 transformer is connected across terminals J_3 and J_4?

10

INDUCTANCE BRIDGES

10.1 BRIDGE VOLTAGE REQUIREMENTS

Inductance values are usually measured with bridge-type instruments. An inductance bridge must be driven by ac voltage, because the null indication is based on reactive values. Most inductance bridges are driven by 1-kHz oscillators. In the first analysis, an inductance bridge is very similar to a capacitance bridge. For example, it is apparent that the bridge in Fig. 10.1 will be balanced if $R_1 = R_2$ and $L_1 = L_2$. However, if the inductors are not the same, a complete null cannot be obtained with this basic bridge arrangement. Thus, a series-resistance inductive bridge is not as simple as a series-resistance capacitive bridge. As explained next, very few inductors can be regarded as ideal in practice, whereas many capacitors can be regarded as ideal.

10.2 CHARACTERISTICS OF TYPICAL INDUCTORS

Although the effective series resistance of a paper, mica, or ceramic capacitor can often be neglected, this is seldom the case for an inductor. Any coil has a winding resistance; although the winding resistance can be minimized by using large-diameter wire, it cannot be eliminated. Therefore, a general-purpose inductance bridge must provide for balancing the effective series resistance of the inductor under test. One of the basic configurations employs resistance connected in series with the standard inductance. Figure 10.2 shows the configuration of this type of bridge. We observe that if the inductors were replaced by capacitors, a familiar type of capacitance bridge would be obtained.

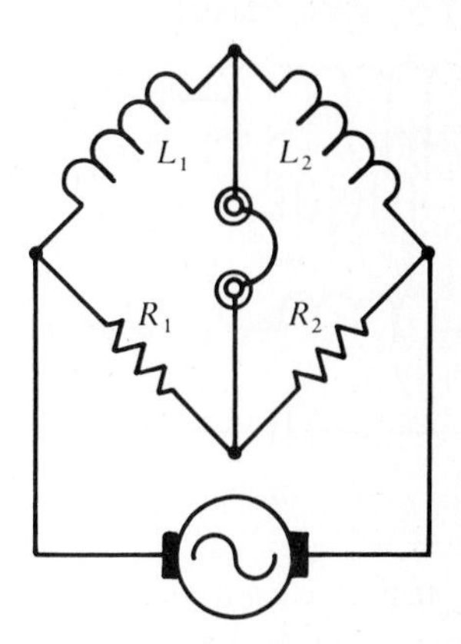

FIG. 10.1 *Basic inductance bridge.*

In Fig. 10.2, R_a and R_b are the ratio arms, and both are adjustable in this example to obtain various R_a/R_b ratios. The unknown inductance is represented by L_x, and its effective series resistance is represented by R_x. This unknown inductance is balanced by adjustment of the standard inductance L_s, and R_x is balanced by adjustment of the standard resistance R_s. The student may show that the equations for balance are as follows:

$$L_x = L_s \frac{R_b}{R_a} \tag{10.1}$$

$$R_x = R_s \frac{R_b}{R_a} \tag{10.2}$$

Any coil also has distributed capacitance, as depicted in Fig. 10.3. Accordingly, the coil becomes self-resonant at some particular frequency. Insofar as bridge operation is concerned, our purpose is to measure an inductance value on the basis of inductive reactance. At its self-resonant frequency, an inductor has an impedance value that is much greater than its inductive-reactance value. Therefore, if we are to obtain accurate inductance measurements with a bridge, the operating frequency must be kept well below the self-resonant frequency of the unknown inductor. It is for this reason that most inductance bridges operate at 1 kHz. However, if we are interested in measuring the inductance of an audio transformer winding, for example, resonance problems will be encountered unless the bridge is operated at a low frequency, such as 60 Hz.

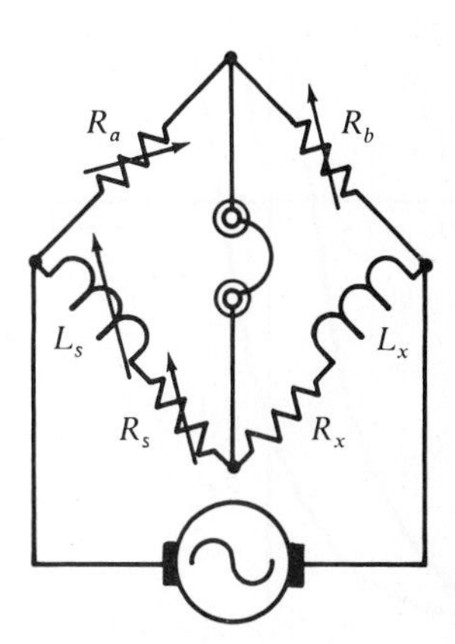

FIG. 10.2 *A series-resistance inductive bridge.*

It might be supposed that resonance problems could be avoided in bridge operation by increasing the operating

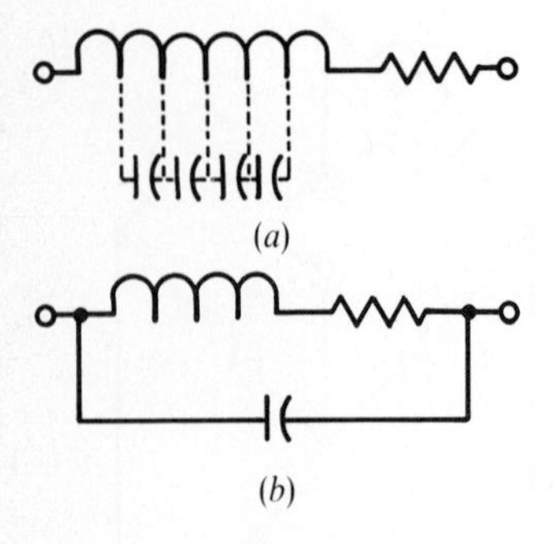

FIG. 10.3 *Equivalent circuits for inductors. (a) Equivalent circuit indicating distributed capacitances; (b) equivalent self-resonant "parallel" circuit.*

frequency well above the resonance value of the unknown. This is not so. The reason is apparent from Fig. 10.3. That is, if a voltage is applied at a high frequency, most of the current is drawn by the distributed capacitance of the coil. In turn, the inductor "looks like" a capacitor at high frequencies. To summarize briefly, a practical inductor is essentially an *LCR* impedance and its inductive component must be measured at a frequency well below its self-resonant frequency.

Thus far, we have considered air-core coils. Bridge operation is comparatively simple when air-core coils are employed, because this type of coil is essentially a linear device. On the other hand, many practical inductors have iron cores; an iron-core inductor is a nonlinear device because its magnetic circuit is nonlinear. Figure 10.4 shows a *B-H* curve and hysteresis loop for an iron core. The inductance of an iron core is greatest at low-flux densities, and is least at high-flux densities. Let us consider how this nonlinear inductance characteristic affects the measurement of an inductance value with a bridge such as depicted in Fig. 10.2.

The bridge is driven by a constant-voltage generator in the example of Fig. 10.2; the current divides into the ratio arms and the inductive arms in accordance with the impedance of the latter. We assume here that both the unknown inductor L_x and the standard inductor L_s have iron cores. As we adjust the value of L_s to find the null point, the impedance of the standard arm changes. In turn, the current through the inductive arms changes. This causes the effective inductance of L_x to shift. Moreover, the hysteresis loop depicted in Fig. 10.4 is associated with a core loss which contributes to the effective series resistances R_x and R_s. Thus, both the inductance and effective series resistance of the unknown shift as we approach the null point.

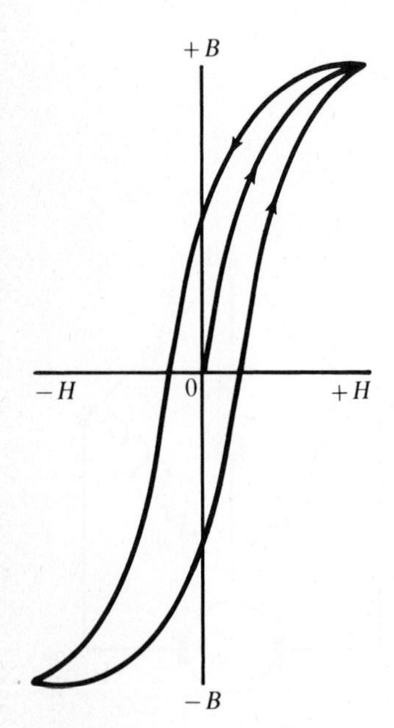

FIG. 10.4 *Nonlinear B-H curve for an iron core.*

In practice, inductance shift and resistance shift introduce a complication in bridge operation which is called *chasing the null* or *sliding null*. That is, the adjustments of L_s and R_s in Fig. 10.2 interact; after we obtain the best partial null by adjusting L_s, we find that R_s changes the current flow and L_s must then be readjusted. As we proceed back and forth between the L_s and R_s adjustments, we eventually obtain a satisfactorily complete null. The readings of the L_s and R_s scales then correspond to the inductance and resistance values for L_x and R_x *at the existing value of current flow.* If the output voltage of the bridge generator is increased or decreased, the original null will disappear, and the bridge must be balanced again with respect to the new values of effective inductance and resistance.

10.3 THE MAXWELL BRIDGE

Although the series-resistance inductive bridge shown in Fig. 10.2 can provide very accurate measurements, it has certain disadvantages that should be clearly recognized. First, stable and accurate standard inductors are comparatively expensive. Second, a standard inductor must be so well shielded that it does not pick up any stray fields and thereby introduce a false null. Third, an iron-core standard inductor will not present its rated value of inductance unless the current flow is precisely adjusted. These complications can be avoided by employing a standard capacitor instead of a standard inductor, as shall be explained.

Figure 10.5 shows the configuration for a Maxwell bridge. Since the standard capacitor draws a leading current, while the unknown inductor draws a lagging current, the reactive arms are not adjacent, but are located oppositely to obtain phase cancellation. The ratio arms are R_a and R_b. It can be shown that the balance equations for the Maxwell bridge are written:

$$L_x = R_a R_b C_s \tag{10.3}$$

$$R_x = R_a \frac{R_b}{R_s} \tag{10.4}$$

We observe that the L_x and R_x indications are frequency independent. However, as explained previously, if L_x is an iron-core inductor, its effective inductance and resistance will be current dependent, and we will have to chase the null. In most commercial bridges, the R_s scale is not calibrated in resistance values, but in Q values. That is, this scale is calibrated in terms of the formula:

$$Q = \frac{\omega L_x}{R_x} = \omega C_s R_s \tag{10.5}$$

Since an omega factor is involved in the Q indication, it is evident that accuracy of the Q measurement depends upon stability of the bridge-driving frequency. The usual driving frequency is 1 kHz. The alert student will also perceive that good waveform is necessary, because a nonsinusoidal wave-

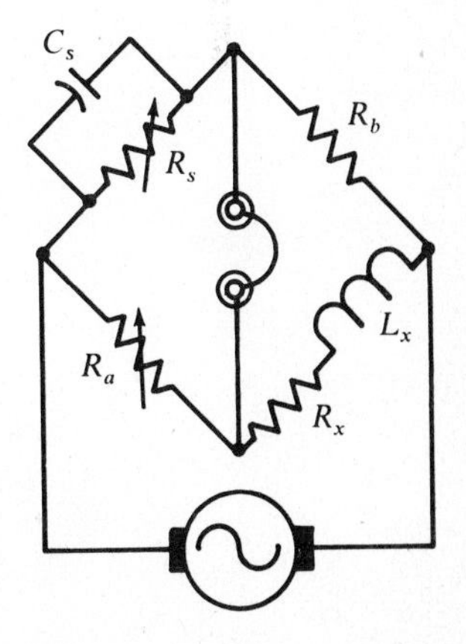

FIG. 10.5 *A Maxwell bridge arrangement.*

form is made of more than one frequency. The chief limitation of the Maxwell bridge is encountered in measuring high Q values, because the required value of R_s becomes so high that residual leakages in the bridge circuit impair the indication accuracy. Therefore a series standard arrangement is used to measure high Q values, as explained next.

10.4 THE HAY BRIDGE

When the standard resistance is connected in series with the standard capacitor as depicted in Fig. 10.6*a*, the arrangement is called a *Hay bridge.* It can be shown that the balance equations for this bridge are written:

$$L_X = \frac{R_A R_B C_S}{1 + (1/Q)^2} \tag{10.6}$$

$$R_X = \frac{R_A R_B}{R_S(Q^2 + 1)} \tag{10.7}$$

Note that the Hay bridge is frequency dependent to a greater extent than the Maxwell bridge, because Formulas (10.6) and (10.7) contain a Q factor in the expression for L_X and R_X. In turn, null chasing is more apparent in operation of the Hay bridge. We recognize that if the bridge driving frequency should drift, the indicated values for L_X and R_X would be inaccurate. Commercial bridges (Fig. 10.6*b*) commonly provide switching arrangements to select a Maxwell configuration for low Q coils, and a Hay configuration for high Q coils. If we attempt to use the Hay bridge to measure low Q values, the value of R_S may be so small that residual resistances in the bridge circuit introduce appreciable error.

10.5 INCREMENTAL INDUCTANCE BRIDGE

Coils and transformer windings are operated under two general conditions in electrical and electronic circuits. That is, an inductor may be energized by ac only, or it may be energized by pulsating dc (a dc voltage with an ac component). For example, the voice coil in a loudspeaker is energized by ac only. On the other hand, a filter choke in a power supply is energized by pulsating dc. As shown in Fig. 10.7, pulsating dc consists of a varying unidirectional current flow—the troughs of the waveform do not cross the zero axis. We shall

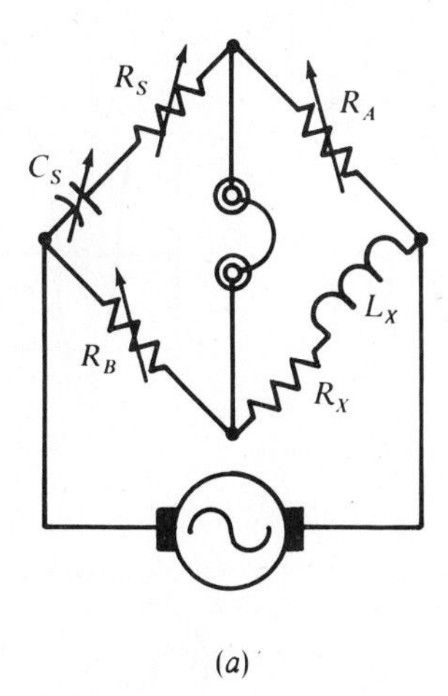

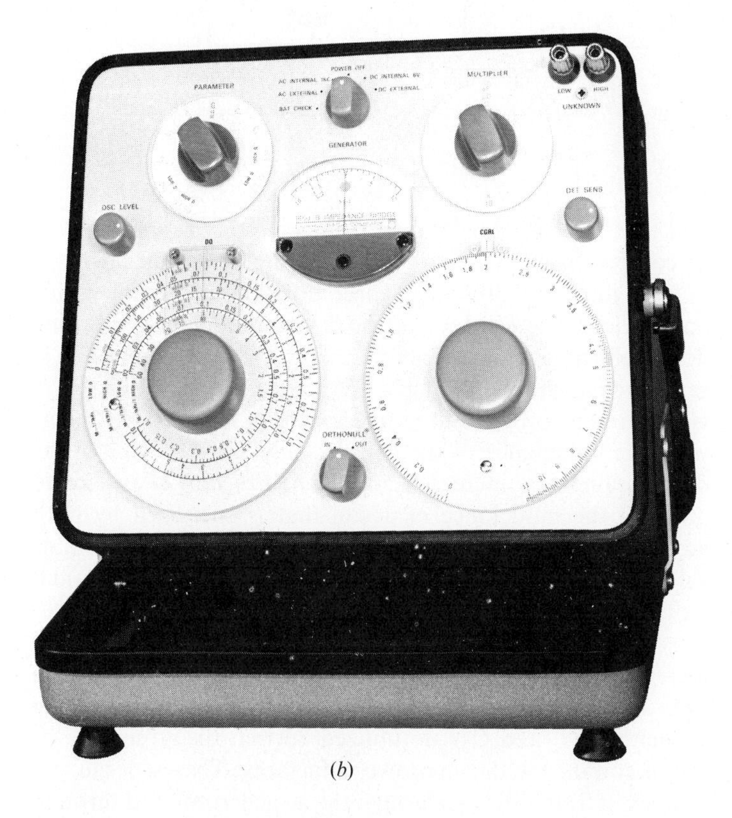

FIG. 10.6 *Hay bridge. (a) Schematic diagram; (b) laboratory bridge that employs both a Hay bridge and a Maxwell bridge. (Courtesy, General Radio Corp.)*

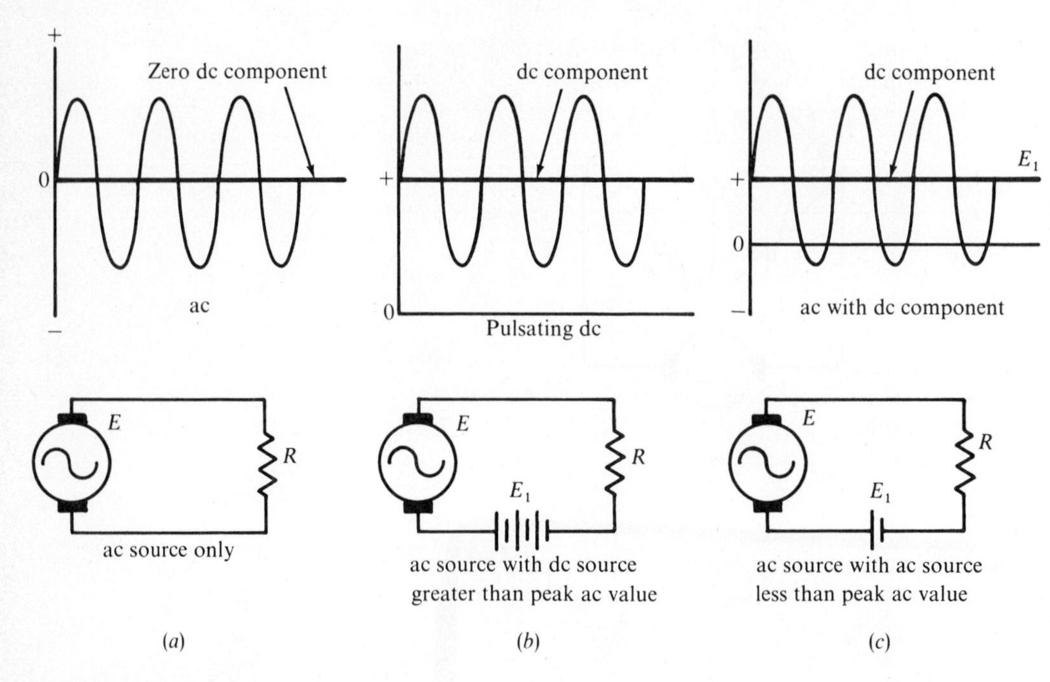

FIG. 10.7 *Alternating voltage (ac). (a) Ac only; (b) pulsating dc; (c) ac with dc component.*

find that the inductance of an iron-core coil decreases as the dc component is increased in a pulsating dc waveform.

Figure 10.8 illustrates three different operating points on a major hysteresis loop as the dc component in a pulsating dc waveform is increased. Note that as the operating point is raised (dc component increased), a greater variation of H is required to produce the same variation in B. Stated otherwise, the inductance of the coil decreases as the value of the dc component is increased. We recognize that if we are to measure the inductance of a coil that is energized by both ac and dc, the bridge that we use must provide adjustable values of ac and dc. This type of bridge is called an incremental inductance bridge.

Although various arrangements can be used to measure incremental inductance, the modified Hay bridge depicted in Fig. 10.9 is usually employed. We observe that blocking capacitors (C_1 and C_2) are utilized, so that the adjustable dc flow is confined to the unknown inductance. That is, dc current is blocked from flowing through the ac generator and through the headphones. An audio-frequency choke is inserted in the bridge to divert the unbalanced ac current through the headphones. The arrangement is usually elaborated by making the ac generator output adjustable, and by connection of an ac current meter in series with the unknown. Thus, the inductance of the coil can be measured under precisely adjusted operating conditions.

From a practical viewpoint, we seldom need to measure an incremental inductance value to a high degree of accuracy. For example, the current demand on a power supply is seldom constant, but changes more or less owing to equipment operating conditions. Moreover, the design of a power supply is somewhat a rule-of-thumb procedure. Therefore, if we measure the incremental inductance of a filter choke to an accuracy of 10 percent, its practical purpose is adequately served. In turn, we can simplify the measurement by using an approximate formula for bridge balance. The exact formula is written:

$$L_x = \frac{R_a R_c C_b}{1 + (1/Q_x)^2} \tag{10.8}$$

Since Q_x is fairly large for most filter chokes, $1/Q_x$ is a rather small fraction; and since the term is squared, its value

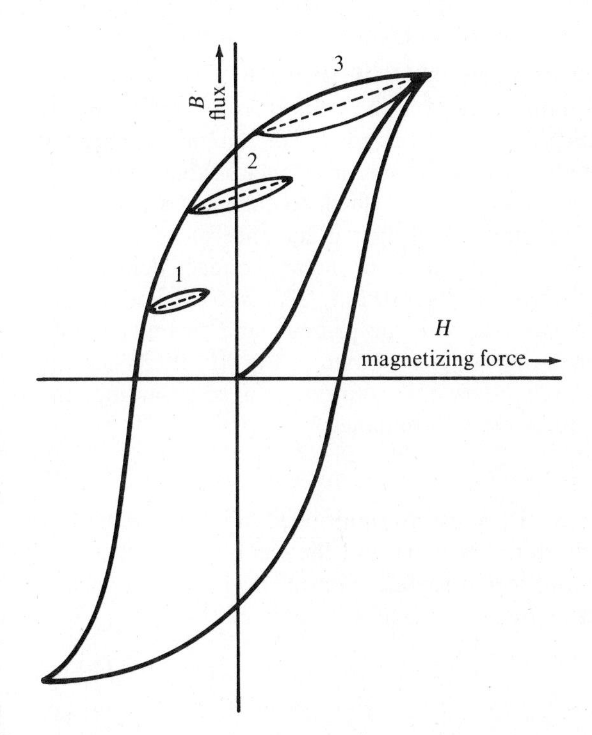

FIG. 10.8 *Minor hysteresis loops at three operating points on a major loop.*

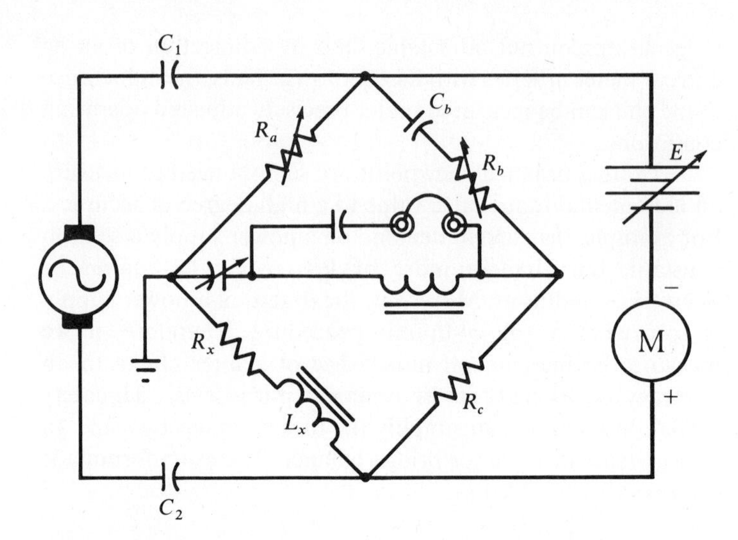

FIG. 10.9 *An incremental inductance bridge with basic Hay configuration.*

is decreased correspondingly. We can often neglect the Q term in practical measurement procedures, and write:

$$L_x \approx R_a R_c C_b \tag{10.9}$$

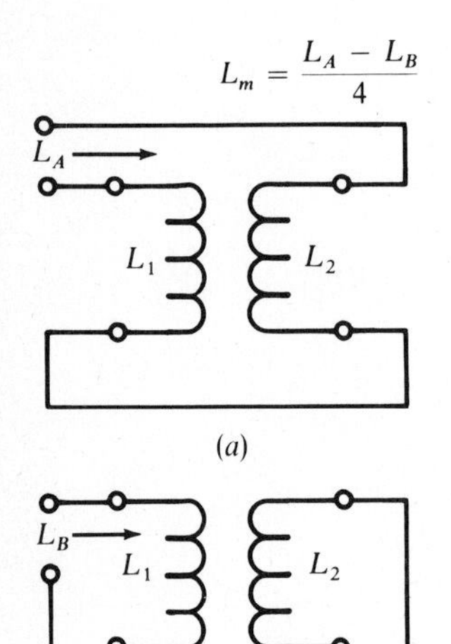

FIG. 10.10 *Measuring the mutual inductance of two coils. (a) Primary and secondary connected in series-aiding; (b) primary and secondary connected in series-opposing.*

10.6 MUTUAL INDUCTANCE BRIDGE

The mutual inductance of a pair of coils is a measure of the extent to which the coils are coupled. We can measure mutual inductance with a conventional inductance bridge if we make a pair of measurements from which to calculate the value of the mutual inductance. The first inductance measurement is made with the primary and secondary connected in series-aiding, as shown in Fig. 10.10*a*. The second inductance measurement is made with the primary and secondary connected in series-opposing, as shown in Fig. 10.10*b*. When the primary and secondary are connected in series-aiding, the measured inductance is formulated:

$$L_A = L_1 + L_2 + 2L_m \tag{10.10}$$

Note that L_m must be multiplied by 2 because it is common to both the primary and the secondary. Next, when the primary and secondary are connected in series-opposing, the measured inductance value is formulated:

$$L_B = L_1 + L_2 - 2L_m \tag{10.11}$$

From the foregoing pair of simultaneous equations, we solve for the value of L_m and obtain:

$$L_m = \frac{L_A - L_B}{4} \tag{10.12}$$

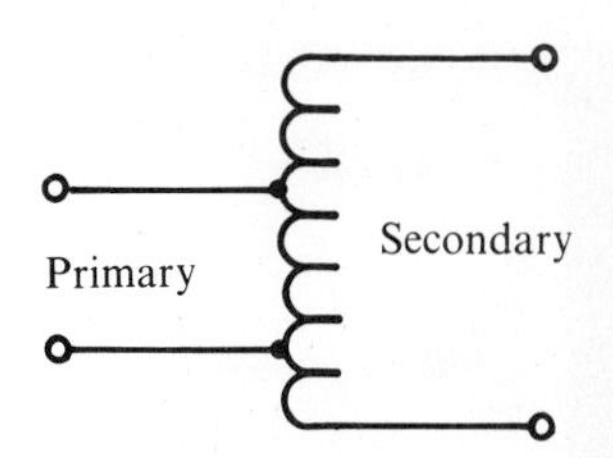

FIG. 10.11 *An autotransformer arrangement.*

In the case of an autotransformer such as depicted in Fig. 10.11, the foregoing method of mutual-inductance measurement is not possible. However, we may employ an inductance bridge to make three measurements from which to calculate the mutual-inductance value. Thus, we measure the primary inductance L_{po} with the secondary terminals open; then we measure the primary inductance L_{ps} with the secondary terminals short-circuited. Finally we measure the secondary inductance with the primary terminals open. In practice, we can usually neglect the resistance of the winding and write the useful approximation:

$$L_m = (L_{po} - L_{ps}) L_{so} \tag{10.13}$$

Specialized mutual-inductance bridges are often used to measure the mutual-inductance values of conventional transformers. For example, the Campbell bridge shown in Fig. 10.12 is in wide use. We recognize that this is a modified series-resistance inductive bridge; the primary of the transformer is connected into the unknown arm, and the secondary is connected in series with the bridge. The switch is first set to position 1, and the primary inductance L_p is measured with the secondary open. Next, the switch is set to position 2, and the changed value of the primary L_d is measured. In other words, L_d is the apparent inductance of the primary when the bridge is nulled with the secondary voltage operating in series with the unbalanced voltage. We write the following formula for calculation of the mutual-inductance value:

$$L_m = \frac{R_a L_p - R_b L_d}{R_a + R_b} \tag{10.14}$$

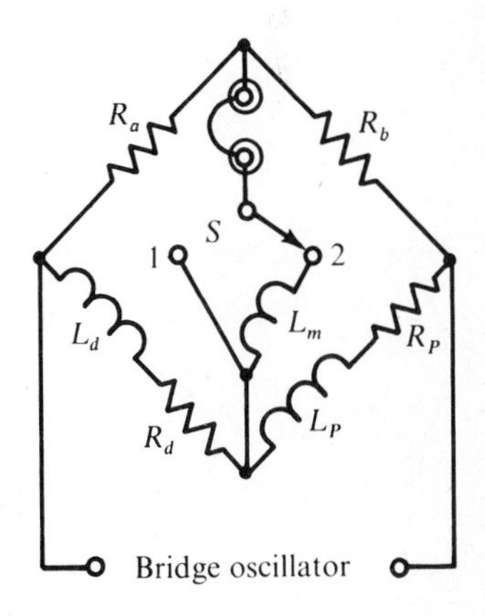

FIG. 10.12 *Configuration of the Campbell mutual-inductance bridge.*

10.7 STANDARD INDUCTORS

The design of standard inductors is a highly specialized subject, and only a brief introduction to the topic can be presented here. An inductance standard should be mechanically rugged and stable; it is usually enclosed in a metal case.

FIG. 10.13 *Appearance of a standard inductor. (Courtesy, General Radio Corp.)*

Figure 10.13 illustrates a standard of self-inductance that is widely used in laboratories. A low-temperature coefficient of inductance is desirable; for example, the inductor shown in Fig. 10.13 has a temperature coefficient of inductance of approximately -25 ppm/°C, between 16° and 32°C. A high Q value is desirable, and the unit in the foregoing example has a Q value of approximately 200 at 2 kHz, and well over 100 at 1 kHz.

To minimize the external magnetic field, and the possibility of the coil picking up stray magnetic fields, a toroidal winding is used in the example of Fig. 10.13. The plan of a toroidal winding is depicted in Fig. 10.14. In comparison to an "air core" unit, a doughnut shaped molybdenum-permalloy core provides a comparatively high Q value. Design engineers minimize distributed capacitance in a standard inductor insofar as practical considerations permit. A low value of distributed capacitance provides a high self-resonant frequency. By way of illustration, the 200-mH standard in Fig. 10.13 has a self-resonant frequency of 39 kHz.

The rated accuracy of an air-core standard inductor is typically ± 0.1 percent, whereas a molybdenum-permalloy core unit might be rated for ± 1 percent accuracy. The student should note that the operating frequency must be chosen considerably below the self-resonant frequency of the inductor, or its apparent inductance will be greater than its actual inductance. For example, if its apparent inductance is not to exceed its actual inductance by more than 1 percent, the operating frequency must not exceed 10 percent of the unit's self-resonant frequency.

Standard inductors in size ranges from 100 μH to 10 H are commonly operated at 1 kHz; litz[1] wire is usually employed to minimize the effective ac resistance. Note that a toroid requires a greater length of wire than a solenoid to obtain a given value of inductance. Thus, litz wire assists in obtaining a high Q value in a toroidal inductor. Ferromagnetic cores provide a substantial increase in Q value; however, the nonlinearity of the magnetic circuit makes it necessary for the operator to take the current flow into account when maximum accuracy is desired. For example, the standard

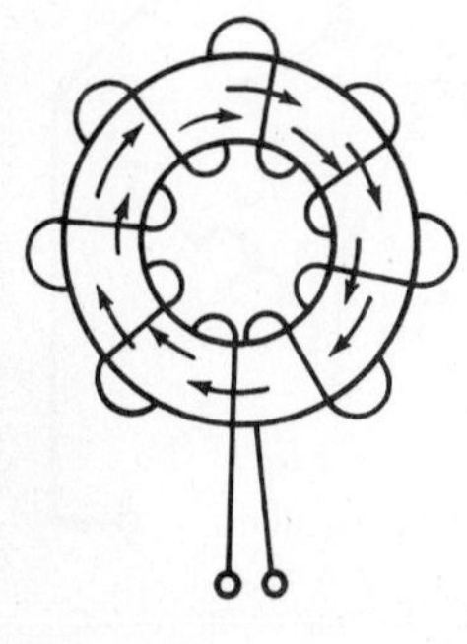

FIG. 10.14 *Toroidal coil construction.*

[1] Litz (litzendraht) wire is fabricated from two or three dozen strands of fine wire braided into cable form. The separate strands are insulated from one another. This construction minimizes both the skin effect and the proximity effect.

inductor illustrated in Fig. 10.13 is rated for an inductance change of 0.24 percent at a current of 10.8 mA.

When a molybdenum-permalloy core is utilized, the direction of the inductance change vs. current depends upon the presence or absence of dc current flow. For example, if the inductor in the foregoing example is energized by ac, the inductance increases with an increase in current until a practical limit of core saturation is reached. On the other hand, if the inductor is energized by pulsating dc, the inductance decreases with an increase in current. Since the inductance change is nonlinear, correction factors must be obtained from charts provided by the manufacturer.

To provide stability against mechanical shock or vibration, standard air-core inductors are typically supported in a mixture of ground cork and silica gel, which is then surrounded by a cast of potting compound contained in an aluminum case. Inductor windings are thermally cycled prior to calibration; this aging process assists in optimizing the thermal stability of the unit. Potting is chiefly useful in maintaining high accuracy under conditions of varying humidity. Standard inductors are rated for maximum voltage, current, and power dissipation, which should never be exceeded. For accurate measurements, an air-core standard inductor should not be operated with a current flow greater than 25 percent of the rated maximum value.

Adjustable inductors, called decade inductors, are occasionally used as inductance standards in bridge measurements. However, this type of standard finds its chief applications in circuit development, equalizers, filters, etc. Continuously variable inductors, called inductometers or variometers, are convenient for use as standards in mutual-inductance bridges, and some other bridge applications. The chief disadvantage of a variable standard is its comparatively low accuracy rating. Figure 10.15 depicts the plan of a variometer. Its inductance value is varied by rotating the inner coil through 180°. If the rotor and stator coils are connected in series, the range of inductance variation is formulated:

$$L_t = L_1 + L_2 \pm 2L_m \tag{10.15}$$

where L_t = total inductance of series-connected coils
L_1 = inductance of stator by itself
L_2 = inductance of rotor by itself
L_m = maximum (series-aiding) mutual inductance

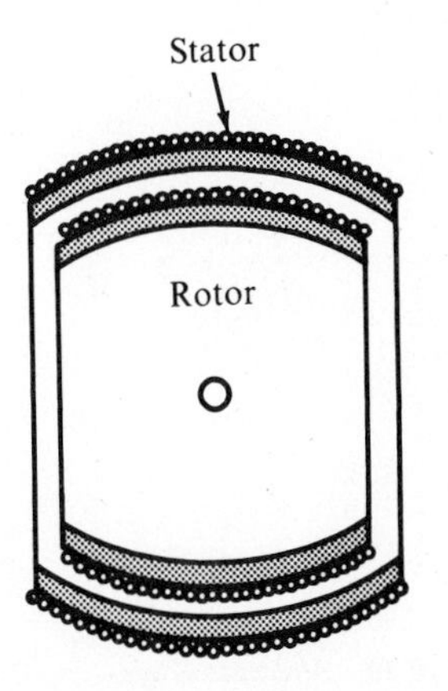

FIG. 10.15 *Plan of a variometer.*

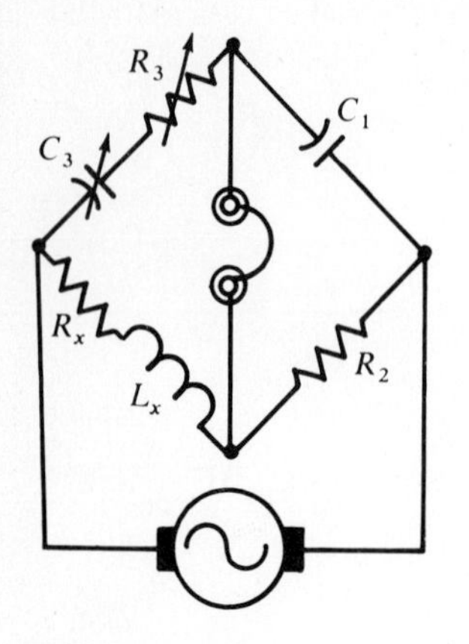

FIG. 10.16 *The basic Owen-bridge configuration.*

If the rotor and stator are connected in parallel, the value of L_t in Formula (10.15) becomes one-fourth the value provided by a series connection. A typical variable inductance standard with a series configuration has a range from 9 to 50 μH. Another unit with larger coils has a range from 90 to 500 mH. Designers use equal values of inductance in both the rotor and stator coils; this detail is not of importance in the series configuration; but in the parallel configuration, inductance equalization avoids losses caused by circulating currents.

10.8 THE OWEN BRIDGE

Commercial inductance bridges often employ the Owen configuration depicted in Fig. 10.16. Although it is comparatively expensive to manufacture, it has definite advantages over less-expensive bridges. This bridge requires the use of a precision variable capacitor C_3. However, it has an operating advantage in that the adjustable parameters R_3 and C_3 are in the same arm; in turn, the Owen bridge eliminates the sliding null encountered in other inductance bridges. Since the operator does not have to chase the null, inductance measurements are speeded up considerably.

The Owen bridge has a very wide range, and the inductance measurement depends basically on the accuracy of R_3 (Fig. 10.16); use of precision-type decade resistors provides high accuracy. The balance equations for the Owen bridge are

$$L_x = C_1 R_2 R_3 \qquad (10.16)$$

$$R_x = R_2 \frac{C_1}{C_3} \qquad (10.17)$$

FIG. 10.17 *A lab-type inductance bridge. (Courtesy, General Radio Corp.)*

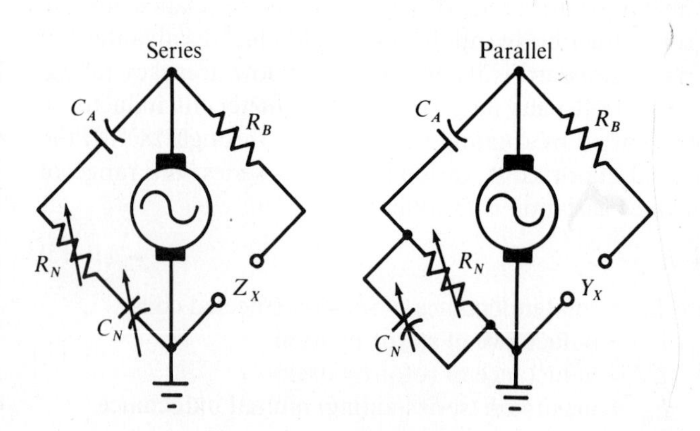

FIG. 10.18 *Alternate Owen-bridge arrangements.*

A commercial high-quality Owen bridge is illustrated in Fig. 10.17. It is rated for an accuracy of 0.1 percent with inductance values as low as 1 μH. Inductance values can be measured down to 0.001 μH, and up to more than 1000 H. The student should note that the resistive and reactive components of an unknown inductor can be measured in terms of resistance in series with reactance, or in terms of an equivalent parallel RL circuit. Thus, the bridge illustrated in Fig. 10.16 can be switched as indicated in Fig. 10.18 to read the unknown parameters either as series RL components, or as parallel RL components. Since most practical inductors can be effectively modeled in terms of series RL components, this function of the Owen bridge finds widest application.

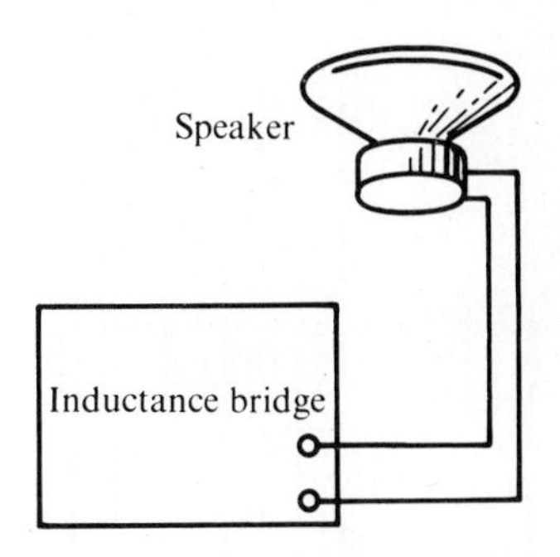

FIG. 10.19 *Measurement of voice-coil inductance and Q value.*

10.9 APPLICATIONS

An inductance bridge is useful for checks of loudspeaker voice-coil impedance, among other routine applications. With reference to Fig. 10.19, the bridge is connected to the voice-coil terminals of the speaker. In a typical example, the inductance measures 100 μH with a Q value of 0.2 at 1 kHz. These values correspond to an impedance of 3+ Ω at 1 kHz; it is customary to rate voice-coil impedance at 1 kHz. Note that the impedance value in this example has a much greater resistive component than an inductive-reactive component. That is, the inductive reactance is only 0.6 Ω, whereas the resistance is 3 Ω. The nominal impedance ratings for various voice coils are 4, 8, and 16 Ω.

Figure 10.20 shows how an inductance bridge is connected to measure the output impedance of an audio amplifier. The output impedance is defined as the impedance value "looking back" into the amplifier from its speaker-output terminals. To obtain an accurate measurement, the amplifier must be turned "on"; however, no input signal is applied to the amplifier. An inductance bridge is employed in this application, because the output impedance has a small inductive component in addition to its resistive component. That is, an output transformer usually has a residual leakage reactance which is inductive and which more than cancels out the inherent capacitances of the output system. The nominal output impedance ratings for audio amplifiers are 4, 8, and 16 Ω.

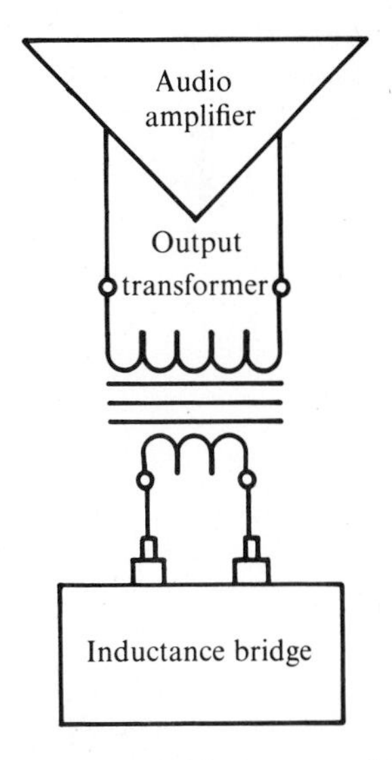

FIG. 10.20 *Measurement of the output impedance of an amplifier.*

Deflection-coil inductance and impedance can be mea-

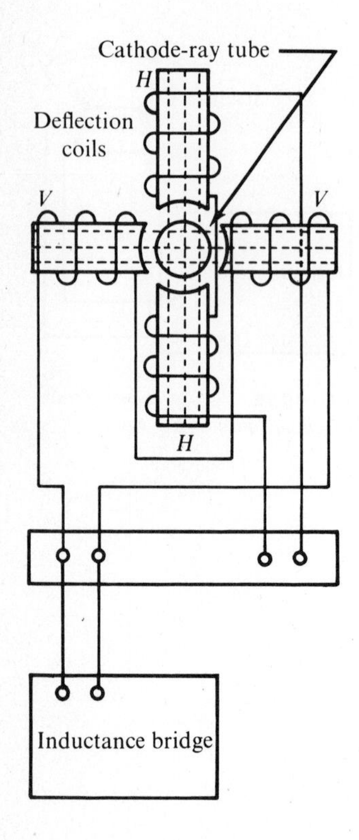

FIG. 10.21 *Measurement of deflection-coil inductance.*

sured for a television yoke as depicted in Fig. 10.21. The horizontal-deflection coils for a small-tube receiver have a typical inductance of 8.2 mH and a winding resistance of 13 Ω. This corresponds to a Q value of 4 at 1 kHz. The vertical-deflection coils in the same yoke have an inductance of 48 mH and a winding resistance of 65 Ω. This corresponds to a Q value of 0.46 at 1 kHz. Inductance and impedance values are the most useful parameters to consider when an exact replacement cannot be made and a selection must be made from stock chokes.

QUESTIONS AND PROBLEMS

1. Why must an inductance bridge be driven by an ac voltage?
2. Why is the series-inductance bridge more complex than the series-capacitance bridge?
3. Suppose the values of the components in the bridge shown in Fig. 10.2 are: R_b = 3 K, R_a = 2 K, L_s = 20 mH, and R_s = 40 Ω. What are the values of the unknown inductance and resistance?
4. Why must the operating frequency of an inductance bridge be much less than the self-resonant frequency of the inductor?
5. Why isn't it possible to overcome the resonant problems by using an operating frequency for an inductance bridge well above the resonant frequency of the inductor?
6. Why is the output voltage level important when measuring an iron-core inductor?
7. Explain the term *sliding null* as it applies to an inductance bridge.
8. What are the disadvantages of the inductance bridge shown in Fig. 10.2?
9. How are the disadvantages of the inductance bridge depicted in Fig. 10.2 overcome?
10. Why is the accuracy of the Q measurement with an inductance bridge dependent upon the frequency stability of the driving voltage?
11. Why are most commercial bridges equipped with a switching arrangement to select a Maxwell configuration or a Hay configuration?
12. What is the effect of an increasing dc current on the inductance of an iron-cored coil through which the current is flowing?

13. What are the characteristics of an incremental inductance bridge?
14. What is the purpose of the capacitors C_1 and C_2 in the bridge circuit shown in Fig. 10.9?
15. Explain the operation of the Hay bridge shown in Fig. 10.9 and the purpose of each component in the circuit.
16. What is the inductance of the unknown inductor shown in the bridge circuit in Fig. 10.9 when $R_b = 10$ K, $R_a = 100\ \Omega$, and $C_b = 1\ \mu F$?
17. Explain the procedure for determining the mutual inductance of two inductors with an inductance bridge.
18. What are the requirements for a standard inductor?
19. What factors are given as maximum ratings of a standard inductor?
20. What is the primary purpose of adjustable or decade inductors?

11

IMPEDANCE BRIDGES

11.1 MEASURING *Q* AND *D* VALUES

An impedance bridge measures the resistive and reactive components of an unknown impedance. Calibrated scales are provided for inductance and capacitance values. The resistive component of an impedance is seldom indicated directly; instead, a scale calibrated in *Q* values is provided for inductive impedances, and a scale calibrated in *D* values is provided for capacitive impedances in typical bridges. A Wheatstone bridge function is nearly always provided for measuring dc resistance values. Many bridges permit a choice of dc or ac drive to the Wheatstone bridge. In turn, the ac resistance of a component can be compared with its dc resistance.

We perceive that an impedance bridge does not measure resultant impedance values; instead, the bridge measures the components of an impedance in terms of capacitance and dissipation values, or in terms of inductance and quality-factor values. That is, the value of the resistive component is usually measured indirectly in terms of Q or D values. The reason for this mode of indication is that practical engineering work is more immediately concerned with Q and D values than with resistive-component values.

The student should note that Q and D values are functions of frequency. In other words, most bridges indicate quality and dissipation values at 1 kHz. Not only are Q and D values functions of frequency, but they are somewhat unpredictable functions, because it is difficult to predict the effective or ac resistance of an inductor or a capacitor at an arbitrary frequency. Therefore, it is necessary to measure a Q or D value at the frequency of interest, and to specify the test frequency in the data report. In general engineering work, it is often sufficient to measure the Q of an inductor at 1 kHz only. The chief exceptions occur in vhf and uhf situations, and special test equipment is required at these high frequencies.

11.2 THE BASIC IMPEDANCE BRIDGE

A configuration for a basic impedance bridge that measures resistance, capacitance, inductance, Q, and D values is shown in Fig. 11.1. To measure resistance, the unknown is connected across the Res. terminals, and the controls are operated as explained subsequently. To measure inductance or capacitance, the unknown is connected across the LC terminals. Switches S_2 and S_3 are ganged, and are shown in their resistance-measuring (Wheatstone) position in Fig. 11.1. Switch S_1 determines the amount of resistance in the ratio arms. The resistors connected to S_{1A} shall be called the R_a group, and the resistors connected to S_{1B} shall be called the R_b group. R_{10} is a precision rheostat with a calibrated scale. Thus, we write the general formula:

$$\frac{R_x}{R_b} = \frac{R_{10}}{R_a} \tag{11.1}$$

or

$$R_x = R_{10}\frac{R_b}{R_a} \tag{11.2}$$

When resistance values are being measured, switch S_5 is set to the position shown in Fig. 11.1. Thus, the galvanometer

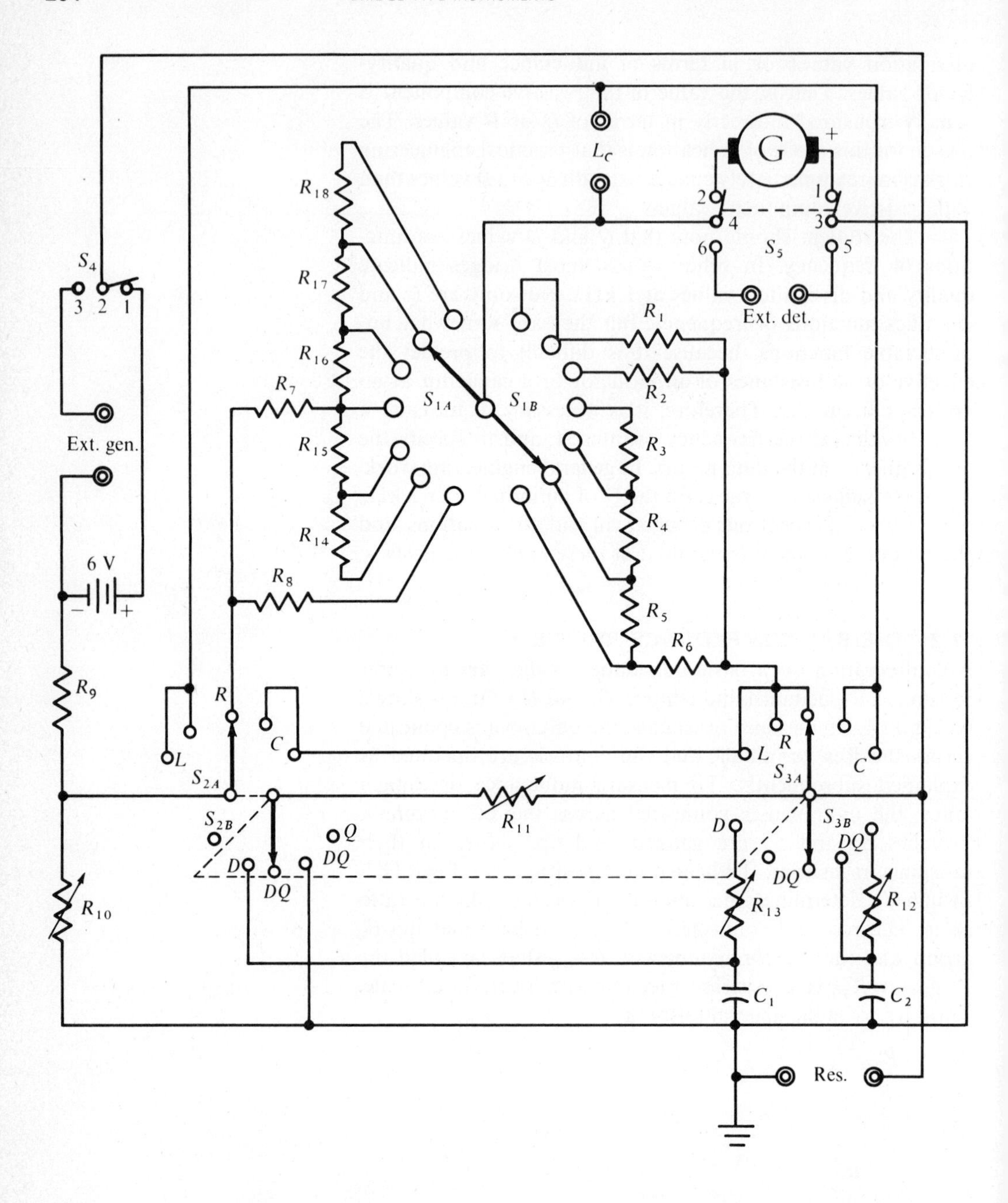

FIG. 11.1 *Schematic diagram of a practical impedance bridge.*

is connected into the bridge circuit for indicating dc balance. Switch S_4 is set to the position shown in Fig. 11.1, thereby energizing the Wheatstone bridge from a 6-V battery. Resistor R_9 is a current-limiting resistor which limits the amount of short-circuit current that can flow through the bridge components. With the switches set to the positions depicted in Fig. 11.1, the resulting Wheatstone bridge circuit shown in Fig. 11.2 is selected. Note that seven ranges are available. This type of bridge has a useful range of approximately 1 Ω to 10 M. Although resistance values as low as 0.01 Ω could be measured in theory, we shall find that the indicating accuracy at the extreme low end of the scale becomes poor. Let us consider this circuit briefly.

The left-hand ratio arm in Fig. 11.2 includes R_7, R_{16}, and

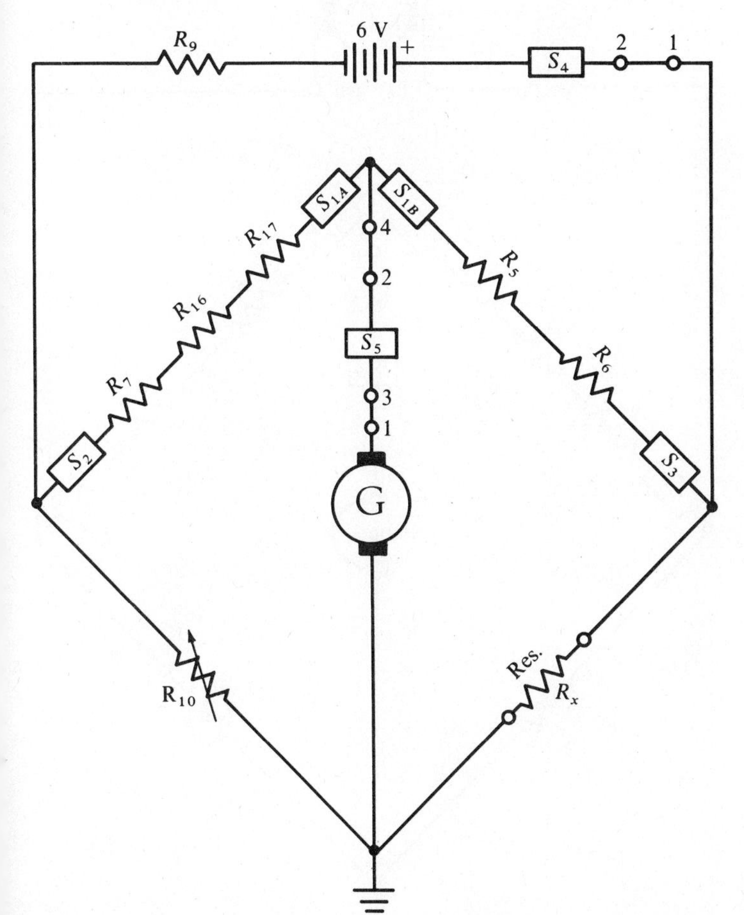

FIG. 11.2 *Wheatsone configuration of the impedance bridge.*

R_{17}; the right-hand ratio arm comprises R_5 and R_6. These ratio arms establish the range of resistance values that are indicated on the calibrated scale; this scale is a part of the standard-resistance assembly R_{10}. As noted previously, R_{10} is a precision rheostat. We observe in Fig. 11.1 that S_4 may be set to its Ext. gen. position, and S_5 may be set to its Ext. det. position, without changing the configuration of the bridge arms. It is accordingly a matter of choice whether the operator drives the Wheatstone bridge from a dc source or from an ac source. However, we will find that the applications for ac drive are quite limited.

When operating the bridge in this example with ac drive, the configuration appears as seen in Fig. 11.3. The external

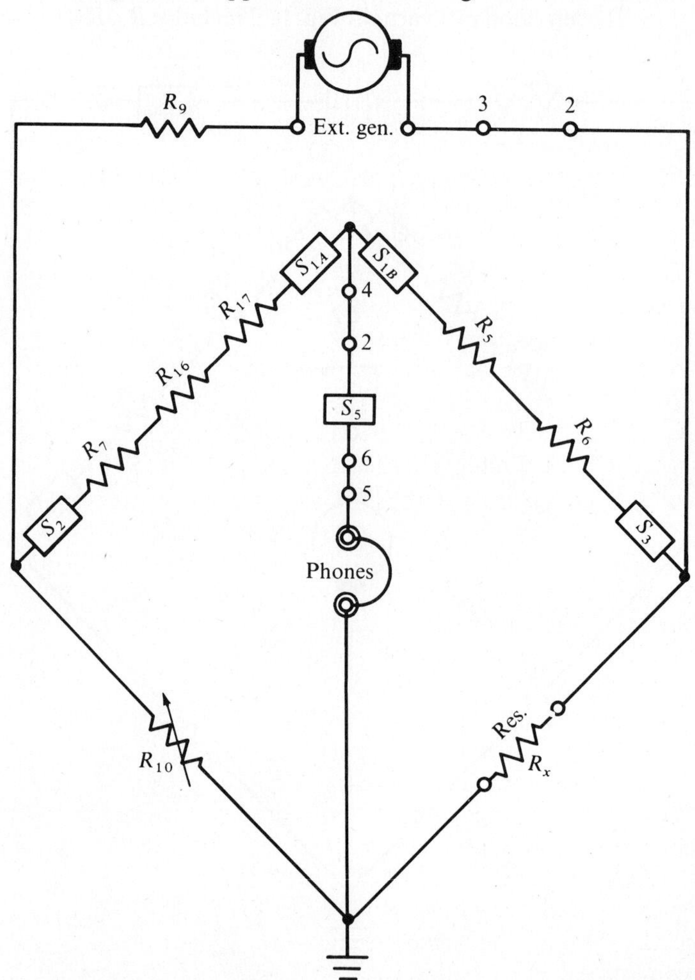

FIG. 11.3 *Wheatstone bridge with ac drive voltage.*

generator is typically an audio oscillator with a frequency range from 20 Hz to 20 kHz. Suppose we wish to measure the internal resistance of a battery at a frequency at 1 kHz. We set the generator in Fig. 11.3 to a frequency of 1 kHz, and connect the battery to the R_x terminals with a large blocking capacitor to prevent the flow of dc from the battery into the bridge. Figure 11.4 shows how the battery and blocking capacitor are connected. Insofar as bridge operation is concerned at 1 kHz, the blocking capacitor has negligible reactance, and the bridge indicates the ac internal resistance of the battery at balance.

FIG. 11.4 *Battery connected in series with blocking capacitor.*

11.3 CAPACITANCE-MEASURING FUNCTION

When the switches in Fig. 11.1 are set to their *C* positions, a resistance-ratio capacitive bridge is provided, as shown in Fig. 11.5. R_{10} operates in one of the ratio arms, and indicates capacitance values on a calibrated scale. R_{13} indicates *D* values. From considerations of symmetry, it is evident that the scale indications are independent of frequency. This conclusion neglects a second-order factor that involves a slight increase in *D* value at higher frequencies of operation. However, over the audio-frequency range, the *D* value of a capacitor is essentially constant.

The bridge depicted in Fig. 11.5 is chiefly useful for checking paper, mica, and ceramic capacitors. That is, it is not provided with a dc polarizing function for checking electrolytic capacitors. Moreover, a bridge of this type does not have extended high-capacitance ranges, and cannot measure capacitance values greater than approximately 10 μF. Like the individual capacitance bridge described in a previous chapter, an impedance bridge of this type can measure capacitance values as low as 10 pF. In theory, a capacitance value as small as 1 pF can be measured—however, we find that the accuracy of indication tends to become poor at the extreme low end of the scale.

Unlike a specialized capacitance bridge, an impedance bridge does not customarily provide a leakage test for capacitors. However, most leaky capacitors will produce a high reading on the *D* scale. A supplementary leakage test can be made by operating the impedance bridge on its Wheatstone function, with the capacitor connected in place of R_x in Fig. 11.2. The chief limitation in leakage measurement is encountered in checking high-voltage capacitors, such as utilized

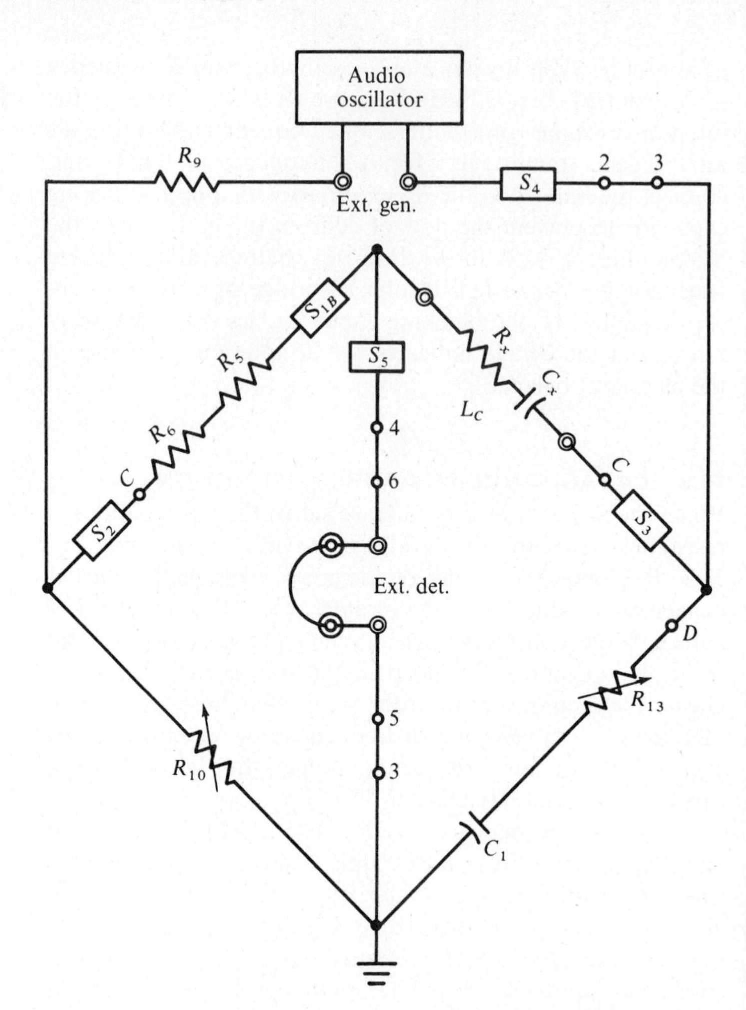

FIG. 11.5 *Circuit for measuring capacitance and dissipation factor.*

in television high-voltage rectifier circuits and damper circuits. That is, a conventional capacitance bridge or impedance bridge applies a comparatively small voltage to the capacitor under test. On the other hand, a high-voltage capacitor sometimes develops a leakage defect that involves a sudden partial breakdown at a particular high-voltage value.

11.4 INDUCTANCE-MEASURING FUNCTION

A choice of two inductance-bridge configurations is provided by the impedance bridge depicted in Fig. 11.1. With S_2 and S_3 set in the L and DQ positions, a Maxwell bridge circuit is provided, as seen in Fig. 11.6. *DQ* means "*D* values or *Q* values." We read the scale in terms of *Q* values when using the inductance function of the bridge. R_{10} is a precision rheostat calibrated in inductance values; R_{11} is also a precision rheostat, and is calibrated in *Q* values. C_2 is a standard capacitor which serves the purpose of a standard inductor, as explained in a previous chapter. When R_{11} and C_2 are connected in parallel, as in Fig. 11.6, the useful range of R_{11} is limited to storage factors (*Q* values) less than 10. The *Q* dial in a basic impedance bridge is typically calibrated from 0.1 to 10. We encounter somewhat less accuracy at the extreme ends of the

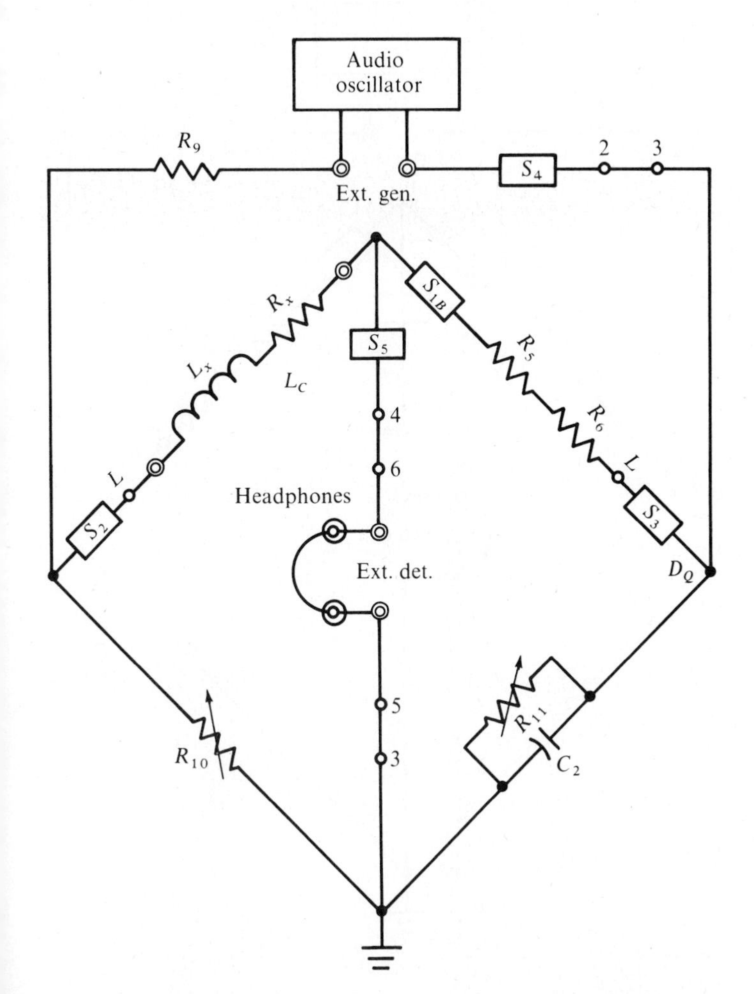

FIG. 11.6 *Maxwell bridge circuit for measuring inductance and storage factor.*

scale. Inductance values from 10 μH to 1 H are usually covered in this type of bridge.

Next, when switches S_2 and S_3 in Fig. 11.1 are set in the L and Q positions, a Hay-bridge configuration is provided, as shown in Fig. 11.7. This arrangement covers the same inductance range as before, but indicates Q values from 10 to 1000. That is, the Q-value indication of the Hay configurations starts where the Q scale of the Maxwell configuration ends. Since R_{10} and R_{12} are not in the same arm of the bridge, a "sliding balance" (or sliding null) is encountered, as explained in a previous chapter. This interaction of the balance

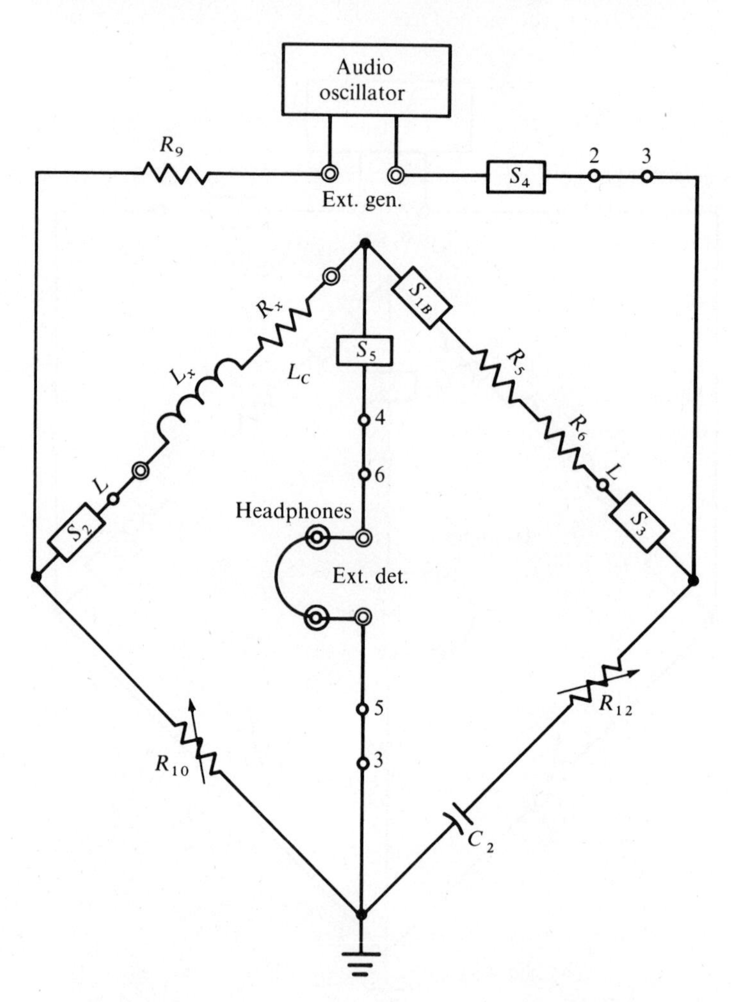

FIG. 11.7 *Hay-bridge circuit for measuring inductance and storage factor.*

controls is particularly prevalent when low Q inductance values are being measured with the Maxwell configuration (Fig. 11-6).

Although an Owen-bridge configuration provides operating convenience in checking low Q inductors, its comparatively high cost rules it out in the design of simple and economical impedance bridges. Note also that the use of headphones as a null detector provides manufacturing economy. Omission of an internal or built-in generator also lowers the cost of the basic bridge. It is instructive to consider the same measuring facilities discussed above, as designed into a somewhat more costly impedance bridge.

11.5 A MODERATELY ELABORATE IMPEDANCE BRIDGE

An impedance bridge with a built-in phase-shift oscillator is depicted in Fig. 11.8. The oscillator circuit is designed to operate in the range from 800 to 1200 Hz. A trimmer capacitor is provided in the phase-shift network so that the operating frequency can be adjusted to 1 kHz. The oscillator output is fed to a level control and thence to a buffer amplifier. This buffer prevents shift in the oscillator frequency caused by changes in load impedance. To obtain optimum power transfer, an impedance-matching transformer (bridge transformer) is employed between the buffer amplifier and the bridge circuit. Figure 11.9 shows a highly stable 1-kHz solid-state bridge oscillator.

The basic bridge arrangements in this example are conventional, as seen in Fig. 11.9. It follows from previous discussion that *CRL* readings will be independent of frequency, whereas readings of D and Q will depend upon the operating frequency. Thus, if the Ext. gen. terminals in Fig. 11.9 are used, a correction factor must be applied to D or Q readings at any frequency other than 1 kHz. With reference to Fig. 11.10, readings on the Q scale must be divided by the operating frequency in kilohertz. For example, if the bridge is operated at 2 kHz, a reading on the Q scale must be divided by 2. Or, if the bridge is operated at 0.5 kHz, a reading on the Q scale must be multiplied by 2. The Hay bridge configuration (Fig. 11.10) is used in conjunction with the Q scale.

Note that the function switch in Fig. 11.11 also provides for inductance measurement with the Maxwell-bridge con-

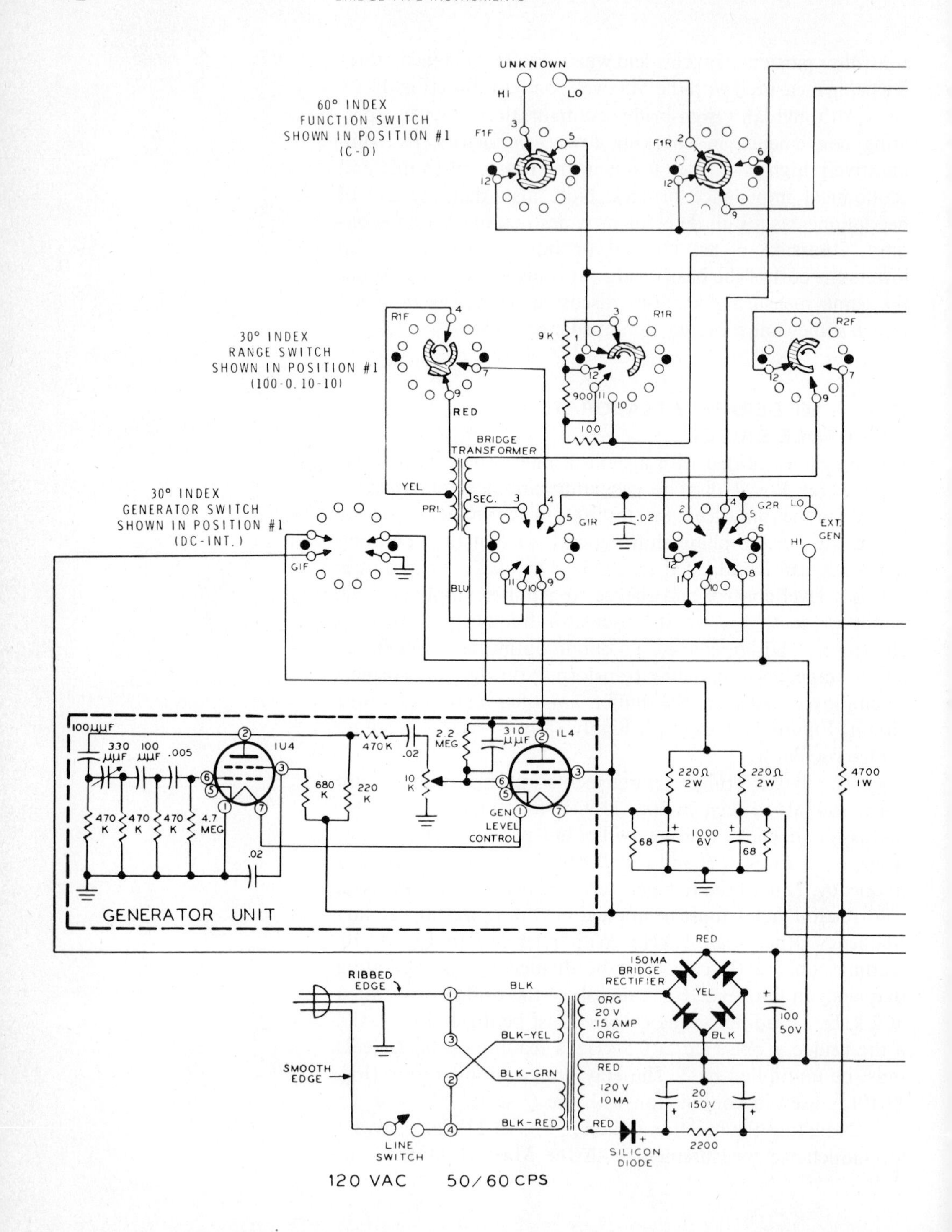
UNKNOWN
HI
LO
60° INDEX
FUNCTION SWITCH
SHOWN IN POSITION #1
(C-D)
F1F
F1R
30° INDEX
RANGE SWITCH
SHOWN IN POSITION #1
(100-0.10-10)
R1F
R1R
R2F
9K
900
100
RED
BRIDGE
TRANSFORMER
YEL
PRI.
SEC.
BLU
30° INDEX
GENERATOR SWITCH
SHOWN IN POSITION #1
(DC-INT.)
GIF
G1R
G2R
.02
LO
EXT.
GEN
HI
100μμF
330 μμF
100 μμF
.005
1U4
470 K
.02
2.2 MEG
310 μμF
1L4
680 K
220 K
10 K
470 K
4.7 MEG
GEN LEVEL CONTROL
220Ω 2W
4700 1W
68
1000 6V
GENERATOR UNIT
RED
150MA
BRIDGE
RECTIFIER
YEL
BLK
100 50V
RIBBED EDGE
BLK
ORG
20 V
.15 AMP
ORG
BLK-YEL
BLK-GRN
RED
120 V
10MA
20 150V
SMOOTH EDGE
BLK-RED
RED
SILICON DIODE
2200
LINE SWITCH
120 VAC 50/60 CPS

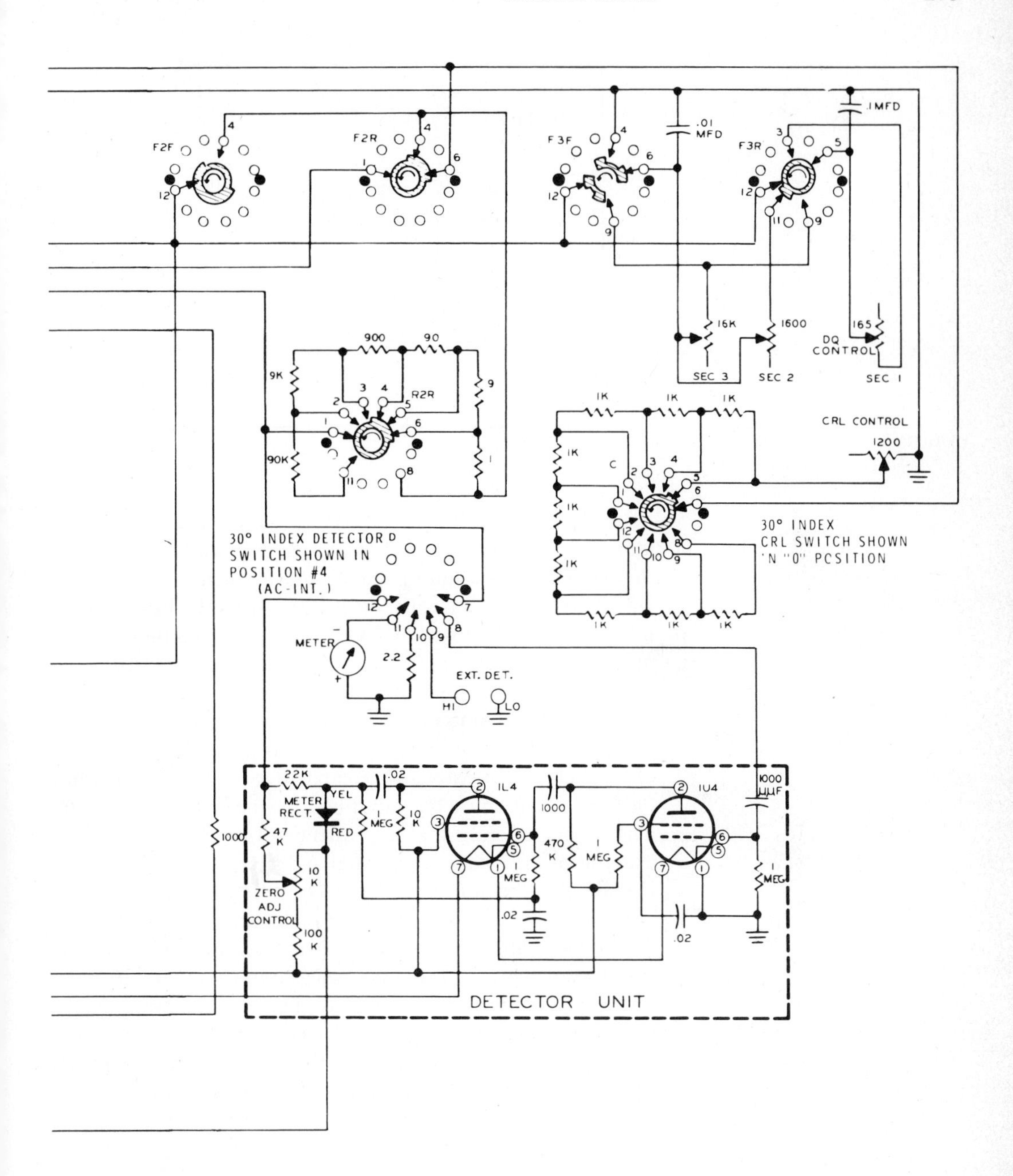

FIG. 11.8 *A moderately elaborate impedance-bridge circuit.* (*Courtesy, Heath Co.*)

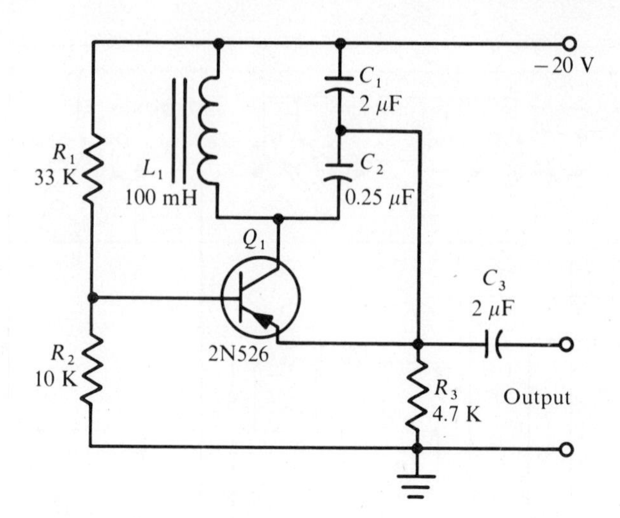

FIG. 11.9 *A stable 1-kHz transistor bridge oscillator.*

figuration shown in Fig. 11.10. Q values are accordingly indicated on another scale, called the DQ scale. Q values on this scale must be multiplied by the frequency in kilohertz, when operating at a frequency other than 1 kHz. For example, if the bridge is operated at 2 kHz, the Q reading on this scale must be multiplied by 2. Or, if the bridge is operated at 0.5 kHz, the scale reading must be divided by 2.

When the function switch is set to its C position, the arrangement shown in Fig. 11.10*b* is provided. This is a comparison, or resistive-ratio capacitive bridge. The reading on the D scale must be multiplied by the frequency in kilohertz, when operating at a frequency other than 1 kHz. For example, if the bridge is operated at 2 kHz, the reading on the D scale must be multiplied by 2. Or, if the bridge is operated at 0.5 kHz, the reading on the D scale must be divided by 2. To facilitate adjustment of the D-value control, the function switch provides for capacitance measurements with a D rheostat value of either 1600 Ω or 16 K. These are identified as the D and DQ switch settings, respectively. If the D value is comparatively small, it is preferable to operate with the 1600-Ω rheostat, because the control setting is less critical.

A 100–0–100 microampere meter is used in this example as a null indicator. To avoid possible damage to the meter before the bridge is adjusted to near-balance, a 2.2-Ω resistor is shunted across the meter. This shunt can be removed by throwing the detector switch to obtain maximum sensitivity in the final balance adjustments. High sensitivity is provided by a two-stage amplifier; output from this amplifier is rectified by a semiconductor rectifier and fed through a zero-adjust control to the meter. If preferred, a pair of headphones may be connected to the Ext. det. terminals and used instead of the meter for null indication.

11.6 BRIDGE ACCURACY RATINGS

Bridge measurements based on null methods are recognized as the most precise way to measure resistance, capacitance, and inductance values. In the first analysis, bridges can be classified into service and laboratory types. Most service-type bridges

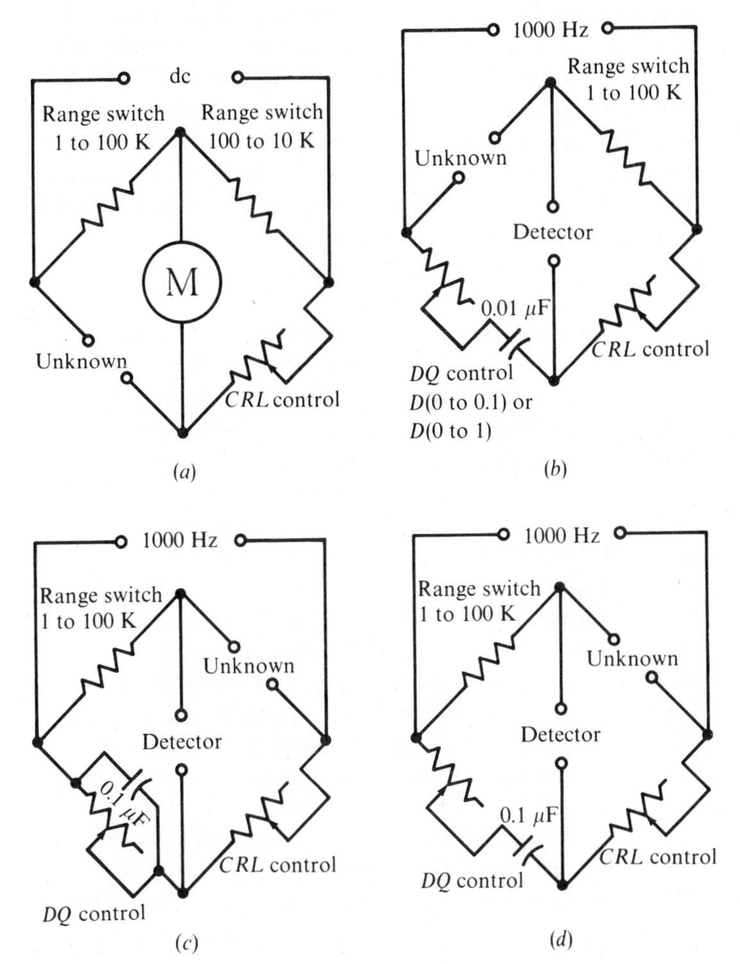

FIG. 11.10 *Basic bridge arrangements for the circuit in Fig. 11.8. (a) Resistance bridge (Wheatstone); (b) capacitance bridge (comparison); (c) Maxwell inductance bridge (Q of 0–10); (d) Hay inductance bridge (Q of 10–1000).*

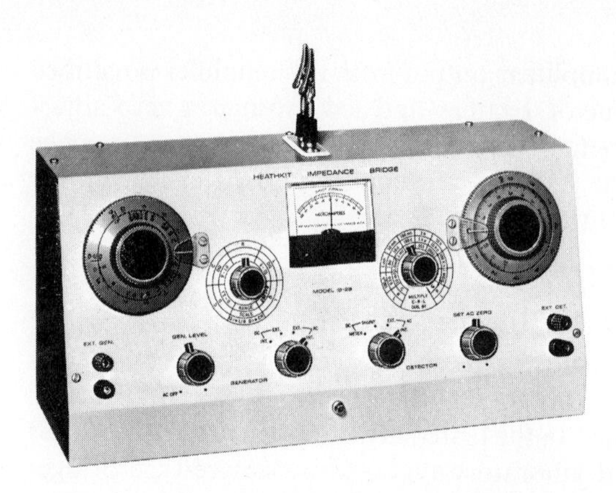

FIG. 11.11 *Front panel of an impedance bridge showing controls. (Courtesy, Heath Co.)*

have no accuracy rating, but have the advantage of convenience in operation. For example, a service-type capacitance bridge may be no more accurate than a service-type VOM, TVM, or signal generator. However, it is much more convenient to measure capacitance values directly with a bridge than to calculate them from measurements made in test circuits. Although service-type *CRL* bridges are less widely used than service-type capacitance bridges, the same consideration of convenience applies to measurement of inductance values in service-type operations.

Laboratory-type bridges are always rated for accuracy; they provide both precise measurements and operating convenience. Three categories of lab-type bridges may be noted, based on relative accuracy and cost. The lowest-priced bridges have a basic accuracy rating of ± 1 percent; medium-priced bridges are rated for ± 0.1 percent accuracy; the highest-priced bridges have an accuracy rating of ± 0.05 percent. As would be anticipated, a lab-type *CRL* bridge does not necessarily have the same accuracy rating on each function. For example, a bridge that is rated for ± 1 percent accuracy in measurement of resistance, capacitance, and inductance values, might be rated for ± 5 percent accuracy in measurement of Q or D values. Similarly, a bridge that is rated for ± 0.05 percent accuracy over most of its ranges might be rated for ± 0.2 percent accuracy on its lowest R and L ranges and on its highest C range.

Most laboratory-type bridges have a basic accuracy rating for 1-kHz operation. The same accuracy rating usually applies over the entire audio-frequency range from 20 Hz to 20 kHz. If this type of bridge is operated at frequencies above

20 kHz, its accuracy is derated, and the maximum operating frequency is typically 100 kHz. Another class of bridges is designed for operation at radio frequencies; for example, a typical rf bridge is rated for ± 2 percent accuracy at frequencies from 400 kHz to 50 MHz. This type of bridge can be operated at frequencies as low as 100 kHz, but is less accurate in this range. Similarly, operation at frequencies somewhat higher than 60 MHz is possible with reduced accuracy.

11.7 PRINCIPLES OF *Q* METERS

A conventional impedance bridge is limited in its ability to measure Q and L values because it is restricted to operation in the audio-frequency range. Although rf bridges are available, they are comparatively costly, and many engineers use Q meters instead. A high-quality Q meter is illustrated in Fig. 11.12. It is designed for operation at frequencies from 1 kHz to 300 MHz. It is instructive to briefly review some basic principles of Q values. The basic formula for the Q value of a coil is written:

$$Q = \frac{\omega L}{R} \tag{11.3}$$

where $\omega = 2\pi f$
L = inductance of coil
R = resistance of coil

If a coil is to be used in a radio receiver over a frequency range from 550 to 1500 kHz, a measure of its Q value at a

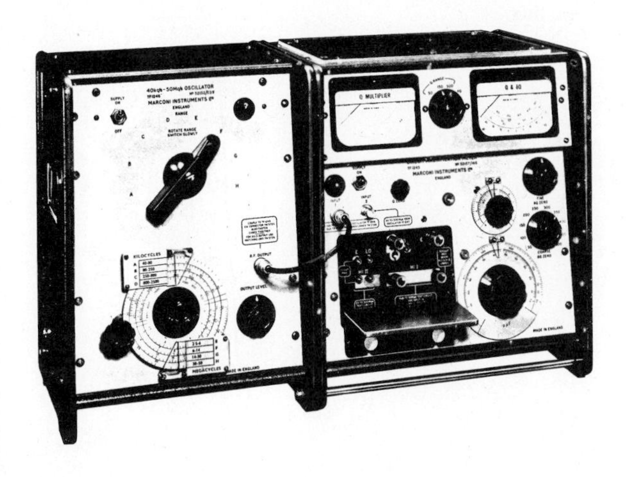

FIG. 11.12 *A high-quality Q meter. (Courtesy, Marconi Instruments)*

frequency of 1 kHz is almost meaningless. In other words, the ac resistance of a coil is a somewhat unpredictable function of frequency. Although the dc resistance of the winding is constant, the ac resistance is a combination of this dc resistance and the apparent resistance due to the *skin effect* depicted in Fig. 11.13*a*. At high frequencies, the current in a conductor becomes redistributed, with least concentration at the center and with maximum concentration at the surface of the conductor. This skin effect is a result of the nonuniform distribution of magnetic flux. That is, the center of the conductor is encircled by a maximum number of flux lines. In turn, the center of the conductor has a higher inductance value than at its surface.

Next, when a pair of conductors are adjacent to each other, a high-frequency current becomes further redistributed due to the *proximity effect*, as depicted in Fig. 11.13*b* and *c*. This increase in apparent resistance, due to restriction of the current channels, is caused by the action of magnetic flux from each conductor upon its adjoining conductor. Moreover, eddy-current losses and dielectric losses contribute to the total increase in apparent resistance, and these losses are also somewhat unpredictable functions of frequency. Therefore, it is not practical to measure the Q value of a coil at 1 kHz, and to attempt calculation of its Q value at 1 MHz, for example. Instead, it is essential to measure the Q value at a frequency of 1 MHz.

With the first analysis of Q-measurement problems, let us observe the basic configuration for a Q meter shown in

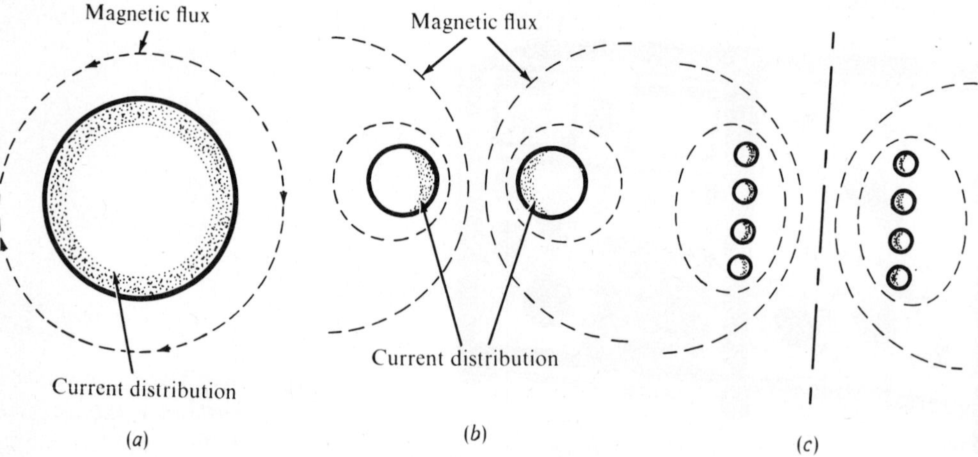

FIG. 11.13 *Skin effect in conductors. (a) Skin effect in a single conductor; (b) proximity and skin effect in a pair of adjacent conductors; (c) proximity and skin effect in a solenoid.*

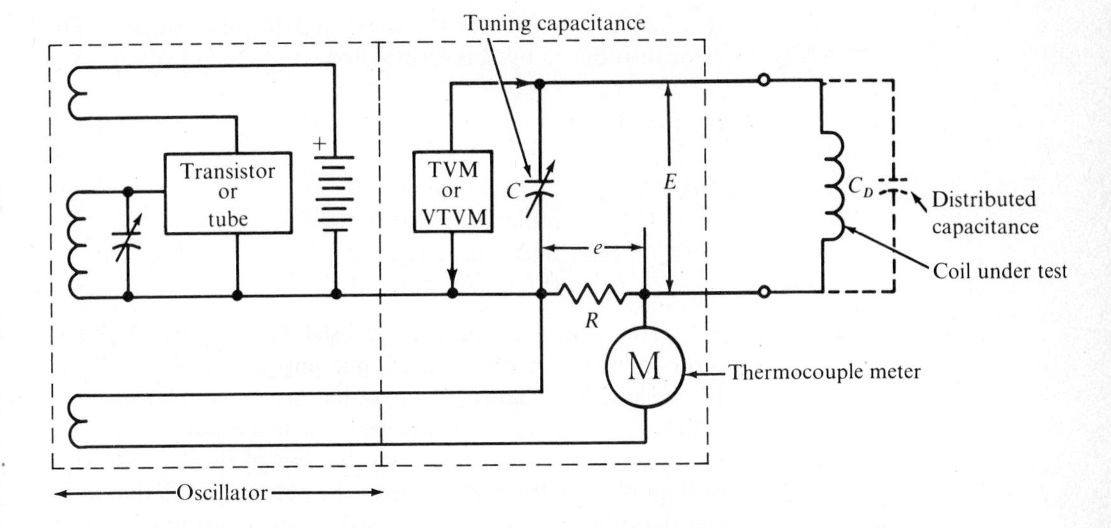

FIG. 11.14 *Simplified diagram of a Q meter.*

Fig. 11.14. A small rf voltage e is coupled into the tuned circuit by means of a resistor R which has a value of a few hundredths of an ohm. The resulting circuit disturbance is negligible. The coil under test is connected across a standard calibrated variable capacitor C. A TVM or VTVM is placed in shunt with C to indicate the voltage magnification of the circuit in response to the series-injected test voltage. After the oscillator is set to the desired frequency of test, the tuning capacitance is then adjusted for maximum indication of E. The Q value can then be read directly from the voltmeter scale. In other words, the voltage magnification of a series-resonant circuit is formulated:

$$Q = \frac{E}{e} \tag{11.4}$$

Note that the thermocouple meter in Fig. 11.14 serves to indicate the value of rf current flow through R. That is, the output from the oscillator is adjusted to obtain the specified value of reference-current flow through R. Thereby, the voltmeter becomes direct-reading in terms of Q values. This measured value of Q is commonly regarded as the Q of the coil. However, certain minor errors are involved in this point of view. First, although the value of R is very small, it reduces the measured Q value slightly, and this effect is greatest when measuring coils with very high Q values, such as ferrite-core antenna coils. Although a TVM has very high-input impedance, this factor also reduces the measured Q value slightly. The error introduced by R is formulated:

$$Q_{\text{true}} = Q_{\text{meas}} \frac{1}{1 + R/R_s} \tag{11.5}$$

where Q_{true} = actual Q value
Q_{meas} = indicated Q value
R = injection resistance
R_s = rf resistance of coil

The same formula can be used to calculate the small error incurred by the shunt input impedance of the TVM. If the shunt impedance is converted into its equivalent series components, the equivalent series resistance can be added to R to obtain a more accurate evaluation of the true Q value. Still another factor which causes the measured Q value to be less than the true Q value is based on the distributed capacitance of the coil under test. Any coil has a small value of capacitance between turns, which can be summed and represented as C_D in Fig. 11.14. This capacitance forms a capacitive voltage divider with respect to the tuning capacitor C. In turn, the TVM indicates the proportion of voltage that drops across C. If we know the value of the distributed capacitance, the associated correction can be made by means of the formula:

$$Q_{\text{true}} = Q_{\text{meas}}\, 1 + \frac{C_D}{C_T} \tag{11.6}$$

where C_D = distributed capacitance of coil
C_T = total shunt capacitance consisting of distributed capacitance, tuning capacitance, and input capacitance of TVM

A reasonably accurate measurement of the distributed capacitance of a coil can be made by means of the frequency-doubling method. The tuning capacitor is first fully meshed, and the oscillator tuned to resonance. We note the resonant frequency and the value of tuning capacitance. Next, the oscillator is set to twice this frequency, and the tuning capacitor again adjusted for resonance. We note the new value of tuning capacitance, and employ the formula:

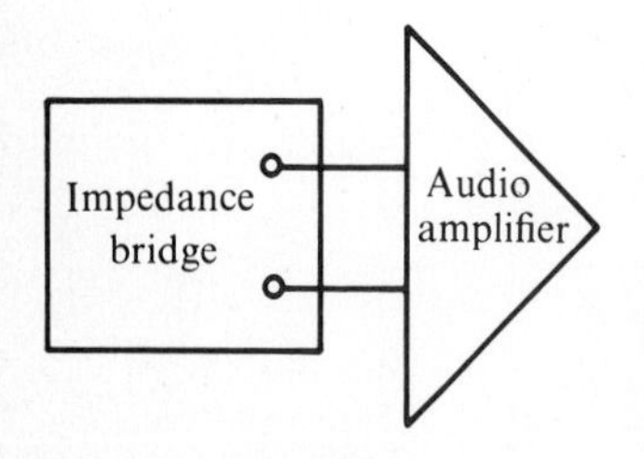

FIG. 11.15 *Measurement of the impedance of an amplifier.*

$$C_D = \frac{C_1 - 4C_2}{3} \tag{11.7}$$

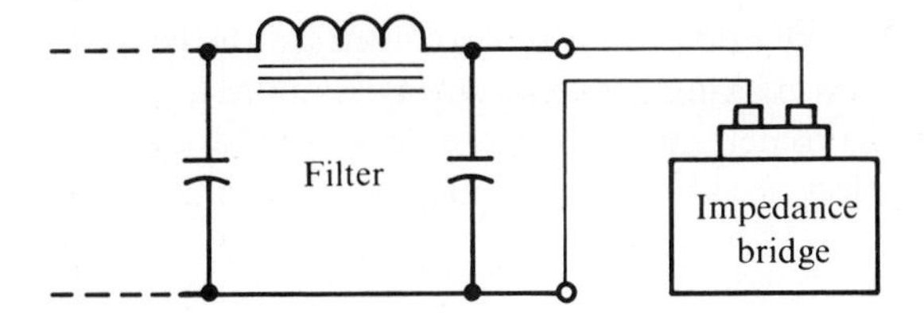

FIG. 11.16 *Measurement of the output impedance of a power-supply filter.*

where C_1 is the first value of tuning capacitance and C_2 is the second value.

11.8 APPLICATIONS

Among the conventional applications for an impedance bridge is the measurement of the input impedance of an audio amplifier, as depicted in Fig. 11.15. This check is useful to determine the optimum transducer impedance; that is, maximum power is transferred to the amplifier when the output impedance of the transducer matches the input impedance of the amplifier. Note that although the normal input impedance of an amplifier is predominantly resistive, it will also have a significant reactive component. This reactive component may be either inductive or capacitive. If the amplifier employs an input transformer, the input impedance will be inductive. On the other hand, if the amplifier employs *RC* input coupling, the input impedance will be capacitive. It is standard practice to measure the input impedance of an audio amplifier at 1 kHz.

Another useful application for an impedance bridge is shown in Fig. 11.16. The output impedance of a power supply is of concern when troubleshooting a "motorboating" symptom in an electronic system. In normal operation, the output impedance has a very low value of capacitive reactance. Since motorboating usually takes place at a low audio frequency, it is most informative to make impedance measurements at approximately the same frequency. This requires that an external generator be used with the bridge; an audio oscillator is the usual driving source that is employed. Note that some impedance bridges have null detectors that are frequency limited; if this difficulty is experienced, an external detector such as an oscilloscope can be utilized.

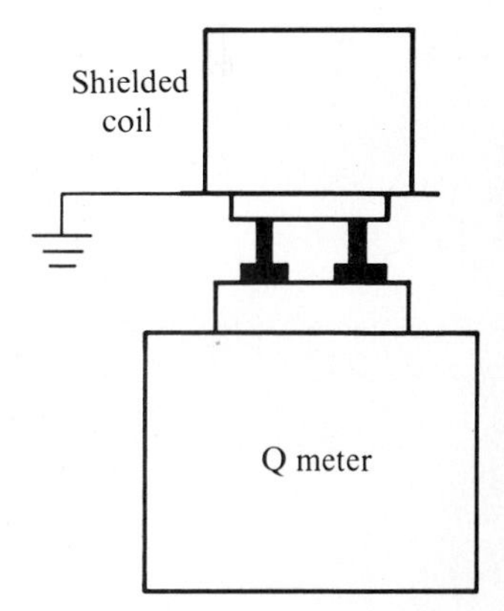

FIG. 11.17 *Measurement of the effect of a shield on the Q value of a coil.*

The Q meter is useful, not only to measure the Q value of a coil, but also to determine the reduction in Q value caused by various types and sizes of shields. With reference to Fig. 11.17, the shield of a coil should always be grounded (to the case of the Q meter) when making a measurement. Similarly, the Q meter can be used to determine the increase in Q value caused by the introduction of various ferromagnetic or ferrite cores.

QUESTIONS AND PROBLEMS

1. How do impedance bridges usually indicate the resistive component of a capacitor or inductor?
2. Why are impedance bridges usually designed to indicate D and Q values rather than the impedance of an inductor or a capacitor?
3. Why is it necessary to measure D and Q values at the frequency at which they are to be used?
4. What is the function of resistor R_9 in the schematic shown in Fig. 11.1?
5. Why is it desirable that the operator be able to drive an impedance bridge circuit with either ac or dc voltage?
6. Why isn't the bridge shown in Fig. 11.5 used to test electrolytic capacitors?
7. What are the results of measuring a leaky capacitor on an impedance bridge?
8. Why does a check for leakage in a high-voltage capacitor require a high value of test voltage?
9. What is the circuit configuration of the bridge circuit shown in Fig. 11.1 when the bridge is connected to measure low Q inductors?
10. Why is the circuit configuration of the bridge circuit shown in Fig. 11.1 changed to a Hay-bridge configuration when high Q inductors are measured?
11. Why isn't the Owen bridge circuit configuration used for checking low Q inductors?
12. What is the purpose of the buffer amplifier in the bridge circuit shown in Fig. 11.8?
13. What is the purpose of connecting a 2.2-Ω resistor across the meter movement in the resistance bridge shown in Fig. 11.9*a*?
14. What is the purpose of the Ext. det. terminals on a resistance-bridge circuit?
15. Describe a method that may be used to measure the internal resistance of a resistance bridge.

ELECTRONIC DISPLAY INSTRUMENTS

OSCILLOSCOPE PRINCIPLES AND THE BASIC OSCILLOSCOPE

12.1 BASIC MEASUREMENTS

An oscilloscope is an electronic display instrument that plots functions of electrical variables and parameters. For example, a voltage variation may be displayed as a function of time; again, a current variation may be displayed as a function of time. Or, a screen pattern might represent voltage variation as a function of current variation. Another type of pattern depicts the variation of one frequency as a function of frequency; in other applications, we display output voltage as a function of input voltage. In vectorscope applications, the magnitude of an ac voltage or current with its associated phase angle is displayed on the screen. Although the oscilloscope is basically a two-dimensional display device, we shall find that it can often process three variables effectively. For example, it is possible to display voltage variations on a time axis, as well as current variations on the same time axis. These and other basic applications for the oscilloscope shall be explained in this and following chapters.

12.2 THE CATHODE-RAY TUBE

A cathode-ray tube (crt) is an electron tube that is specially designed to display electrical data in terms of curves or patterns on a phosphorescent screen. The basic crt consists of four principal parts: an evacuated glass envelope, an electron gun for producing a stream of electrons, electrostatic or electromagnetic structures for deflecting the electron beam, and a phosphorescent screen for converting the kinetic energy of the electron beam into light energy. The working parts of a crt are enclosed in a glass envelope with a high vacuum to permit the emitted electrons to move freely from one end of the tube to the other.

Commercial cathode-ray tubes are fabricated in various sizes. Screen diameters range from 1 to 20 in. or more. The air pressure exerted on the screen of even a moderately sized crt is considerable; for example, the face of a 5-in. screen must withstand an air pressure of about 300 lb. That is, atmospheric pressure is approximately 16 psi, and a 5-in. circular screen has an area of 19.6 in.^{2}. The electron gun, which emits electrons and forms them into a beam, consists of a heater, a cathode, a control grid, a focusing anode, and an accelerating anode. We note these parts in Fig. 12.1, and their relative

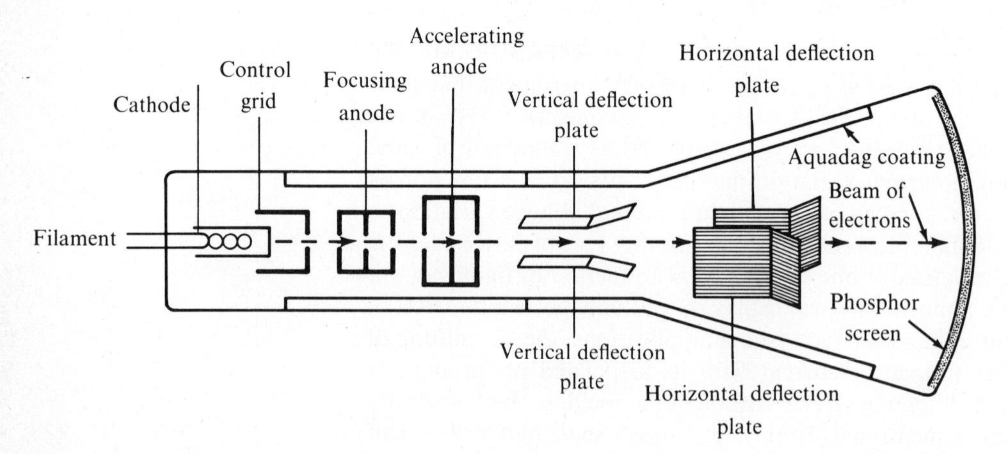

FIG. 12.1 *A typical cathode-ray tube (crt).*

positions inside the glass envelope. In smaller types of cathode-ray tubes, connections to the various electrodes are brought out through pins in the base of the tube. Larger types and medium-sized high-performance tubes operate at very high voltages, and these leads are usually brought out through the sides of the glass envelope.

In general-purpose tubes, the cathode is not connected to the heater; in special-purpose tubes, the cathode is connected to the heater internally. The heater element is energized in most applications by alternating current. It heats the cathode, which is basically a nickel cylinder. A layer of barium and strontium oxide is deposited on the end of the cathode cylinder to obtain high emission of electrons at moderate temperatures. We observe in Fig. 12.1 that the accelerating anode is a cylinder. A diaphragm is mounted inside this cylinder, and has a small hole at its center. Since the accelerating anode is operated at a high positive voltage, it attracts the electrons that are emitted by the cathode. Voltages applied to the accelerating anode range from 250 to 10,000 V or more.

Cathode-ray tubes in the 3- to 5-in. screen sizes operate from 1000 to 2000 V in simple oscilloscopes. However, a 5-in. crt operates at much higher voltages in the more sophisticated types of instruments. The accelerating voltage causes the electrons to travel at a high velocity, which is proportional to the square root of the potential difference. For example, a 1000-V potential difference will produce an electron velocity of about 10,000 mi/s.

The grid is cylindrical in shape and partially surrounds the cathode. It is biased negative with respect to the cathode and controls the density of the electron beam. If the negative bias is sufficient, the crt is cut off. This bias control is commonly called the *intensity control.* (See Fig. 12.2.) It is evident from Fig. 12.1 that some means must be provided for focusing the electron beam to a small spot on the screen. Note that the focusing anode and accelerating anode form an *electrostatic lens* which brings the electron beam into spot focus on the screen. This focusing action is depicted in Fig. 12.3. Let us briefly consider the lens action produced by the electrostatic field between the two electrodes.

We shall consider a typical example in which the focusing anode operates at a potential of 1200 V, and the accelerating anode at 2000 V. Thus, there is an 800-V potential difference which produces a strong electrostatic field. The operator can

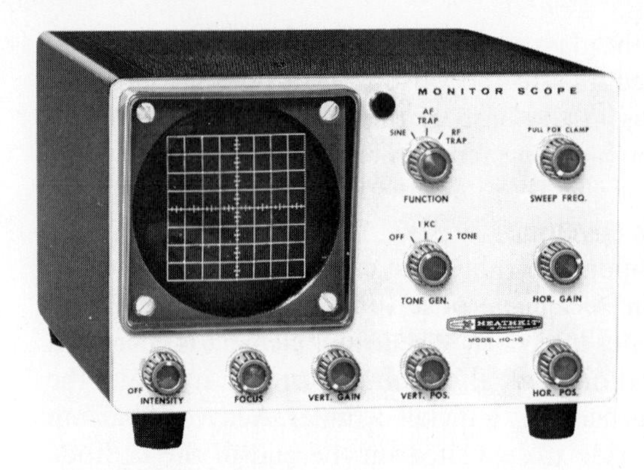

FIG. 12.2 *A simple type of oscilloscope.*

vary the field strength by changing the operating voltage on the focusing anode. An electron passing through these anodes has two forces acting upon it; the high accelerating voltage attracts the electron and speeds it up in a forward direction, and the electrostatic field between the two electrodes tends to deflect the electron. The end result is that all the electrons entering the lens area tend to come together at a point called the focal point.

Light is produced where the electrons strike the phosphorescent screen. The screen is a deposit of semitransparent phosphor substance; willemite (zinc silicate) is commonly used, and produces a green light output. Screen substances are rated for persistency, or the time that the screen continues to glow after the electron beam has passed a given point. Low-persistency screens used in cathode-ray tubes for oscilloscopes usually have a decay time of less than 0.1 s. Low persistency ensures that a changing pattern will not appear confused and overlapped.

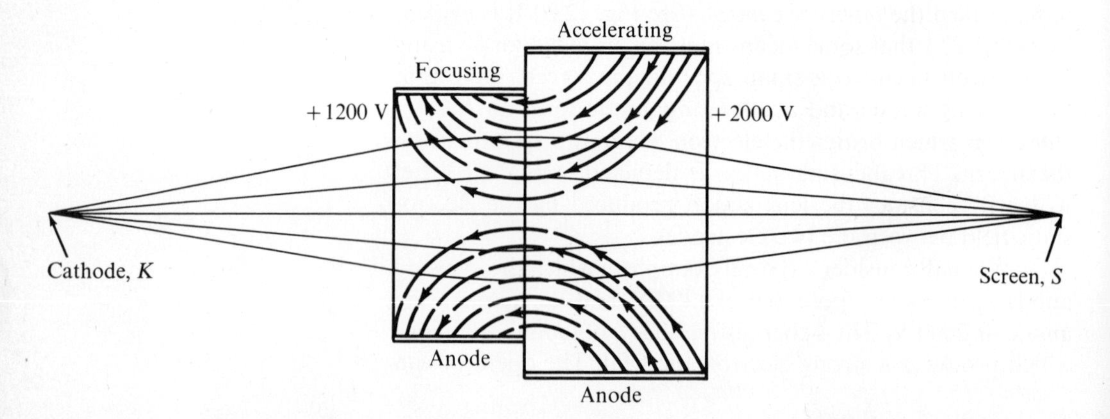

FIG. 12.3 *Focusing action on the electron beam.*

If electrons were not removed from the screen, a negative charge would build up until no more electrons could strike it. Similarly, if electrons were not returned to the cathode, a positive charge would build up on this electrode until no more electrons could be emitted. Since the electrons arrive at the screen with very high velocity, as noted previously, secondary electrons are emitted. In turn, these secondary electrons are collected by the Aquadag coating depicted in Fig. 12.1. This is a graphite coating that covers the inside surface of the envelope nearly up the screen. The coating is connected to the cathode, although it does not contact the screen. Thus, secondary electrons emitted by the screen when it is struck by the electron beam are attracted to the comparatively positive Aquadag coating, and the electric circuit is thereby completed or closed.

Two pairs of deflection plates are provided in the electrostatic type of crt, as seen in Fig. 12.1. The pairs of deflection plates are mounted at right angles to each other to provide electron-beam deflection along the vertical axis and along the horizontal axis of the screen. Figure 12.4 shows the basic principles involved in beam deflection. Electrons are deflected in accordance with Coulomb's law, with attractive force exerted by a positive plate and repulsive force exerted by a negative plate. Next, if the battery in Fig. 12.4*b* were replaced by an ac generator, we perceive that a horizontal line would be displayed on the screen. Or, if the battery in Fig. 12.4*c* were replaced by an ac generator, a vertical line would be displayed on the screen.

12.3 OPERATION OF THE CATHODE-RAY TUBE

The deflection sensitivity of a crt is defined as the amount of voltage that must be applied between a pair of deflection plates to obtain 1 in. of spot deflection. The deflection factor is defined as the distance (expressed as a fraction of an inch) that is produced when 1 V is applied between a pair of deflection plates. There is a tendency to use these two terms interchangeably, inasmuch as a deflection sensitivity is often expressed as the value of voltage required per inch of deflection. Note also that the amount of deflection may be stated in either inch units or centimeter units. Although the deflection sensitivity and deflection factor are defined in terms of dc (or equivalent peak-to-peak) volts, we often find that deflection ratings are expressed in rms volts.

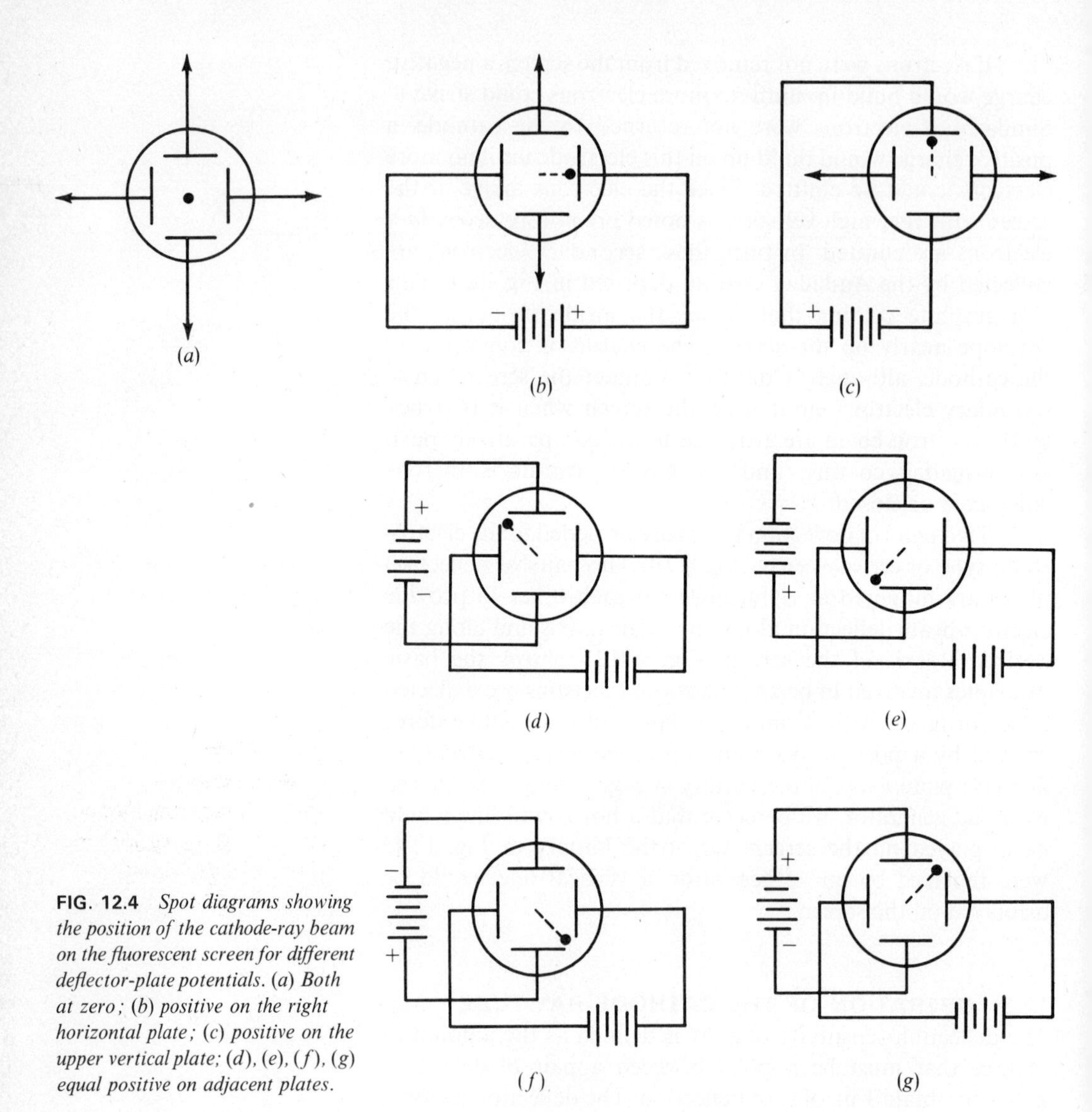

FIG. 12.4 *Spot diagrams showing the position of the cathode-ray beam on the fluorescent screen for different deflector-plate potentials. (a) Both at zero; (b) positive on the right horizontal plate; (c) positive on the upper vertical plate; (d), (e), (f), (g) equal positive on adjacent plates.*

When a crt is operated at comparatively high accelerating voltage, the electrons in the beam move faster and they remain for a shorter period of time between the deflection plates. The beam thus becomes more difficult to deflect, or, it becomes "stiffer." In a conventional crt, from 35 to 150 V dc of potential difference are required to deflect the spot 1 in. on the screen. A 5-in. crt with 1200 V applied to the accelerating anode typically requires a deflection voltage of more than 200 dc (or peak-to-peak) volts for full-screen deflection. Or, if 2000 V of accelerating potential are applied in this example, about 350 dc (or peak-to-peak) volts of deflection potential will be required for full-screen deflection.

With reference to Fig. 12.1, we note that the "vertical" deflection plates are farthest from the screen. These plates produce a vertical deflection of the spot on the screen. Since they are in a position where the electrons are moving a bit slower, a deflection voltage applied to these plates has somewhat greater effect on beam deflection. On the other hand, the "horizontal" deflection plates are nearest to the screen. Electrons move somewhat faster through these plates, and a deflection voltage applied to these plates has a lesser effect on beam deflection. In other words, it requires a somewhat greater value of horizontal deflection voltage to produce the same amount of beam deflection that is produced by a given value of vertical deflection voltage.

12.4 THE BASIC OSCILLOSCOPE

Figure 12.5 shows how the basic circuitry associated with a crt is commonly arranged. Somewhat elaborated arrangements for improved performance shall be considered subsequently. Voltages for operation of the crt are obtained from two power-supply sections, called the low-voltage power supply and the high-voltage power supply. In this example, +420 V and −780 V, respectively, are provided. The cathode voltage is fixed at −720 V. In turn, the grid-bias voltage is varied between −720 and −780 V by the 50-K intensity control. Note that the position of the contactor arm on the 250-K potentiometer, denoted the *focus* control, varies the voltage applied to the focusing anode. This value is adjustable between −320 and −590 V in this example.

Although it would be possible to operate the cathode at ground potential, and to operate the deflection plates and screen at a high positive voltage, this would impose a shock hazard to the operator. In practical scope design, the cathode is maintained at a comparatively high negative voltage, while the deflection plates and screen operate at $B+$ potential in most cases. Thus, the amount of electrostatic charge that can build up on the face of the crt is limited below a value that would impose a shock hazard. In some scopes, the deflection plates and screen are operated at an average potential of zero.

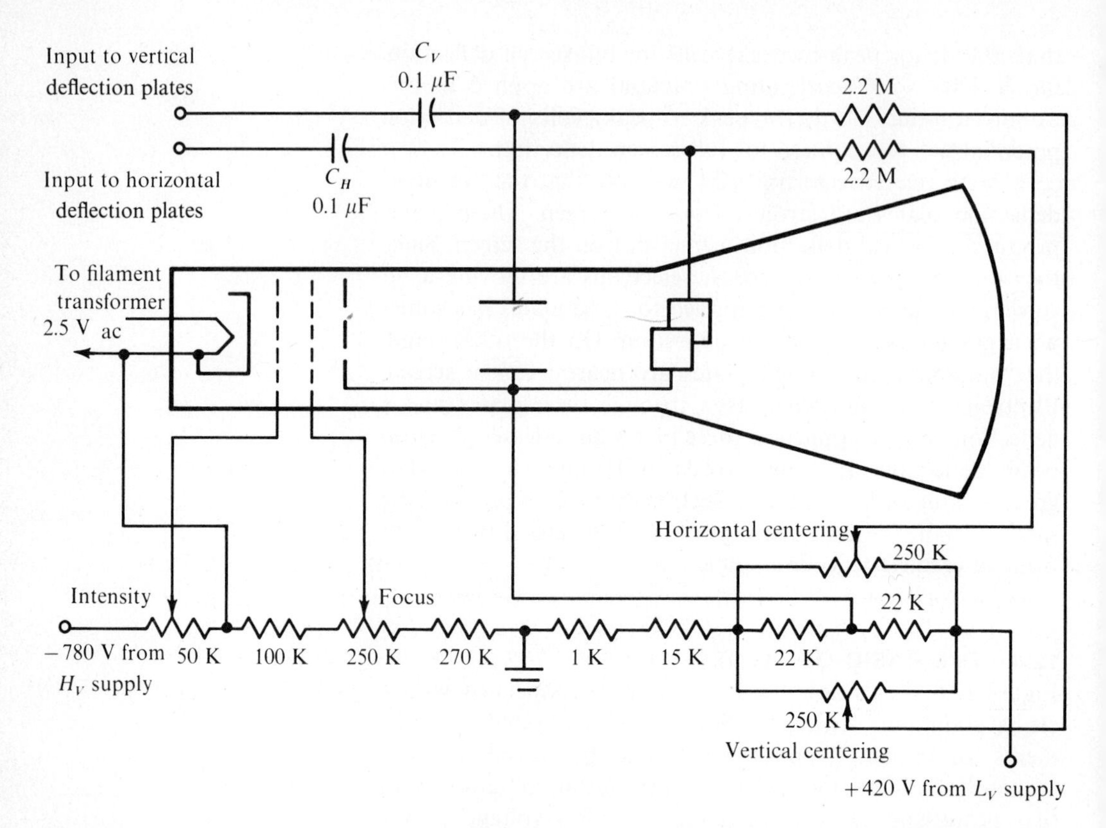

FIG. 12.5 *Typical control circuits for a crt.*

We observe in Fig. 12.5 that the lower vertical deflection plate and the left-hand horizontal deflection plate are connected internally to the accelerating anode and maintained at + 275 V. The contactor arm on the 250-K potentiometer is called a *vertical centering* control. It provides bias voltages between +120 and +415 V, which permits the operator to bring the spot to rest at any point on the vertical axis that might be chosen. The right-hand horizontal deflection plate is connected to a similar *horizontal centering* control, which permits the operator to bring the spot to rest at any point on the horizontal axis that might be chosen.

In this simple arrangement, the deflection plates are ac-coupled to the deflection voltage sources via C_V and C_H. The 2.2-M resistors permit the dc centering voltages to reach the deflection plates, while draining away a negligible amount of the applied ac deflection voltages. The time constant of each of these coupling circuits is 0.22 s; thus a 60-Hz deflection voltage is not excessively attenuated by C_V and C_H. The upper frequency limit extends well into the megahertz range. At usual brightness, the beam current is typically 250 μA. Thus, if the potential difference between the cathode and screen is 1000 V, the electron beam delivers 0.25 W of energy to the screen. Nevertheless, a spot or line appears rather bright, because the radiated light energy stems from a very small area.

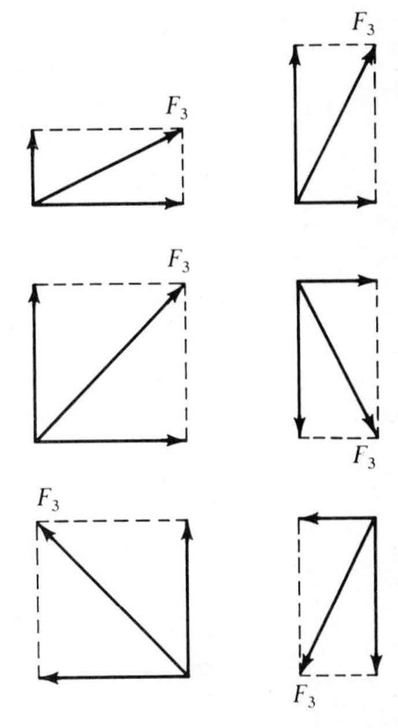

FIG. 12.6 *Vector representation of two forces at right angles.*

12.5 BASIC OSCILLOSCOPE OPERATION

It is instructive to consider the pattern characteristics that result from application of sine-wave voltages to the vertical and horizontal input terminals depicted in Fig. 12.5. That is, we shall analyze the basic bidirectional deflection processes. The simultaneous application of vertical and horizontal deflection forces causes a resultant motion of the electron beam, as depicted in Fig. 12.6. If the vertical and horizontal deflection voltages are in phase with each other, a diagonal line is displayed on the crt screen. The length and slope of the displayed line will depend on the relative values of the two deflection voltages, as is apparent from Fig. 12.6.

Next, let us suppose that the vertical and horizontal deflection voltages are 90° out of phase with each other. In such case, equal deflection voltages of the same frequency produce a circular pattern, as shown in Fig. 12.7. This is a

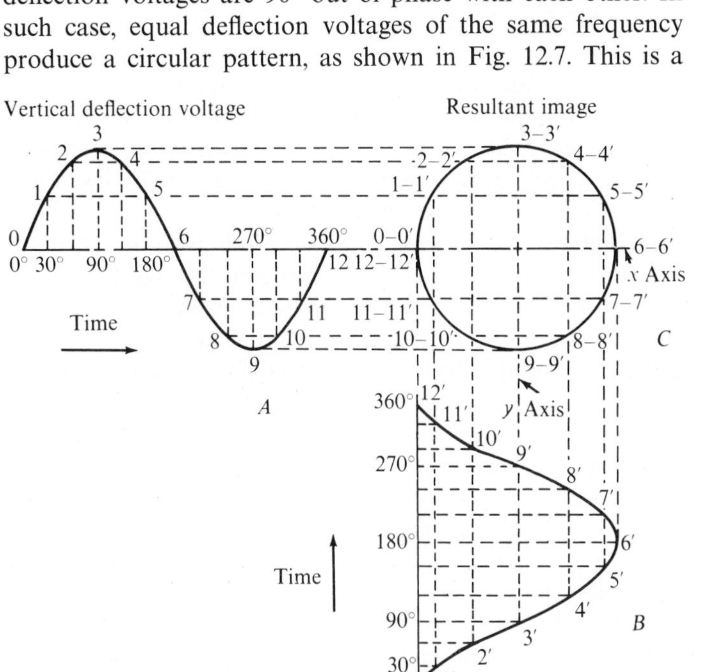

FIG. 12.7 *Production of a circular sweep by two sine waves of equal frequency and amplitude but 90° out of phase.*

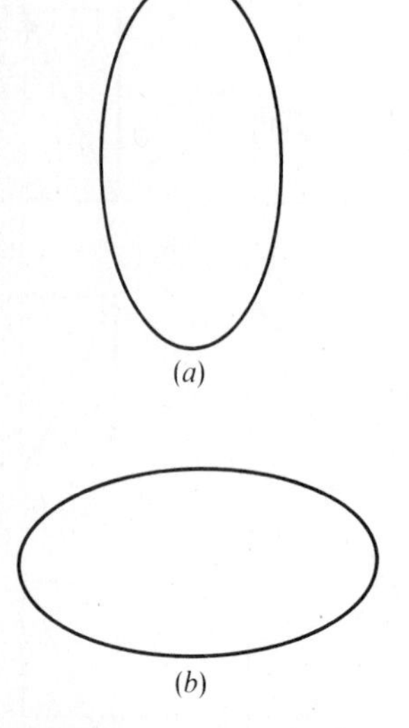

FIG. 12.8 *Horizontal and vertical deflection frequencies equal. (a) Vertical voltage greater than horizontal voltage; (b) horizontal voltage greater than vertical voltage.*

simple example of a *Lissajous figure*. If the vertical deflection voltage is unequal to the horizontal deflection voltage, a vertical or horizontal ellipse is displayed, as depicted in Fig. 12.8. The reason for the formation of an ellipse follows from a comparison of Fig. 12.8 with Fig. 12.7. In the limit (when one of the deflection voltages is zero), the pattern collapses into a vertical or a horizontal line.

Returning to the case of equal deflection voltages of the same frequency, let us observe the result of varying the phase between the two voltages. As seen in Fig. 12.10, progressive phase variation causes the pattern to vary from a straight diagonal line through diagonal ellipses of different eccentricities to a circle, and then through another series of ellipses to a straight diagonal line once more. It can be shown that the phase difference between the vertical and horizontal deflection voltage is formulated as illustrated in Fig. 12.9.

When voltages having different frequencies are applied to the vertical and horizontal deflection plates, crossover types of Lissajous figures are produced. For example, Fig. 12.11 shows the range of patterns that are formed by deflection voltages having a 1:2 frequency ratio, and with various phase differences between them. This is commonly called a "bow-tie" pattern. Note that if the deflection frequencies are almost but not quite in exact 1:2 ratio, the pattern will "writhe" on the screen as it goes through the progressive phases illustrated in Fig. 12.11. Lissajous patterns are widely used to check the calibration of a generator against a frequency standard at various harmonic intervals.

Figure 12.12 shows a Lissajous figure for deflection frequencies having a 2:3 ratio. To obtain the ratio of vertical and horizontal deflection frequencies from any Lissajous pattern, count the number of horizontal tangent points, and divide this number by the number of vertical tangent points. Note that this method is straightforward so long as the pattern contains visible crossovers. However, in the example of Fig. 12.11*c*, the crossover is masked by trace coincidence; that is, the two horizontal tangent points fall together in this special case of phase relations. The beginning scope technician must be on guard in such situations to avoid jumping to false conclusions.

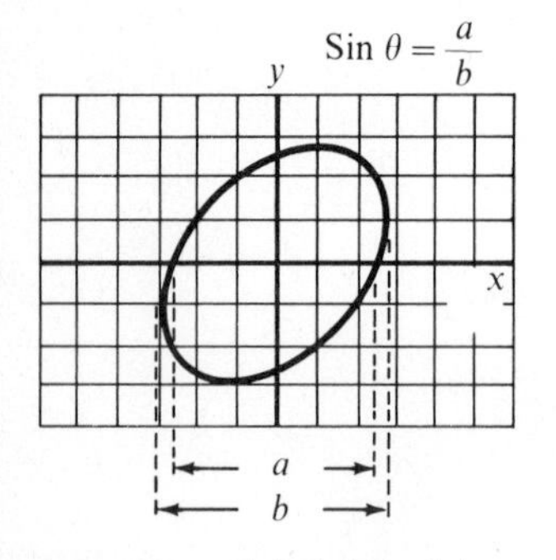

FIG. 12.9 *Calculation of phase angle from the ratio of dimensions a and b.*

In color television test procedures, we work with more complex forms of Lissajous figures which are called *vectorgrams*. This simplest form of vectorgram is developed by

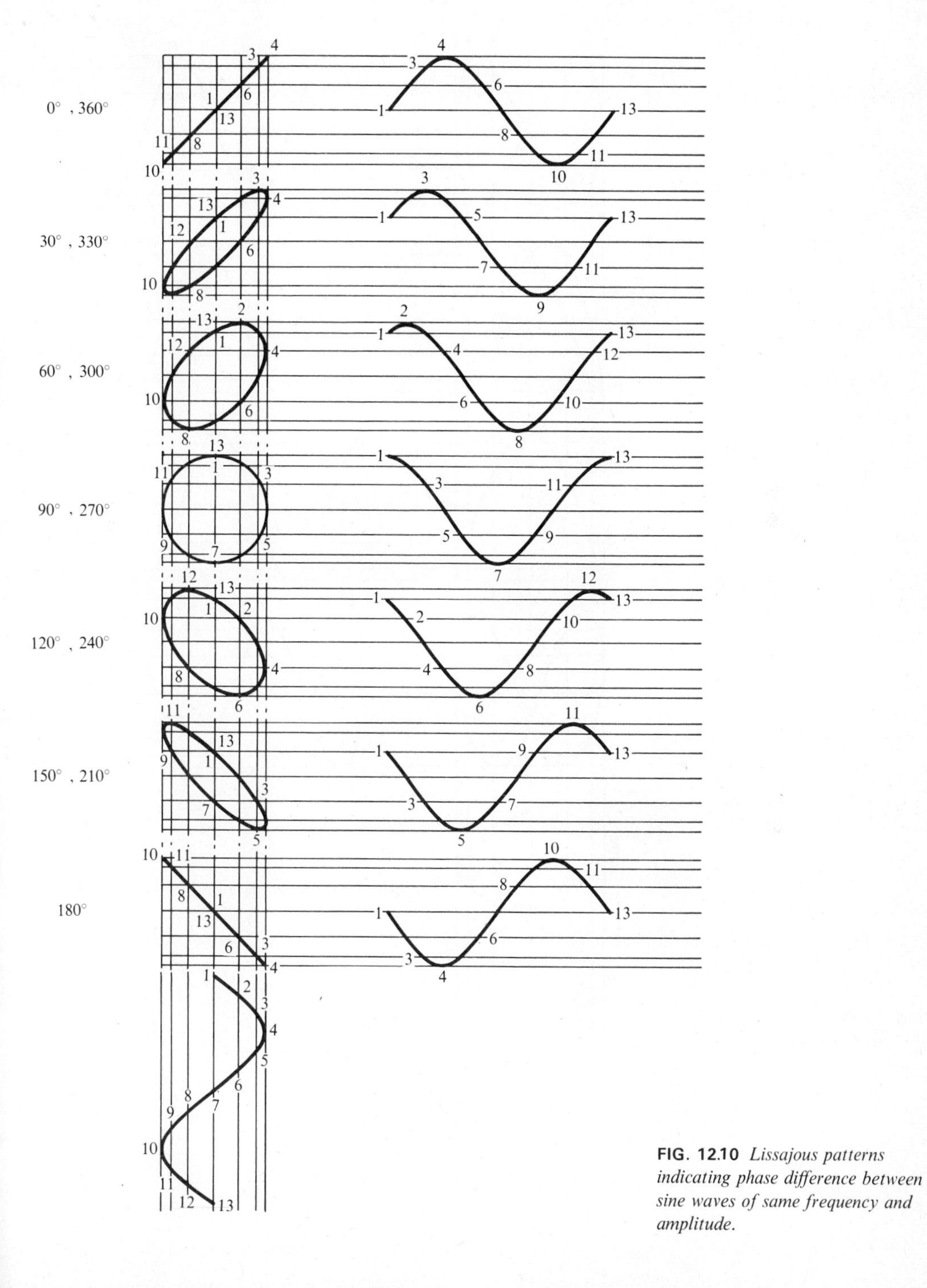

FIG. 12.10 *Lissajous patterns indicating phase difference between sine waves of same frequency and amplitude.*

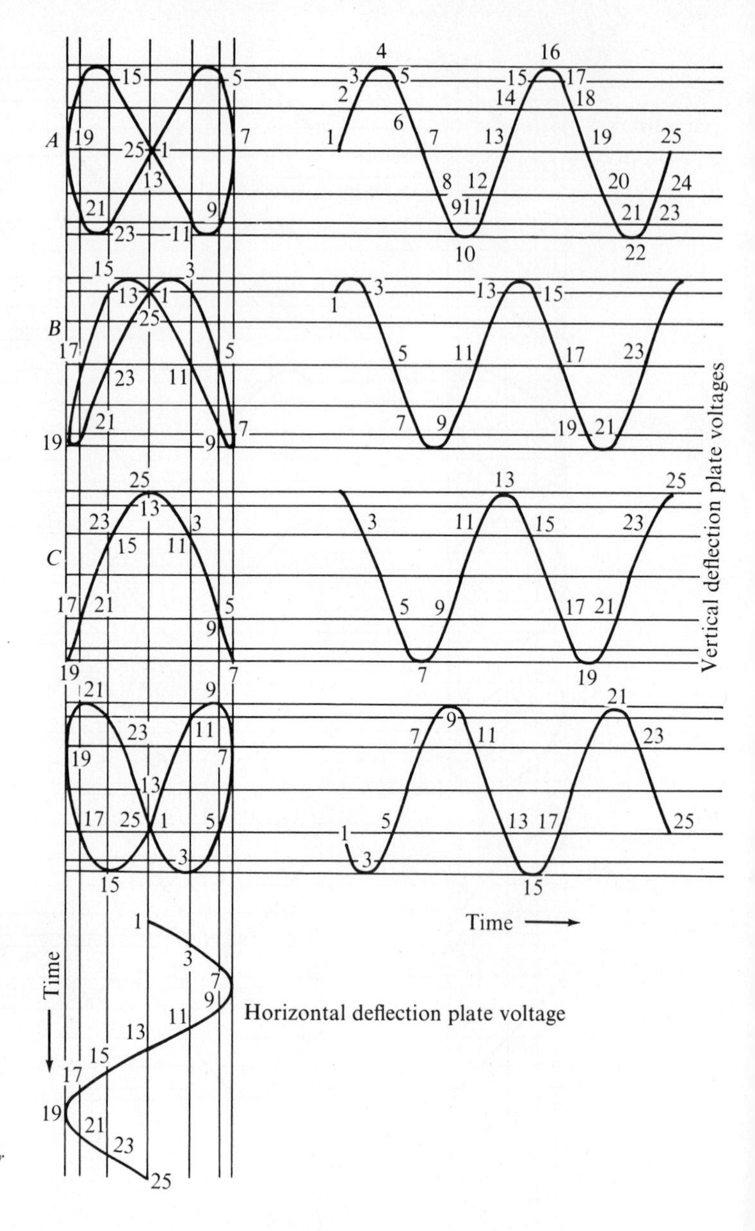

FIG. 12.11 *Lissajous patterns for 1:2 frequency ratio.*

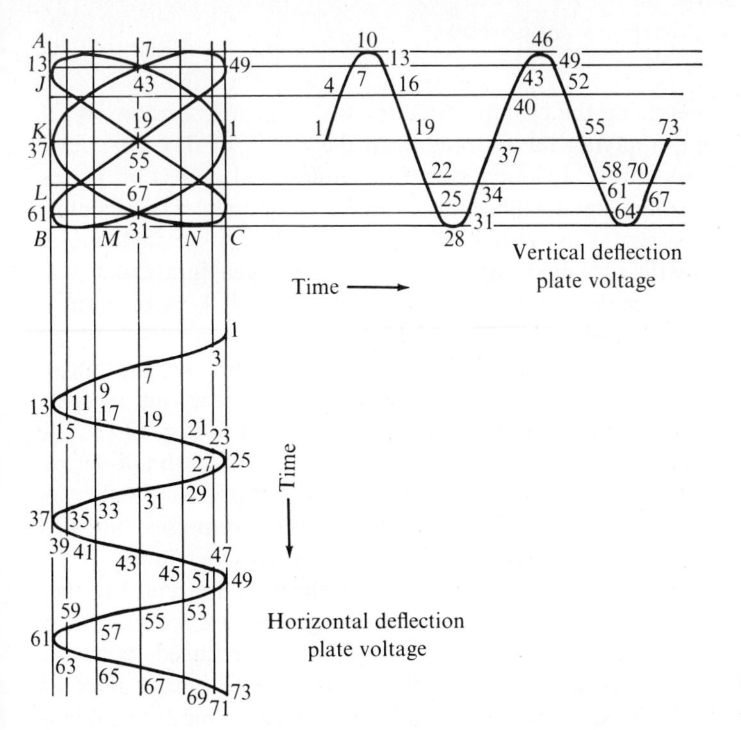

FIG. 12.12 *Lissajous patterns for 2:3 frequency ratio.*

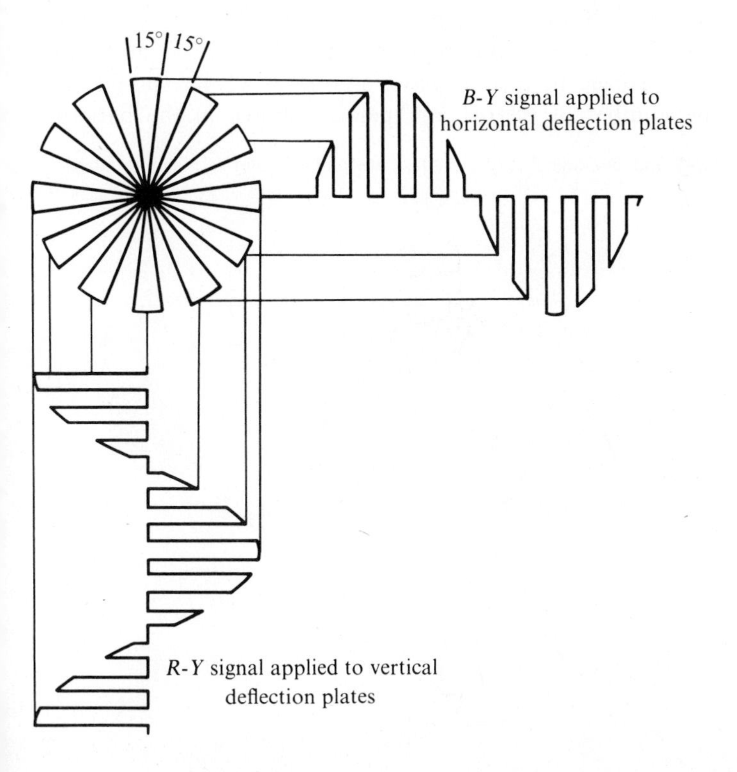

FIG. 12.13 *Development of an ideal vectorgram from keyed-rainbow R-Y and B-Y signals.*

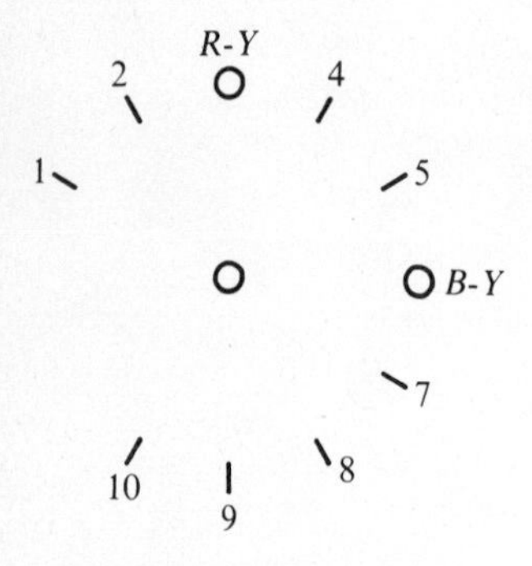

FIG. 12.14 *Typical graticule for a vectorscope.*

energizing the chroma circuitry with a keyed-rainbow signal, and applying the outputs from the *R-Y* and *B-Y* channels,[1] respectively, to the vertical and horizontal deflection plates. In somewhat idealized format, the development of the resulting vectorgram is visualized in Fig. 12.13. A vectorscope is usually provided with a transparent screen (graticule) with phase calibrations as depicted in Fig. 12.14. A vectorgram is viewed as shown in Fig. 12.15.

The alert reader will perceive that the vectorgram in Fig. 12.15 is depicted with an inner circular contour, whereas the vector patterns come to a central point in Fig. 12.13. The diameter of the inner circle is a measure of the *R-Y* and *B-Y* channel bandwidth. Let us see why this is so. The keyed-rainbow signal consists of a series of positive pulses, followed by a series of negative pulses, and so forth, as seen in Fig. 12.13. Now, if the pulse train has passed through a circuit that has subnormal bandwidth, partial integration of the waveform occurs. That is, the pulse amplitude is reduced, with the result that the baseline of the pulse train becomes curved as exemplified in Fig. 12.16. In turn, an inner boundary circle is formed in the vectorgram.

The vectorgrams represented in Figs. 12.13 and 12.16 are idealized in various ways, among which we should note the number of vectors, or "petals," shown in the patterns.

[1] *R-Y* and *B-Y* signals are explained in standard color television textbooks.

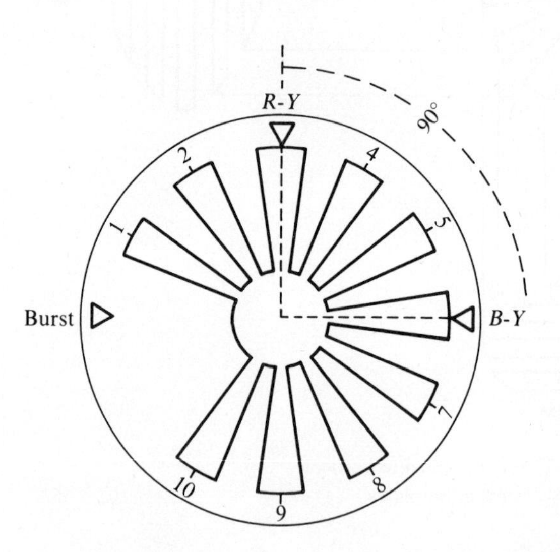

FIG. 12.15 *Keyed-rainbow signal segmental vector display. This is the ideal shape.*

FIG. 12.16 *Vectorgram produced by circuits having limited bandwidth.*

Although the vectors are normally 30° apart, we usually observe 10 vectors in practice, instead of 12. In other words, Fig. 12.15 is representative of actual practice. A keyed-rainbow generator normally supplies 11-pulse trains, separated by horizontal sync pulses. The eleventh pulse in each train is normally blanked in the receiver circuitry, as depicted in Fig. 12.17. The foregoing vectorgrams have been idealized in certain other respects, also, in the interest of clarity.

Now, let us consider the result of unequal *R-Y* and *B-Y* waveform voltages—this is the rule rather than the exception in actual practice. The *R-Y* voltage is usually somewhat greater than the *B-Y* voltage, producing an elliptical envelope as shown in Fig. 12.18. The phase angle between the vertical and horizontal deflection voltages is 90° in Fig. 12.18, as in the previous examples. Although successive vector petals are separated by 30°, this does not *seem* to be true. The reason for this discrepancy is seen in Fig. 12.19; compression of the horizontal deflection interval makes some angles appear to be greater than 30°, and others appear to be less than 30°.

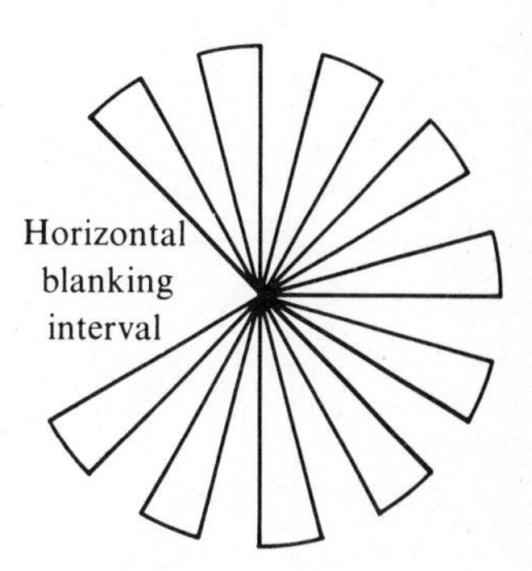

FIG. 12.17 *Diagram showing the normally blanked-out sector of the vectorgram.*

When the demodulation angle is incorrect, the so-called $R\text{-}Y$ and $B\text{-}Y$ voltages then have a phase difference other than 90°. Figure 12.20 depicts the result of a phase difference less than 90°. As would be expected, an inclined elliptical outline is displayed. Although the successive pulses in the deflection signals are separated by 30° in this example, the angles between

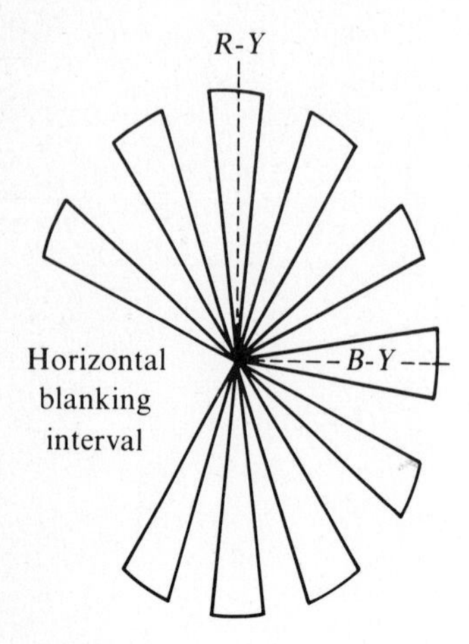

FIG. 12.18 *Diagram showing R-Y voltage greater than B-Y voltage.*

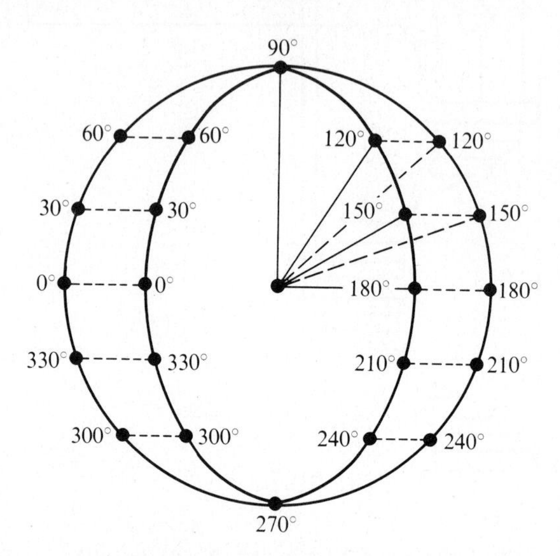

FIG. 12.19 *Apparent change in angle values due to unequal deflection voltages.*

the vector petals vary from less than 30° to more than 30° as we proceed around the circumference. Only the more basic aspects of vectorgram analysis can be presented here. Interested students are referred to specialized texts on the subject of using scopes in color television servicing.

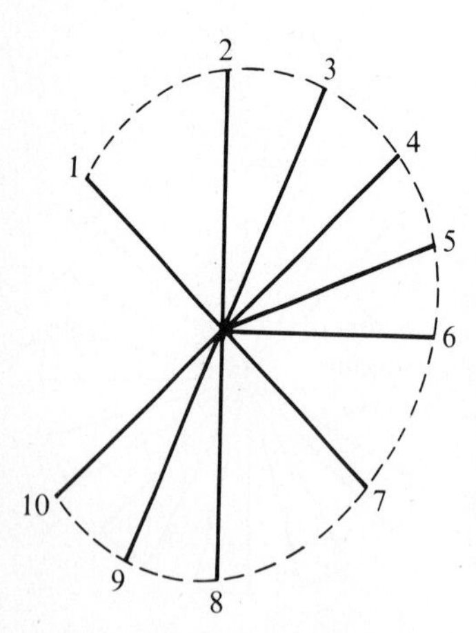

FIG. 12.20 *Basic vectorgram characteristics when R-Y and B-Y voltages have less than 90° phase difference.*

12.6 APPLICATIONS

One of the basic applications for the oscilloscope is phase-angle measurement in electrical power work. For example, the phase angle of an induction motor can be measured as depicted in Fig. 12.21. The voltage applied to the motor is coupled to the vertical deflection plates in the crt, and the current drawn by the motor is coupled to the horizontal deflection plates in the crt. Note that the current is actually applied by means of the voltage drop across a series resistor R; since voltage and current are in phase for a resistance, we thereby effectively apply the current phase to the horizontal deflection plates. In turn, an elliptical pattern is observed on the crt screen. This pattern is evaluated as shown in Fig. 12.22;

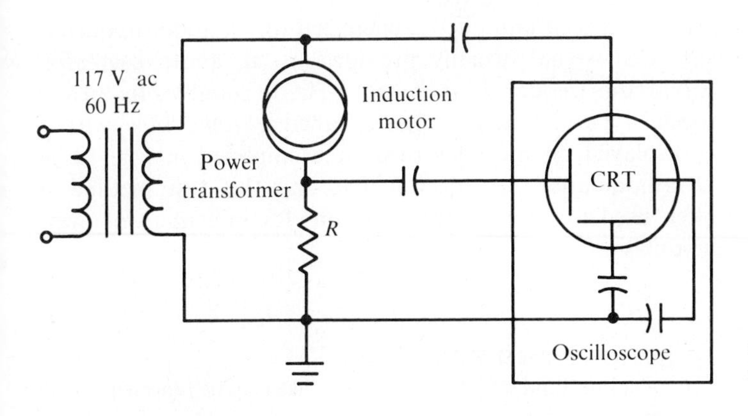

FIG. 12.21 *Measurement of phase angle for an induction motor.*

the phase angle between current and voltage is equal to the arcsin of the ratio A/B.

Another basic application for the oscilloscope is modulation percentage measurement in amplitude-modulated radio transmitters. For example, the test setup shown in Fig. 12.23 can be utilized. The modulated rf signal is coupled to the vertical deflection plates of the crt, and the audio modulating voltage is coupled to the horizontal deflection plates of the

$$\theta = \sin^{-1}\frac{A}{B}$$

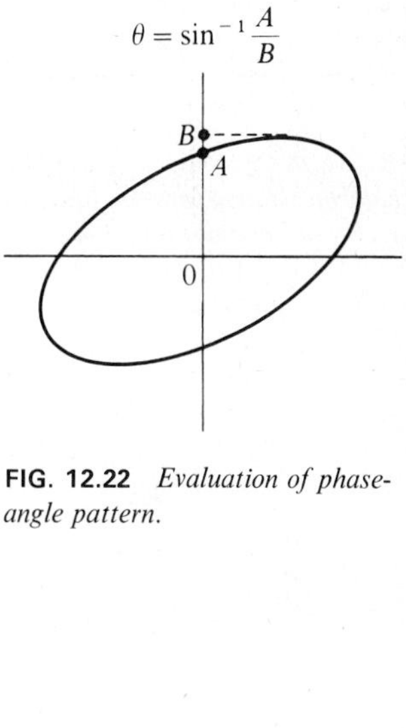

FIG. 12.22 *Evaluation of phase-angle pattern.*

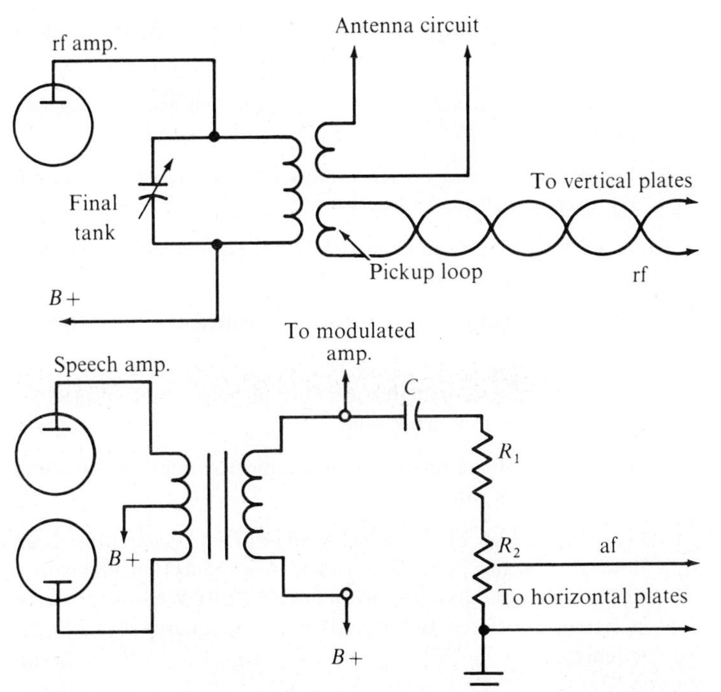

FIG. 12.23 *Measurement of percentage of modulation.*

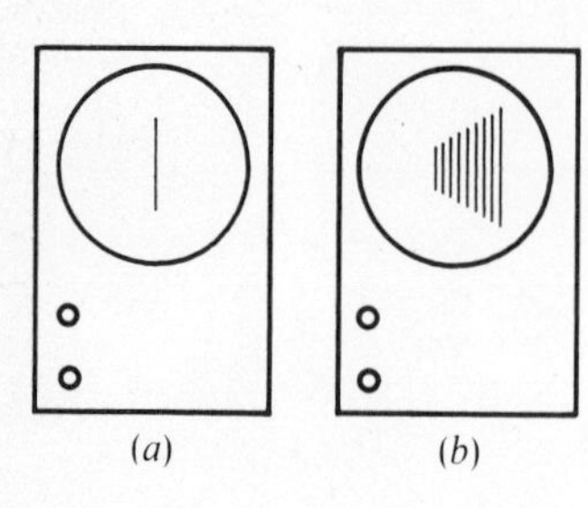

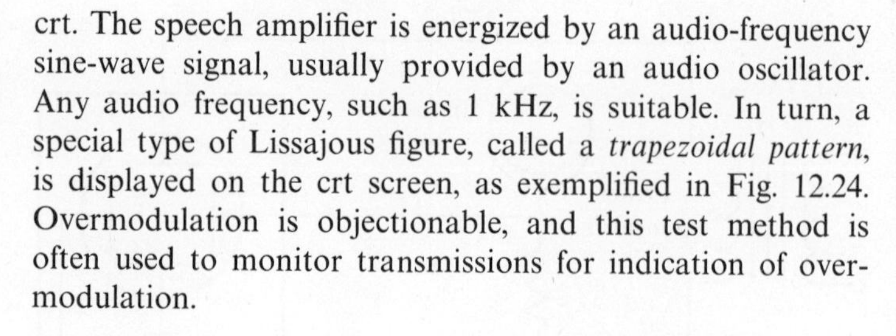

crt. The speech amplifier is energized by an audio-frequency sine-wave signal, usually provided by an audio oscillator. Any audio frequency, such as 1 kHz, is suitable. In turn, a special type of Lissajous figure, called a *trapezoidal pattern*, is displayed on the crt screen, as exemplified in Fig. 12.24. Overmodulation is objectionable, and this test method is often used to monitor transmissions for indication of overmodulation.

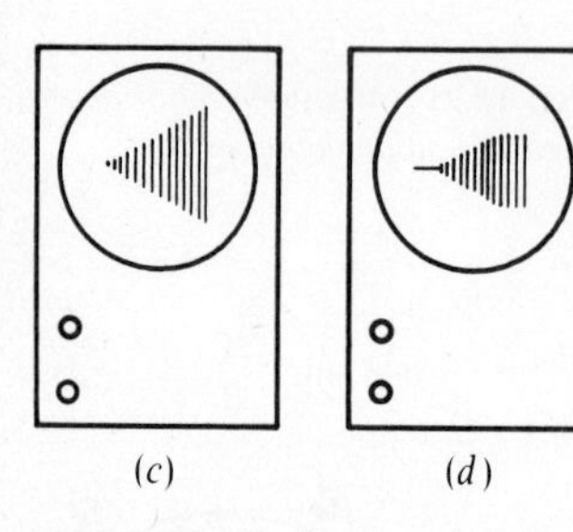

FIG. 12.24 *Evaluation of trapezoidal modulation patterns. (a) Zero percent modulation; (b) 50-percent modulation; (c) 100-percent modulation; (d) overmodulation.*

QUESTIONS AND PROBLEMS

1. List four fundamental parameters that may be represented by an oscilloscope.
2. Sketch a cathode-ray tube and label the most important parts.
3. What is the purpose of the accelerating voltage in a crt?
4. Explain the purpose of the *intensity control* and its operation on an oscilloscope.
5. How is the electron beam focused into a spot on the screen of a crt?
6. Explain the term *persistency* as it applies to a crt.
7. What is the purpose of the Aquadag coating along the inside of a crt envelope?
8. Define the term *deflection sensitivity* as it is used to describe the rating of a crt.
9. If a crt has a deflection sensitivity of 100 V/cm, what voltage is required to deflect the beam 4 in.?
10. In a crt, which pair of deflection plates has the least sensitivity? Explain your answer.
11. Why is the cathode of a crt usually operated at a high negative voltage?
12. What is the approximate value of the power that an electron beam imparts to a crt screen?
13. What two voltage relationships are most commonly measured with Lissajous patterns?
14. The application of the signal voltages from two points in an amplifier to the horizontal and vertical deflection plates of a crt results in the pattern shown in Fig. 12.10. What is the phase relationship of the two signals if $a = 4$ squares and $b = 2.5$ squares?

15. The input and output signals of an amplifier form a pattern such as depicted in Fig. 12.10. What is their phase relation if $a = 2$ squares and $b = 3.2$ squares?

16. What name is applied to the Lissajous patterns that are used to evaluate test procedures in color television?

17. What is the purpose of the vectorgram type of display on a crt?

18. What is the shape of the Lissajous pattern seen on a crt when the signal applied to the vertical deflection plates is 90° out of phase and of greater amplitude as compared to the signal applied to the horizontal deflection plates?

13

GENERAL-PURPOSE OSCILLOSCOPES

13.1 TIME-BASE REQUIREMENTS

A general-purpose oscilloscope consists of the basic oscilloscope discussed in the previous chapter, with provision for a vertical deflection amplifier, a horizontal deflection amplifier, and a time base. The plan of a general-purpose scope is depicted in Fig. 13.1. Deflection amplifiers provide indication sensitivity and permit tests in low-level circuits. For example, a vertical amplifier may produce 1 in. of beam deflection for a 0.02-V input signal. A time base permits the operator to display voltage or current variations in time. This is a very useful facility, because the majority of our formulas employed in ac-circuit analysis are expressed as functions of time.

We will recall that the basic formula for a sinusoidal voltage is written:

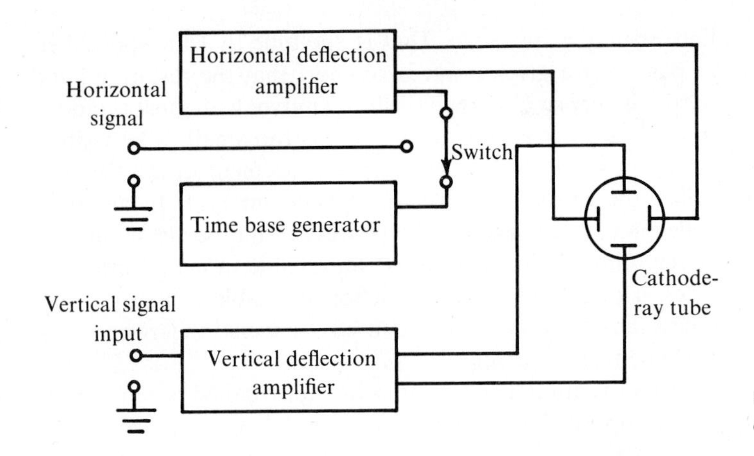

FIG. 13.1 *Block diagram of an oscilloscope.*

$$e = E \sin \omega t \tag{13.1}$$

Again, the basic formula for an exponential voltage is written:

$$e = \epsilon^{-bt} \tag{13.2}$$

From these elementary examples, it is apparent that a large proportion of general-scope applications will require a time base that produces a horizontal movement of the electron beam at a constant velocity, so that successive horizontal intervals are proportional to elapsed time. The horizontal axis is commonly chosen to represent time, in accordance with our convention of choosing the abscissa of a graph to represent time.

13.2 PRINCIPLES OF TIME BASES

The most widely used time bases generate an output voltage which is used to move the beam across the crt screen in a straight horizontal line, and then to return the beam quickly back to its starting point. The comparatively slow movement of the spot from left to right will appear as a solid line, provided that its rate of motion exceeds the threshold of persistence of vision. Below this threshold limit, the observer does not perceive a line, but instead perceives a moving spot. This left-to-right, or forward, movement of the beam is called the *trace* interval, or the forward trace.

On the other hand, the comparatively rapid movement of the spot from right to left will appear as a thin and dim line, or may be invisible. That is, the time that the spot takes to pass over a given point is so short that the resulting light output is very small. When the beam moves with great rapidity, the light output at any given point is too small to be visible. This rapid right-to-left, or reverse, movement of the beam is called the *retrace* interval, or flyback interval. In an ideal deflection system, the retrace process would be accomplished in zero time. Of course, this is impossible in actual practice. Thus, a retrace line may be definitely visible in many cases, unless auxiliary circuits are used to *blank* the retrace, as explained subsequently.

Conventional deflection action (also called sweep action) is accomplished by applying a *sawtooth* voltage across the horizontal deflection plates. A sawtooth waveform rises at a uniform rate to a peak value, and then quickly drops back to zero. As depicted in Fig. 13.1, the sawtooth waveform produced

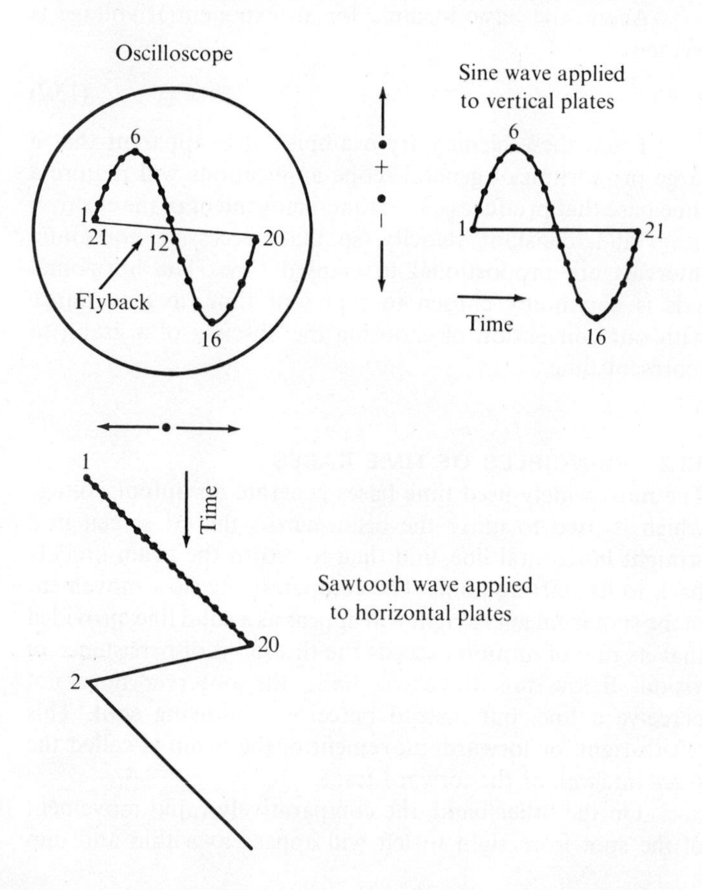

FIG. 13.2 *Development of a sine-wave pattern on the face of a crt.*

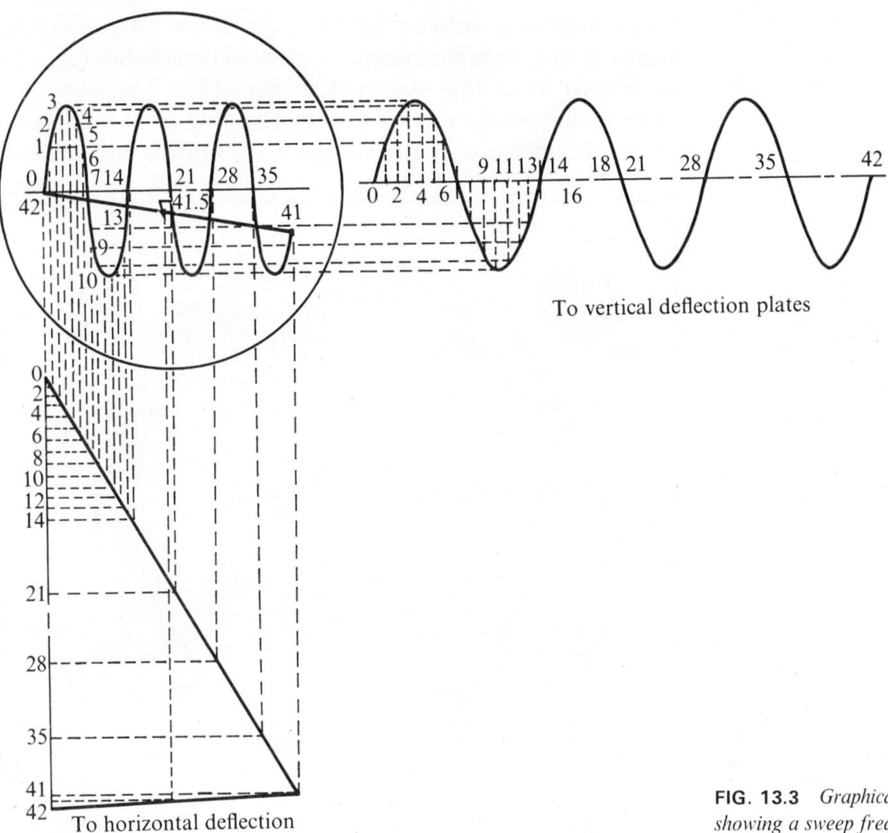

FIG. 13.3 *Graphical representation showing a sweep frequency of one-third the signal frequency.*

by the time base generator is stepped up through the horizontal deflection amplifier before it is applied to the crt. Figure 13.2 shows how a sine waveform is displayed on the crt screen when a sinusoidal voltage is applied to the vertical deflection plates, and a sawtooth voltage is applied to the horizontal deflection plates. The alert reader will perceive that the period of the sawtooth waveform in Fig. 13.2 is the same as the period of the sinusoidal waveform, and that a small portion of the sine-wave pattern is lost on flyback.

Next, let us consider the display process that results when the time-base period is three times that of the vertical input signal. In such case, the displayed pattern is developed as shown in Fig. 13.3. Note that almost three cycles of the vertical input waveform are displayed; as before, a small portion of a cycle is lost on flyback. Because the spot must travel a greater distance to trace out three cycles than to trace out one cycle per sweep, the pattern depicted in Fig. 13.3 is dimmer than that depicted in Fig. 13.2. Therefore, the operator adjusts the intensity control (see Fig. 13.4) as required to obtain adequate pattern brightness. On the other hand,

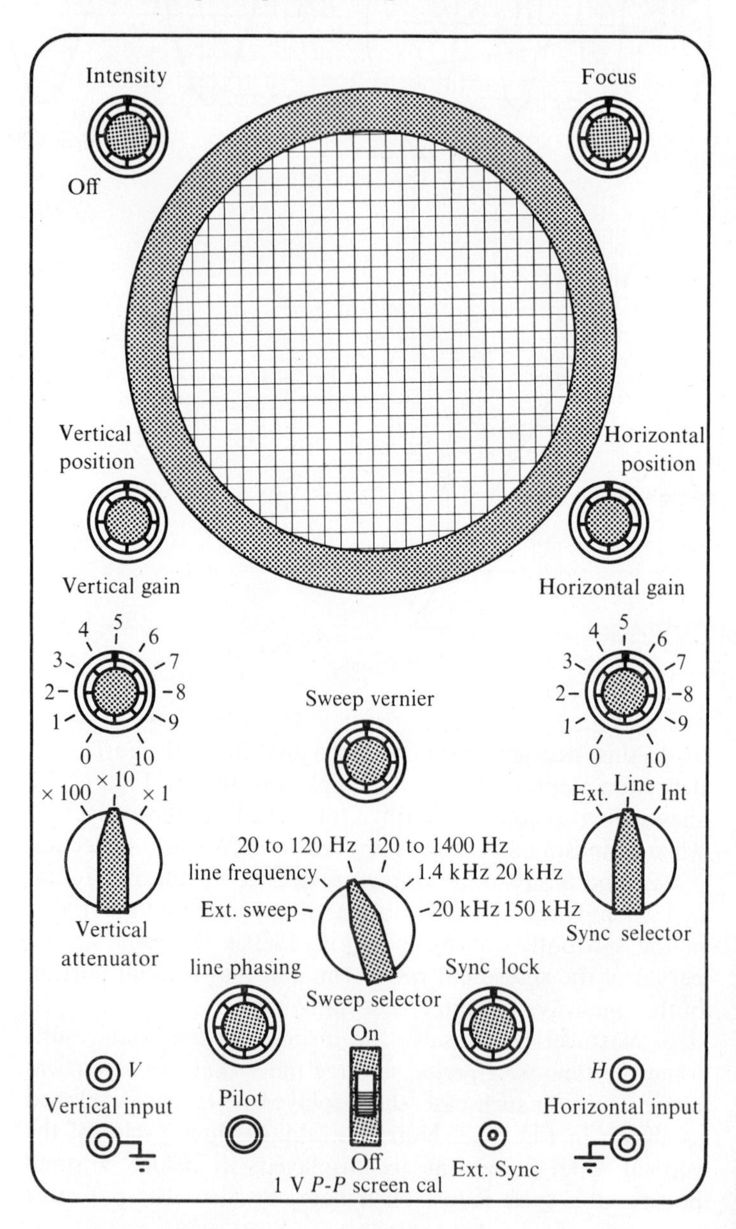

FIG. 13.4 *Front panel plan of a general-purpose oscilloscope.*

the intensity control must not be advanced to the point that the spot burns the screen.

At this point, it is instructive to consider the configuration for a simple time base, as shown in Fig. 13.5. In this arrangement, a dc voltage is applied to an RC integrator circuit. In turn, the capacitor charges exponentially as depicted in Fig. 13.6. When the capacitor voltage reaches the firing potential of the four-layer diode, the diode suddenly conducts and discharges the capacitor rapidly. When the capacitor voltage falls to the cutoff potential of the diode, conduction stops and the capacitor starts to charge again. Note that the value of R in Fig. 13.5 is many times greater than the internal resistance of the four-layer diode in its conduction state. Therefore, a semi-sawtooth waveform is generated.

To avoid drain-off of dc current in Fig. 13.5, a blocking capacitor C_1 is employed to pass the ac component only from the time base to the utilization circuitry. We recognize that if the value of the series resistor R is varied, the time constant of the integrator is changed and the repetition rate of the sawtooth waveform is changed. Similarly, the repetition rate can be changed by varying the value of C. An increase in the time constant lowers the repetition rate. Note that the repetition rate can also be changed by varying the value of the dc source voltage; that is, if the supply voltage is increased, the operating frequency is increased, as depicted in Fig. 13.7.

Another important characteristic is changed by increase of the dc supply voltage in Fig. 13.5. It follows from the relations depicted in Fig. 13.7 that the linearity of the semi-sawtooth waveform is increased, because the firing and

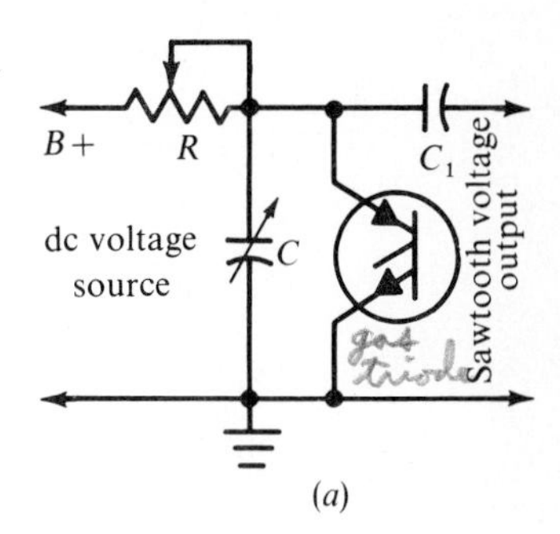

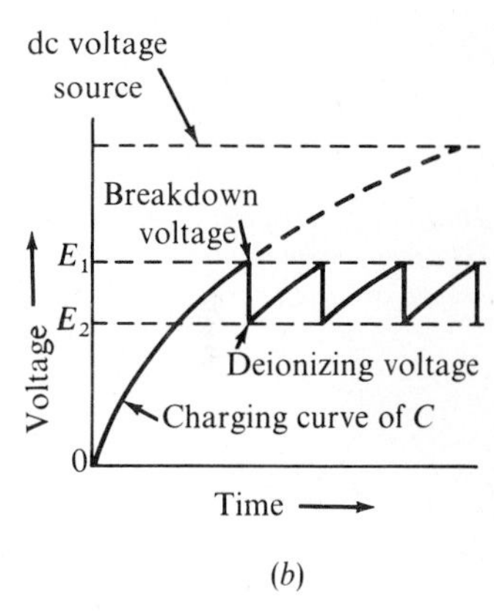

FIG. 13.5 *Circuit diagram (a) and voltage-time characteristics (b) of a four-layer diode sawtooth generator.*

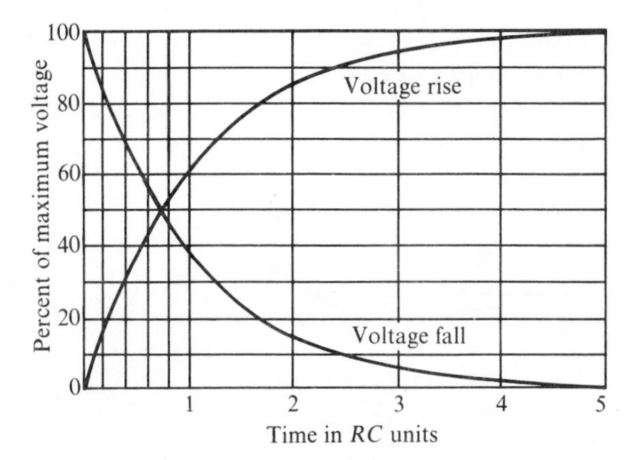

FIG. 13.6 *Universal RC time-constant chart.*

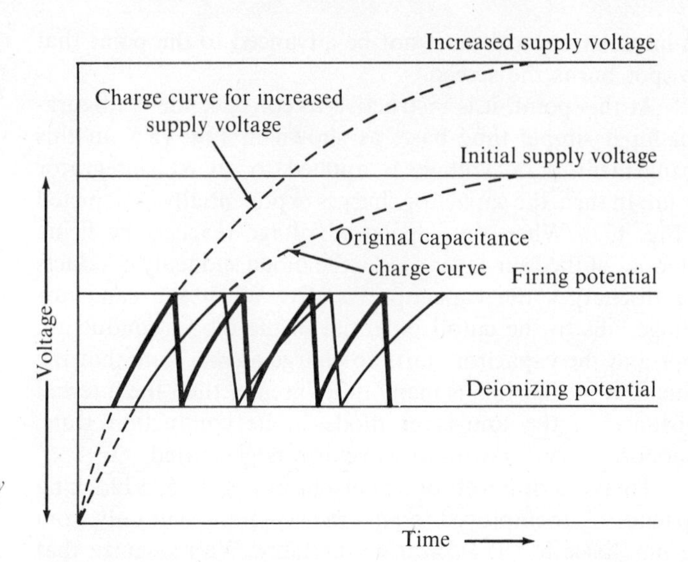

FIG. 13.7 *Frequency and linearity changes caused by variations of supply voltage.*

deionizing potentials are then farther down on the total charge curve. In other words, to improve the linearity of the time base arrangement in Fig. 13.5, and to maintain the same repetition rate, we can increase the $B+$ value and increase the time constant of the charging circuit accordingly. This is the simplest method of linearizing an exponential waveform. Of course, complete linearization cannot be obtained by this simple approach.

13.3 VACUUM-TUBE TIME BASES

The great majority of oscilloscopes in present use employ vacuum tubes. The triode sawtooth generator depicted in Fig. 13.8 utilizes discharge-tube action. Since a triode contains no gas, its plate resistance can be reduced to the required low value only by driving its grid positive to initiate flyback.

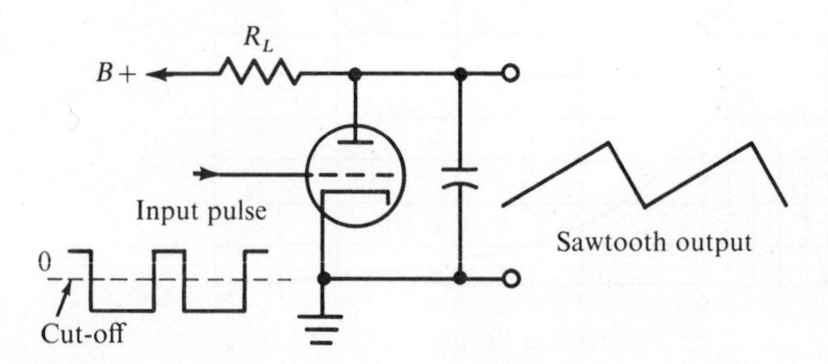

FIG. 13.8 *Vacuum tube employed as a discharge tube and actuated by a pulse.*

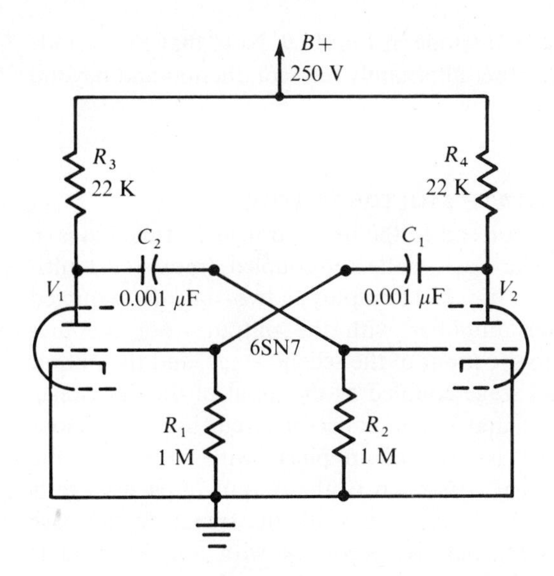

FIG. 13.9 *Circuit of a plate-coupled multivibrator with typical component values.*

The rectangular grid waveform shown in the diagram biases the grid negatively while the capacitor charges, and then rises as a positive pulse to rapidly discharge the capacitor through the resulting low plate resistance of the tube.

Although a separate discharge tube is used in the more elaborate types of scopes, most general-purpose scopes employ a tube that does double duty as a nonsinusoidal oscillator and as a discharge tube. It is instructive to briefly review the operation of the symmetrical plate-coupled multivibrator depicted in Fig. 13.9. Figure 13.10*a* and *b* shows the plate and grid waveforms, respectively, for the first triode; Fig. 13.10*c* and *d* shows the plate and grid waveforms, respec-

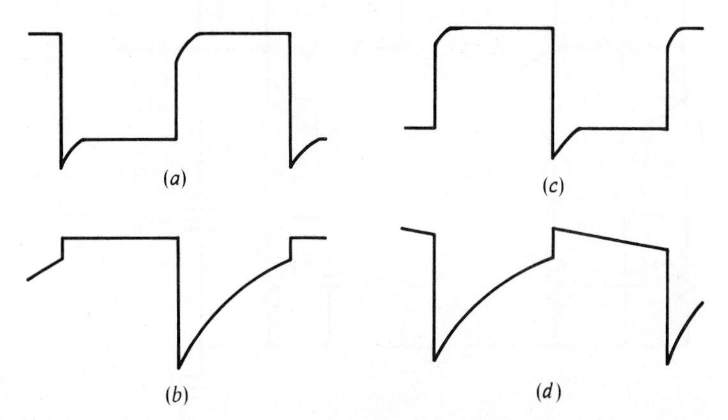

FIG. 13.10 *Waveforms produced by a symmetrical multivibrator.*

tively, of the second triode in Fig. 13.9. Note that each triode is driven by the other alternately into conduction and beyond cutoff.

13.4 SOLID-STATE MULTIVIBRATOR

There is a marked trend to the use of transistor time bases in oscilloscopes. The basic collector-coupled transistor multivibrator shown in Fig. 13.11 employs a two-stage *RC*-coupled common-emitter amplifier with the output from the first stage coupled to the input of the second stage, and the output from the second stage coupled to the input of the first stage.

Since the signal in the collector circuit of a common-emitter amplifier is reversed in phase with respect to the input of that stage, a portion of the output of each stage is fed to the other stage in phase with the signal on the base electrode. This regenerative feedback with amplification is required for oscillation. Bias and stabilization are established identically for both transistors.

Circuit operation is as follows:

1. Because of the variation in tolerances of the components, one transistor will conduct before the other, or will conduct more heavily than the other in Fig. 13.11.

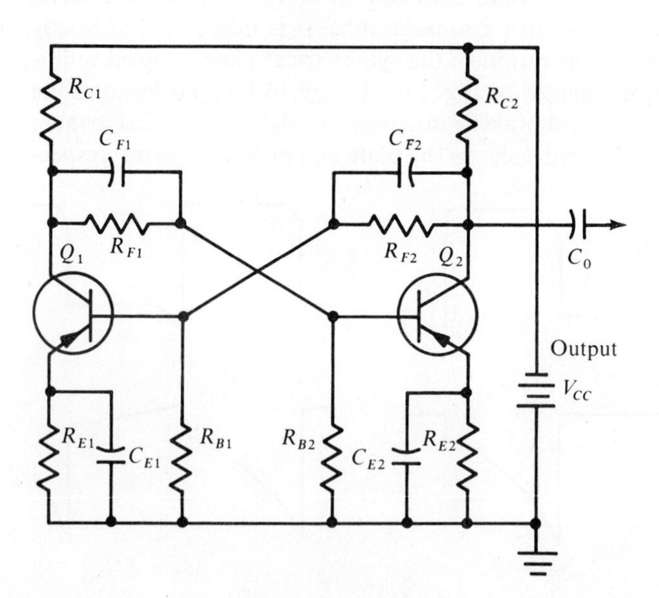

FIG. 13.11 *Transistor multivibrator.*

2. Assuming transistor Q_1 is conducting more heavily than transistor Q_2, more current i_{b1} will flow in the base circuit of transistor Q_1 than in the base circuit of Q_2. Collector current i_{c1} in transistor Q_1 increases rapidly, causing collector voltage V_{c1} (Fig. 13.12), and voltage at the junction of resistors R_{C1} and R_{F1} to decrease (become more positive). This increasing positive voltage is applied through capacitor C_{F1} to the base of transistor Q_2.

3. As base voltage v_{b2} of transistor Q_2 becomes more positive, the forward bias decreases, resulting in a rapid decrease in base current i_{b2} and collector current i_{c2} in transistor Q_2. Collector voltage V_{c2}, and thus the voltage at the junction of resistors R_{C2} and R_{F2} becomes more negative. This negatively increasing voltage is fed back through capacitor C_{F2} to the base of transistor Q_1, increasing the forward bias.

4. The processes in (1) and (2) continue until a point is reached where base voltage v_{b2} of transistor Q_2 is made so positive with respect to the emitter that transistor Q_2 is cut off (reverse bias is applied) and transistor Q_1 is saturated (total dc voltage V_{CC} appears across resistor R_{C1}). That is, the current through transistor Q_1 increases steadily as the current through transistor Q_2 decreases steadily until transistor Q_2 is cut off. Point A, Fig. 13.12, represents this action. This entire action happens so quickly that capacitor C_{F1} does not get a chance to discharge and the increased positive voltage at the collector of transistor Q_1 appears entirely across resistor R_{B2}.

5. During the period from A to B, Fig. 13.12, collector current i_{c1} and collector voltage V_{c1} remain constant and capacitor C_{F1} discharges through resistor R_{F1}. As capacitor C_{F1} discharges, more of the previously increased positive voltage at the collector of transistor Q_1 appears across capacitor C_{F1} and less across resistor R_{B2}. This decreases the reverse bias on the base of transistor Q_2. This action continues until the time at point B, Fig. 13.12, is reached; and forward bias is reestablished across the base-emitter diode of transistor Q_2.

6. Transistor Q_2 conducts. As collector current i_{c2} in transistor Q_2 increases, the collector voltage V_{c2} becomes less negative or more positive. This voltage, coupled through capacitor C_{F2} to the base of transistor Q_1, drives it more positive and causes a decrease in current flow through transistor Q_1. The resulting increased negative voltage at the collector of transistor Q_1 is coupled through capacitor C_{F1}

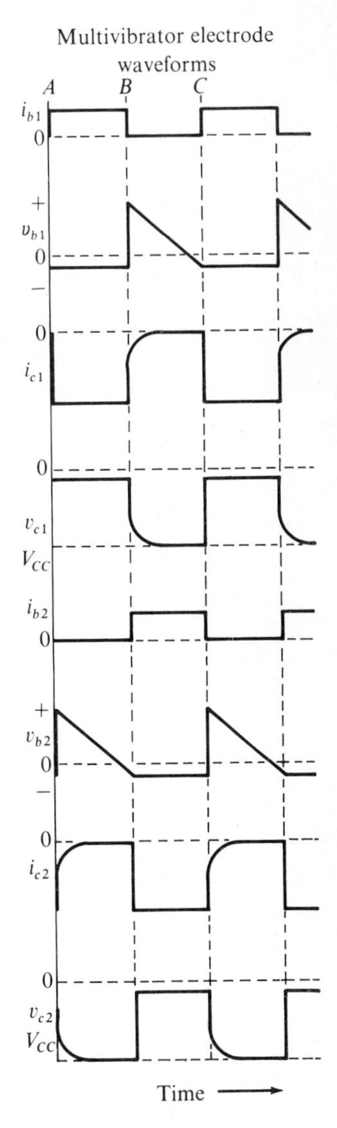

FIG. 13.12 *Transistor multivibrator waveforms.*

and appears across resistor R_{B2}. The collector current of transistor Q_2 therefore increases. This process continues rapidly until transistor Q_1 is cut off. Transistor Q_1 remains cut off (and transistor Q_2 conducts) until capacitor C_{F2} discharges through resistor R_{F2} enough to decrease the reverse bias on the base of transistor Q_1, point C, (Fig. 13.12). The cycle is repeated in (1) above.

7. The oscillating frequency of the multivibrator is usually determined by the values of resistance and capacitance in the circuit. In the collector-coupled multivibrator of Fig. 13.11, collector loads are provided by resistors R_{C1} and R_{C2}. Base bias for transistor Q_1 is established through voltage divider resistors R_{B1} and R_{F1}. Base bias for transistor Q_2 is established through voltage divider resistors R_{F1} and R_{B2}. Stabilization is obtained with emitter-swamping resistor R_{E1} for transistor Q_1, and resistor R_{E2} for transistor Q_2. Emitter capacitors C_{E1} and C_{E2} are ac bypass capacitors.

8. The output signal is coupled through capacitor C_0 to the load. This output waveform, which is essentially square, may be obtained from either collector. To have a sawtooth output, a capacitor is usually connected from collector to ground for development of the output voltage. This principle will be explained subsequently.

13.5 ASYMMETRICAL MULTIVIBRATOR

Next, it is evident that since C_2 is much smaller than C_1 in Fig. 13.13, the time constant of the Q_2 base circuit will be much shorter than that of Q_1. In turn, an asymmetrical or rec-

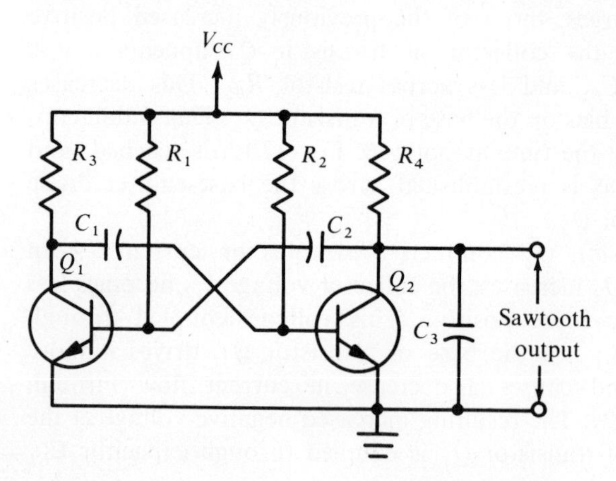

FIG. 13.13 *Practical circuit of an asymmetrical multivibrator used as a sweep generator.*

tangular waveform will be generated. For example, let us consider the operation of the generator depicted in Fig. 13.13, with C_3 disconnected from the circuit temporarily. Instead of producing the symmetrical waveforms seen in Fig. 13.12, the asymmetrical waveforms shown in Fig. 13.14*a* through *d* are obtained. However, before this multivibrator can be used as a time base, the waveform in Fig. 13.14*c* must be shaped into a semi-sawtooth.

The required waveshaping function is provided by capacitor C_3 in Fig. 13.13. With this capacitor connected to the collector of Q_2, the waveform in Fig. 13.14*c* becomes modified into a practical sawtooth shape, as shown in Fig. 13.14*e*. That is, although multivibrator operation continues as before, the collector voltage on Q_2 cannot rise abruptly when the transistor is in its cutoff mode (Fig. 13.13). The rise of collector voltage is determined by the integrator action of R_{C2} and C_3. Next, when the base of Q_2 is suddenly driven positive by Q_1, C_3 discharges with great rapidity because Q_2 acts as a discharge transistor at this time. Insofar as sawtooth-wave generation is concerned, we may regard Q_2 as an electronic switch with a long "off" period and a very short "on" period.

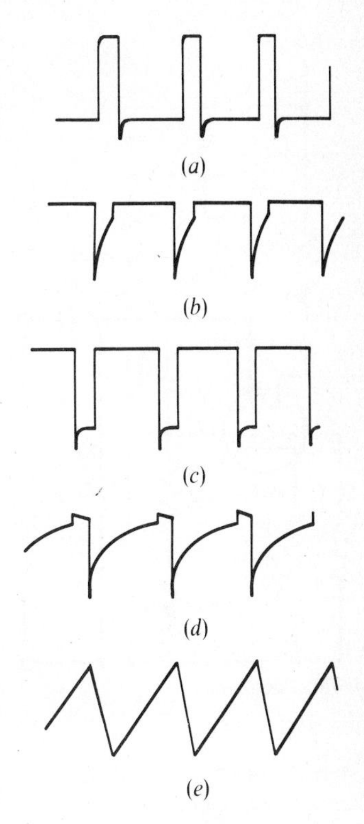

FIG. 13.14 *Collector, base, and output-voltage oscillograms obtained from an asymmetrical multivibrator.*

General-purpose scopes often employ the cathode-coupled multivibrator type of time base depicted in Fig. 13.15. This circuit employs a duotriode with a common-cathode resistor. The output of V_1 is coupled via C_1, R_6, and R_1 to the input of V_2. In turn, the output of V_2 is coupled to the input of V_1 through the common-cathode resistor. If C_2 were disconnected we would obtain a rectangular waveform across the output terminals. However, with C_2 shunted across the output path, the plate of V_2 cannot rise suddenly to the $B+$ value when V_2 is cut off. Instead, C_2 charges exponentially through R_4 and R_5. Let us consider the circuit details briefly.

The coarse frequency adjustment C_1 in Fig. 13.15 is designed as a bank of fixed capacitors with a switch; these capacitors range in value from 100 pF to 0.5 μF. When the capacitance of C_1 is changed, the coupling circuit has a different time constant, which changes the repetition rate of the multivibrator. In turn, the repetition rate of the sawtooth output waveform is changed. For example, if the capacitance of C_1 is increased, the repetition rate is slowed down. The switches for capacitor banks C_1 and C_2 are ganged on the

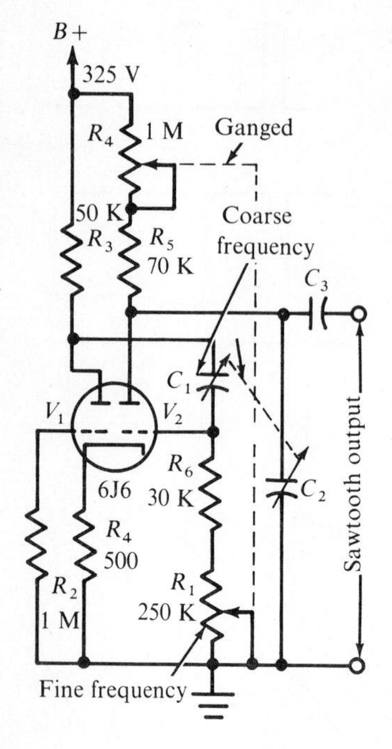

FIG. 13.15 *Typical cathode-coupled multivibrator sweep generator.*

same shaft; this provides an appropriate value for C_2 at various repetition rates, so that exponential curvature in the sawtooth waveform is minimized.

Next, fine (vernier or continuous) frequency control is provided by potentiometers R_1 and R_4 in Fig. 13.15. These potentiometers are ganged, and their resistance ranges are designed to maintain the amplitude of the sawtooth output reasonably constant at various repetition rates. The practical limits on the repetition rates obtainable from this type of sawtooth oscillator are approximately 15 Hz and 80 kHz. Since from 10 to 15 harmonics must be passed to retain a reasonable facsimile of a sawtooth waveform (see Fig. 13.16), it follows that the horizontal deflection amplifier should have a bandwidth of at least 1 MHz. Otherwise, exponential curvature in the 80-kHz sawtooth waveform will be aggravated by frequency distortion in passage through the horizontal amplifier. Portable oscilloscopes may use transistors in an emitter-coupled multivibrator to generate a sweep.

13.6 TIME-BASE SYNCHRONIZATION

In the operation of a general-purpose scope, the same pattern is "redrawn" at a rapid rate on the screen, as long as the particular waveform is displayed. If a single excursion, or two excursions or more, is retraced in such a manner that the pattern always occupies the same position on the screen, the eye sees a stationary image. This is true whether we are viewing a power-frequency waveform at 60 Hz, or a chroma tv waveform at 3.58 MHz. That is, the human eye provides persistence of vision to a lower limit of about 16 events per second.

We know that if a single excursion of a signal is to be displayed, the sweep frequency must be equal to the signal frequency, or the sweep frequency may be any exact submultiple of the signal frequency. Unless strict time coincidence is maintained, the pattern "walks" or "runs" on the screen, as depicted in Fig. 13.17. If the pattern moves from left to right on the screen, the sawtooth frequency is too high; if the pattern moves from right to left on the screen, the sawtooth frequency is too low. Therefore, it is necessary to employ circuit means for locking the sawtooth repetition rate to that of the vertical input signal. Let us see how this is accomplished.

Figure 13.18 shows the plan of a general-purpose oscilloscope, with the synchronizing section included. Note that

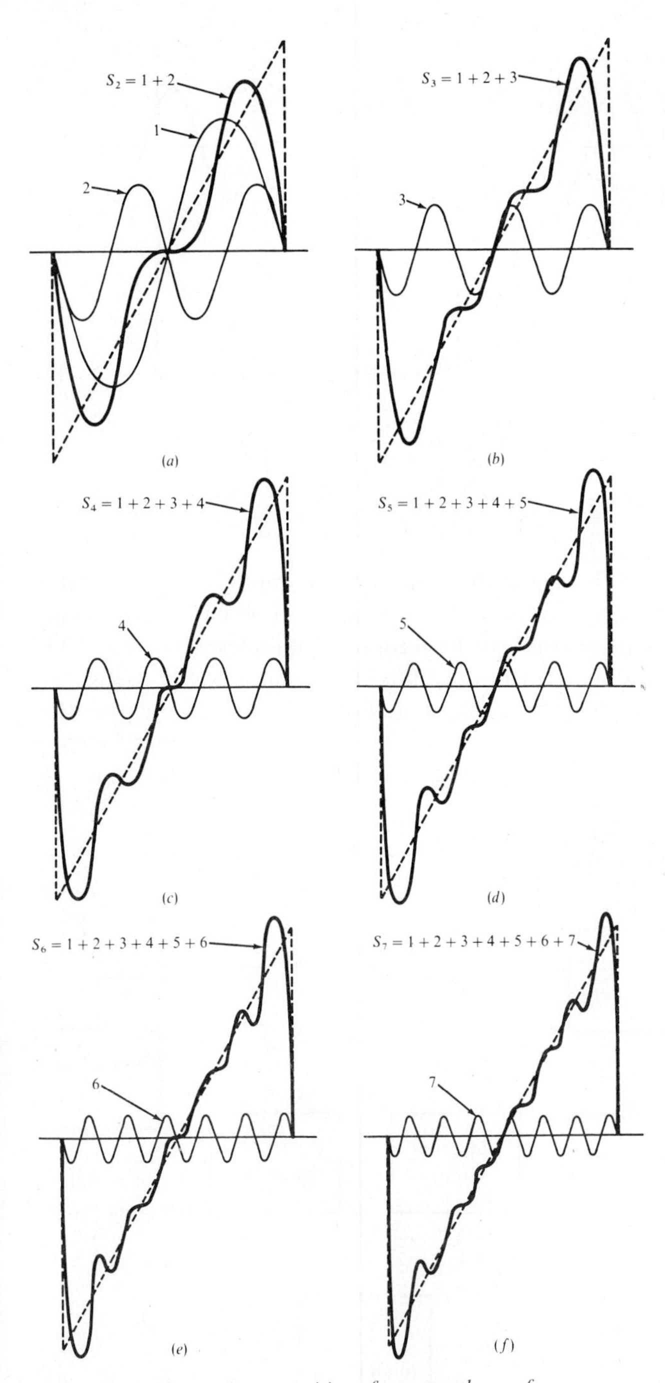

FIG. 13.16 *Harmonic composition of a sawtooth waveform.*

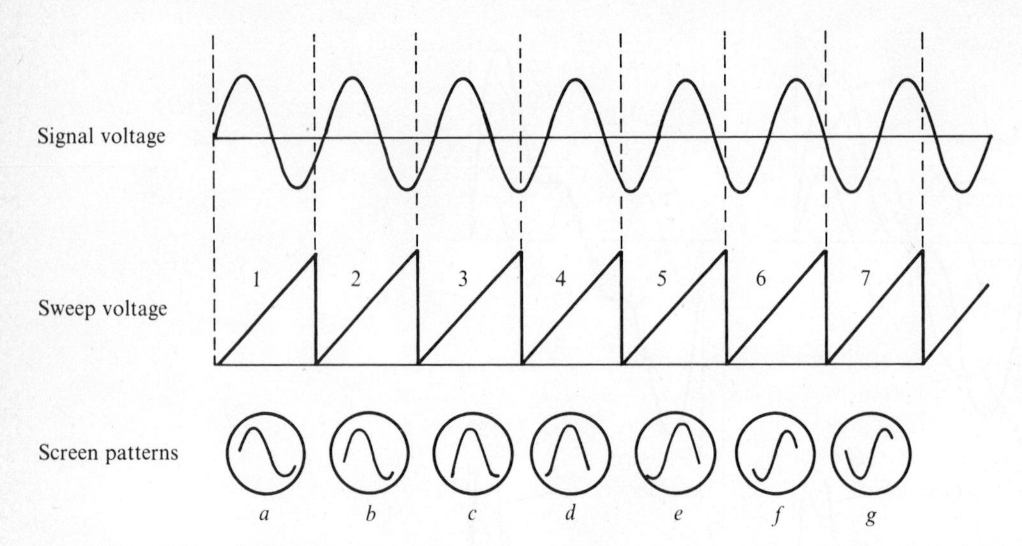

FIG. 13.17 *Drifting of pattern from left to right on cathode-ray screen when sweep frequency is slightly higher than signal frequency.*

some of the signal from the vertical amplifier is processed by the sync section in order to lock the time base. This method is termed *internal synchronization.* With reference to Fig. 13.19, the sync signal is injected into the grid circuit of V_1 in Fig. 13.9. In turn, the tube is triggered into conduction a bit earlier than would be the case in its free-running mode, as shown in Fig. 13.19. When a scope employs a sync signal clipper, the

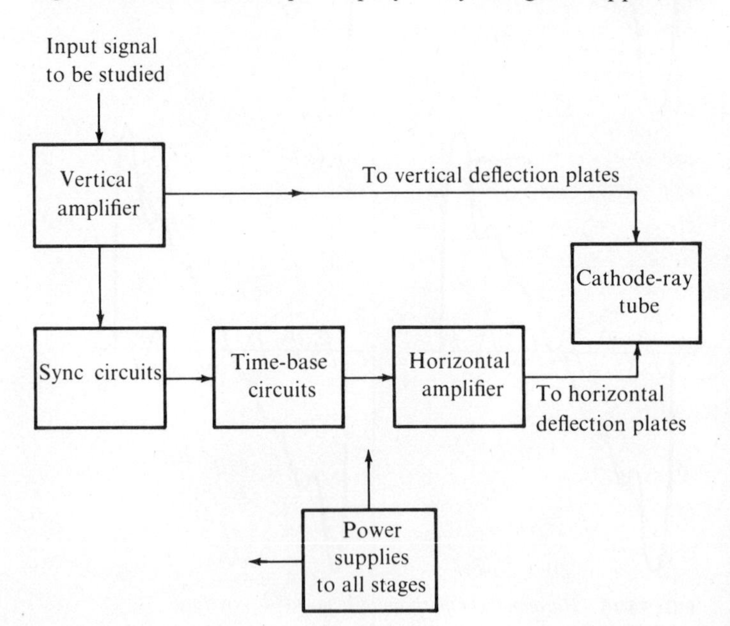

FIG. 13.18 *Simplified block diagram of basic oscilloscope.*

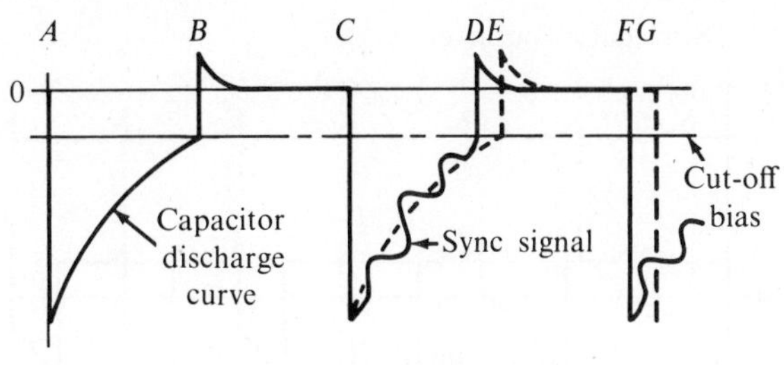

FIG. 13.19 *Grid voltage of a multivibrator showing effect of a sine-wave sync signal.*

clipper may be followed by a differentiating circuit. In turn, "spikes" or pulses are applied to the multivibrator circuit. This elaboration provides more stable sync lock over a wide range of complex waveforms. Figure 13.20 shows how a multivibrator is locked by differentiated pulses in its fundamental and third submultiple modes of operation.

13.7 OSCILLOSCOPE AMPLIFIERS

Various types of amplifiers are employed in oscilloscopes. From the most basic viewpoint, we can classify amplifiers into narrowband and wideband types. These categories are not sharply defined, and tend to overlap. However, it is generally agreed that an amplifier falls into the narrowband classification if its high-frequency response does not extend to 3.58 MHz (the subcarrier frequency in color television). Conversely, any amplifier that has full response up to 3.58 MHz will be classified as a wideband amplifier. Most utility-type wideband scopes provide full response through 4.5 MHz, so that intercarrier sound circuitry of a television receiver can be checked. Intercarrier sound circuitry is explained in all standard television receiver textbooks. All scope amplifiers provide full response down to at least 60 Hz.

Oscilloscope amplifiers are also classified into high-gain and low-gain types. Although high gain is desirable in general-purpose scopes, it may not be economically feasible to provide more than moderate gain, particularly when extended high-frequency response is a dominant consideration. For example, it is common practice to provide a vertical deflection sensitivity in the order of 15 to 30 mV/in. for a service-type scope that has frequency response to 4.5 MHz. On the other hand, a scope capable of precision pulse analysis and having frequency response to 15 MHz provides a vertical

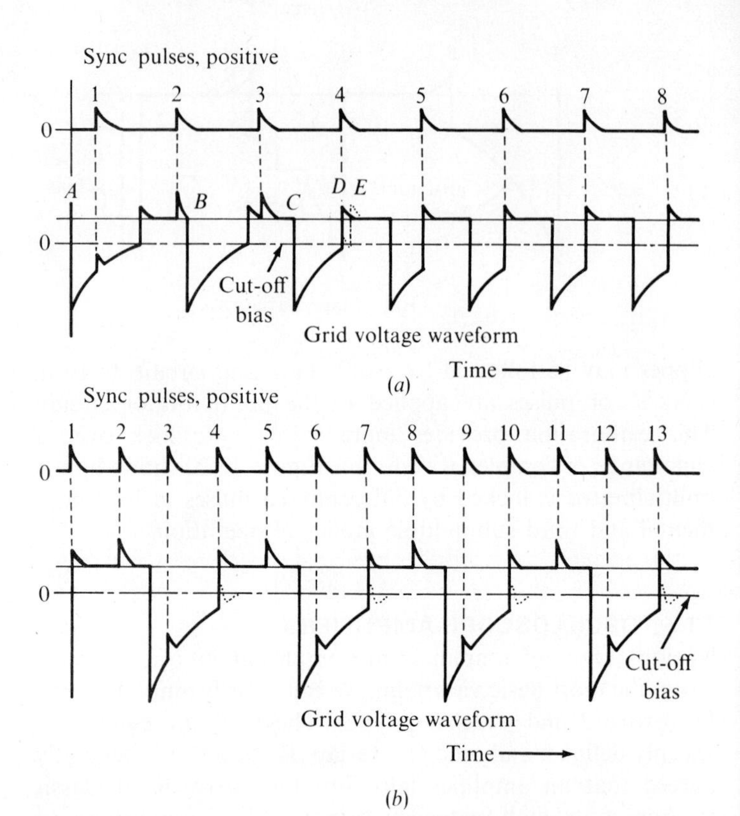

FIG. 13.20 *Synchronization of a multivibrator by means of sharply peaked pulses.*

deflection sensitivity of 125 mV/in., approximately. Since high gain and extended high-frequency response are conflicting design factors, various compromises become necessary.

We commonly group oscilloscope amplifiers also into ac-coupled and dc-coupled types. Although dc-coupled amplifiers are comparatively costly, they provide the advantage of full response down to zero frequency (dc). If a scope is provided with a dc-coupled vertical amplifier, it will display the dc component along with the ac component in a pulsating dc waveform, or in an ac waveform with a dc component. Details of these considerations are explained subsequently. Another advantage of a dc-coupled amplifier is absence of low-frequency phase shift and resulting waveform distortion when a pulse train with a low repetition rate is being displayed. A dc-coupled amplifier is comparatively elaborate, because the designer must provide for operating stability and freedom from drift.

Another classification of oscilloscope amplifiers is

made into single-ended and double-ended (push-pull) types. Most general-purpose scopes have two vertical-input terminals, one of which is grounded to the case of the instrument; this is called a single-ended arrangement. On the other hand, some general-purpose scopes and many specialized scopes have three vertical-input terminals; one terminal is grounded, and the other two may be driven in push-pull (180° out of phase), if desired. This is sometimes called a balanced-input or double-ended arrangement. It has several practical advantages in testing procedures which are detailed subsequently.

13.8 BASIC OSCILLOSCOPE AMPLIFIERS

The oscilloscope is a test instrument which must display an exact duplicate waveform of the signal or signals applied to its input; in other words, it must not cause any distortion of the applied signals. This is accomplished through the use of linear Class *A* amplifiers. The oscilloscope also must have a high input impedance to reduce the loading effect on the circuit or equipment being tested.

The function of the horizontal and vertical amplifiers in the oscilloscope is to amplify the signal applied to them with minimum distortion. These amplified signals are then applied to the horizontal and vertical deflection plates in the crt, which will cause horizontal and vertical deflection of the electron beam.

The horizontal amplifier channel is shown in block diagram form in Fig. 13.21. It consists of the horizontal attenuator, horizontal cathode follower, the first and second direct-coupled amplifiers, and the cathode-ray tube. During the following discussion, reference to this block diagram will

FIG. 13.21 *Block diagram of the horizontal channel of an oscilloscope.*

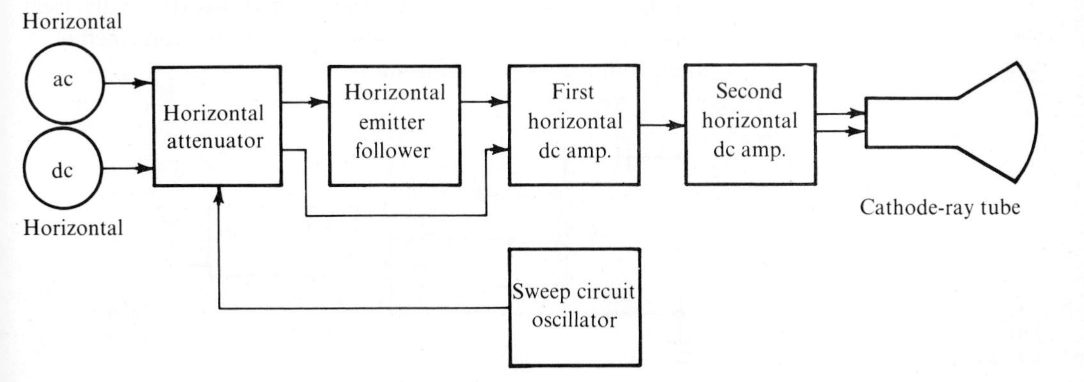

aid in your understanding of the basic operation of the following circuits.

There are three possible inputs to the horizontal channel: two external inputs and one internal. When using an oscilloscope a linear time base is generally desired. In this case the signal fed into the horizontal channel would come from a sweep generator located inside the oscilloscope. In special cases a signal other than a sawtooth is used for horizontal deflection, and the external inputs would be used. The selection of these various inputs is made with the horizontal attenuator switch. All inputs to the channel, except the external dc input, are fed to the horizontal cathode-follower stage. The purpose of the horizontal amplifier is to increase the strength of the horizontal signal to achieve adequate lateral (horizontal) deflection of the crt beam.

The purpose of the horizontal attenuator is to select the desired type and adjust the strength of the signal fed into the horizontal channel.

Since the input signal can vary considerably in amplitude, a means of coarse as well as fine amplitude adjustment is desirable. This is usually accomplished by the use of a step-switch attenuator and potentiometer. A basic step-switch attenuator is shown in Fig. 13.22.

The tapped resistor is merely an ac voltage divider where the output voltage (input to the horizontal amplifier) depends on the position of switch S_1. The taps occur at calibrated intervals so the amount of attenuation or reduction of the input is known by the switch position. As the frequency applied to this network increases, the stray and input capacitance (represented by C_s) causes a shunting effect across the tapped portion of the voltage divider. This shunting action changes the resistance and impedance ratios originally used in calibration. For this reason, most attenuators include a high-frequency-compensating network.

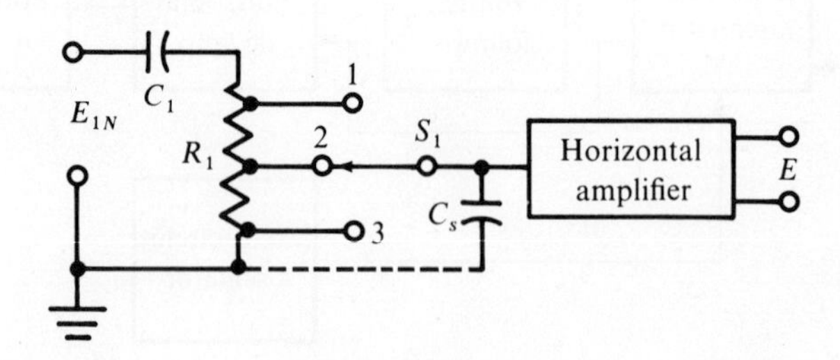

FIG. 13.22 *Basic step-switch attenuator.*

A typical step attenuator with high-frequency compensation is shown in Fig. 13.23. With switch S_{102} is moved to the ac-1 position, R_{128} is shorted out, and the sawtooth voltage is fed directly into the horizontal cathode follower stage. The ac-1 position would be used for very weak signals applied to the horizontal input ac terminal.

When switch S_{102} is in the ac-10 position, the input is reduced by a factor of 10; and in the ac-100 position it is

FIG. 13.23 *Horizontal input attenuator.*

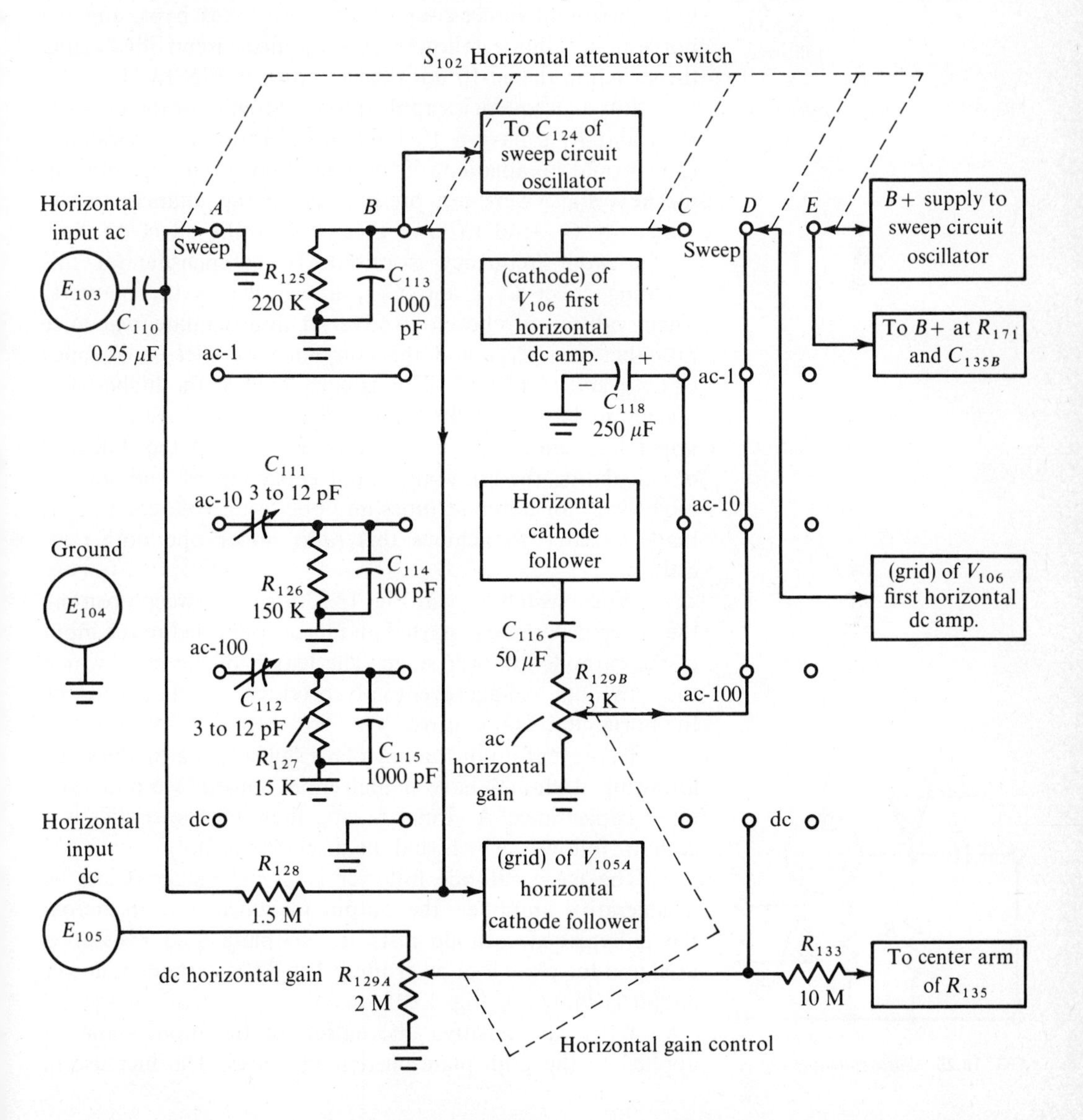

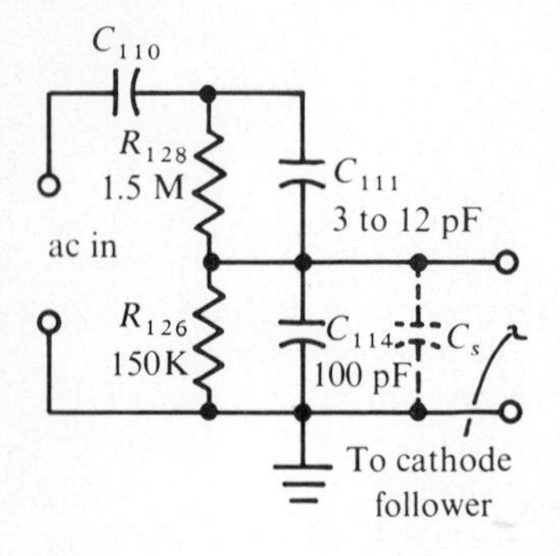

FIG. 13.24 *Simplified attenuator.*

reduced by a factor of 100. This means that in the ac-10 position one-tenth of the input is fed to the next stage, and in the ac-100 position, one-hundredth of the input is used.

With the horizontal attenuation control in the dc position, the connection to the horizontal ac input is removed, and the horizontal dc input jack E_{105} is connected to the horizontal gain control which can be used to adjust the magnitude of the dc voltage input before it is applied to the first horizontal direct-coupled amplifier, thus bypassing the horizontal cathode follower. A simplified circuit illustrating the switch in the ac-10 position is shown in Fig. 13.24.

For all frequencies applied to the attenuator the resistors R_{128} and R_{126} have a 10:1 resistance ratio. This resistance ratio would be adequate if the stray and input capacities of the next stage were not present. These capacitances, represented by C_s, tend to have greater shunting effect on R_{126} as the input frequency is raised. To compensate for this shunting effect C_{111} and C_{114} are added to the network. Their values are chosen to give an approximate reactance ratio between C_{111} and the combined parallel capacitance of C_{114} and C_s of 10:1. This is calculated at the highest frequency to be passed. In the circuit shown in Fig. 13.23 the upper frequency limit is 500 kHz. Because of the difficulty of calculating the stray and input capacitances, and since it might vary for different tubes and operating voltages, C_{111} is made variable to achieve this ratio under operating conditions.

When switch S_{102} in Fig. 13.23 is in the sweep position, the sweep generator is started and its output is fed to the input of the cathode-follower stage. The length of the trace which this sawtooth voltage eventually produces can be varied by the horizontal gain control.

To prevent shunt-loading of the step attenuator, the stage following it should have a high-input impedance and low-input capacitance. A circuit having these characteristics is a cathode follower, which shall be discussed next.

The basic cathode follower is a single-stage, Class A, degenerative amplifier, the output of which appears across the unbypassed cathode resistor. No plate load resistor is used and the plate is at ac ground. The basic cathode follower circuit is shown in Fig. 13.25.

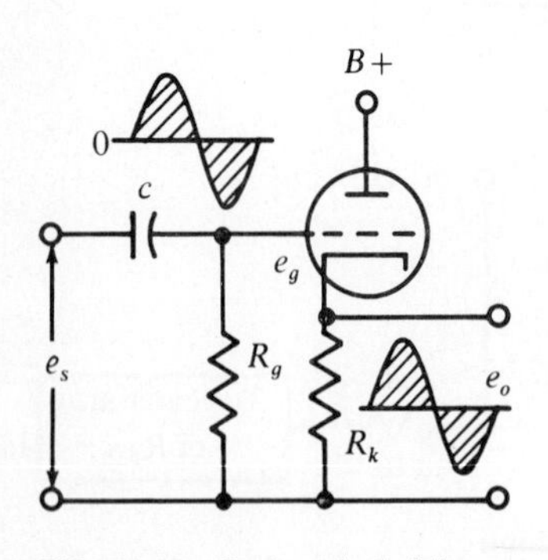

FIG. 13.25 *Basic cathode follower.*

When the positive alternation of the input signal is applied to the grid, plate current increases. The increase in plate current will cause a greater voltage drop across the cathode resistor. Since the plate current will vary sinusoidally, the voltage drop across the cathode resistor, R_{R_k}, will also vary sinusoidally as shown in Fig. 13.25. The output will be a sine wave, the reference level of which is the quiescent cathode voltage.

Equating a series loop from the grid to the cathode, the ac voltage changes are series-opposing. In Fig. 13.25 the following is therefore true:

$$e_g = e_s - e_o \tag{13.3}$$

where e_g = ac grid to cathode voltage

e_s = ac input signal

e_o = ac output signal

The tube plate current is controlled by e_g. If the output voltage change e_o were to equal the input voltage change e_s, the grid voltage change e_g would be zero, resulting in no original change in plate current. This is an impossible condition. Therefore, the ac output voltage from a cathode follower must always be less than the input voltage. For this reason, the voltage gain of a cathode follower is less than unity although it will subsequently be shown that the circuit is capable of a power gain. As the name implies, the output voltage follows the input voltage. It not only has the same waveform, but also the same instantaneous polarity. It should be noted that the voltage gain of less than unity only applies to the ac voltages and not to the dc component of the output.

The circuit shown in Fig. 13.25 can be represented by the equivalent circuit shown in Fig. 13.26.

where μ = amplification factor

r_p = ac plate resistance

e_g = as shown in Eq. (13.3)

According to Ohm's law,

$$i_p = \frac{\text{voltage applied}}{\text{total resistance}}$$

Then:

$$i_p = \frac{\mu e_g}{r_p + R_k} = \frac{\mu(e_s - e_o)}{r_p + R_k} \tag{13.4}$$

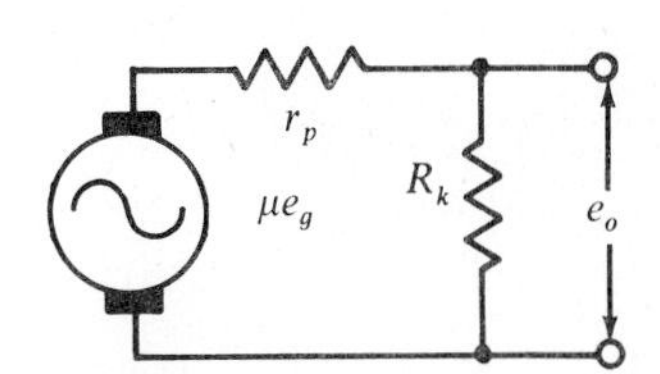

FIG. 13.26 *Equivalent circuit of a cathode follower.*

since:

$$e_o = e_{R_k} = i_p R_k$$

Substituting in Eq. (13.4) gives

$$i_p = \frac{\mu(e_s - i_p R_k)}{r_p + R_k}$$

This equation can be solved for i_p as follows:

$$i_p(r_p + R_k) = \mu e_s - i_p R_k$$

$$i_p(r_p + R_k) + \mu i_p R_k = \mu e_s$$

$$i_p(r_p + R_k + \mu R_k) = \mu e_s$$

$$i_p = \frac{\mu e_s}{r_p + R_k + \mu R_k} = \frac{\mu e_s}{r_p + R_k(\mu + 1)}$$

since

$$e_o = i_p R_k = \frac{\mu e_s R_k}{r_p + R_k(\mu + 1)}$$

The voltage gain of the amplifier is equal to

$$\text{Voltage gain} = \frac{e_o \text{ (voltage out)}}{e_s \text{ (voltage in)}}$$

and therefore,

$$\text{Voltage gain} = \frac{e_o}{e_s} = \frac{\mu e_s R_k}{r_p + R_k(\mu + 1)} \div e_s$$

$$= \frac{\mu e_s R_k}{r_p + R_k(\mu + 1)} \times \frac{1}{e_s}$$

$$= \frac{\mu R_k}{r_p + R_k(\mu + 1)} \qquad (13.5)$$

Upon examination of Eq. (13.5) it can be seen that the denominator will always be greater than the numerator; thus the voltage gain of the cathode follower will always be less than one.

In comparing the output-voltage formula of a cathode follower to a conventional triode *RC*-coupled amplifier, the tube used as a cathode follower appears to have an amplification factor equal to $\mu/(\mu + 1)$, and an ac plate resistance equal to $r_p/(\mu + 1)$. From this comparison another equivalent circuit

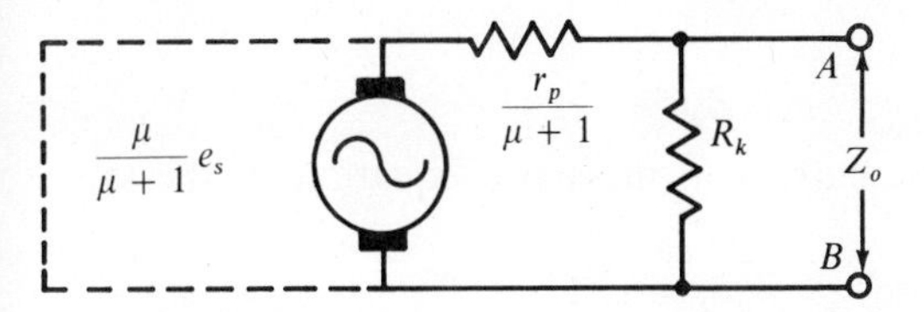

FIG. 13.27 *Modified equivalent circuit of a cathode follower.*

may be drawn that more closely represents the cathode follower. This modified equivalent circuit is illustrated in Fig. 13.27.

The impedance Z_o "looking back" from points A and B in Fig. 13.27 consists of a parallel network composed of the ac plate resistance $\mu r_p/(\mu + 1)$ and the cathode resistor. The constant-voltage generator may be considered a short circuit. Therefore, the following equation may be used to determine the value of the output impedance Z_o.

Using the product-over-the-sum formula for parallel impedance:

$$Z_o = \frac{\dfrac{r_p}{1 + \mu} \times R_k}{\dfrac{r_p}{1 + \mu} + R_k} = \frac{r_p R_k}{r_p + (1 + \mu)R_k} \tag{13.6}$$

In general, this equation shows that the output impedance Z_o of a cathode follower is less than R_k and is usually resistive in nature. This makes the cathode follower capable of feeding a low impedance load.

The input impedance of a cathode follower is high and the effective input capacitance is low when compared with like values of a conventional amplifier. Both characteristics result from the degenerative action that occurs due to the unbypassed cathode resistor. The equivalent circuit for the input impedance is shown in Fig. 13.28.

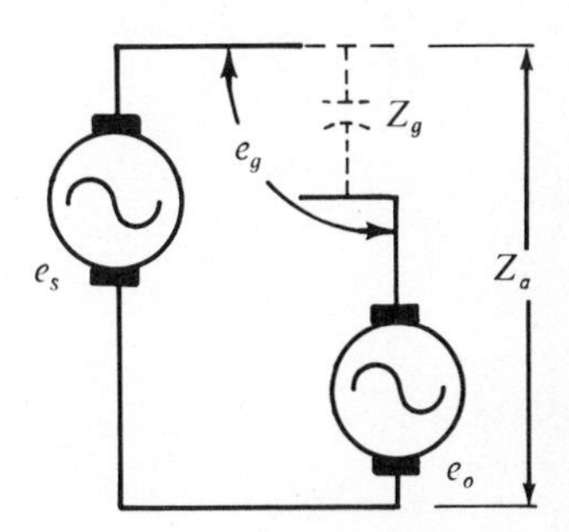

FIG. 13.28 *Input impedance.*

The apparent input impedance as seen by the generator is Z_a. The input signal is represented by e_s. According to Ohm's law,

$$I_s = \frac{e_s}{Z_a}$$

Since the output voltage opposes the input voltage, the voltage applied between the control grid and cathode equals $e_s - e_o$. According to Ohm's law,

$$I_g = \frac{e_s - e_o}{Z_g} \quad \text{or} \quad \frac{e_g}{Z_g}$$

Since the current in these two equations is the same,

$$\frac{e_s}{Z_a} = \frac{e_g}{Z_g} \quad \text{and} \quad Z_a e_g = e_s Z_g$$

therefore,

$$Z_a = \frac{e_s Z_g}{e_g}$$

Dividing through by e_s, we obtain

$$Z_a = \frac{Z_g}{e_g/e_s}$$

Substituting Eq. (44.1) for e_g yields

$$Z_a = \frac{Z_g}{(e_s - e_o)/e_s} = \frac{Z_g}{1 - (e_o/e_s)}$$

$$\frac{e_o}{e_s} = \text{voltage gain } A$$

therefore,

$$Z_a = \frac{Z_g}{1 - A} \tag{13.7}$$

In Eq. (13.7) the denominator will always be less than one so the apparent input impedance A_a, as seen by a source, will always be larger than the input impedance of a conventional amplifier.

The reduced input capacitance results from the fact that degeneration reduces the amplitude of the ac component of the grid-to-cathode voltage, or in effect, increases the input impedance, and thus causes less current to flow through the tube capacitances.

One of the principal advantages of a cathode follower is that it can be used to match a high impedance to a low impedance. Thus it can take the voltage developed across a high impedance and supply a low impedance load with only a slightly less voltage but with a correspondingly large increase in current. One or more of the circuit elements of a cathode follower may be varied to achieve a more precise impedance match if the match is critical.

When tubes having a high mutual conductance are used, the low value of output impedance extends the amplification into the upper range of frequencies because the shunting effects of interelectrode and distributed capacitances are proportionately smaller. The low-frequency response is improved by allowing the dc component of cathode current to flow in the load, thus avoiding the use of a series blocking capacitor.

The degenerative effect caused by the unbypassed cathode resistor increases the input impedance. Thus less shunting effect is offered to the previous stage, and a better overall frequency response is produced.

As stated before, the input and output voltages have the same instantaneous polarity. When pulses are used, it may be necessary to feed a positive- or a negative-going pulse to a load without polarity inversion. The cathode follower could thus serve two purposes: to prevent polarity inversion and to afford an impedance match.

Circuit stability is also improved, as in regular amplifiers, by degenerative feedback. Specifically, this type of circuit counteracts amplitude distortion occurring within the tube, the effect of plate supply voltage variations, aging of tubes, production of harmonics, and other undesirable effects that occur within the stage.

However, these advantages are achieved at the expense of an overall reduction in voltage gain. Normally, the voltage gain is slightly less than unity, but the circuit is capable of producing a gain in power.

Because of its high input impedance and low output impedance, the cathode follower can be used to match the impedance of a step attenuator to a variable gain control. A typical cathode follower used in the horizontal channel of an oscilloscope is shown in Fig. 13.29. It should be noted that the heavy black line represents signal flow and not necessarily current flow.

Any voltage applied to the control grid pin 2 will appear slightly reduced across the cathode resistor R_{132}. This signal is capacitively coupled to the horizontal ac gain control. Blocking capacitor C_{116} blocks the dc component of the output and passes only the ac signal. Because of the impedance characteristics of the cathode follower, the gain control (possessing a small value of resistance, 3 K) has a negligible loading effect on the input. The gain control provides fine adjustment of the signal fed into the first horizontal direct-

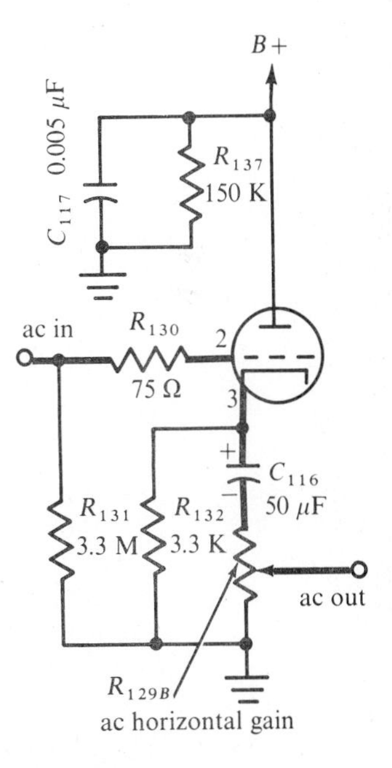

FIG. 13.29 *Horizontal cathode follower.*

coupled amplifier. It must be remembered that the step switch attenuator and the horizontal gain control together set the amplitude of the ac signal fed into the first horizontal amplifier.

Under normal operating conditions the output of a cathode follower stage is practically free of amplitude distortion. However, if the input signal swings the grid voltage too far negative, the output waveform will be cut off and any further increase in negative grid potential will have no corresponding change in plate current. To prevent this distortion the attenuator should be at the maximum attenuation position before applying a signal of unknown amplitude to the horizontal channel.

In the practical cathode follower circuit in Fig. 13.29, the input resistor is in parallel with the apparent input impedance discussed previously. The actual input impedance can be no higher than the value of this grid resistor, but the loading effect on this resistor is much less in a cathode-follower stage than in a conventional amplifier.

The horizontal cathode follower feeds the first direct-coupled amplifier.

13.9 SOLID-STATE CONFIGURATIONS

Circuit action in solid-state configurations can be discussed to good advantage by employing equivalent circuits. There are various types of equivalent circuits, and it will be helpful to use a comparatively simple arrangement in which a transistor is represented by a resistive T network. As in the case of any equivalent circuit, an equivalent T network has definite limitations, and we must avoid exceeding the competency of our representation. Most solid-state oscilloscope circuits utilize the common-emitter configuration, although the common-collector (emitter follower) configuration is also employed. The common-base configuration is used occasionally. From the standpoint of circuit analysis, we shall find it helpful to start with the common-base configuration.

We recall from our electronic circuit course that the input resistance of a transistor changes as the load resistance is varied. In other words, as the load is varied in a common-base stage, the emitter draws a changing amount of current from the driving source. Ohm's law states the value of input resistance to a transistor stage as

$$R_{in} = \frac{V_{in}}{I_{in}} \qquad (13.8)$$

With reference to Fig. 13.30, we note that the input resistance (ac resistance) is 35 Ω when a collector load resistance of 10K is used in a common-base configuration with a typical junction transistor. In this example, the generator has an internal resistance of 1K, which is representative of general conditions of operation. Next, Fig. 13.31 shows how the ac input resistance varies with the value of collector load resistance in the common-base (cb) configuration. We may assume that the transistor in part (*b*) of the figure is represented by the equivalent T network in part (*c*). Thus, R_e denotes the emitter resistance, R_b denotes the base resistance, and R_c denotes the collector resistance. It is evident that if we increase the value of R_L in Fig. 13.30 *c*, the value of R_{in} must also increase.

Next, let us consider the output resistance of a transistor in the cb configuration. In Fig. 13.30, we observe that an ac output resistance of 1 M is typical when the generator internal resistance is 1 K. With reference to Fig. 13.32, we note that the input resistance to the transistor changes considerably when the generator internal resistance is varied, and that the roc (rate of change) becomes comparatively slow for

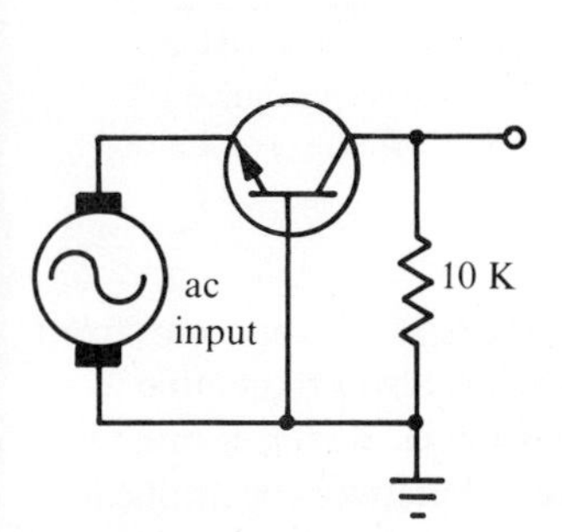

Voltage gain: 380 times
Current gain: 0.98
Power gain: 26 dB
Input resistance: 35 Ω
Output resistance: 1 M
(For generator internal resistance of 1 K)

(*a*)

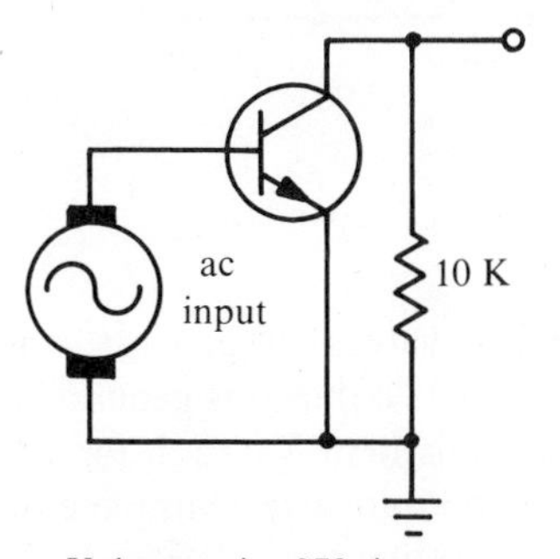

Voltage gain: 270 times
Current gain: 35 times
Power gain: 40 dB
Input resistance: 1.3 K
Output resistance: 50 K
(For generator internal resistance of 1 K)

(*b*)

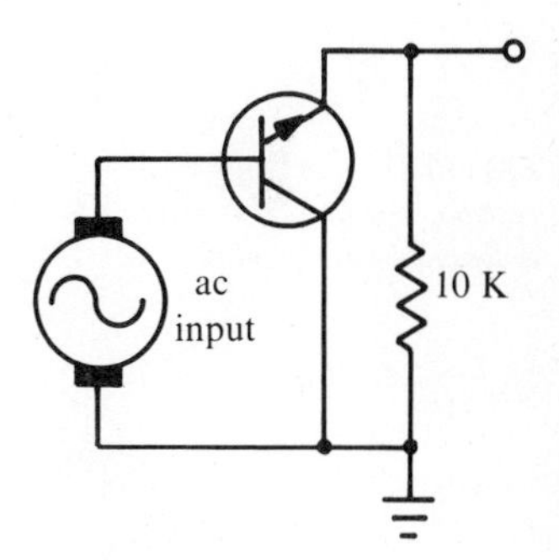

Voltage gain: 1
Current gain: 36 times
Power gain: 15 dB
Input resistance: 350 K
Output resistance: 500 Ω
(For generator internal resistance of 1 K)

(*c*)

FIG. 13.30 *Comparison of typical characteristics of a triode transistor in the three basic amplifier configurations.*

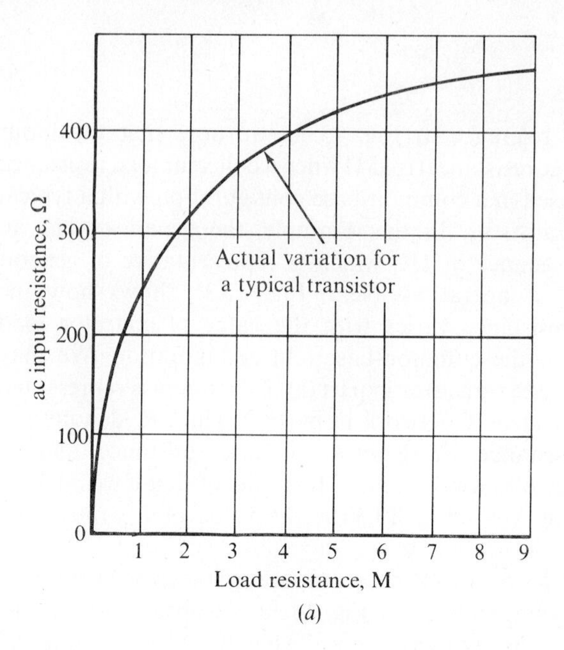

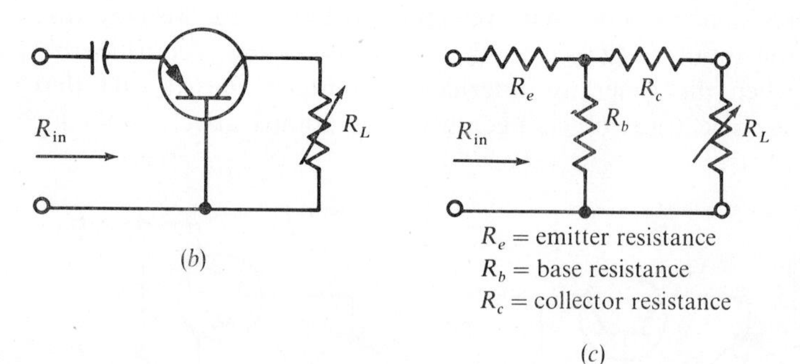

FIG. 13.31 *Variation of input resistance with load resistance for a common-base amplifier.*

higher values of generator resistance. It is apparent from Fig. 13.32 *c* that this general variation is also predictable from the equivalent T circuit for a transistor. However, it must not be supposed that a simple equivalent T network is applicable to all transistor circuits. For example, Fig. 13.33 shows the typical variation of input resistance with load resistance for a common-emitter configuration. However, if we inspect the equivalent T network in part (*c*), we would predict that the input resistance would increase as the load resistance increases. To anticipate subsequent discussion, we shall find that another equivalent circuit is commonly employed, which is in better agreement with variational analysis of the ce configuration.

Figure 13.34 shows that an equivalent T network does not account for the decrease in output resistance of a transistor in the ce configuration when the generator resistance is increased. That is, if the value of R_g is increased in part (*c*), the value of R_{out} would also be increased. Thus, the equivalent T network is not applicable in this case. On the other hand, the equivalent T circuit predicts an increase of input resistance for the common-collector (cc) configuration when the load resistance is increased, as shown in Fig. 13.35. Thus, for the

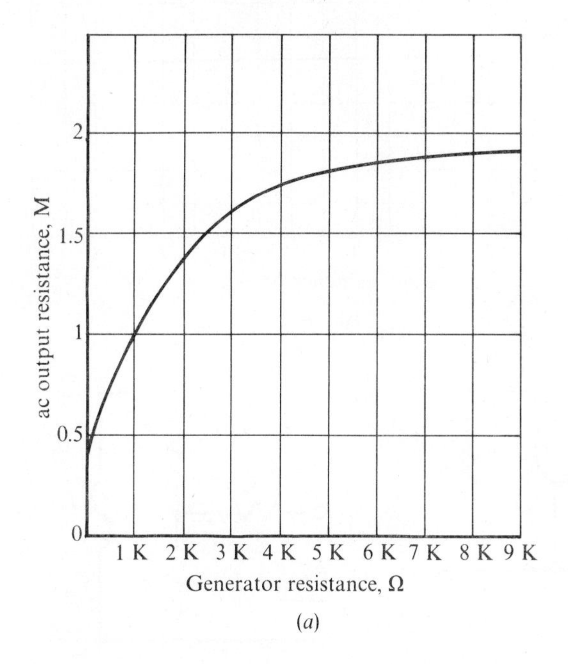

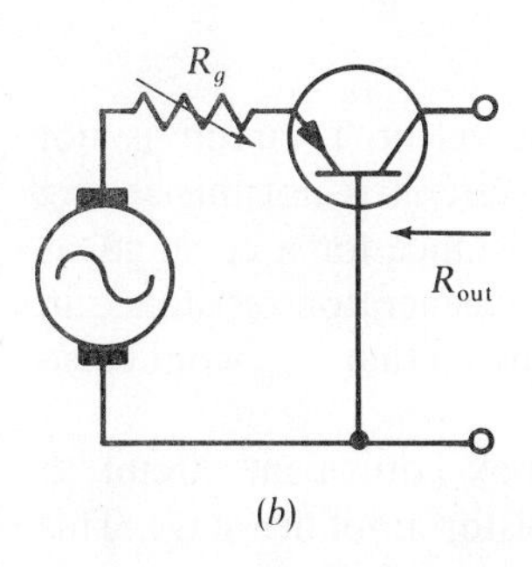

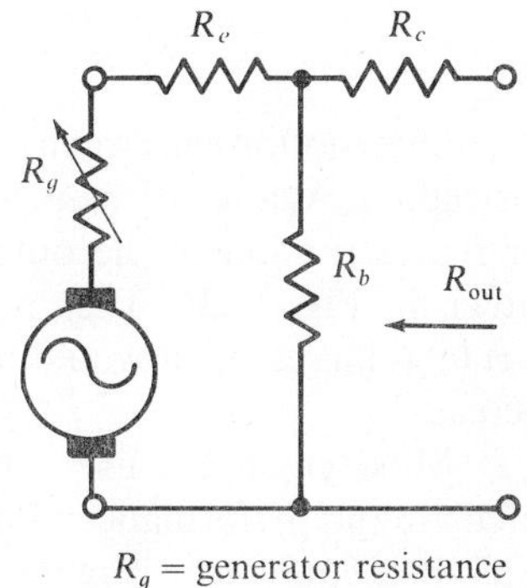

FIG. 13.32 *Variation of output resistance with generator resistance for a common-base amplifier.*

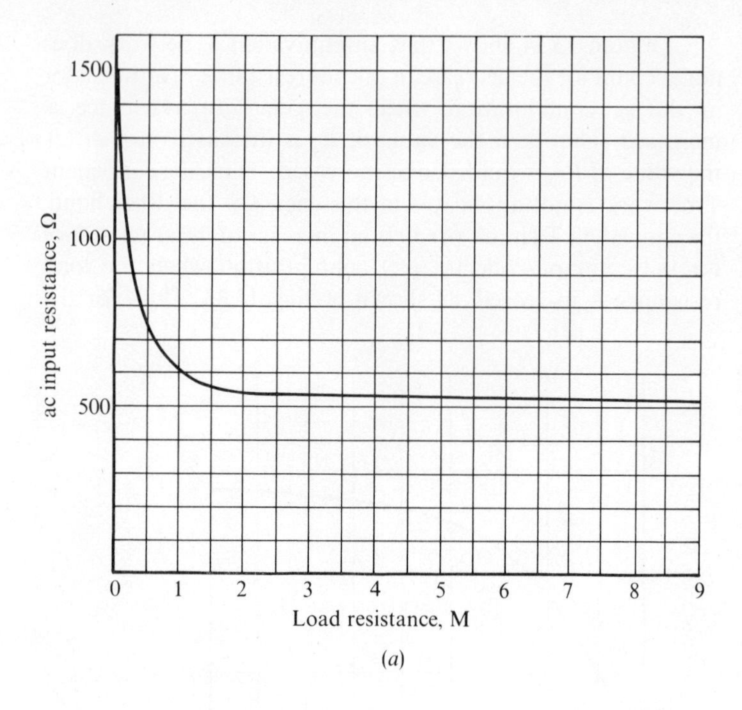

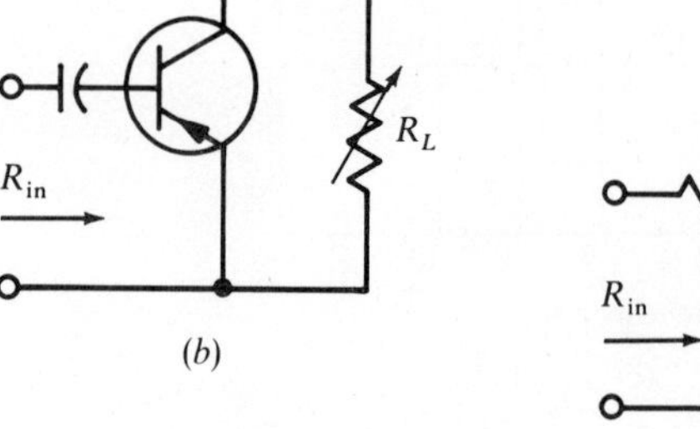

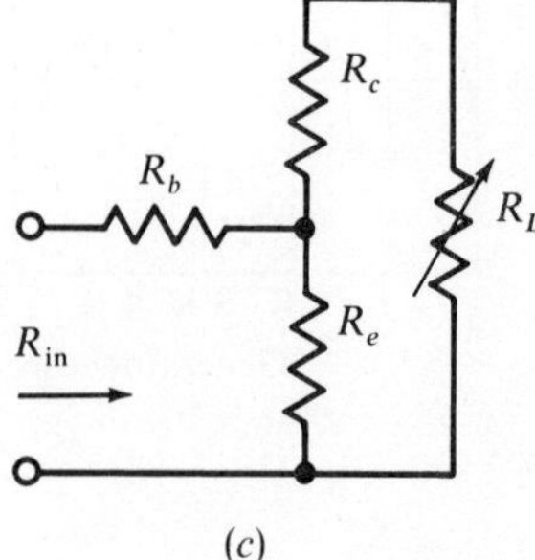

FIG. 13.33 *Variations of input resistance with load resistance for a common-emitter amplifier.*

cc (emitter-follower) circuit, an equivalent T circuit is not misleading. Again, the equivalent T circuit is not misleading for representation of the output resistance for a cc stage, as shown in Fig. 13.36. That is, as the generator resistance in part (*c*) is increased, it would be predicted that R_{out} would also increase.

Most engineers use a black-box equivalent circuit to calculate the performance of a transistor amplifier stage. This black-box equivalent is merely an unspecified active network with a pair of input terminals and a pair of output terminals.

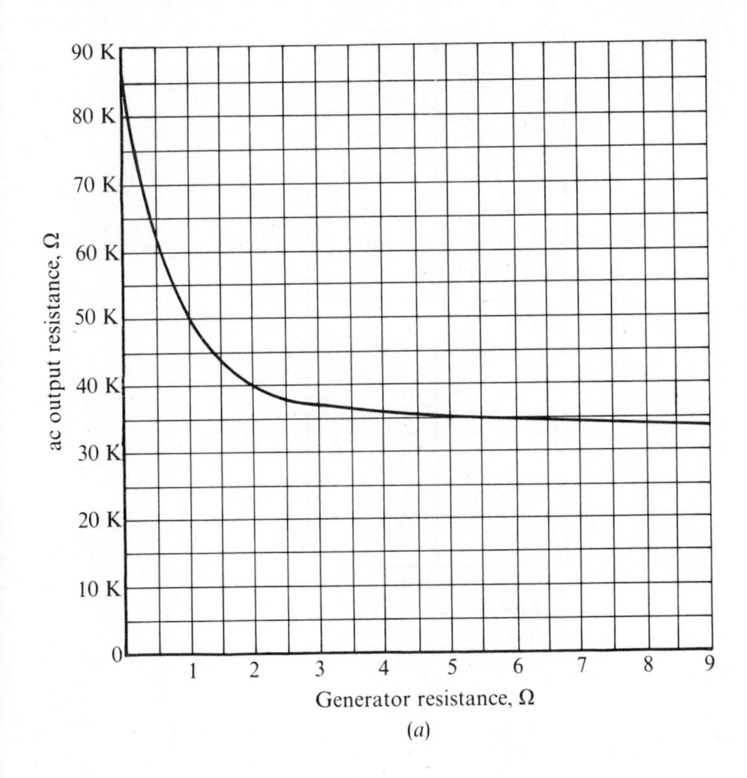

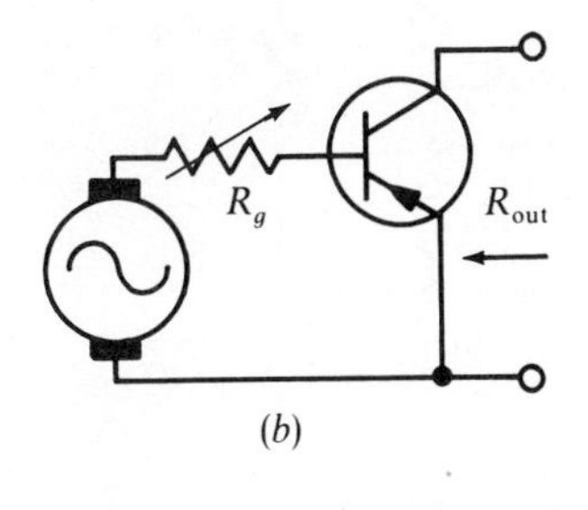

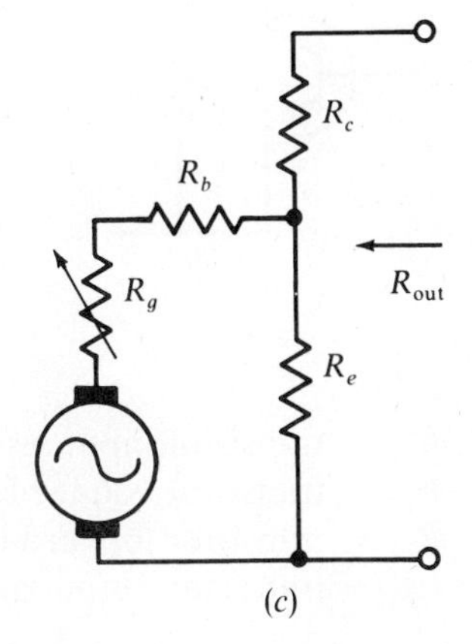

FIG. 13.34 *Variation of output resistance with generator resistance in a common-emitter amplifier.*

It is characterized by means of hybrid parameters, published in transistor manuals. Hybrid parameters are referred to the cb configuration, from which equations for the ce and cc configurations can be derived. The four basic hybrid parameters are:

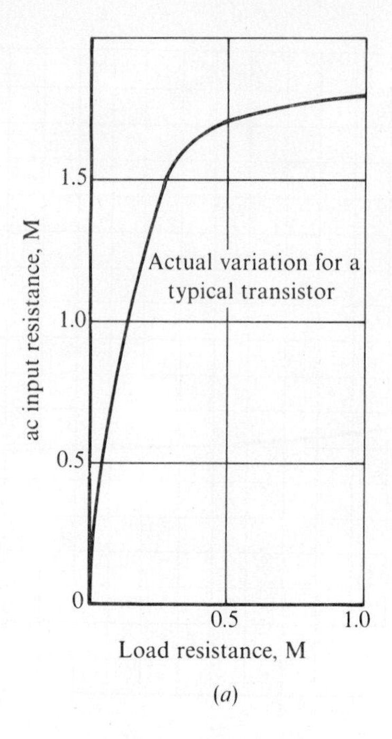

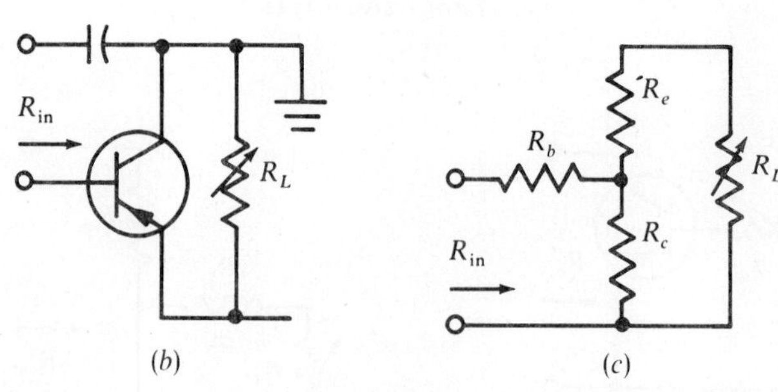

FIG. 13.35 *Variation of input resistance vs. load resistance for a common-collector amplifier.*

h_{11} = transistor input resistance
h_{12} = transistor voltage-feedback ratio
h_{21} = transistor forward-current ratio
h_{22} = transistor output admittance

Let us briefly consider the common-collector configuration as an example of the application of hybrid parameters. We will stipulate that the following notation indicates the transistor characteristics of concern to us in the cc configuration:

h_{ic} = transistor input resistance
h_{rc} = transistor voltage-feedback ratio
h_{fc} = transistor forward-current gain
h_{oc} = transistor output admittance

In turn, the derived expressions are written:

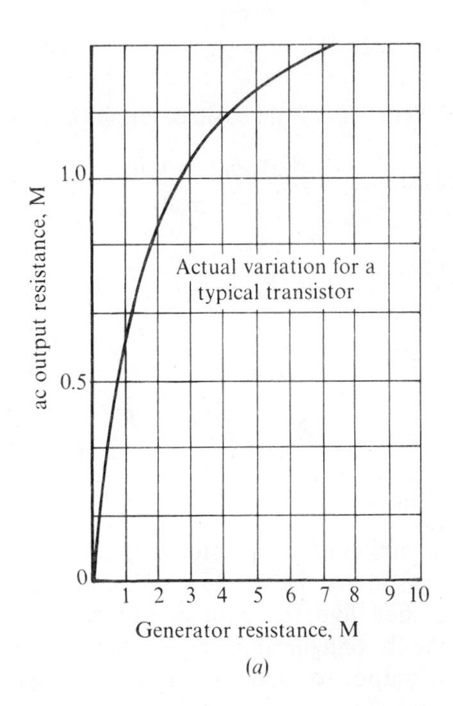

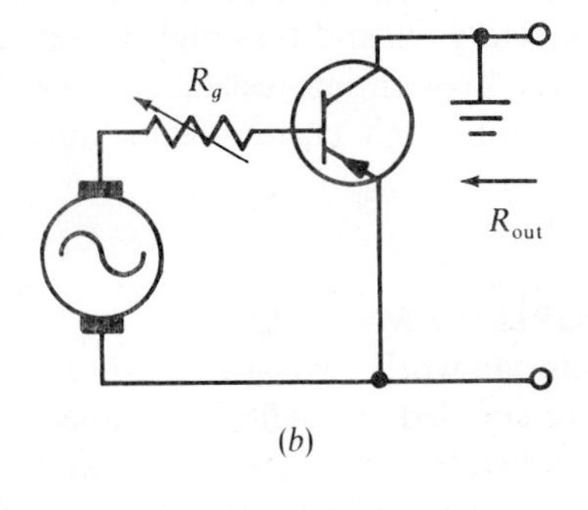

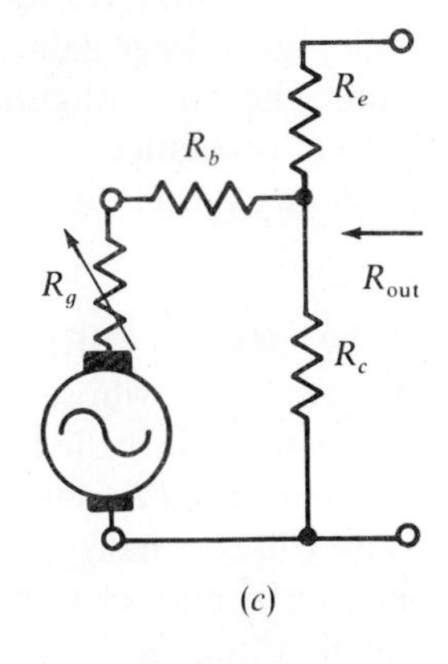

FIG. 13.36 *Variation of output resistance vs. generator resistance for a common-collector amplifier.*

$$h_{ic} = \frac{h_{11}}{1 + h_{21}}$$

$$h_{rc} = 1$$

$$h_{fc} = \frac{-1}{1 + h_{21}}$$

$$h_{oc} = \frac{h_{22}}{1 + h_{21}}$$

Finally, the parameters of the basic cc circuit are written:

$$R_i = \text{input resistance} = \frac{R_L(h_{ic}h_{oc} - h_{rc}h_{fc}) + h_{ic}}{1 + R_L h_{oc}}$$

$$A_e = \text{voltage gain} = \frac{-R_L h_{fc}}{R_L(h_{ic}h_{oc} - h_{rc}h_{fc}) + h_{ic}}$$

$$A_i = \text{current gain} = \frac{h_{fc}}{1 + R_L h_{oc}}$$

$$R_o = \text{output resistance} = \frac{h_{ic} + R_g}{h_{ic}h_{oc} - h_{rc}h_{fc} + R_g h_{oc}}$$

$$A_p = \text{power gain} = A_i A_e$$

Thus, the action of a transistor amplifier circuit can be analyzed and calculated. From the standpoint of a simplified summary, we shall find the data in Fig. 13.30 to be very useful. Thus, the cb configuration has very low input resistance, and very high output resistance, with high voltage gain, no current gain, and high power gain. The ce configuration has moderate input resistance and moderate output resistance, with high voltage gain, high current gain, and very high power gain. The cc configuration has high input resistance, low output resistance, no voltage gain, high current gain, and moderate power gain.

13.10 BASIC DIRECT-COUPLED AMPLIFIER

An amplifier whose coupling networks consist of direct connections is called a direct-coupled amplifier. A direct-coupled amplifier is used to amplify dc voltage changes as well as ac voltage changes. The use of direct-coupled amplifiers in the horizontal channel of an oscilloscope permits an ac signal voltage to be superimposed on a dc reference voltage, thus providing a method of centering or positioning the pattern on the face of the crt. This method of centering shall be discussed later.

The simplest form of direct-coupled amplifiers consists of a single tube or transistor with a grid resistor across the input and a load connected in the plate circuit (or collector), as shown in Fig. 13.37*a*. The dc voltage change to be amplified

FIG. 13.37 *Comparison of direct input and capacitance input to a direct-coupled amplifier.*

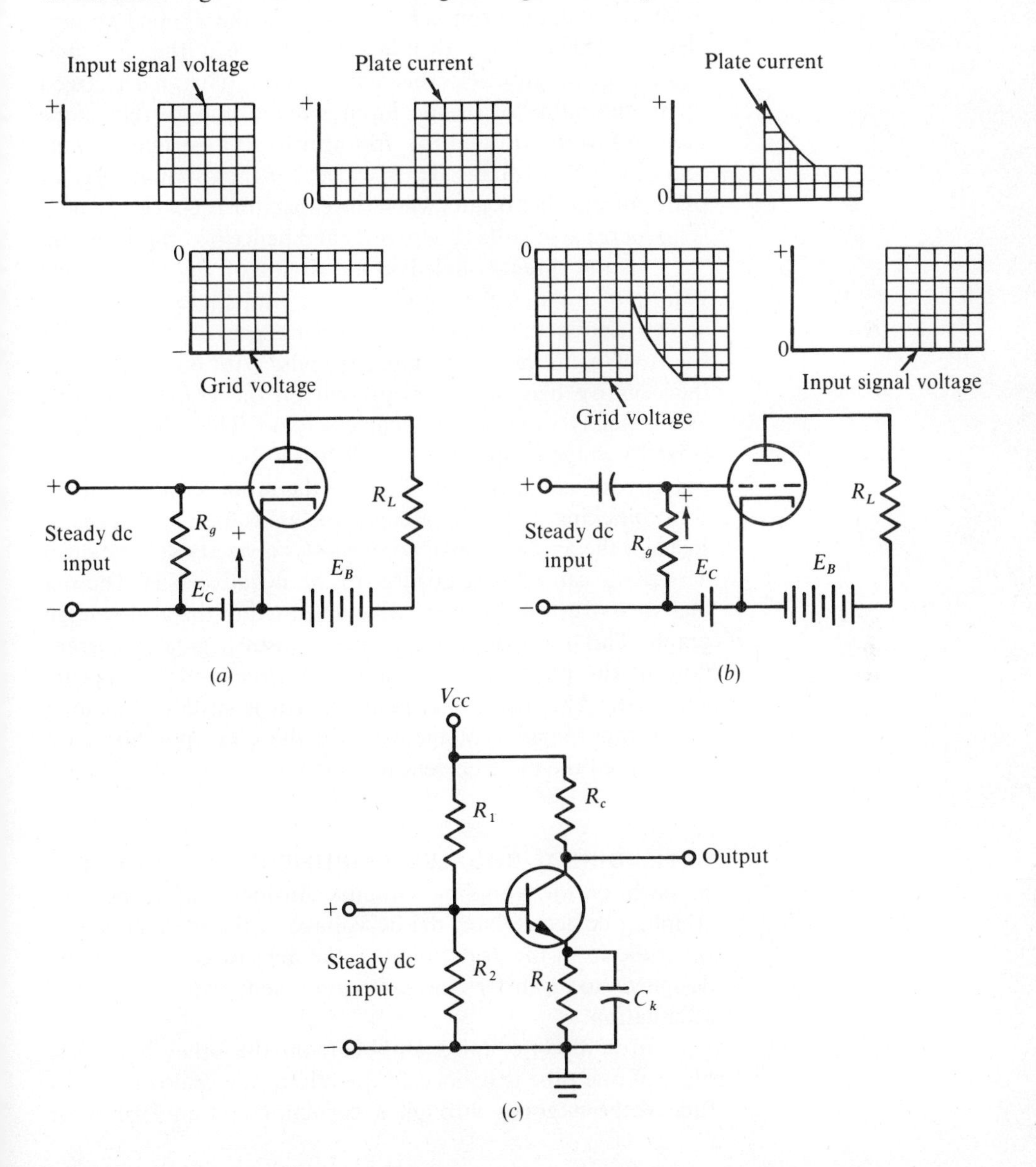

is applied directly to the grid of the amplifier tube; thus, direct coupling is required in the input circuit. A capacitive input circuit is also shown to indicate how the capacitor changes a pulsating-dc voltage to an ac signal.

In the capacitor input circuit of Fig. 13.37*b*, graphs of the signal voltage, grid voltage, and plate current are shown. The applied dc voltage charges the capacitor, and momentarily the voltage drop across R_g equals the applied voltage change. This voltage then appears between the grid and cathode of the tube. However, when the capacitor is charged up to the value of the dc input voltage, the current stops flowing through R_g and the grid returns to its original value, that of the bias voltage. Thus, except for the original surge of plate current that occurs when the capacitor is charging, there is no increase in voltage across R_L and hence no amplification.

In the direct-coupled input circuit of Fig. 13.37*a*, the graphs of input signal, grid voltage, and plate current are shown above the circuit. The input signal is like that in Fig. 13.37*b*, but here the similarity ends. With no input signal the negative bias voltage is present on the grid of the tube and a steady value of plate current flows. This action causes a fixed voltage drop across R_L. When a direct voltage of the polarity indicated is applied across the input terminals, there is no blocking action by a capacitor as in the previous case. Instead, the applied signal continues as a steady voltage drop across R_g, canceling a portion of the negative bias. The net bias then drops to the new value indicated in the grid voltage graph. This reduction in grid bias causes a greater current flow in the plate circuit, and thus a greater drop appears across R_L. The increase in plate current is sustained as long as the input signal voltage exists at the corresponding level that caused the plate current to increase.

13.11 DIRECT-COUPLED AMPLIFIERS IN CASCADE

In each of the coupling circuits considered thus far, the coupling device isolates the dc voltage in the plate circuit of one tube from the grid circuit of the next tube; but they are designed to transfer the ac component with minimum attenuation.

In a direct-coupled amplifier, on the other hand, the plate of one tube is connected directly to the grid of the next tube without going through a capacitor, a transformer, or any similar coupling device. This arrangement presents a voltage distribution problem. Since the plate of a tube must have a positive voltage with respect to its cathode, and the grid of the next tube must have a negative voltage with respect to its cathode, it follows that the two cathodes cannot operate at the same potential. Proper voltage distribution is obtained by a voltage divider, as shown at points *A*, *B*, *C*, *D*, and *E* in Fig. 13.38.

In this amplifier the plate of V_1 is connected directly to the grid of V_2. The grid of V_1 is returned to point *A* through R_{g1}. The cathode of V_1 is returned to point *B*. The grid bias for V_1 is developed by the voltage drop between points *A* and *B* of the voltage divider. The plate of V_1 is connected through its plate-load resistor, R_L, to point *D* on the divider. R_L also serves as the grid resistor for V_2.

Since the plate current from V_1 flows through R_L, a certain amount of the supply voltage appears across R_L. The amount of voltage developed across R_L must be allowed for in choosing point *D* on the divider. Point *D* is so located that approximately half the available voltage is applied to the plate of V_1. The plate of V_2 is connected through a suitable output load, *R*, to point *E*, the most positive point on the divider. Since the voltage drop across R_L may place too high a negative bias on the grid of V_2, it may be necessary to connect the cathode of V_2 at point *C*, which is negative with respect to point *D*, in order to lower the bias on the grid of V_2

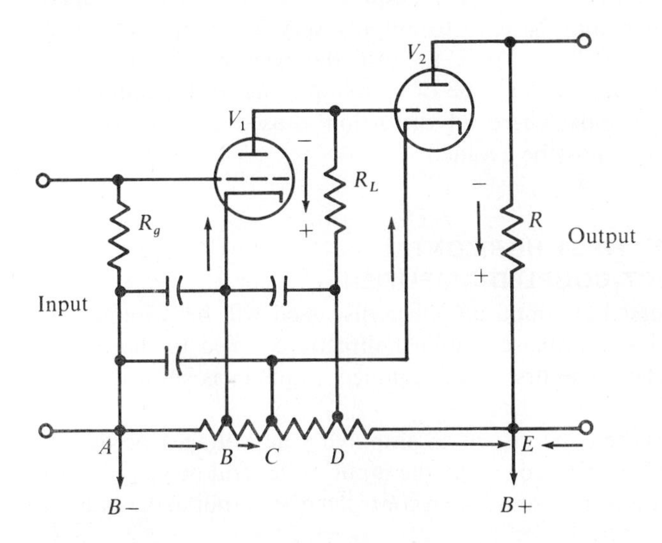

FIG. 13.38 *Direct-coupled amplifier.*

(since the voltages across R_L and CD are in opposition). Point C, together with the value of R, determines the proper voltage for V_2.

The entire circuit is a complex resistance network that must be adjusted carefully to obtain the proper plate and grid voltages for both tubes. If more than two stages are used in this type of amplifier, it is difficult to achieve stable operation. Any small changes in the voltages of the first tube will be amplified and will thus make it difficult to maintain proper bias on the final tube connected into the circuit. Because of the instability thus encountered, direct-coupled amplifiers are practically always limited to two stages. Furthermore, the power-supply voltage must be twice that required for one stage.

One method of supplying the range of voltage needed is to use a power supply that provides approximately equal amounts of both positive and negative voltages with respect to ground. This arrangement allows cascading without necessitating excessively high plate supply voltages. In Fig. 13.38, either point C or point D might properly be tied to ground potential.

When the tube voltages are properly adjusted to give Class A operation, the circuit serves as a distortionless amplifier whose response is uniform over a wide frequency range. This type of amplifier is especially effective at the lower frequencies because the impedance of the coupling elements does not vary with the frequency. Thus a direct-coupled amplifier may be used to amplify very low frequency variations in voltage. Also, because the response is practically instantaneous, this type of coupling is useful for amplifying pulse signals where all distortion caused by the coupling elements must be avoided.

13.12 FIRST HORIZONTAL DIRECT-COUPLED AMPLIFIER

The first horizontal amplifier discussed will be a frequency-selective paraphase amplifier directly coupled to the output amplifier. The first direct-coupled amplifier is shown in Fig. 13.39.

When the circuit is amplifying ac voltages or the internally generated sweep, the input to the grid of V_{106A} comes from the ac horizontal gain control in the output of the cathode

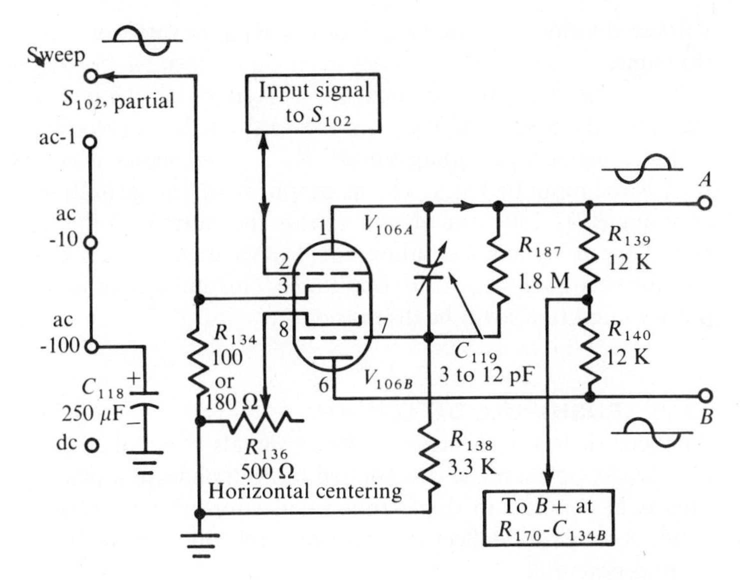

FIG. 13.39 *First direct-coupled amplifier.*

follower stage. In all the ac positions the cathode resistor R_{134} is bypassed to prevent degeneration. When the circuit is amplifying dc voltages the input to the grid of V_{106A} comes from the direct-coupled horizontal gain control.

The bias for V_{106A} is equal to the voltage dropped across R_{134}. The bias for V_{106B} is the voltage dropped across R_{136}. Under normal static conditions the dc plate voltage of V_{106A} would be equal to the dc plate voltage of V_{106B}. This would result in both deflection plates having equal positive dc voltages which would place the beam in the center of the crt screen. By adjusting the horizontal centering control R_{136} the resultant bias on V_{106B} can be changed. This causes a change of V_{106B} dc plate voltage. The dc output voltage at point B can be changed in respect to the dc output voltage at point A, thereby providing a method of positioning the crt beam at any desired horizontal position on the crt screen. It should be remembered that a paraphase amplifier depends on an ac voltage divider network for proper operation. The operation of this amplifier is as follows: The horizontal ac signal is coupled to the grid of V_{106A}. An amplified ac signal appears on the plate of V_{106A}. This is the ac signal present at output A as shown in Fig. 13.39. C_{119}, R_{187}, and R_{138} form an ac voltage divider. At low and medium frequencies the resistance ratio between R_{187} and R_{138} primarily determines the

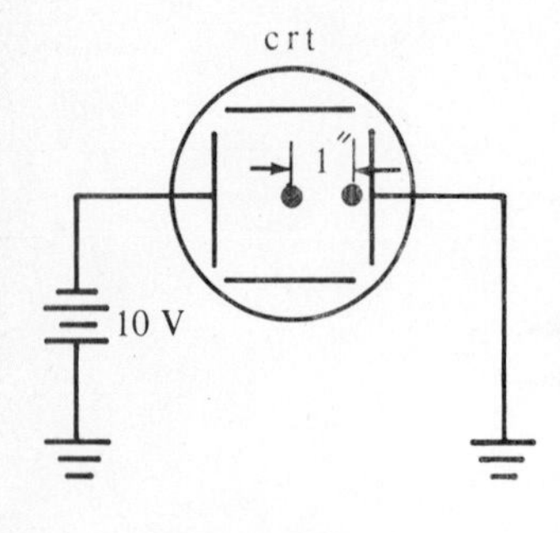

FIG. 13.40 *Single-ended deflection.*

voltage division. The ac signal across R_{138} is very small at this time.

As the frequency applied to the horizontal channel is increased the reactance of C_{119} starts to shunt R_{187}, resulting in more signal developing across R_{138}. This signal across R_{138} is the input to V_{106B}. The ac output of this stage (output at point B) is 180° out of phase with the output of V_{106A}. Both output signals are riding equal dc reference levels but are not necessarily equal in amplitude. This unbalance in ac output amplitudes will be discussed later.

13.13 PUSH-PULL DEFLECTION

Balanced deflection is achieved when signals of equal amplitude but opposite phase are applied to both deflection plates. This is in contrast to single-ended deflection where a signal is applied to one deflection plate and the other plate is at ground potential.

Most oscilloscopes employ push-pull amplifiers in the horizontal deflection channel to achieve push-pull deflection. In push-pull deflection using push-pull amplifiers, the amplitude of the driving signal required is one-half the amplitude of a signal required for single-ended deflection systems. In Fig. 13.40 where a dc voltage is used, the same results would be obtained if an ac signal were used.

In Fig. 13.40 a 10-V signal is applied to the left deflection plate and the right deflection plate is grounded. This deflects the crt beam 1 in. to the right. The same amount of deflection can be produced by applying one-half this voltage—but of different polarity—to both plates, as illustrated in Fig. 13.41. This opposite polarity driving signal may be achieved by using a push-pull amplifier output stage, or a paraphase amplifier.

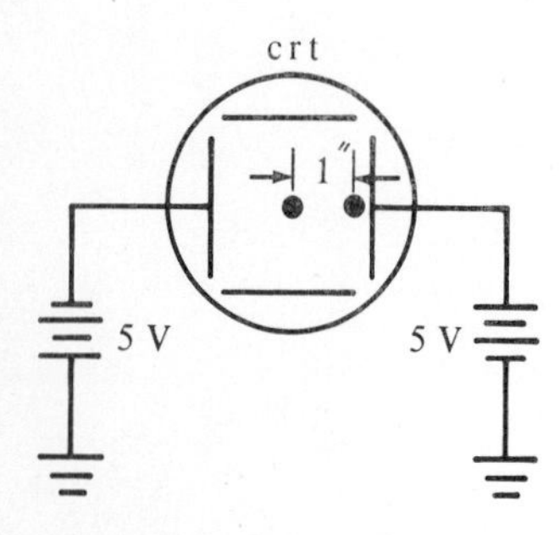

FIG. 13.41 *Push-pull deflection.*

13.14 SECOND HORIZONTAL DIRECT-COUPLED AMPLIFIER

A push-pull amplifier, with a paraphase amplifier driving it, is illustrated in Fig. 13.42. This is not a conventional push-pull amplifier but is sometimes referred to as a cathode-coupled push-pull amplifier.

The two control grids of the push-pull amplifier are directly coupled to the plates of the paraphase amplifier.

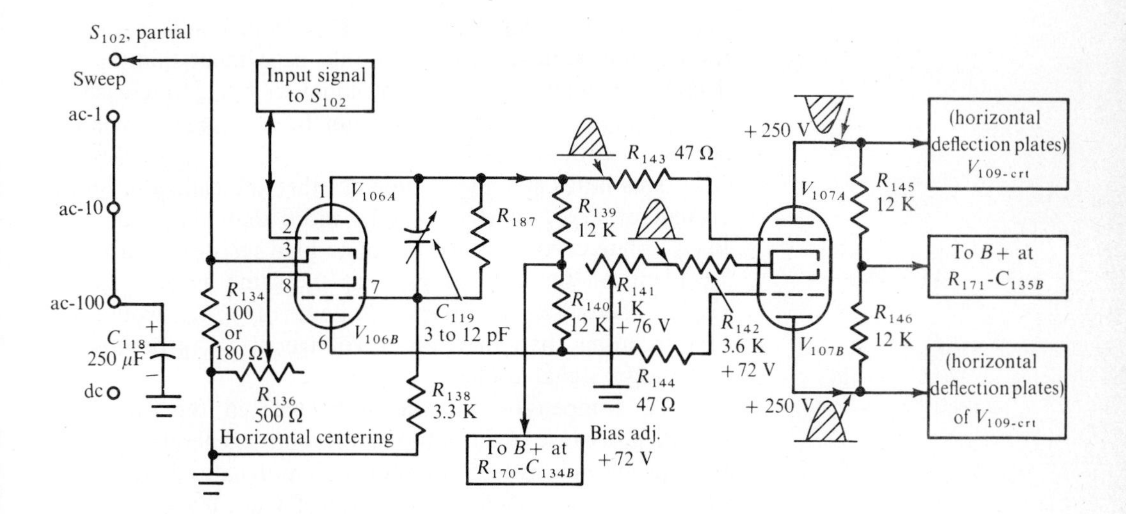

FIG. 13.42 *First and second direct-coupled stages.*

so it can couple frequencies down to zero.

Each grid is approximately 72 V positive in respect to ground. The cathodes are tied together and connected to a common resistive network composed of R_{141} and R_{142}. Since the plate load resistors are identical, the quiescent currents through the tubes will be equal. The sum of these two currents will provide a cathode voltage of approximately 76 V, or approximately minus 4-V bias for each tube.

The bias control R_{141} is provided so that compensation for tolerances in resistors and electron tubes may be made in order to maintain the proper bias on the final stage. Proper adjustment of R_{141} is obtained when the voltage drop across the plate load resistors R_{145} and R_{146} is 90 V, with the electron beam centered horizontally on the crt screen. The plates of V_{107A} and V_{107B} are connected directly to the horizontal deflection plates of the crt.

The ac driving signal is directly coupled to the grid (pin 6) of V_{107A}. For simplicity, a positive alternation of the horizontal ac signal at the plate of V_{106A} (Fig. 13.42) will be used.

With a positive-going signal applied, the plate current of V_{107A} will increase, producing a negative-going alternation of plate voltage and causing a positive voltage to be felt on the cathode. Since the cathode is common to both stages, this rise will affect V_{107B}. The grid of V_{107B} is effectively at ac ground. The positive cathode voltage causes the same effect as a negative control grid. This action results in a decrease in the plate current of V_{107B} and tends to reduce the cathode voltage drop. It should be evident that $V_{107B's}$ decrease in plate current is not equal to $V_{107A's}$ increase. If this were the case, there would not be a change of voltage across the cathode resistors.

This latter statement indicates that the change in plate current of V_{107A} must always be more than the change in $V_{107B's}$ plate current. Since the plate load resistors and tubes are identical, the gain of V_{107A} is the same as the gain of V_{107B}. Therefore, the output voltage of V_{107A} would be slightly larger than the output voltage of V_{107B} because it has a larger signal applied.

To compensate for this unbalance in the push-pull amplifier, because of the cathode follower degenerative action and the degenerative action of the common cathode resistors, some signal is coupled into the grid of V_{107B} from the paraphase amplifier. The amount of signal coupled is just enough to compensate for the losses in gain.

As the input frequency to the first and second direct-coupled amplifiers is increased, a loss of gain occurs that is due to the shunting effects of stray and interelectrode capacitance. To correct for this, more signal is coupled into V_{106B} by the shunting effect of C_{119}.

By not operating the push-pull amplifiers with driving signals of equal amplitude, at low and medium frequencies, the deflection sensitivity is less than maximum. The overall deflection sensitivity can be improved by approaching a condition of equal driving signals as the input frequency is increased. This counteracts the loss of gain due to shunting capacity at high frequencies and keeps the deflection sensitivity of the horizontal channel uniform over the frequency range of from 1 Hz to 500 kHz.

Since the circuit is push-pull, any $B+$ variation caused by fluctuating line voltage has little or no effect on the centering of the crt beam. This is because a dc voltage change on one deflection plate is accompanied by an equal voltage change on the other deflection plate.

13.15 HORIZONTAL CHANNEL USED TO INDICATE DC VOLTAGES

Figure 13.43 shows the input circuitry to the horizontal channel when the attenuator switch S_{102} is in the dc position.

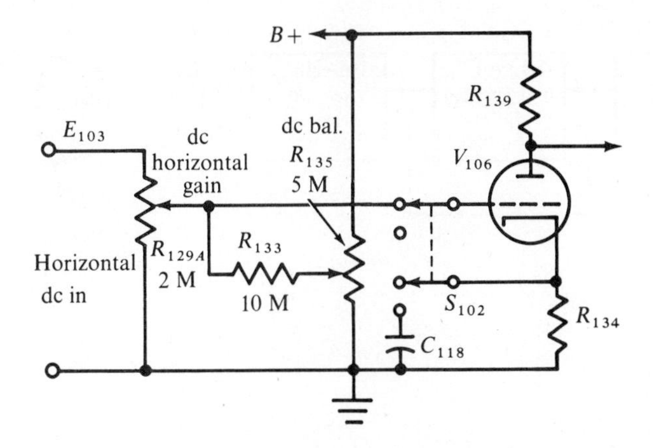

FIG. 13.43 *Direct-coupled input circuit.*

When the horizontal channel is used to indicate dc voltages, the input circuit to the first direct-coupled amplifier is modified and becomes a direct-coupled input circuit. When the horizontal attenuator switch S_{102} is in the dc position, a negative contact potential is developed on the grid of V_{106A} owing to the high impedance of the horizontal dc gain control R_{129A}. This condition is canceled by the selection of a positive potential by the horizontal dc balance control R_{135}, which is in series with R_{133}.

Improper adjustment of R_{135} will result in a shift of the electron beam with various settings of the horizontal gain control R_{129A}. Proper adjustment of the horizontal balance control is made when rotation of the horizontal gain control has no effect on the position of the horizontal trace.

The dc voltage to be measured is developed across the horizontal gain control R_{129A}. This potentiometer is used as a variable voltage divider. The voltage selected by the position of this control helps to bias the first direct-coupled amplifier V_{106A}. This bias controls the dc plate voltage of V_{106A}. As a result of direct coupling, V_{106A} in turn controls the push-pull amplifier V_{107A} which positions the crt beam in proportion to the dc input voltage.

13.16 VERTICAL AMPLIFIERS

The vertical amplifier channel is shown in block diagram form in Fig. 13.44. There are two possible inputs, both of which are external. In most cases, the signal to be viewed on the oscilloscope is applied to the vertical deflection amplifier.

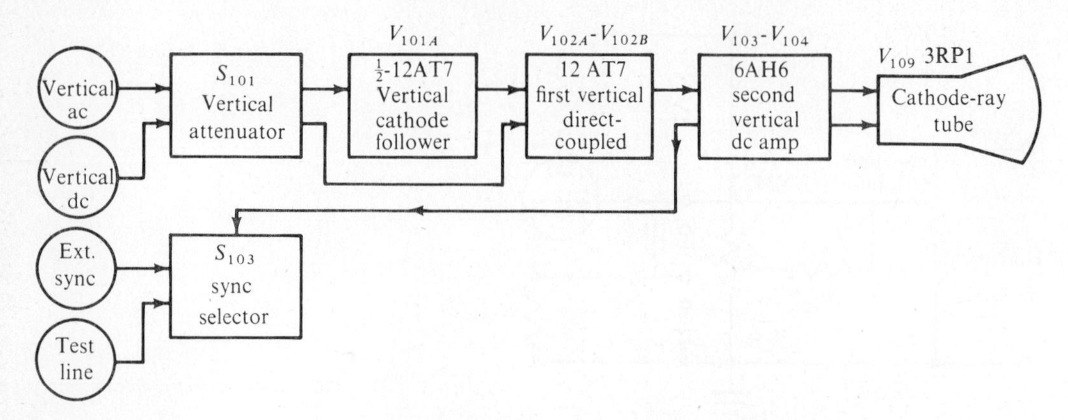

FIG. 13.44 *Block diagram of vertical channel.*

The specific signals that would be applied to the vertical channel are dependent on the type of equipment under test. The vertical channel provides two useful outputs. One is to the crt for beam deflection and the other is sent to the sync selector for the purpose of synchronizing the sweep generator with the vertical signal.

The ac input signal is coupled through the vertical attenuator to the vertical cathode follower. The dc input signal is coupled through the attenuator but bypasses the vertical cathode follower.

13.17 VERTICAL ATTENUATOR

As in the horizontal channel the purpose of the vertical attenuator is to select the type and strength of signal fed into the vertical amplifier channel. The vertical amplifiers are fixed-gain amplifiers and the output amplitude is adjusted by adjusting the input. The most common type of vertical attenuator employs a step-switch control. An ideal attenuator usually has a variable potentiometer for five amplitude adjustments. This potentiometer control is usually low in resistance and would tend to load down the attenuator if it were connected directly across the attenuator. For this reason, the potentiometer should have an impedance matching device between it and the attenuator. This impedance matching is accomplished by the use of a cathode follower stage.

13.18 PRACTICAL ATTENUATOR

Figure 13.45 shows a practical vertical attenuator. Section *A* of switch S_{101} selects the voltage divider to be used, and couples its output into the grid of the cathode follower stage. In the ac-1 position the ac signal receives no attenuation.

With S_{101} in positions ac-10 and ac-100, the input is attenuated by a factor of 10 and 100, respectively. The voltage divider networks have capacitors added for high frequency compensation as in the horizontal channel. With S_{101} in the dc position the ac input terminal J_{101} is grounded through C_{105}. In all the ac positions section C of switch S_{101} connects C_{108} across the cathode resistor of the first direct-coupled amplifier and prevents cathode degeneration. Section D of switch S_{101} applies the output of the cathode follower stage to the first direct-coupled amplifier in the ac positions. In the dc position the input from terminal E_{102} is applied to the dc vertical gain control through section D of switch S_{101} of the first vertical direct-coupled amplifier.

FIG. 13.45 *Vertical input attenuator.*

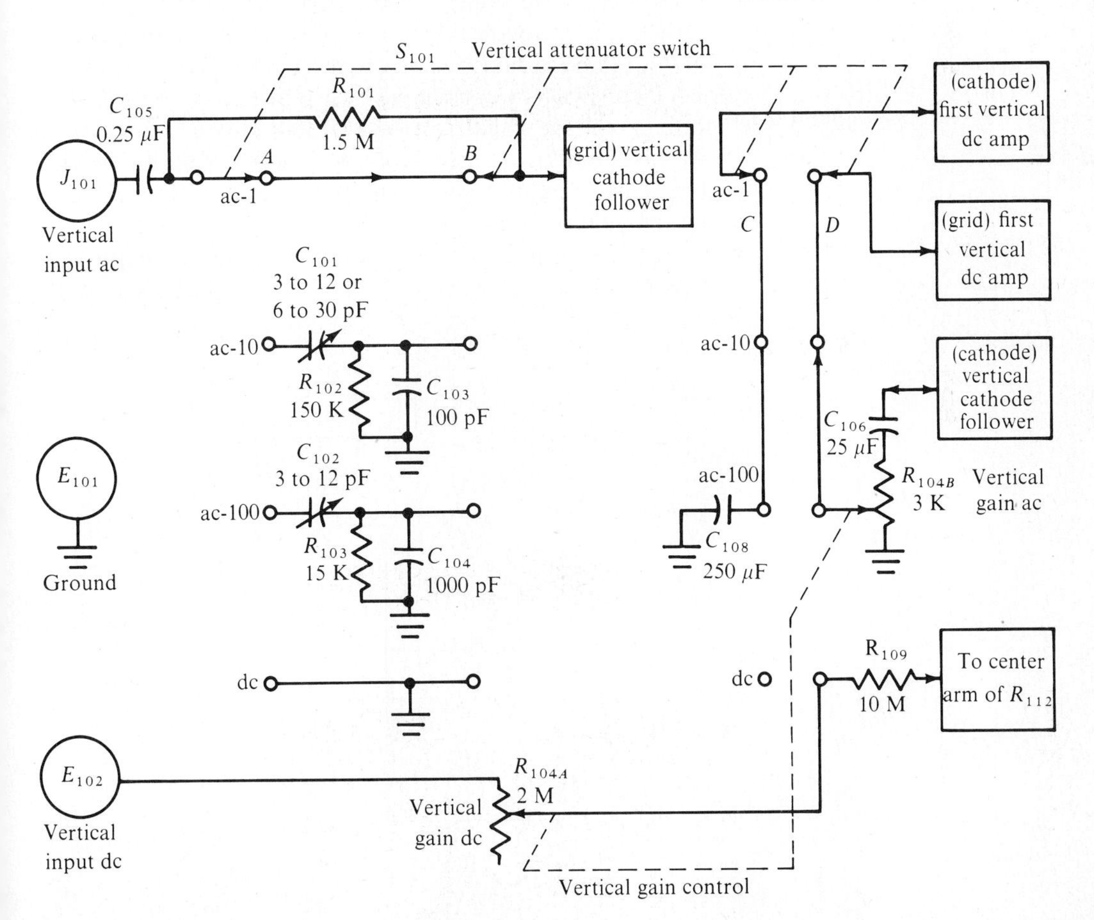

13.19 VERTICAL CATHODE FOLLOWER

Figure 13.46 shows a practical vertical channel cathode follower. In operation it is identical to that of the horizontal cathode follower discussed previously. Its main purpose is to provide an impedance match between the low impedance of the ac gain control R_{104B} and the attenuator network. The output of the cathode follower stage is passed through switch S_{101} section D in Fig. 13.38 to the grid of the first direct-coupled amplifier.

13.20 FIRST DIRECT-COUPLED VERTICAL AMPLIFIER

Figure 13.47 shows the first and second direct-coupled vertical amplifiers.

The operation of the first vertical direct-coupled amplifier is essentially the same as the horizontal amplifier discussed earlier. One primary difference, however, is the bias control of V_{102B}. The cathode pin 8 is connected through R_{189} to $B+$.

This resistor and the vertical positioning control, R_{111}, form a voltage divider network. The voltage drop across R_{111} is a resultant of the sum of the divider current and normal tube current. This entire network provides a wider range of bias voltage available to V_{102B} when R_{111} is varied.

Capacitor C_{109} and R_{113} form a frequency-sensitive ac voltage divider network. As the frequency applied to the

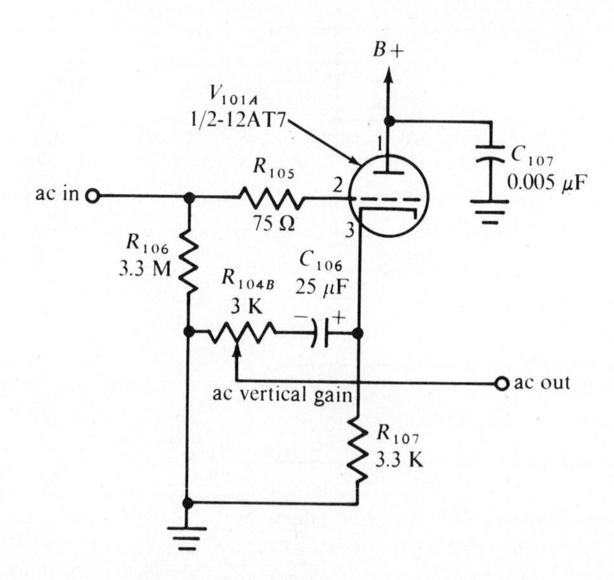

FIG. 13.46 *Vertical cathode follower.*

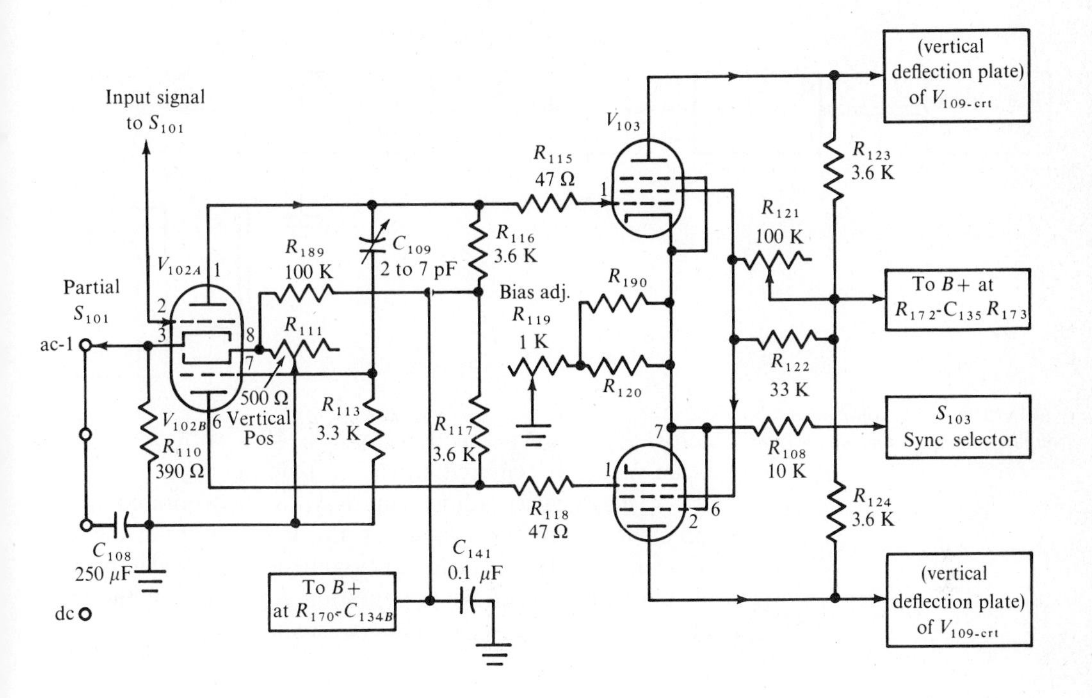

FIG. 13.47 *Vertical amplifiers.*

vertical amplifiers is raised, more signal is developed across R_{113}. The ac outputs of the first direct-coupled amplifier are 180° out of phase but not necessarily equal in amplitude. The first direct-coupled amplifier is directly coupled to the push-pull amplifier.

13.21 SECOND DIRECT-COUPLED AMPLIFIER

The operation of the push-pull vertical amplifier is similar to the operation of the horizontal push-pull amplifier as discussed above. Pentode tubes are used primarily to increase the frequency response of the vertical amplifier. This particular circuit has an upper frequency response of approximately 2 MHz. Since the push-pull amplifiers are cathode-coupled, a driving signal is developed across the common-cathode resistor. This vertical signal is coupled from the cathodes of the push-pull amplifier for the purpose of synchronizing the sweep generator to the signal being viewed.

The bias adjustment potentiometer R_{119} is provided to compensate for tolerances in resistors and electron tubes. Proper adjustment of R_{119} is obtained when the voltage drop

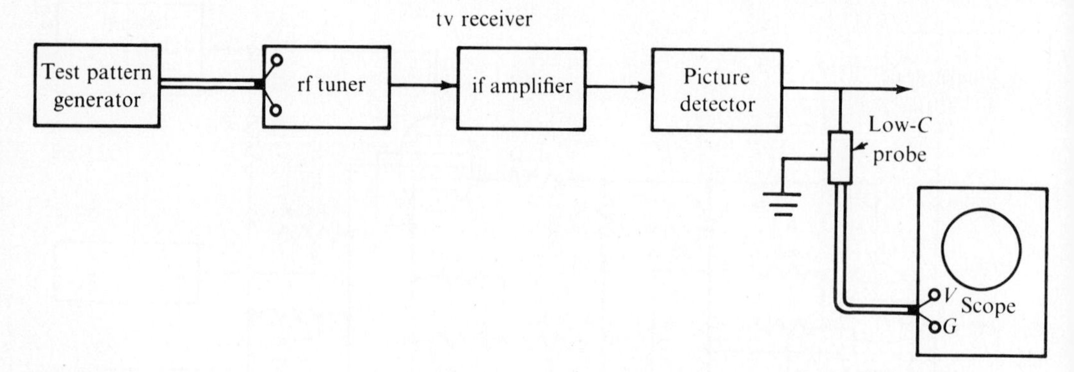

FIG. 13.48 *Video waveforms display at output of picture detector.*

across R_{123} or R_{124} in this circuit is 45 V with the electron beam vertically centered on the cathode-ray tube.

The linearity adjustment R_{121} is incorporated in the circuit to adjust the voltage on the screen grids of the final push-pull amplifier. The linearity control is adjusted to provide an undistorted vertical presentation regardless of the trace placement (vertically) on the crt screen.

13.22 APPLICATIONS

One of the basic applications for an oscilloscope in television work is to check the video waveform at the output of the picture detector, as shown in Fig. 13.48. It is preferable to utilize a test pattern generator to energize the receiver, because a steady waveform is thereby provided. The oscilloscope should have a vertical amplifier with frequency response to

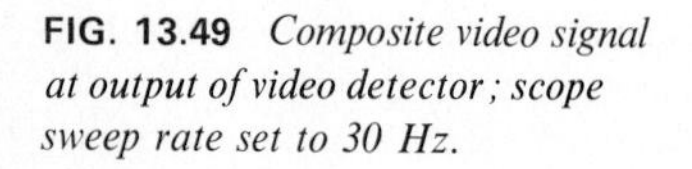

FIG. 13.49 *Composite video signal at output of video detector; scope sweep rate set to 30 Hz.*

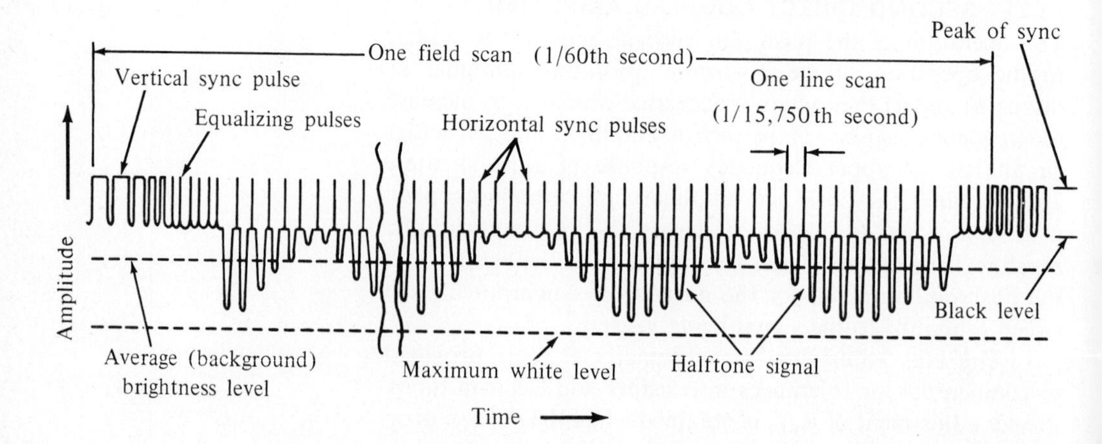

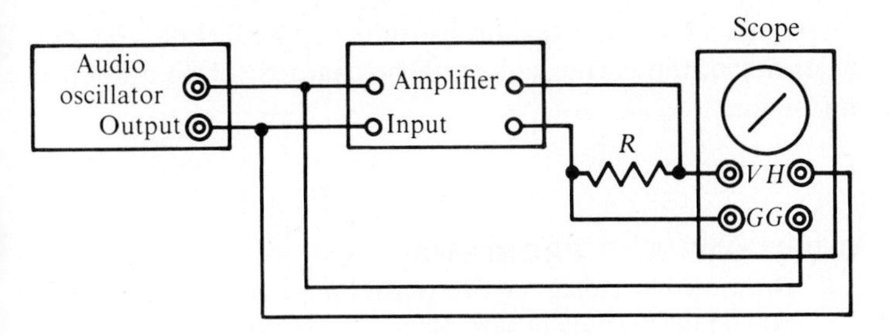

FIG. 13.50 *Check of audio-amplifier distortion.*

at least 1.5 MHz, and preferably to 4 MHz. When the oscilloscope is operated at a horizontal deflection rate of 30 Hz, a waveform similar to that illustrated in Fig. 13.49 is displayed, provided that the receiver is in normal operating condition.

Figure 13.50 shows a basic application for an oscilloscope in audio-frequency work. This test setup develops a Lissajous figure of input voltage vs. output voltage. The scope display reveals the presence of amplitude distortion, phase shift, even-harmonic distortion, even-harmonic distortion with phase shift, odd-harmonic distortion, and odd-harmonic distortion with phase shift. As shown in Fig. 13.51, an amplifier develops a straight-line Lissajous figure, provided that its operation is distortionless. Note that a high-quality oscilloscope is required; that is, if either or both of the oscilloscope

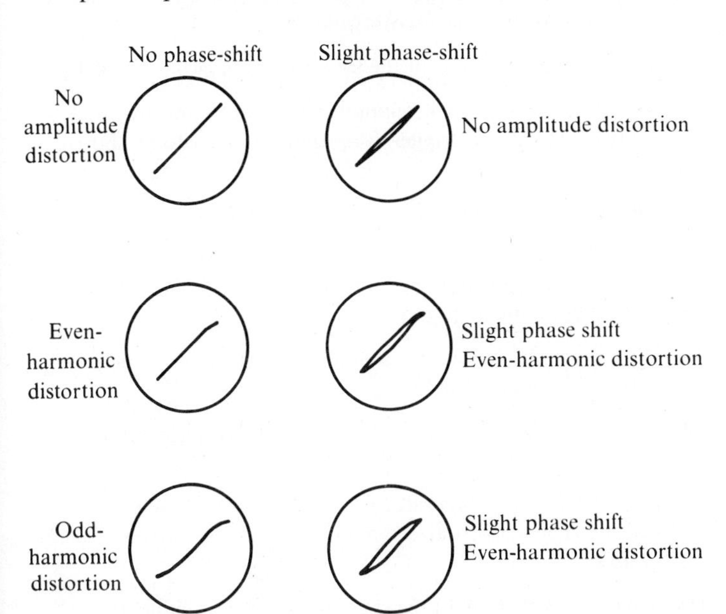

FIG. 13.51 *Evaluation of typical audio test patterns.*

amplifiers are nonlinear or introduce phase shift, the deficiencies of the oscilloscope will be charged to the amplifier under test.

QUESTIONS AND PROBLEMS

1. Draw the block diagram of a general-purpose oscilloscope and explain the purpose of each block.
2. What is the name applied to a line on a cathode-ray tube that is caused by the right-to-left movement of the electron beam?
3. Why is it desirable that the crt be "blanked" during the retrace of the electron beam?
4. What are two desirable characteristics of a sawtooth waveform when used to develop the horizontal deflection on a crt?
5. Describe the movement of the electron beam resulting from application of the input signals shown in Fig. 13.2.
6. What are the time relationships of the waveforms depicted in Fig. 13.3 that produce the resultant waveform on the crt?
7. What factors control the frequency of the sawtooth waveform in the circuit shown in Fig. 13.5?
8. Look up the advantages of the thyratron-sawtooth generator over the gas-diode sawtooth generator.
9. Explain the operation of the sawtooth generator in Fig. 13.10.
10. What is the necessary relationship between the free-running frequency and the trigger frequency of the circuit shown in Fig. 13.10?
11. Describe one cycle of operation for the sweep generator shown in Fig. 13.15.
12. What is the purpose of potentiometers R_1 and R_4 in the circuit diagram in Fig. 13.15?
13. Why must the horizontal amplifier of an oscilloscope have a substantial bandwidth?
14. Why must the sweep frequency of an oscilloscope be equal to or an exact submultiple of the signal frequency?
15. How is the sweep generator of an oscilloscope synchronized in the internal synchronization position of the sweep selector switch?
16. What is the purpose of the sync-phase inverter circuit?

17. What is the noticeable result of too large a synchronization waveform?
18. What is the purpose of a semiconductor limiter in the sync section of a general-purpose oscilloscope?
19. What is the purpose of a differentiator circuit following a clipper in the sync section of a general-purpose oscilloscope?
20. Why do most attenuators employ high-frequency compensation?
21. Why is the horizontal cathode follower not used when measuring dc voltages?
22. What would be the output of a cathode follower if the cathode resistor were bypassed?
23. What is the phase relationship between input and output signals of a cathode follower?
24. Why are direct-coupled amplifiers usually limited to two stages?
25. Can the voltage gain of a cathode follower ever be greater than unity? Why?
26. What is the primary advantage of push-pull deflection compared to single-ended deflection systems?
27. How is positioning of the crt beam accomplished?
28. What are the three possible inputs to the horizontal deflection channel?
29. What is a requirement of a power supply for a direct-coupled amplifier?
30. Why are pentode tubes used in the vertical push-pull amplifiers?
31. What type of circuit is generally used to achieve push-pull deflection?
32. What are the current requirements of an output stage utilizing electrostatic deflection?
33. Why should the input impedance of the vertical deflection channel be high?
34. Why couldn't a sync signal be taken from the cathodes of the vertical push-pull amplifier if both tubes had equal grid signals?

14

TRIGGERED-SWEEP OSCILLOSCOPES

14.1 CIRCUIT REQUIREMENTS

The time-base generators for oscilloscopes previously discussed have been free-running oscillators. The repetition rate of this type of sweep oscillator depends both upon the time constant of the oscillator circuit, and upon the frequency of the injected synchronizing voltage. Although precise horizontal deflection rates are not required in many general-purpose applications, the more sophisticated scope applications require precise control of the sweep timing with respect to the signal under test. These comparatively demanding requirements are met by triggered time bases.

High-speed, precisely timed sweeps provide data of fundamental importance in waveform analysis. For example, one of the basic characteristics of a square wave or pulse is its rise time, as depicted in Fig. 14.1. This measurement can be made only with a triggered-sweep scope. Another basic characteristic of a pulse is its width, expressed in microseconds of duration. Precise width measurements require accurately calibrated sweeps. Adequate expansion of narrow pulses for width measurement is impossible without the use of high-speed triggered sweeps. Another important capability of a triggered-sweep scope is its flexible triggering facilities. That is, a small portion of a complex waveform can be "picked out" and displayed on the crt screen for examination of detail. For example, the color burst in a color-bar waveform (Fig. 14.2) can be selected and expanded to occupy the full screen of a triggered-sweep scope.

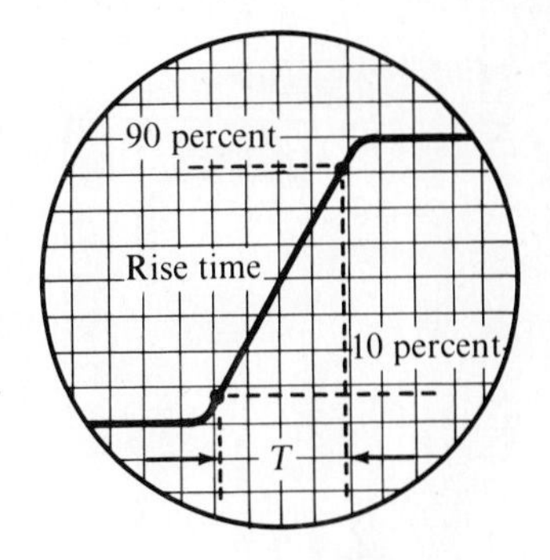

FIG. 14.1 *Measurement of rise time.*

Although a triggered-sweep oscillator is not always calibrated, the vast majority of sophisticated scopes provide time-base controls calibrated in seconds per centimeter. For example, a typical instrument employs a time base with 22 calibrated steps ranging from 0.2 μs/cm to 2 s/cm. A magnifier control expands the horizontal trace 5 times in length, thereby extending the deflection speed to 0.04 μs/cm. The rated accuracy of the time base in this example is ± 5 percent. Although a vernier sweep control is provided to "fill in" between each of the 22 steps, this continuous control is uncalibrated. It is difficult to calibrate a continuous control accurately at reasonable cost, and a large number of calibrated step intervals provide the required flexibility for quantitative tests.

Most triggered-sweep scopes also provide a calibrated vertical input attenuator. This control is calibrated in peak-to-peak volts (or dc volts) per centimeter. A typical attenuator provides nine steps, calibrated from 0.05 V/cm to 20 V/cm. The rated accuracy of the vertical system in this example is

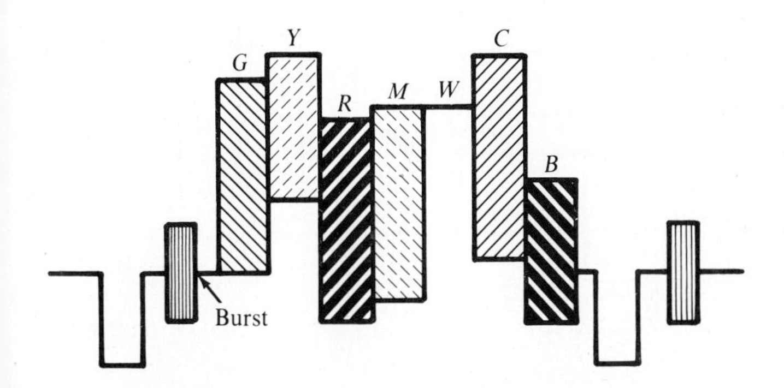

FIG. 14.2 *A color-bar waveform.*

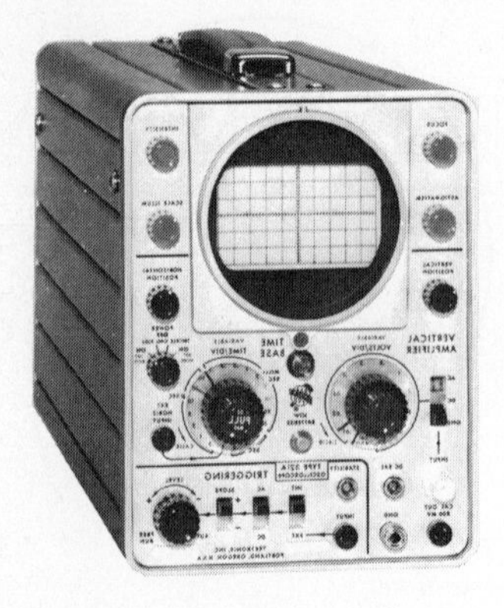

FIG. 14.3 *A typical triggered-sweep oscilloscope. (Courtesy, Tektronix, Inc.)*

±3 percent. When used with a 10:1 low-capacitance probe, the vertical attenuator range is from 0.5 V/cm to 200 V/cm. Although a vernier vertical gain control is provided to fill in between each of the nine steps, this continuous control is uncalibrated. The screen graticule for the crt is ruled in centimeter intervals vertically and horizontally, so that the amplitude and period of a waveform can be read directly from the control settings.

14.2 CIRCUIT OPERATION

A typical triggered-sweep oscilloscope is illustrated in Fig. 14.3, and a basic block diagram is shown in Fig. 14.4. Signals to be displayed on the crt are applied to the vertical input connector. The signal is amplified by the vertical amplifier section and is also converted to a push-pull signal for driving the vertical deflection plates of the crt. A trigger-pickoff stage in the vertical amplifier section feeds a sample of the vertical signal to the time-base trigger circuit for actuating the time-base circuitry. Note that the time-base section can also be actuated from an external signal source.

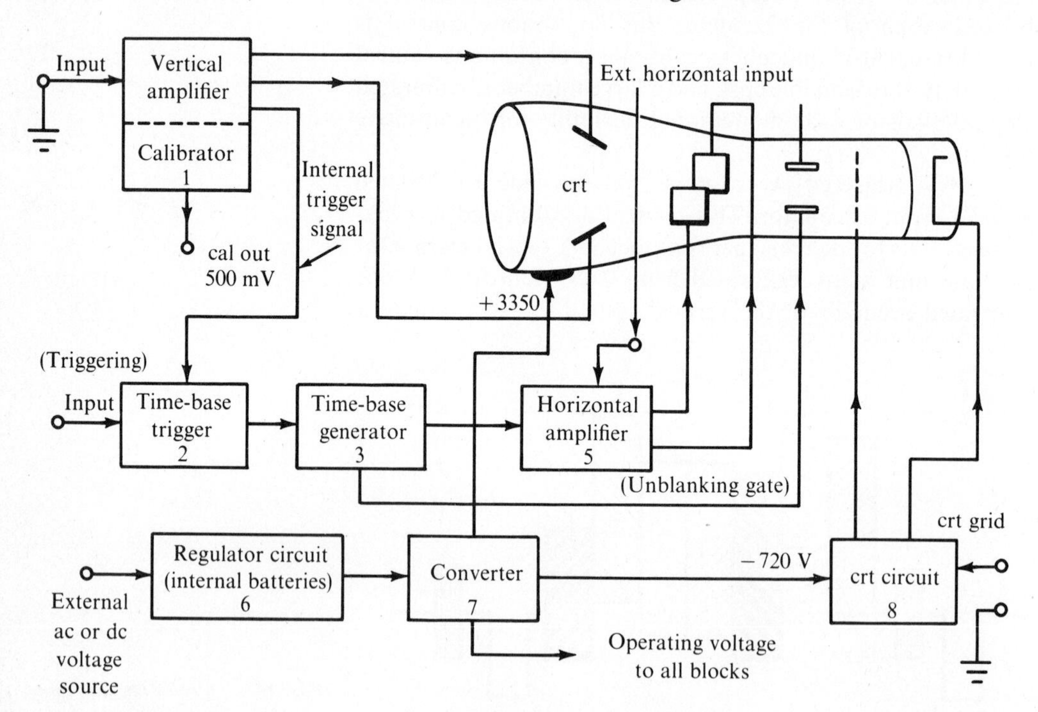

FIG. 14.4 *Block diagram of a triggered-sweep oscilloscope. (Courtesy, Tektronix, Inc.)*

The calibrator circuit depicted in Fig. 14.4 generates a square-wave voltage that has an accurate amplitude. This waveform is used to check the gain of the vertical system, and to check the compensation of low-capacitance probes. It will be found that this square-wave source has limited utility in conventional square-wave testing procedures, because its rise time is comparatively slow, and its repetition rate is not adjustable. In this example, the rise time is 1 ms and the repetition rate is 2 kHz. The square-wave signal is made available at an output connector on the front panel of the oscilloscope for convenience in testing the low-capacitance probe that is commonly employed.

Next, the time-base trigger circuit depicted in Fig. 14.4 produces an output pulse which initiates the sweep signal generated by the time-base generator circuit. This time-base generator section produces a linear sawtooth output waveform when actuated by the time-base trigger circuit. Note that the time-base circuit also develops an unblanking gate signal which unblanks (enables) the crt so that a visible trace can be presented on the screen. This unblanking gate signal is coincident with the sawtooth output from the time-base generator. From the time-base generator, the sawtooth voltage waveform is fed to the horizontal amplifier section. In turn, the horizontal amplifier steps up the sawtooth amplitude and produces a push-pull output signal to drive the horizontal deflection plates of the crt.

The regulator circuit in Fig. 14.4 provides a regulated 10-V output to the converter circuit. This regulated output voltage can be provided either from an ac line, external batteries, or internal batteries. Note that the regulator circuit also contains a battery-charging circuit for use with rechargeable internal batteries. The converter circuit produces positive and negative accelerating potentials for the crt, and the low-voltage power values required by the instrument.

14.3 CALIBRATOR SECTION

A detailed block diagram showing the calibrator section in relation to the vertical amplifier is seen in Fig. 14.5. The schematic diagram shown in Fig. 14.6 should also be referred to in the following discussion. The calibrator circuit produces a square-wave output voltage with a precise amplitude. It consists of a grounded emitter amplifier which is overdriven

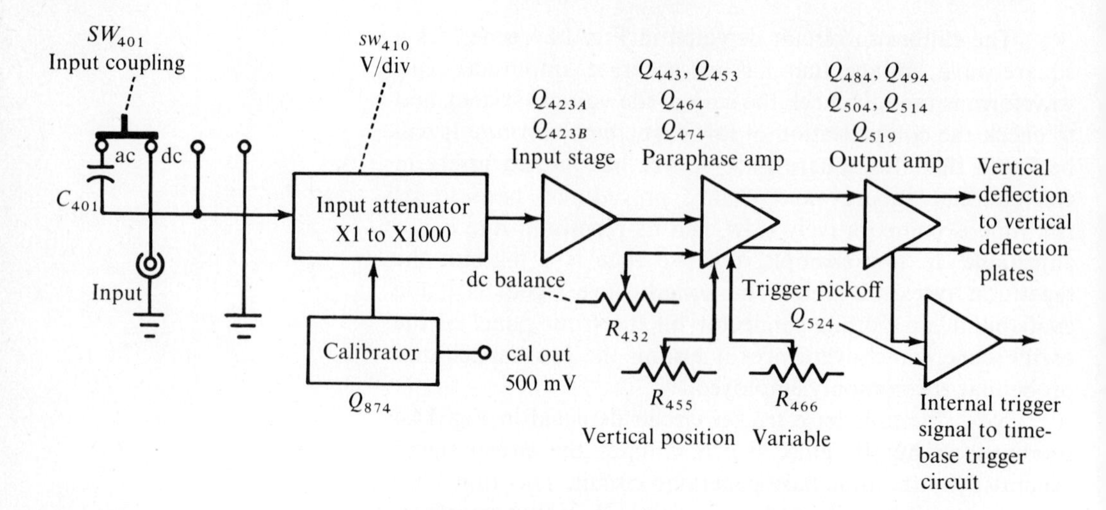

FIG. 14.5 *Vertical amplifier and calibrator detailed block diagram.*

by a signal from the converter circuit. This drive signal for the calibrator circuit is a 100-V 2-kHz sine wave. In turn, the sine-wave voltage is applied to the base of Q_{874} through a pair of networks that include C_{871}-R_{871}. These networks assist in shaping the driving current waveform, so that a good square wave is processed. When the drive signal is more negative than approximately ± 0.5 V, Q_{874} becomes reverse biased and its collector rises in a positive direction toward the +10-V supply through R_{881}. Zener diode D_{881} clamps the collector at approximately +5.1 V.

When the drive signal to Q_{874} in Fig. 14.6 goes positive above +0.5 V, approximately, Q_{874} becomes forward biased. The transistor is overdriven by the high amplitude base signal, and is quickly driven into saturation. The collector of Q_{874} drops negative to the emitter level (nearly zero volts). This produces the negative portion of the output square wave. Diode D_{882} is reverse biased by the collector level of Q_{874}, which effectively disconnects all voltage to the output. Therefore, the negative portion of the output square wave drops to zero volts.

Note that the output level in Fig. 14.6 is determined by the voltage divider R_{882}-R_{884}-R_{886}-R_{888} between the collector of Q_{874} and ground. The voltage level at the junction of R_{884}-R_{886} is connected to the calibrate output 500-mV jack on the front panel of the instrument. The calibrate amplifier adjustment R_{884} is set to provide an accurate 500-mV peak-to-peak square-wave output level at this jack. The voltage level at the junction of R_{886}-R_{888} is internally connected to the vertical amplifier in the Cal 4 Div position of the volts/division switch to provide a quick check of the basic vertical amplifier gain. The amplitude of this internal calibrator signal is 40 mV peak to peak.

14.4 VERTICAL ATTENUATOR

A vertical input connector can be ac-coupled, dc-coupled, or internally disconnected. When the input coupling switch *SW*401 is in its dc position, the input signal is coupled directly to the input attenuator stage. Next, in the ac position, the input signal is connected to capacitor C_{401}. This capacitor prevents the dc component of the input signal from being applied to the vertical amplifier. In the Gnd position, the signal path is open and the input of the amplifier is connected to ground. This provides a ground reference without the necessity for disconnection of the applied signal from the input connector.

We observe that the input attenuator is a part of the vertical amplifier input network in Fig. 14.6. These attenuators are switched into the circuit singly or in pairs to produce the vertical deflection factor noted on the front panel for the particular setting (such as 0.1 V per division). The deflection factor is understood to be stated in peak-to-peak or dc values. These attenuators are frequency-compensated voltage dividers. The theory of compensated dividers was previously explained in the VTVM chapter. Each attenuator includes an adjustable series capacitor to provide optimum response for the high-frequency components of the signal and an adjustable shunt capacitor for optimum response of the low-frequency components. Each attenuator section is designed to present the same input *RC* characteristics (1 M and 35 pF in this example).

Note that the signal from the input attenuator in Fig. 14.6 is connected to the input stage through a network comprising C_{422}-C_{423}-R_{422}-R_{423}-R_{424}. The input resistance of this stage is established by R_{422} and is part of the attenuation network on all positions of the vertical attenuator. R_{423} limits the current that can be supplied to drive the input stage. Q_{423A}, a field-effect transistor, is employed as a source follower to provide a high-input impedance with respect to the incoming signal, and a low-output impedance for driving the following stage. Diodes D_{423} and D_{424} protect Q_{423A} by

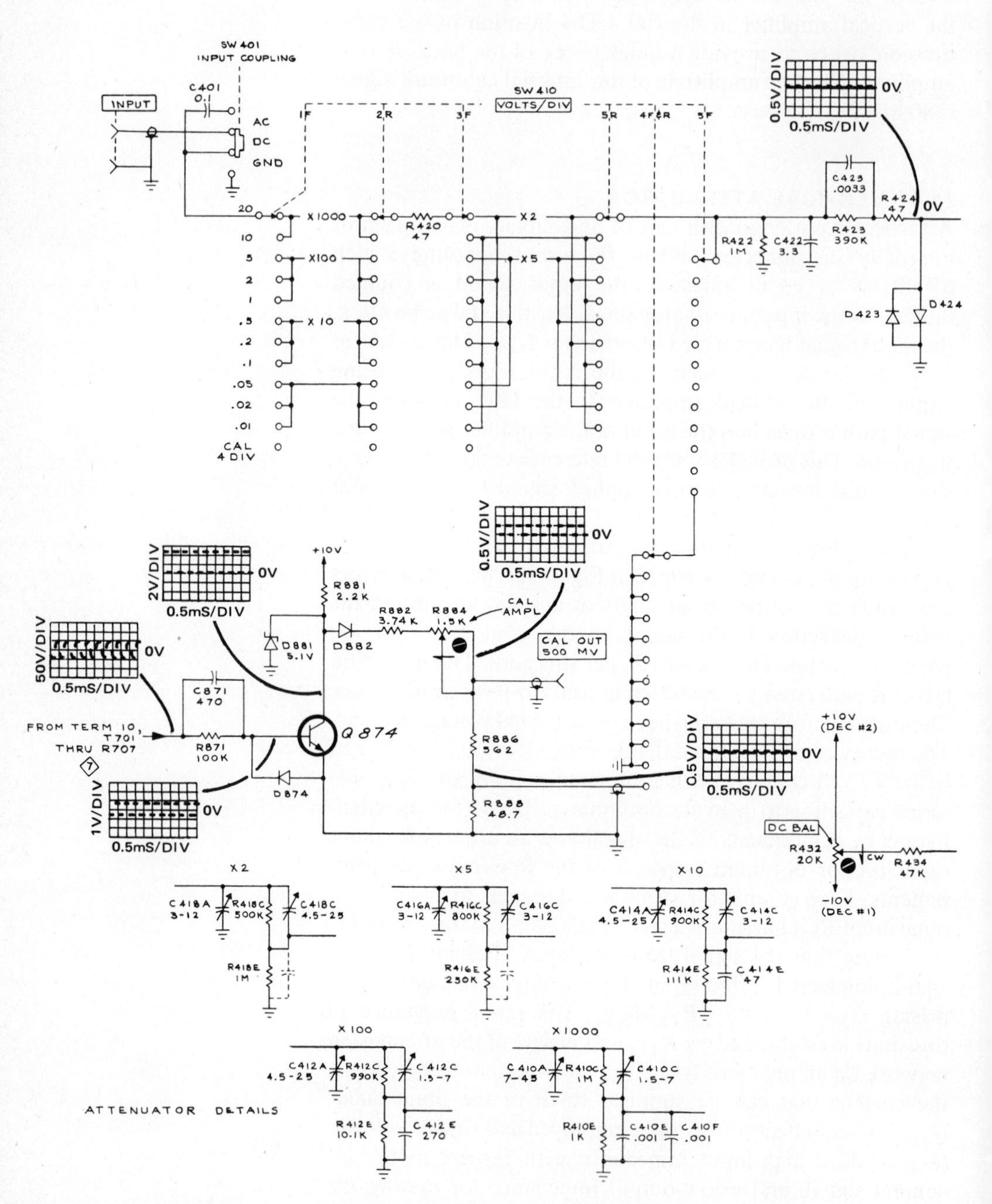

FIG. 14.6 *Vertical amplifier and calibrator schematic. (Courtesy, Tektronix, Inc.)*

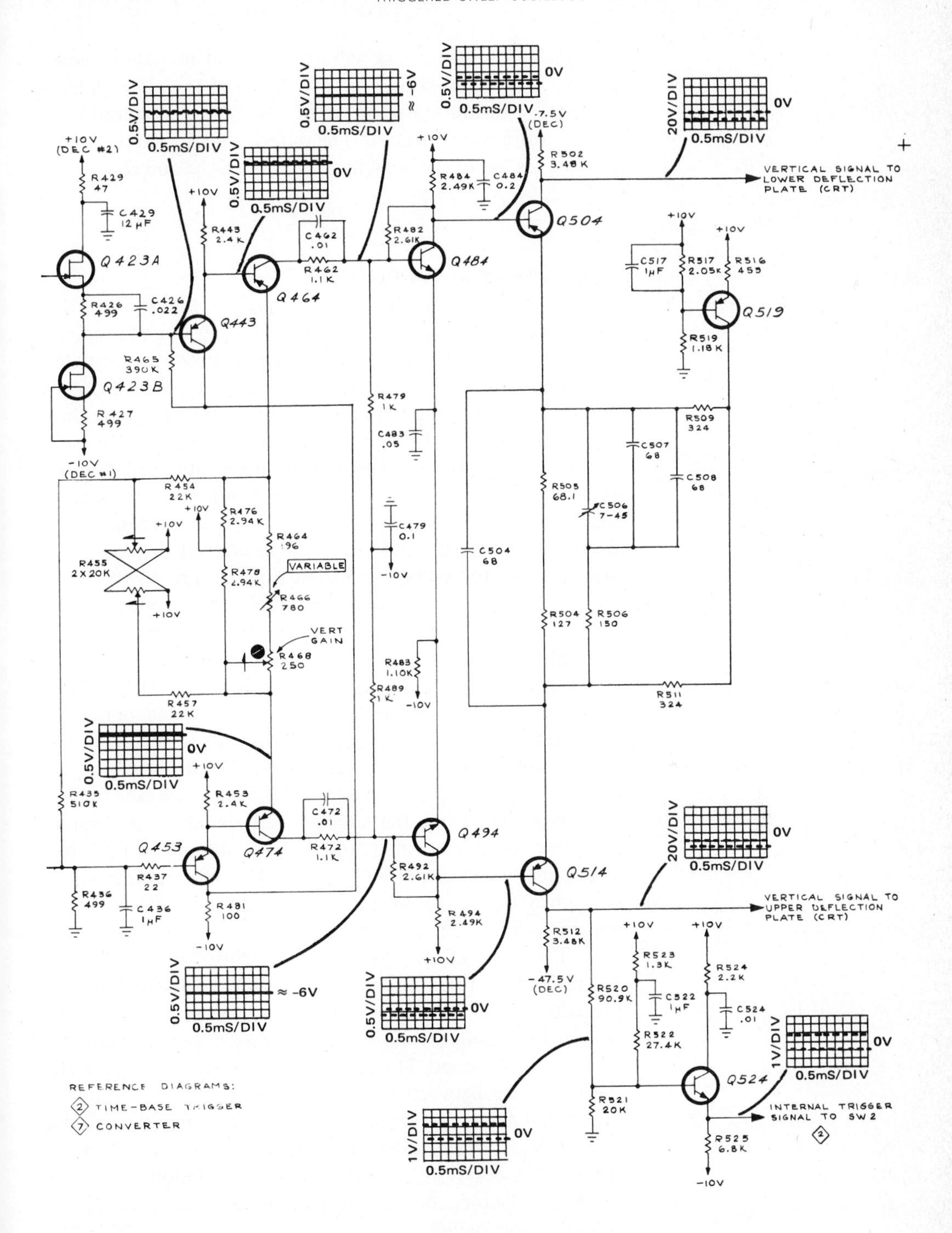
VERTICAL SIGNAL TO LOWER DEFLECTION PLATE (CRT)
VERTICAL SIGNAL TO UPPER DEFLECTION PLATE (CRT)
INTERNAL TRIGGER SIGNAL TO SW2
VARIABLE
VERT GAIN
Q423A
Q423B
Q443
Q464
Q484
Q504
Q519
Q453
Q474
Q494
Q514
Q524
0.5V/DIV
0.5mS/DIV
20V/DIV
1V/DIV
REFERENCE DIAGRAMS:
2 TIME-BASE TRIGGER
7 CONVERTER

limiting the peak-to-peak voltage swing at its gate to about 1.2 V. Q_{423B}, the other field-effect transistor, provides a practical constant-current source for Q_{423A} in addition to temperature compensation. The output signal from the source of Q_{423A} is fed to the next stage through C_{426} and R_{426}.

14.5 VERTICAL AMPLIFIER

Next, the signal from the input stage in Fig. 14.6 is coupled to the base of paraphase amplifier Q_{464}-Q_{474} through the emitter follower Q_{443}. This emitter follower provides a very low impedance drive to Q_{464} (about 20 Ω). Q_{464} and Q_{474} are connected as a common-emitter phase inverter (paraphase amplifier) to convert the single-ended input signal to a push-pull output signal. The push-pull output is obtained from the single-ended input signal in the following manner. Assume that the signal voltage at the base of Q_{464} is increasing. This produces a corresponding decrease in current through Q_{464}, and its collector voltage goes negative toward the collector supply voltage. At the same time, the emitter of Q_{464} goes positive and this change is applied to the emitter of Q_{474} through R_{464}, R_{466}, and R_{468}. As far as signal changes are concerned, Q_{474} is connected as a grounded-base stage so that it operates as the emitter-driven section of the paraphase amplifier. The positive-going signal applied to the emitter of Q_{474} forward-biases Q_{474}, and its collector voltage goes positive by about the same amount that the collector of Q_{464} went negative. Thus, the single-ended signal at the base of Q_{464} has been amplified and is available as a push-pull signal at the collectors of Q_{464} and Q_{474}. This output signal is applied to the output amplifier stage through C_{462}-R_{462} and C_{472}-R_{472}.

The gain of the paraphase amplifier in Fig. 14.6 is determined by the amount of emitter degeneration which is present. As the resistance between the emitters of Q_{464} and Q_{474} is increased, the emitter degeneration increases and the stage gain is lessened. The vertical gain adjustment R_{468} varies the resistance between the emitters of Q_{464}-Q_{474} to control the overall gain of the vertical amplifier. Note that the variable control R_{466} also changes the amount of resistance between the emitters of Q_{464}-Q_{474} and provides continuously variable deflection factors between the calibrated settings of the volts/division switch.

Next, we observe that the vertical position control R_{455} in Fig. 14.6 varies the dc emitter current of Q_{464} and Q_{474}; although this produces only a small voltage change at the emitters of Q_{464} and Q_{474}, it produces a voltage change at the collectors which establishes the position of the display on the crt screen. We perceive that the dc-balance adjustment R_{432} sets the base level of Q_{474} through the emitter follower Q_{453}. This adjustment is set to establish the same dc voltage level at the emitter of Q_{474} as at the emitter of Q_{464}. Since the emitters of Q_{464} and Q_{474} are at the same dc voltage level, no current flows through the variable control R_{466}. This configuration prevents the dc level of the display from shifting when the variable control is turned.

Two individual push-pull amplifier stages are employed in the output amplifier, Fig. 14.6. The first stage of amplification, Q_{484}-Q_{494}, utilizes collector-to-base feedback via R_{482} and R_{492}. This negative feedback provides linear amplification from this stage. The signal at the collectors of Q_{484} and Q_{494} connects to the bases of Q_{504} and Q_{514}. Thus, Q_{504}-Q_{514} provides the final amplification of the vertical deflection signal before it is applied to the crt. High-frequency compensation (explained in greater detail subsequently) is provided by C_{504}-C_{506}-C_{507}-C_{508}-R_{504}-R_{505}-R_{506}. Capacitor C_{506} is adjustable to provide optimum high-frequency response. Transistor Q_{519} operates as a constant-current source for the emitters of Q_{504}-Q_{514}. This constant current is applied at the junction of R_{509}-R_{511}. Note that the output signal at the collector of Q_{504} is fed to the lower vertical deflection plate in the crt, and the signal at the collector of Q_{514} is fed to the upper vertical deflection plate.

14.6 PICKOFF POINT FOR TIME-BASE GENERATOR

Next, let us consider the trigger-pickoff configuration. This signal is obtained at the collector of Q_{514}, in Fig. 14.6, and is fed to the trigger-pickoff stage, Q_{524}, through divider R_{520}-R_{521}. Thus a sample of the vertical amplifier signal provides internal triggering. Q_{524} is connected as an emitter follower to provide isolation between the vertical amplifier circuit and the time-base trigger circuit. It also minimizes loading of the output amplifier stage while providing a low output impedance to the time-base trigger circuit. The output from this stage is applied to the time-base trigger circuit through the triggering source switch SW_2.

14.7 TIME-BASE TRIGGER SECTION

Next, let us consider the operation of the time-base trigger section in Fig. 14.7. Trigger pulses are produced by the time-base trigger circuit to initiate the time-base generator circuit. These trigger pulses are derived either from the internal trigger signal picked off from the vertical amplifier, or from an external signal applied to the triggering input jack. Controls are provided in this circuit to select trigger level, slope, ac or dc coupling, source, and mode, as shall be explained. A schematic diagram of the time-base trigger section is shown in Fig. 14.8. The triggering source switch SW_2 selects either an internal or an external signal. When SW_2 is set to its external position, the input resistance is about 90 K, established by R_2. Note that C_2 is a high-frequency-compensating capacitor that maintains good trigger waveform.

The trigger coupling switch SW_8 (Figs. 14.7 and 14.8) permits selection of the portion of the trigger signal from which the trigger pulse is derived. In the dc position of the coupling switch, SW_8 bypasses capacitor C_8 and the trigger signal is coupled directly to the slope-comparator stage. This position provides equal coupling for all trigger signals from dc to 6 MHz, in this example. When the coupling switch is set to its ac position, SW_8 is open and the trigger signal must be coupled by C_8. This capacitor blocks the dc component of the trigger signal and attenuates ac signals below about

FIG. 14.7 *Detailed block diagram of time-base trigger circuit.*

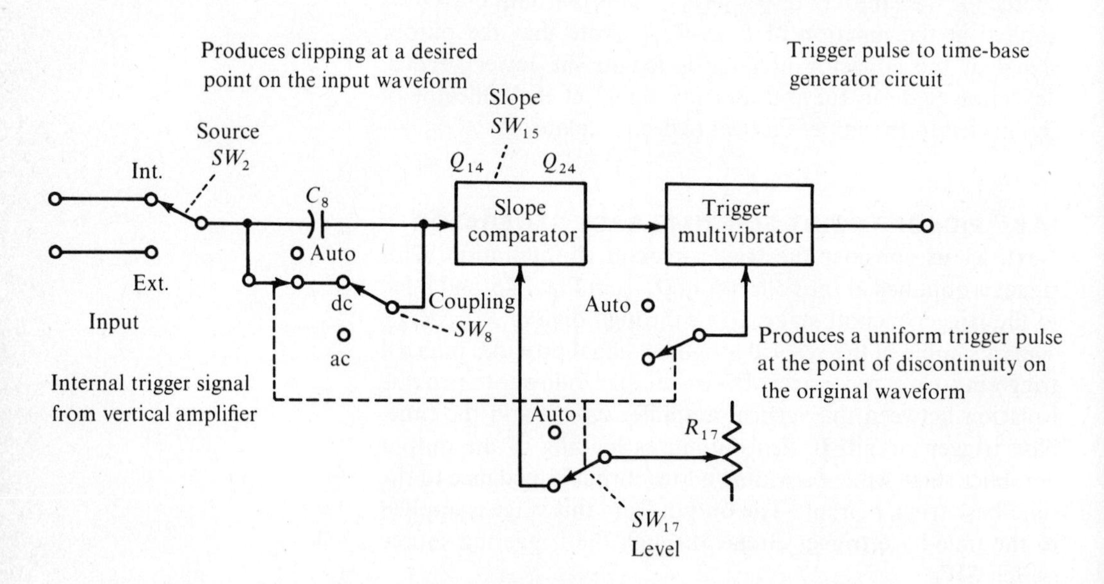

600 Hz in the internal-source switch position; in the external-scource switch position, ac signals below about 16 Hz are attenuated.

When the triggering level control (Figs. 14.7, 14.8) is turned fully counterclockwise to its automatic position, SW_{17} (ganged with the level control) opens the dc path around C_8. All trigger signals must be coupled by this capacitor in the automatic trigger mode; in other words, only ac trigger coupling is provided in this mode.

Next, let us observe the action of the slope comparator. Transistors Q_{14} and Q_{24} in Fig. 14.8 are connected as a difference amplifier (comparator) to provide selection of the slope and level at which the sweep is triggered. Figure 14.9 depicts the effect of the triggering level and slope controls on the displayed pattern. The output signal from the slope comparator stage is always obtained from the collector of Q_{24} in Fig. 14.8, and the sweep is started from the positive-going portion of the signal at this point. To provide selection of the trigger slope, the slope switch SW_{15} is employed. This switch applies the trigger signal to the base of Q_{14} for positive slope triggering, or to the base of Q_{24} for negative slope triggering.

For positive slope triggering in Fig. 14.8, the trigger signal drives the base of Q_{14}, and a reference voltage from the level control, R_{17}, is applied to the base of Q_{24} through R_{19}. Resistor R_{23} establishes the emitter current of Q_{14} and Q_{24}. Capacitor C_{23} is connected across R_{23} to improve high-frequency response of the stage. In this configuration, the transistor with the most negative base controls conduction to the comparator. For example, with a positive-going trigger signal, Q_{14} conducts until its base is raised more positive than the base of Q_{24}. Then, Q_{24} becomes reverse biased and the decreased current flow through R_{23} produces a smaller voltage drop; in turn, the emitters of both Q_{14} and Q_{24} go more positive. This more positive voltage at the emitter of Q_{24} represents a forward bias, since the base is held at the voltage set by the level control. Thus the collector current of Q_{24} increases to produce a positive-going output signal from this stage. Notice that the output signal from this stage is in phase with the input signal. Therefore, the sweep is started on the positive portion of the trigger input signal.

We observe that the level control R_{17} in Fig. 14.8 sets the base level of Q_{24} for positive slope triggering. This in turn determines the level on the trigger signal at which the

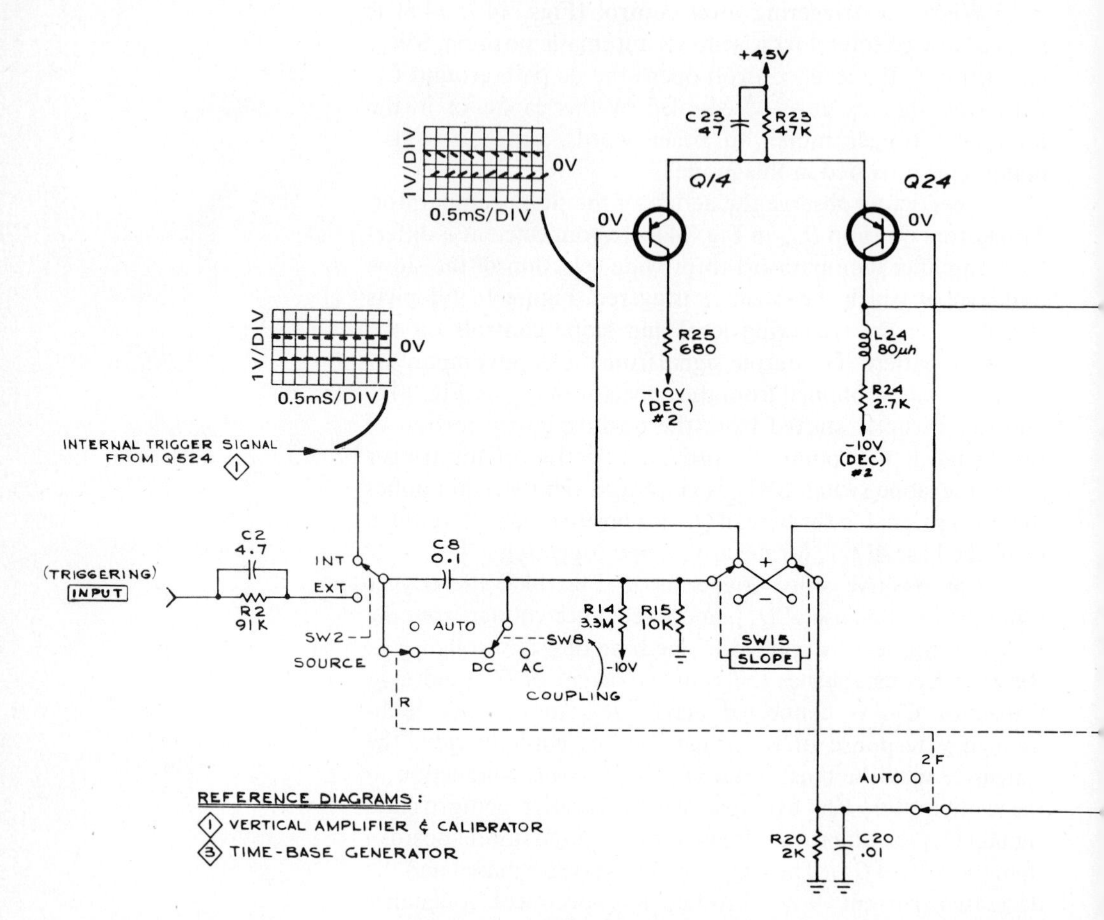

FIG. 14.8 *Time-base trigger circuit. (Courtesy, Tektronix, Inc.)*

comparator switches. With the level control set near midrange and a positive-going signal applied, Q_{14} conducts until the applied trigger signal raises the base of Q_{14} more positive than the base of Q_{24} (the transistor with the most negative base always controls conduction of the comparator). Q_{14} is then reverse biased to produce a negative-going output at the collector of Q_{24} as explained above. Q_{14} remains off, and Q_{24} conducts until the applied trigger signal drops the voltage at the base of Q_{14} more negative than the base of Q_{24}. Then the circuit returns to its original state. Now, assume that the level control is turned clockwise to produce a display which

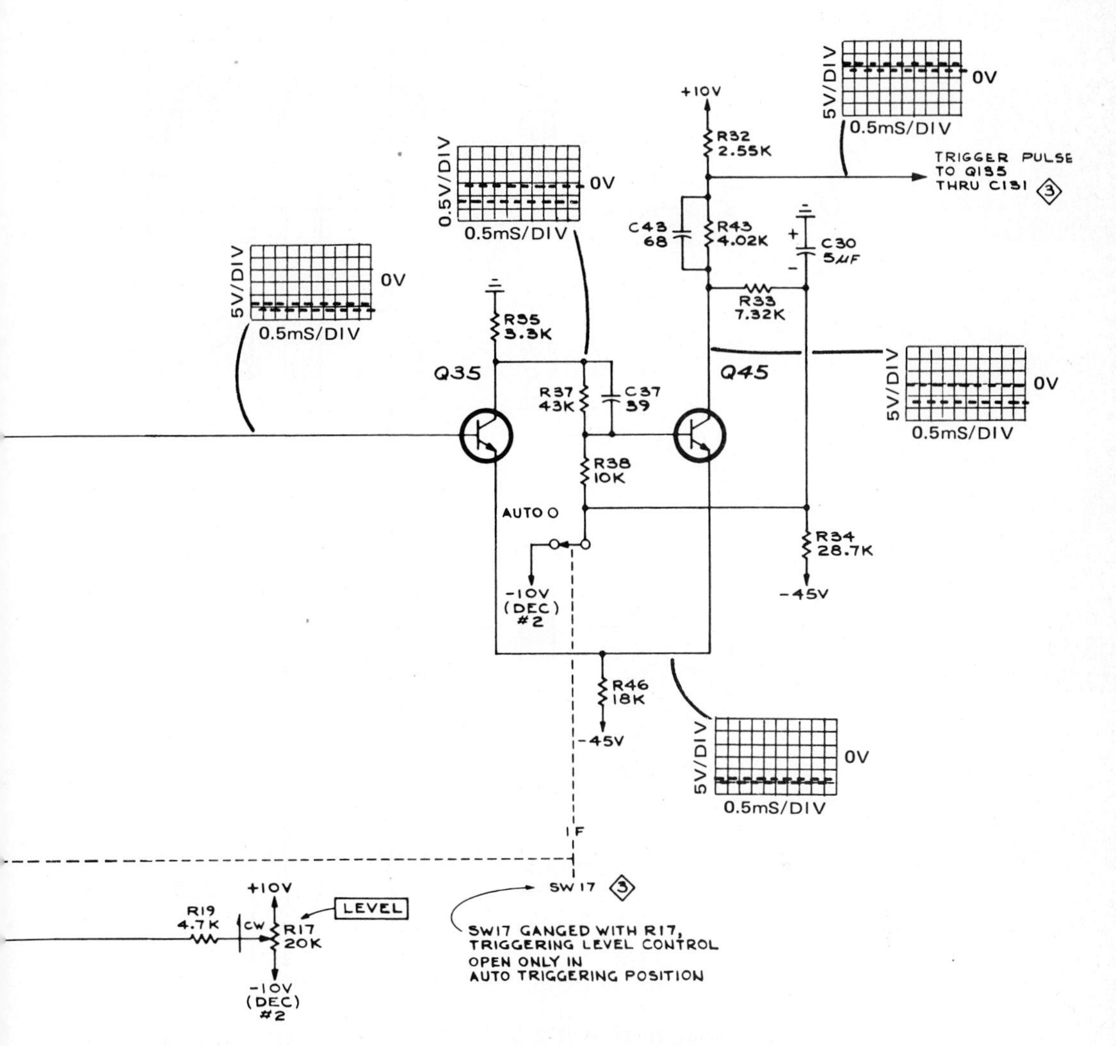

starts at a more positive level. A more positive level is established at the base of Q_{24} by the level control. The trigger signal must now rise more positive to make the base of Q_{14} more positive than the base of Q_{24}. Thereupon, Q_{14} is reverse biased to produce the positive-going output from this stage. The resultant crt display starts at a more positive point on the displayed signal since the sweep is started later. When the level control is turned counterclockwise (toward its negative end), the effect is opposite; the resultant crt display then starts at a more negative point on the trigger signal.

To start the sweep on the negative slope of the trigger

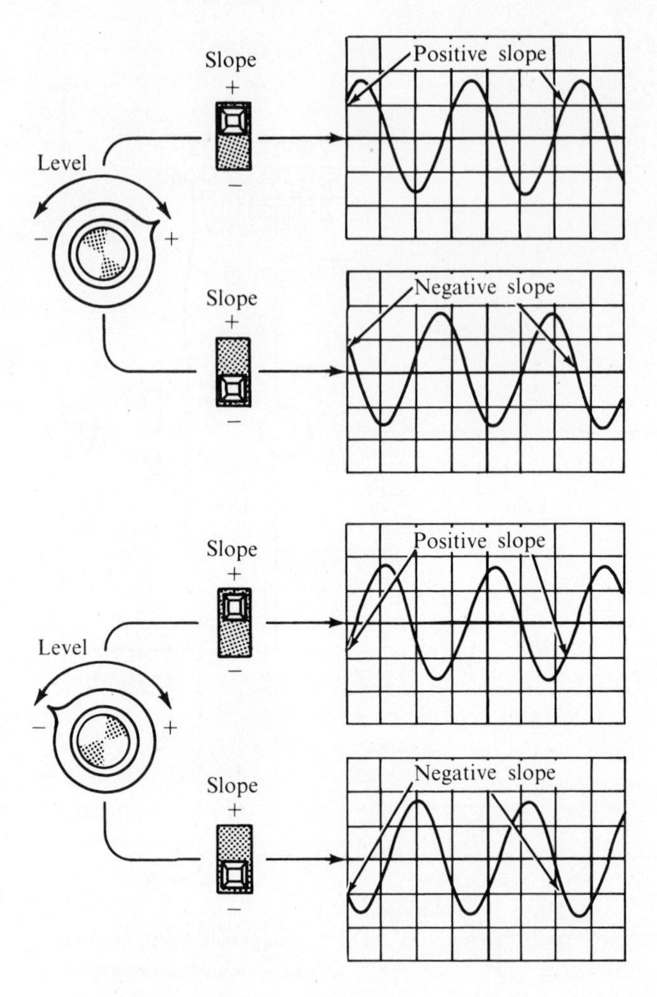

FIG. 14.9 *Effect of triggering level control and slope control switch on the crt display.*

signal, slope switch SW_{15} in Fig. 14.8 reverses the connections to the bases of Q_{14} and Q_{24}. In this situation, the trigger signal is applied to the base of Q_{24} and the reference voltage level from the level control is applied to the base of Q_{14}. The level control establishes the point at which the comparator switches, as explained previously for positive slope triggering. Assume that Q_{24} is conducting with a positive-going signal applied. As the base of Q_{24} goes more positive than the level established at the base of Q_{14} by the level control, Q_{24} becomes reverse biased. The collector current of Q_{24} decreases to produce a negative-going change at its collector. Note that this signal is 180° out of phase with the applied trigger signal.

The sweep is always started on the positive-going portion of the signal at the output of this stage. Therefore, since there is 180° phase shift through the slope comparator stage for negative slope triggering, the sweep is triggered on the negative-going slope of the trigger input signal.

In the automatic (fully clockwise) position of the trigger level control in Fig. 14.8, SW_{17} (ganged with the level control) disconnects the level control R_{17} from the circuit. In this triggering mode, the triggering level is set very near the zero-volt level by R_{20}.

Next, let us consider the operation of the trigger multivibrator stage in Fig. 14.8. We start by stipulating that the level control is switched into operation. The output from the slope comparator stage is applied to the trigger multivibrator stage, Q_{35} and Q_{45}. These transistors are connected as a Schmitt bistable multivibrator when the level control is in its variable range. To understand the operation of this circuit, assume that the level control is set near its midrange, that the slope switch is set to +, and that the circuit is ready to respond to a trigger signal. These conditions produce a negative output voltage level from the slope comparator stage, which reverse-biases Q_{35}. When Q_{35} is off, its collector voltage rises positive toward the collector supply voltage (ground level). The quiescent level on the collector is determined by voltage divider R_{35}-R_{37}-R_{38} between ground and −10 V. This divider also sets the quiescent level at the base of Q_{45} positive enough so that Q_{45} is forward biased. The collector of Q_{45} goes negative to establish an output level of about +3 V at the junction of C_{43}-R_{32}-R_{43}.

When a trigger signal applied to the time-base trigger circuit in Fig. 14.8 produces a positive-going output from the slope comparator stage, Q_{35} is forward biased. The collector of Q_{35} goes negative and the voltage at the base of Q_{45} is also driven negative through dividers R_{37}-R_{38}. At the same time, the emitters of both Q_{35} and Q_{45} rise positive, following the positive voltage at the base of Q_{35}. In turn, Q_{45} is quickly cut off and its collector rises positive toward the supply voltage through C_{43}-R_{32}-R_{43}. Capacitors C_{37} and C_{43} provide high-frequency compensation. This positive-going transition at the junction of C_{43}-R_{32}-R_{43} is applied to the time-base generator circuit through C_{31} in order to start the sweep. The circuit remains in the foregoing condition until the output from the slope comparator stage drops negative and Q_{35} becomes reverse biased. Then, the base of Q_{45} rises positive and Q_{45}

is turned on. The output level from this stage drops negative to the quiescent level of about +3 V.

Let us next consider the operation of this section with the level control in the automatic position. When the level control in Fig. 14.8 is in its automatic (fully counterclockwise) position, SW_{17} (ganged with the level control) disconnects the -10 V level from the divider R_{35}-R_{37}-R_{38}. Note that R_{34} is connected to this divider, and the negative level for the divider is now -45 V. The charge rate of C_{30} also affects the base level of Q_{45}. In the automatic triggering mode, the trigger multivibrator can operate in one of two modes, dependent upon the trigger signal.

When there is no trigger signal present in Fig. 14.8, or if the repetition rate of the trigger signal is less than about 50 Hz, the trigger multivibrator stage operates as an astable multivibrator with C_{30} determining the repetition rate. To understand the operation of the circuit under this condition, assume that no trigger signal has been applied and that Q_{35} has just turned off. When Q_{35} is off, Q_{45} is on and its collector drops negative. This places less voltage across divider R_{33}-R_{34}, and the voltage at the junction of R_{33} and R_{34} starts to go negative at the charge rate of C_{30}. This voltage is applied to the base of Q_{45} through R_{38} to produce a similar negative-going change at the emitter of Q_{45}. Since the emitters of Q_{35} and Q_{45} are connected together, the emitter of Q_{35} goes negative also at the rate determined by the charge rate of C_{30}. The base level of Q_{35} is about -10 V, since Q_{24} is off when there is no trigger applied in the automatic mode.

Next, the emitter of Q_{35} in Fig. 14.8 continues to go negative with the charging of C_{30} until it drops sufficiently negative to forward-bias Q_{35} (the charge time of C_{30} is about 10 ms). Thereupon, Q_{35} comes on and its collector drops negative. The base of Q_{45} goes negative also, as determined by divider R_{37}-R_{38} between the collector of Q_{35} and the junction of C_{30}-R_{33}-R_{34}. We note that Q_{45} is reverse biased, and its collector goes positive to produce a trigger pulse from the time-base generator circuit. Now, there is more voltage across divider R_{33}-R_{34} and the voltage at the junction of R_{33} and R_{34} goes positive at the discharge rate of C_{30}. As C_{30} discharges, the negative voltage to the divider R_{37}-R_{38} slowly rises more positive and the base of Q_{45} also rises positive until Q_{45} is forward biased (the discharge time of C_{30} is about 10 ms). At this time, Q_{45} turns on and its emitter rises positive to reverse-bias Q_{35} by way of the common-emitter connection. The cycle begins again as C_{30} starts to recharge. As long as the instrument is in its automatic mode and no trigger signal is applied, or the signal repetition rate is too low, the trigger multivibrator stage free-runs at a 50-Hz rate to automatically retrigger the sweep and provide a reference trace.

Whenever a trigger signal with a repetition rate higher than about 50 Hz is applied to the base of Q_{35} in Fig. 14.8, the trigger multivibrator stage operates in much the same manner as described with the level control in its variable region. The sweep is triggered at the average voltage level of the trigger signal (trigger signal is always ac-coupled in the automatic trigger mode). In turn, the switching rate of the trigger multivibrator stage when it is triggered at a repetition rate faster than 50 Hz is such that C_{30} does not charge or discharge enough to affect the base level of Q_{45}. Under triggered conditions, the voltage drop across R_{34} is such that divider R_{35}-R_{37}-R_{38} is connected between about the same voltages as when the level control is in its variable region.

14.8 TIME-BASE GENERATOR SECTION

Proceeding to the time-base generator section, this subsystem produces a sawtooth voltage which is amplified by the horizontal amplifier section to provide horizontal beam deflection in the crt. This output signal is generated on command (trigger pulse) from the time-base trigger circuit when the level control is in any position except free-run. When the level control is set fully clockwise to its free-run position, the sweep has a repetition rate selected by the time division switch. The time-base generator also produces an unblanking gate to unblank the crt during the sweep time. Figure 14.10 shows a detailed block diagram of the time-base generator circuit. A schematic diagram of this section is depicted in Fig. 14.11.

We observe that the trigger pulse from the time-base trigger circuit is applied to the time-base generator circuit through C_{131}. This trigger pulse is a fast-rise square wave with a repetition rate and duty cycle determined by the trigger multivibrator stage. The duty cycle is equal to the pulse width divided by the waveform period (Fig. 14.12). C_{131} and its associated resistive branch differentiate this square wave, thereby producing narrow positive- and negative-going pulses. The process of differentiation is depicted in Fig. 14.13. The

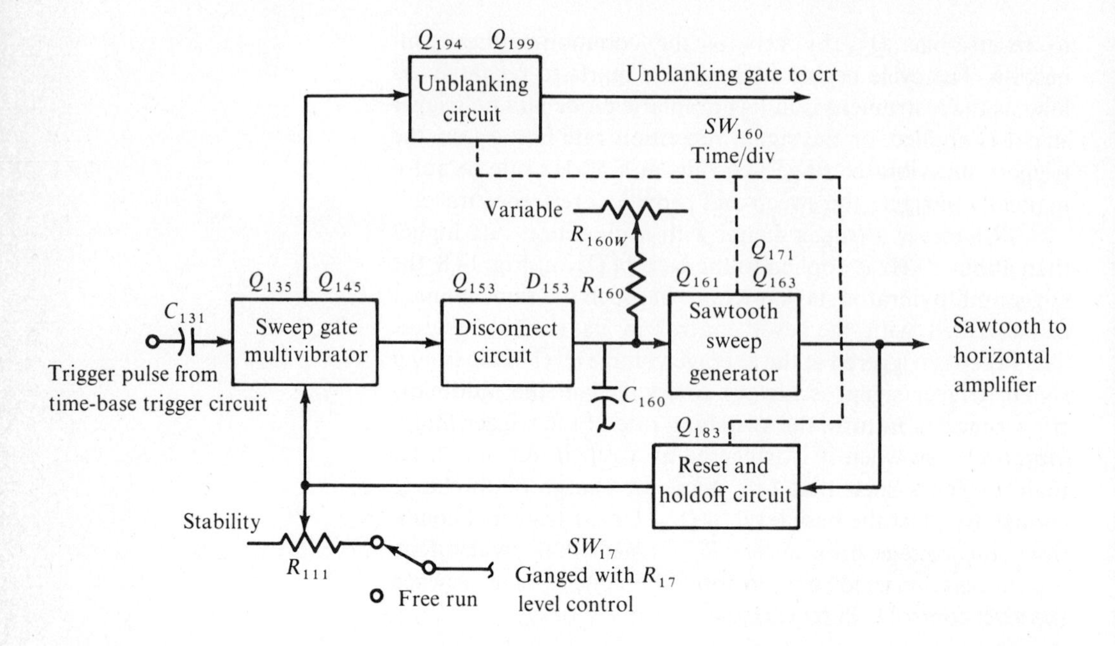

FIG. 14.10 *Block diagram of time-base generator circuit.*

negative-going pulses are shunted by D_{131} to prevent them from turning off Q_{135} during the sweep time. Q_{135} and Q_{145} are connected as a Schmitt bistable multivibrator. Quiescently, Q_{135} is off and Q_{145} is conducting. This produces a negative level at the collector of Q_{145} which forward-biases D_{149} and D_{150}. The function of these diodes shall be explained under the disconnect circuit topic.

Next, we observe that the collector current of Q_{145} in Fig. 14.11 flows through divider R_{147}-R_{148}. Diode D_{148} clamps the junction of R_{147} and R_{148} at about -0.5 V to limit the collector current of Q_{145}. With Q_{135} off, its collector rises positive. This signal is applied to the unblanking circuit. Note that the stability adjustment, R_{111}, sets the quiescent base level of Q_{135} so it is near the forward-bias point. The positive-going trigger pulse applied to the base of Q_{135} turns Q_{135} on, and its collector goes negative. This produces a negative-going signal to the unblanking circuit through R_{134} and R_{135}. It also applies less voltage across divider R_{141}-R_{143}, which results in a more negative voltage at the base of Q_{145}. This reverse-biases Q_{145}, and its collector rises positive to reverse-bias D_{149} and D_{150}. Capacitor C_{141}, connected between the collector of Q_{135} and the base of Q_{145}, improves the response time of this circuit. Diodes D_{144} and D_{145} protect

Q_{135} and Q_{145} from exceeding their base-emitter breakdown voltage when they are reverse biased. The circuit remains in this state with Q_{135} on and Q_{145} off until it is reset by the reset-and-holdoff circuit at the end of the sweep. This reset action is explained in detail subsequently.

Now, let us consider the action of the unblanking circuit in Fig. 14.11. The negative-going signal at the collector of Q_{135} in response to the trigger is applied to the base of Q_{194} through R_{134} and R_{135}. This forward-biases Q_{194}, and as a consequence of the circuit configuration, Q_{194} attempts to go into saturation. However, D_{194} prevents Q_{194} from going into full saturation by clamping the collector of Q_{194} about 0.2 V more negative than its base. With the collector of Q_{194} clamped in this manner, the unblanking gate has a fast trailing edge as well as a fast leading edge. The meaning of leading and trailing edges, with their associated rise and fall times, is depicted in Fig. 14.14. The level at the collector of Q_{194} in Fig. 14.11 during conduction is about +8.4 V. This voltage is applied to the blanking deflection plate of the crt to allow a display to be presented; the unblanking action is described in greater detail subsequently. Note that the collector level of Q_{194} is also applied to the base of Q_{199} through R_{197}. When Q_{194} is on, the positive level at its collector raises the base of Q_{199} sufficiently positive to reverse-bias the transistor.

At the end of the sweep, Q_{135} is cut off and its collector rises positive. This permits the base of Q_{194} to go positive toward +10 V, and Q_{194} becomes reverse biased. Its collector drops negative to approximately −22 V, as established by the divider R_{194}-R_{196}-R_{197}. This voltage level is applied to the blanking deflection plate in the crt to display the beam off-screen. Since D_{194} prevents Q_{194} from going into full saturation when the transistor is turned on during the sweep interval, Q_{194} can in turn be cut off rapidly at the end of the sweep. Then, the voltage connected to the blanking deflection plate quickly drops negative and the beam is deflected off-screen. With Q_{194} off, the voltage at the base of Q_{199} is determined by divider R_{194}-R_{196}-R_{197}. The voltage at the junction of R_{196} and R_{197} is such that Q_{199} is forward biased. Note that Q_{199} is on when Q_{194} is off, and vice-versa. Q_{199} is employed to prevent a change in the loading on the power supply when Q_{194} is turned off. This is the only function of Q_{199}.

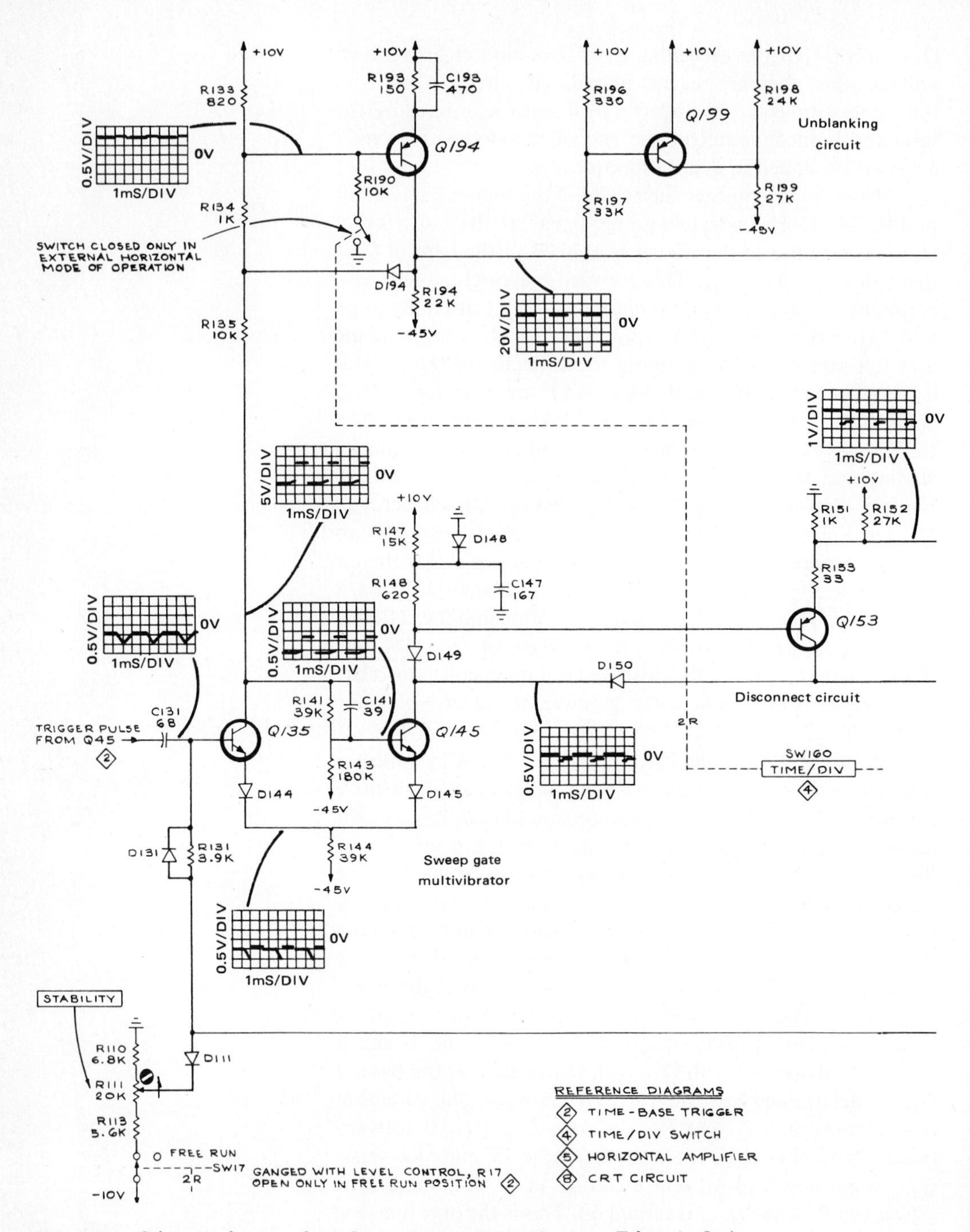

FIG. 14.11 *Schematic diagram of time-base generator circuit.* (*Courtesy, Tektronix, Inc.*)

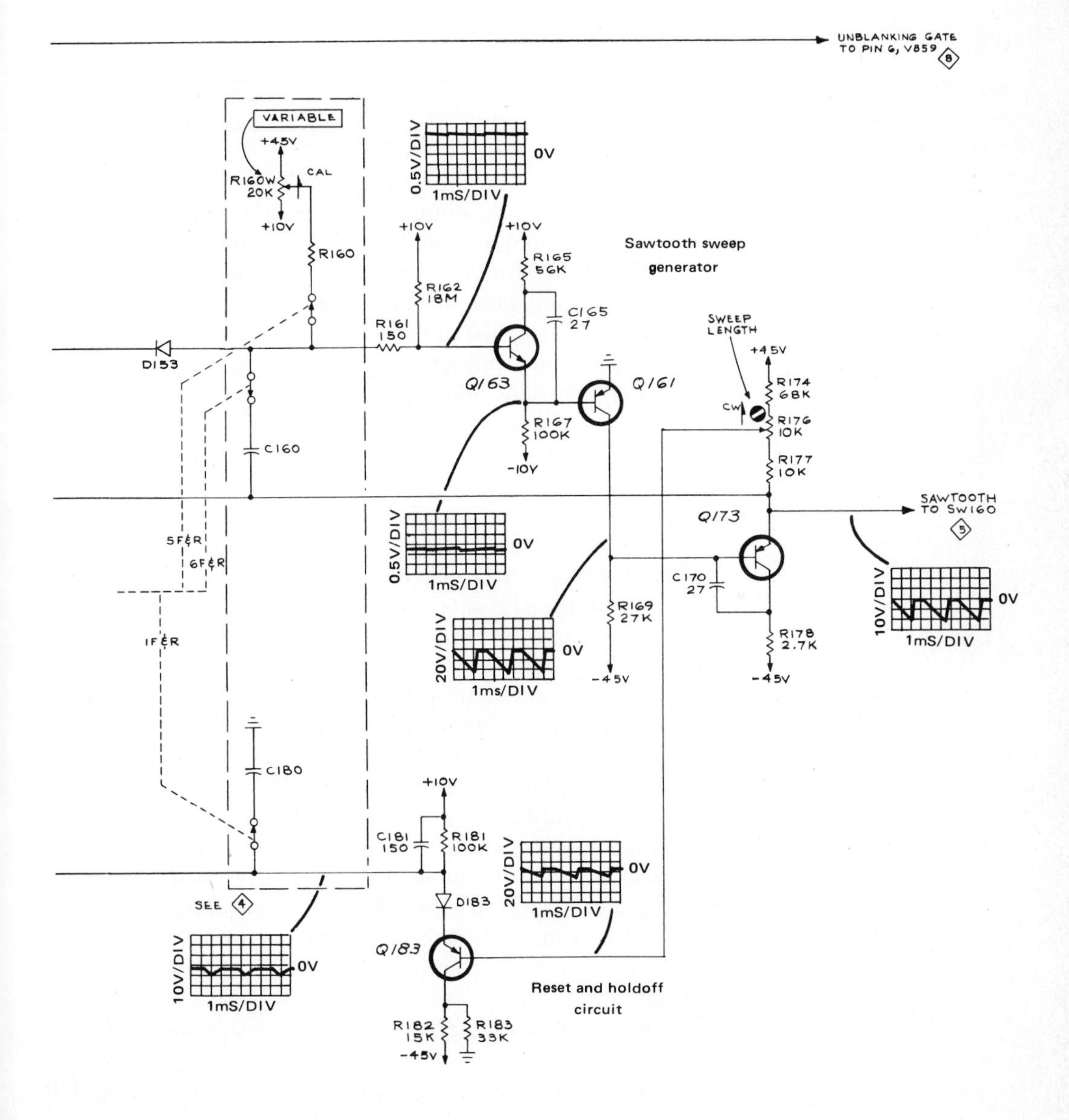
UNBLANKING GATE TO PIN 6, V859
VARIABLE
+45V
CAL
R160W 20K
+10V
R160
0.5V/DIV
0V
1mS/DIV
+10V
+10V
R165 56K
R162 18M
Sawtooth sweep generator
C165 27
R161 150
SWEEP LENGTH
+45V
D153
Q163
Q161
R174 68K
CW
R176 10K
R167 100K
-10V
R177 10K
C160
SAWTOOTH TO SW160
Q173
0.5V/DIV
0V
1mS/DIV
5F&R
6F&R
C170 27
R169 27K
10V/DIV
0V
1mS/DIV
IF&R
20V/DIV
0V
1ms/DIV
R178 2.7K
-45V
-45V
C180
+10V
C181 150
R181 100K
20V/DIV
0V
1mS/DIV
SEE 4
D183
10V/DIV
0V
1mS/DIV
Q183
Reset and holdoff circuit
R182 15K
R183 33K
-45V

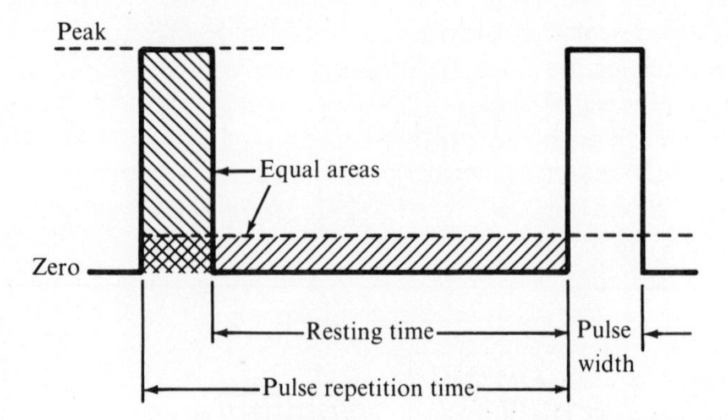

FIG. 14.12 *The duty cycle of the waveform is the ratio of pulse width to pulse-repetition time.*

When the time-division switch in Fig. 14.11 is set for the external horizontal mode of operation, the base of Q_{194} is connected to ground through R_{190}. This forward-biases Q_{194} and unblanks the crt beam. Therefore, an external signal can be applied to the input of the horizontal amplifier, as explained in detail subsequently. We find that most oscilloscope applications utilize the time-base generator to energize the horizontal amplifier. However, there are some applications in which it is desired to drive the horizontal amplifier from an external signal source, with the time-base generator inoperative.

Next, let us consider the operation of the disconnect circuit in Fig. 14.11. The disconnect circuit is comprised of Q_{153} and the disconnect diode D_{153}. Quiescently, Q_{153} is forward biased by the negative level at the collector of Q_{145} in the sweep-gate multivibrator stage. Q_{153} conducts heavily when in this quiescent state before a trigger pulse is applied, and the transistor goes into saturation. However, the action of diodes D_{149} and D_{150} prevent the collector of Q_{153} from reaching the full saturation level. Diode D_{149} is a silicon type with a forward voltage drop of 0.6 V, and diode D_{150} is a germanium type with a forward voltage drop of 0.4 V. In the quiescent state of this circuit, the voltage drop of diodes D_{149} and D_{150} clamps the collector of Q_{153} about 0.2 V more negative than its base and prevents it from reaching full saturation. Since a transistor requires more time to turn off from the saturated state in the absence of clamping, this configuration permits Q_{153} to turn off quickly when a trigger pulse is applied. Disconnect diode D_{153} is quiescently conducting current through Q_{153}, R_{153}, timing resistor R_{160}, and variable control $R_{160}W$ (when the variable control is not set to its calibrate position). The emitter current of Q_{153} sets the level at the cathode of D_{153}, which in turn determines the quiescent conduction of the sawtooth sweep generator stage to establish the starting point of the sweep. Also, the conduction of D_{153} prevents timing capacitor C_{160} from charging in this quiescent state.

Next, when a trigger pulse is applied in Fig. 14.11, the output level from Q_{145} goes positive and Q_{153} is quickly cut off. When Q_{153} cuts off, D_{153} is quickly reverse biased, since

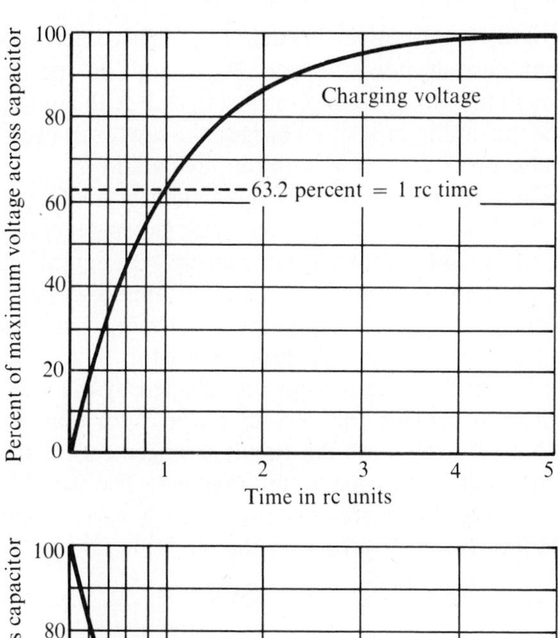

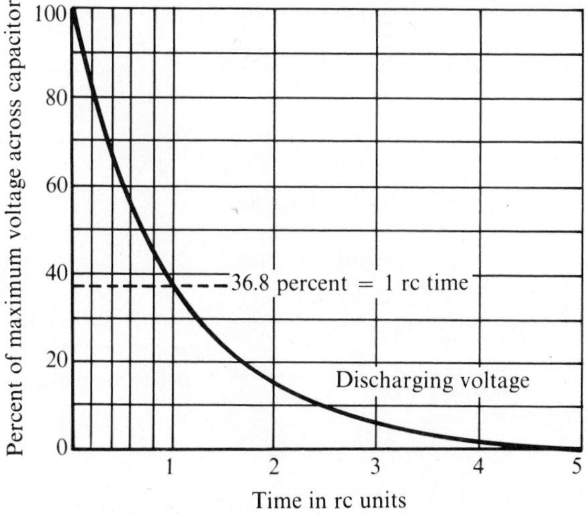

FIG. 14.13 *Universal RC time constant chart. (a) Charge curve; (b) discharge curve.*

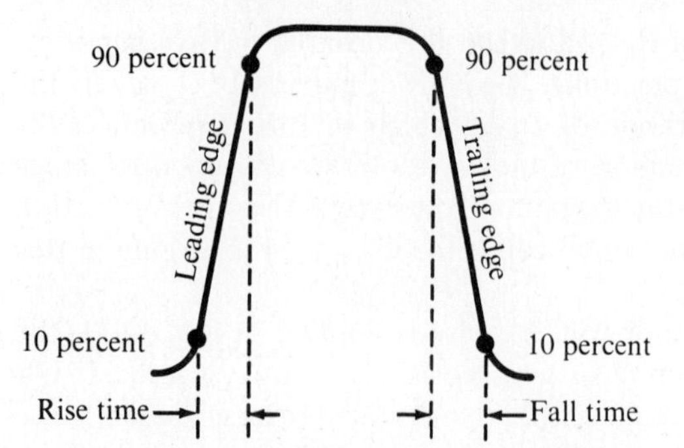

FIG. 14.14 *Leading and trailing edges with rise and fall time intervals.*

its cathode is driven positive through R_{152}. This interrupts the quiescent current flow through D_{153}, and the timing current through the timing resistor begins to charge the timing capacitor. As the timing capacitor charges, the sawtooth sweep generator stage produces a sawtooth output signal.

Proceeding to the sawtooth sweep generator configuration in Fig. 14.11, we may note that the basic sweep generator circuit is a Miller integrator (Miller run-up) configuration. When the quiescent current flow through the disconnect circuit is interrupted by the sweep-gate pulse, timing capacitor C_{160} begins to charge through timing resistor R_{160}. The timing capacitor and timing resistor are selected by the time-division switch to provide the various sweep rates that are available. Figure 14.15 shows the circuit arrangement for the time-division switch in the horizontal amplifier. The variable control R_{160W} in Fig. 14.11 provides continuously variable but uncalibrated sweep rates by varying the charging current to the timing capacitor.

FIG. 14.15 *Horizontal amplifier block diagram.*

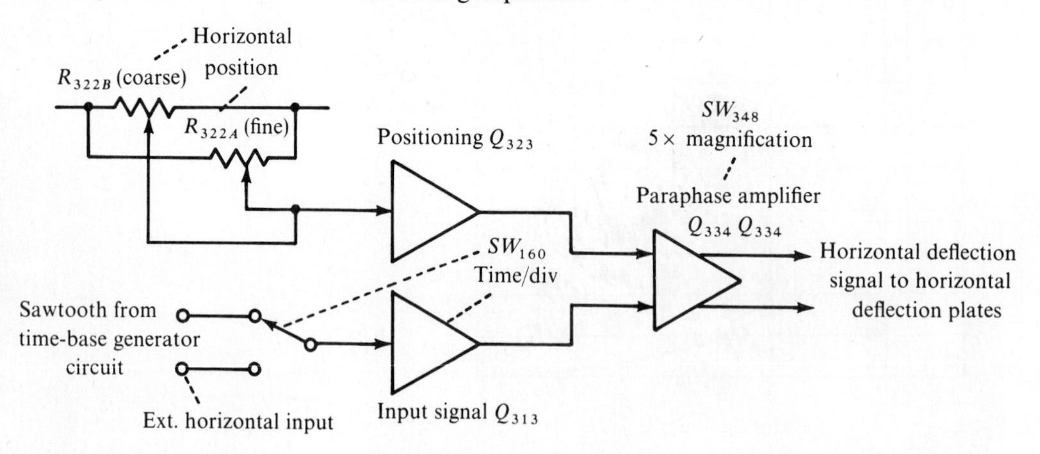

As the timing capacitor begins to charge positive toward the voltage applied to the timing resistor in Fig. 14.11, the base of Q_{163} rises positive also. This produces a positive-going change at the emitter of Q_{163} which is coupled to the base of Q_{161}. In turn, Q_{161} amplifies and inverts the voltage change at the emitter of Q_{163} to produce a negative-going sawtooth output through emitter follower Q_{173}. The signal at the emitter of Q_{173} is applied to the horizontal amplifier circuit through the time-division switch and is also connected back to the negative end of timing capacitor C_{160}. This feedback to the timing capacitor maintains a constant charging current for C_{160} to produce a linear sawtooth output waveform. The actual voltage change at the base of Q_{163} is very small compared to the sawtooth output waveform. This output waveform continues to go negative until the circuit is reset through the reset-and-holdoff circuit. Note that the divider R_{174}-R_{176}-R_{177} in the emitter circuit of Q_{173} determines the level at which the sweep is reset. Thus, the sweep-length adjustment R_{176} is set to provide the standardized sweep length.

We shall now observe the complete operation of the reset-and-holdoff circuit in Fig. 14.11. Notice that the negative-going sawtooth from the emitter of Q_{173} is applied to the base of Q_{183} through R_{177} and R_{176}. The sweep-length adjustment R_{176} determines the dc level and amplitude of the signal applied to the base of Q_{183} to determine the length of the sweep. This negative-going sawtooth at the base of Q_{183} produces a similar negative-going change at the emitter of Q_{183}, which charges holdoff capacitors C_{180} and C_{181} through D_{183}. The level at the anode of D_{183} is also applied to the base of Q_{135} through R_{131}. During the sweep interval, Q_{135} in the sweep-gate multivibrator stage is on, and Q_{145} is off. The negative-going signal from Q_{183} drives the base of Q_{135} negative toward cutoff as the sawtooth produced by the time-base generator circuit goes negative.

Next, observe that D_{111} in Fig. 14.11 disconnects the stability control from the circuit as the negative-going voltage reverse-biases the diode. As the base of Q_{135} is driven negative, the emitter of Q_{135} goes negative also, and this change is coupled to the emitter of Q_{145} through D_{144} and D_{145}. This action continues until the emitter of Q_{145} drops sufficiently negative to forward-bias Q_{145} (the voltage at the base of Q_{145} is held constant by divider R_{141}-R_{143} from the collector of Q_{135} to the -45-V supply). Q_{145} then takes control of the emitter current and the negative-going sawtooth applied to the base of Q_{135} cuts Q_{135} off in order to reset the sweep-gate multivibrator stage. The level on the negative-going sawtooth at which this stage is reset can be varied with the sweep-length adjustment R_{176}. When R_{176} is turned clockwise, the sawtooth must go more negative before the sweep-gate multivibrator is reset, which produces a longer sweep trace. The action is opposite when R_{176} is turned counterclockwise to produce a shorter sweep trace.

As the sweep-gate multivibrator is reset, Q_{145} goes into conduction and its collector drops negative rapidly, which forward-biases Q_{153} and D_{153} in Fig. 14.11. Timing capacitor C_{160} discharges rapidly through D_{153}, R_{153}, and Q_{153}. As C_{160} discharges, the base of Q_{163} goes negative also to produce the retrace portion of the sweep. Since the resistance of the discharge path for C_{160} is less than that of the charging path (through the timing resistor), the retrace portion of the sweep has a steeper slope which quickly returns the crt beam to the left side of the screen. The unblanking pulse produced by the unblanking circuit ends when the sweep-gate multivibrator stage is reset; hence, this retrace is not visible on the screen. At this time the disconnect circuit and the sawtooth sweep generator stages have returned to their quiescent condition, and can produce another sweep sequence as soon as the sweep-gate multivibrator stage is ready to process the next trigger pulse.

Note that as the anode of D_{183} in Fig. 14.11 went negative along with the negative-going sawtooth applied to the base of Q_{183}, the holdoff capacitor charged negative also. However, when the circuit is reset to produce the retrace portion of the sawtooth signal, the emitter of Q_{183} rises positive with the retrace, but the anode of D_{183} is clamped by the charge level on the holdoff capacitor C_{180}. D_{183} is reverse biased to disconnect the positive-going retrace from the holdoff capacitor and the base circuit of Q_{135}. This action blocks incoming trigger pulses for a period of time to establish a holdoff (lockout) period, which allows all circuits to return to their quiescent state before the next sweep sequence is produced.

The holdoff time is determined by the discharge rate of the holdoff capacitor C_{180}. As the holdoff capacitor discharges through R_{181} in Fig. 14.11, the base of Q_{135} rises positive also through R_{131}. When the holdoff capacitor has discharged to the level where D_{111} is forward biased, the voltage level at the base of Q_{135} is again determined by the setting of the stability control. Now the circuit is ready to accept the next positive-going trigger pulse. Holdoff capacitor C_{180} is changed in value by the setting of the time-division switch to provide the correct holdoff time for the sweep rate that is employed. (See Fig. 14.15.) Holdoff capacitor C_{181} is connected into the circuit at all times. For sweep rates of 5 μs and faster, C_{180} is disconnected and C_{181} determines the holdoff time of the circuit.

When the triggering level control is turned fully clockwise to its free-run position, SW_{17} (ganged with R_{17}) disconnects the negative voltage level from the stability control R_{111}. Thus D_{111} and R_{111} have no control on the quiescent level at the emitter of Q_{135} as described previously. Instead, the base of Q_{135} continues to rise positive as the holdoff capacitor discharges, until Q_{135} is forward biased. The result is that the time-base generator is retriggered at the end of each holdoff period to produce a free-running sweep. This sweep free-runs at the rate determined by the setting of the time-division switch and produces a bright reference trace at fast sweep rates. This is in contrast to the 50-Hz repetition rate in the automatic triggering mode.

14.9 HORIZONTAL AMPLIFIER

To provide the required level of horizontal deflection voltage, the sawtooth voltage from the time-base generator circuit is stepped up through the horizontal amplifier. Switching facilities are provided whereby an external signal can be applied to the horizontal amplifier for special types of displays. A horizontal magnifier function is also provided, as explained below. Figure 14.15 shows a detailed block diagram for the horizontal amplifier in this example. A schematic diagram of this section is depicted in Fig. 14.16.

We observe that the input signal for the horizontal amplifier is selected by the time-division switch in Fig. 14.16. In all positions of this switch except external-horizontal (note arrow to external-horizontal input jack), the input signal is the sawtooth produced by the time-base generator circuit. This sawtooth is applied to the base of Q_{313} through C_{311}-R_{311}. These components are part of the compensated

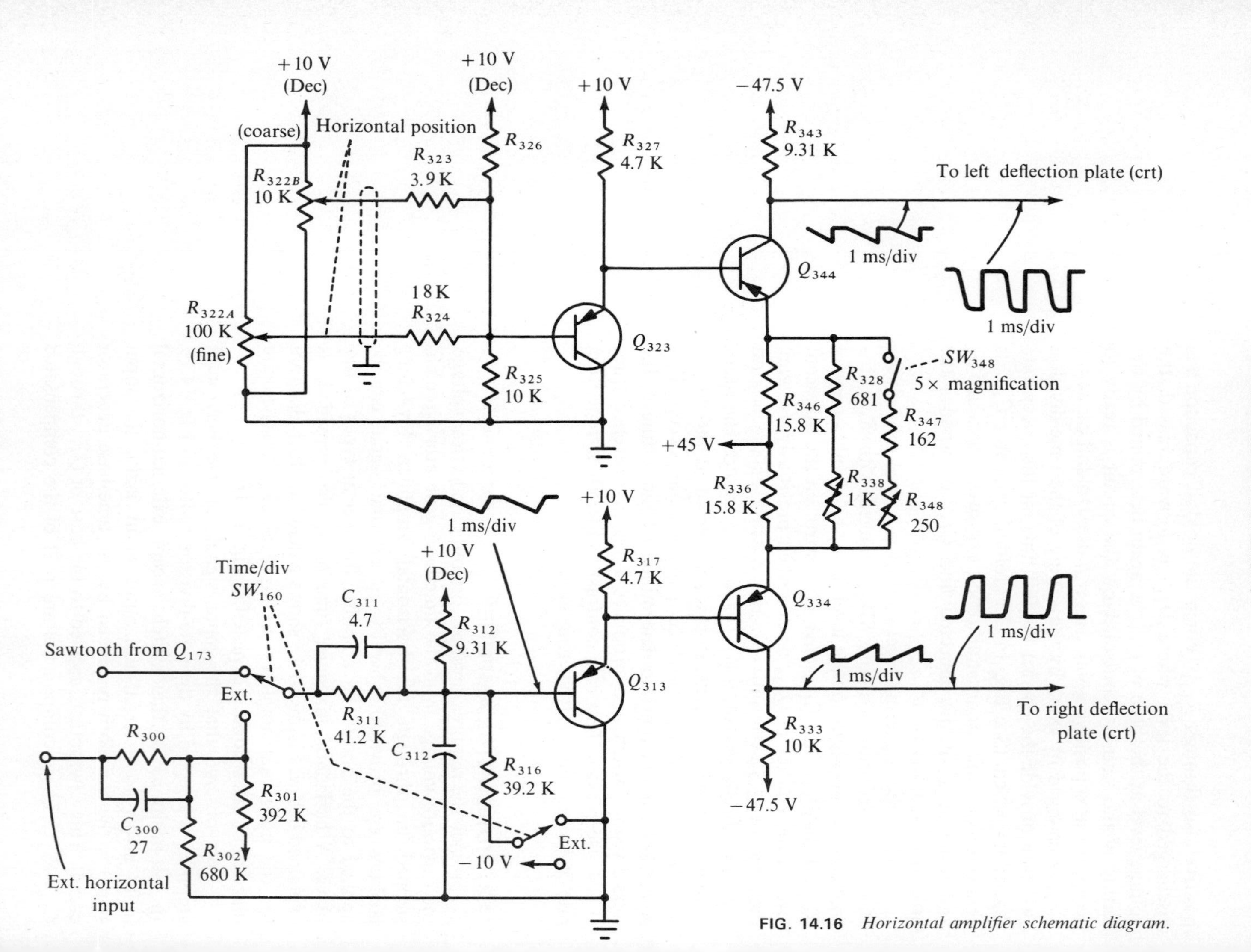

FIG. 14.16 *Horizontal amplifier schematic diagram.*

divider network C_{311}-C_{312}-R_{311}-R_{312}-R_{316} which establishes the input resistance for the horizontal amplifier circuit as well as the correct dc voltage level at the base of Q_{313}.

For external-horizontal-mode operation, an external signal from the external-horizontal input jack in Fig. 14.16 is employed to drive the horizontal amplifier. The network C_{300}-R_{300}-R_{301}-R_{302} is added to the divider network mentioned previously to provide the desired input impedance for external-horizontal mode operation (approximately 100 K and 30 pF). When the time-division switch is set to its external-horizontal position, SW_{160} connects R_{316} to -10 V instead of ground. This voltage is applied to divider R_{312}-R_{316}, and along with divider R_{301}-R_{302} sets the base level of Q_{313} so that the display is near the center of the screen for external-horizontal-mode operation when the horizontal position control is set to midrange.

Next, we observe that the horizontal position control $R_{322}A$ and $R_{322}B$ in Fig. 14.17 sets the dc voltage level at the base of Q_{323}, which determines the horizontal position of the crt display. Q_{323} is connected as an emitter follower (ef), and the voltage at its emitter is applied to the base of Q_{344} in the paraphase amplifier stage. The horizontal position control is

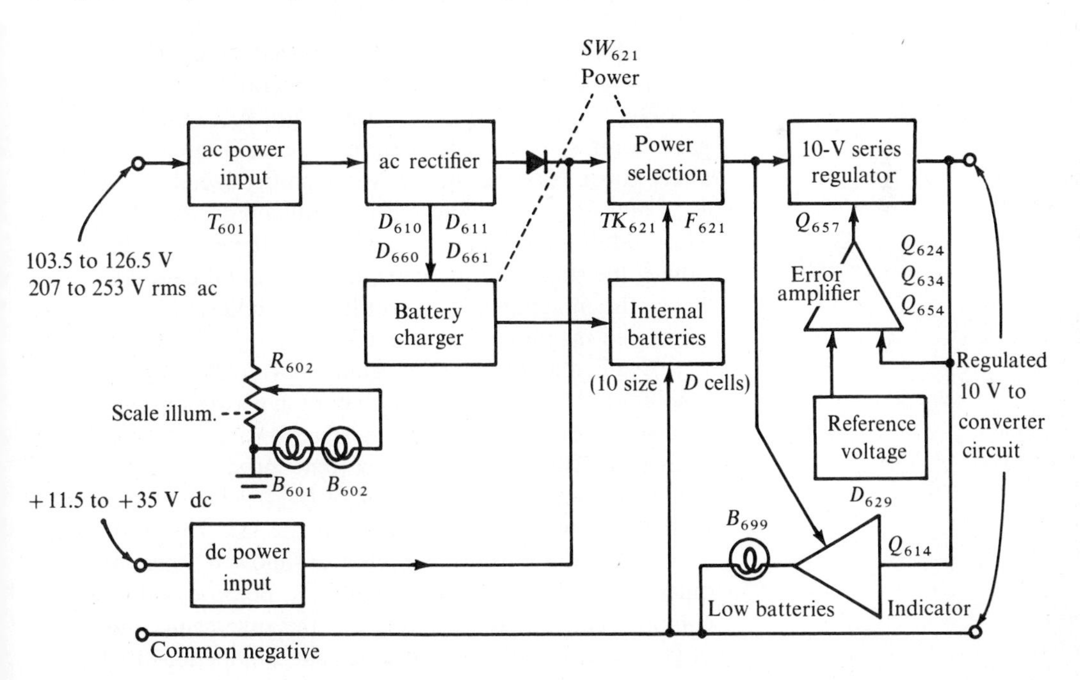

FIG. 14.17 *Regulator circuit block diagram.*

a dual-range control which provides a combination of coarse and fine adjustment in a single control. When this control is rotated, fine control $R_{322}A$ provides positioning for a range of about 60° of rotation. Then, after the fine range is exceeded, the coarse control $R_{322}B$ provides rapid positioning of the trace.

Note that Q_{334} and Q_{344} make up a pair of transistors connected as a paraphase amplifier in Fig. 14.16. This section converts the single-ended input signal into a push-pull output signal. In addition, this stage provides a X5 magnifier and adjustments to set the normal and magnified gain excursions. Note that the input signal from Q_{313} is applied to the base of Q_{334}, and it produces equal but opposite output signals at the collectors of Q_{334} and Q_{344}. This circuit action occurs as follows: The negative-going sawtooth (or negative-going external-horizontal signal) applied to the base of Q_{334} from Q_{313} forward-biases Q_{334} and the current through Q_{334} increases. This increase in current produces a positive-going sawtooth at the collector of Q_{334}, which drives the right-hand deflection plate in the crt.

Simultaneously, the increase in current through Q_{334} in Fig. 14.17 produces a negative-going sawtooth at its emitter. The emitters of Q_{334} and Q_{344} are coupled together through two degeneration networks. For normal sweep action, R_{328} and R_{338} control the emitter degeneration between Q_{334} and Q_{344} to set the gain of this stage. R_{338}, the horizontal gain control, is adjusted to provide calibrated horizontal sweep rates. Next, when the X5 magnifier switch is actuated, R_{347} and R_{348} are connected in parallel with R_{328} and R_{338}. This additional conductance decreases the emitter degeneration of the stage and increases the horizontal gain five times. R_{348}, the magnifier gain control, is adjusted to provide calibrated magnified sweep rates.

Note that the negative-going sawtooth coupled to the emitter of Q_{344} through the degeneration network in Fig. 14.16 reverse-biases Q_{344}, and the current through the transistor decreases. This decrease in current produces a negative-going sawtooth at the collector of Q_{344}, which is connected to the left-hand deflection plate in the crt. The positioning level from the positioning emitter-follower stage sets the quiescent-level bias at the base of Q_{344}. Because of the emitter coupling between Q_{344} and Q_{334}, the quiescent conduction of Q_{344} also establishes the quiescent conduction of Q_{334} to fix the starting point of the sweep, or the quiescent position of an external horizontal display.

14.10 REGULATOR CIRCUIT

Regulation of the power supply is required for stable operation of the oscilloscope. Figure 14.17 shows a detailed block diagram of the regulator circuit in this example. It provides a regulated +10-V output to the converter circuit from a power source; a stable low-ripple output voltage is produced from a variety of sources. A schematic diagram of the regulator circuit is depicted in Fig. 14.19. Alternating-current power may be supplied to the primary of transformer T_{601}. The secondary of T_{601} provides three power outputs. Diodes D_{610} and D_{611} are connected to one secondary winding to form a full-wave rectifier which provides rectified voltage to the power selection stage through D_{620}. C_{612} provides filtering action. Diodes D_{660} and D_{661} are also connected to this secondary winding to form a full-wave rectifier which is in series with the D_{610}-D_{611} rectifier to supply an output of approximately +32 V. This voltage is filtered by C_{660} and is connected to the battery-charger stage for use in recharging storage cells.

The other secondary winding of T_{601} in Fig. 14.19 is connected to the scale-illumination control R_{602}. In turn, the current through the scale-illumination lights, B_{601} and B_{602}, is controlled by the scale-illumination control for illuminating the graticule lines. Notice that the power switch is located after this stage. Therefore, power is applied to the secondary of T_{601} whenever the instrument is connected to an ac voltage source. Alternatively, a dc power source may be utilized. A potential in the range from +11.5 to +20 V is suitable. Note that D_{620} is reverse biased by the applied

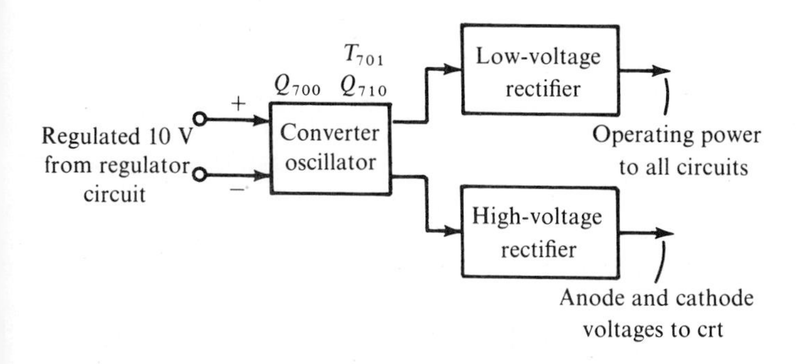

FIG. 14.18 *Converter block diagram.*

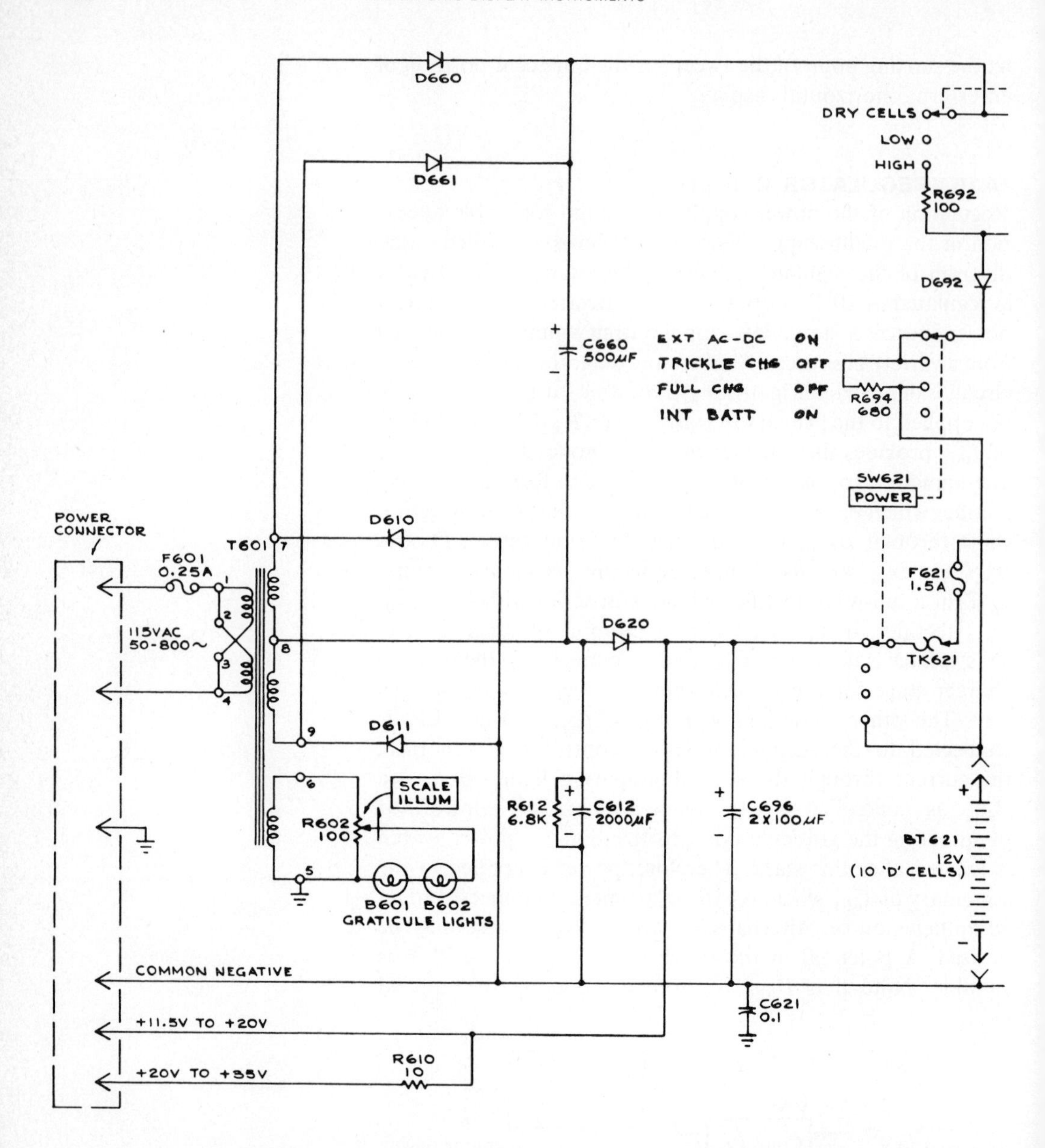

FIG. 14.19 *Schematic of the regulator. (Courtesy, Tektronix, Inc.)*

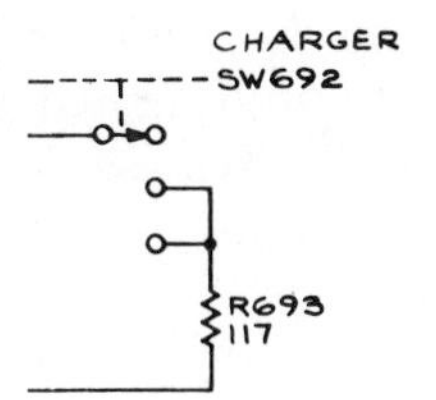

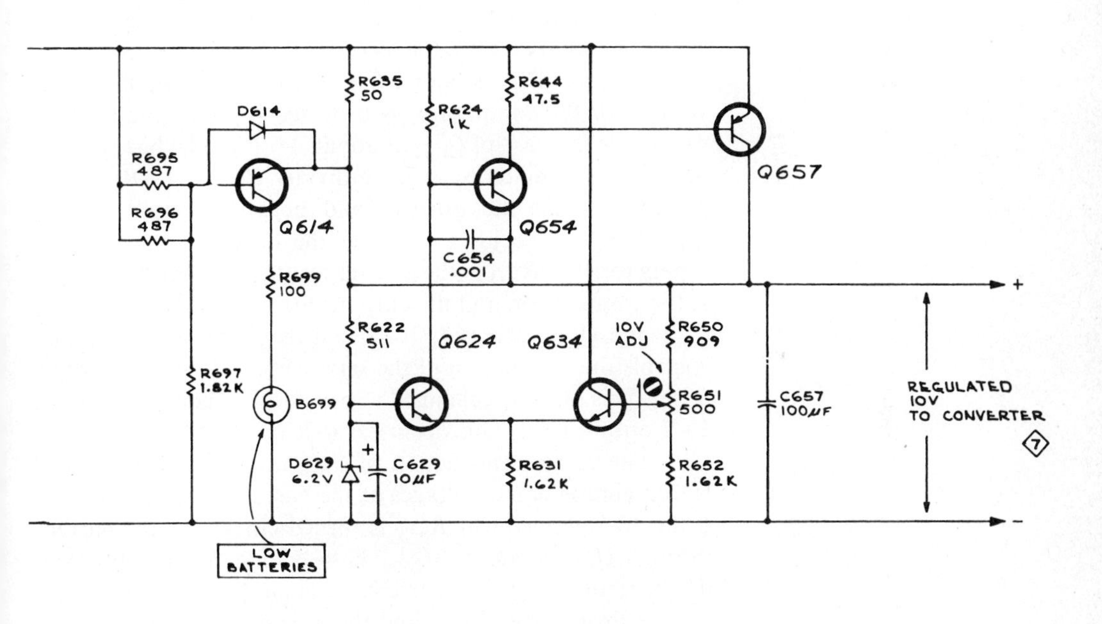

REFERENCE DIAGRAMS:

7 CONVERTER

voltage to disconnect the secondary circuit of T_{601} and the battery charger stage for this mode of operation. If desired, operation is also possible from internal batteries. Ten size *D* cells are used in this mode; secondary-type cells can be recharged from the internal battery charger.

Thermal cutout TK_{621} in Fig. 14.19 opens the power circuit automatically if the ambient temperature becomes excessively high. Overload protection is provided by fuse F_{621}. Diodes D_{660}-D_{661} provide charging current for secondary batteries. Two charging rates are available.

Next, let us observe the operation of the +10 V series regulator, error amplifier, and reference voltage stages. The series regulator can be compared with a variable resistor with a value that changes automatically to maintain a constant output voltage. The error amplifier stage provides this change by varying the conduction of the series regulator stage.

We observe that the error-amplifier stage, Q_{624} and Q_{634} in Fig. 14.19, is connected as a comparator. Reference voltage for the comparator is provided by zener diode D_{629}, which sets the base of Q_{624} at about +5.1 V. The base level of Q_{634} is determined by voltage divider R_{650}-R_{651}-R_{652} connected between the positive and negative output of this supply. R_{651} is adjustable to set the output voltage of this supply to 10 V. R_{631} is the emitter resistor for both transistors in the comparator, and the current through it divides between Q_{624} and Q_{634}. The output current from the error amplifier controls the conduction of the series regulator stage (through Q_{654}). This output is changed to provide a constant low-ripple 10-V output level. This occurs as follows:

The comparator action of Q_{624} and Q_{634} in Fig. 14.19 is to maintain equal voltages at the bases of both transistors. If the 10-V adjustment R_{651} is turned clockwise, the current through Q_{634} increases (Q_{634} base goes more positive than the base of Q_{624}), the current through Q_{624} produces less voltage drop across R_{624}, and the base of Q_{654} goes positive. The emitter of Q_{654} drives the base of Q_{657} positive to decrease the current through the load, thereby decreasing the output voltage of the supply. This places less voltage across divider R_{650}-R_{651}-R_{652} and the divider action returns the base of Q_{634} to about +5 *V*. A similar but opposite action takes place when R_{651} is turned counterclockwise so that the base of Q_{634} is more negative than the base of Q_{624}. The 10-V adjustment is set to provide a 10-V level at the output of the supply.

The output voltage of the supply in Fig. 14.19 is regulated

to provide a constant voltage to the load (converter circuit) by feeding a sample of the output voltage back to series regulator Q_{657}. For example, assume that the output voltage increases because of a change in load or an increase of input voltage. This produces a more positive voltage at the collector of Q_{657}, which is connected across the voltage divider R_{650}-R_{651}-R_{652} to produce a more positive level at the base of Q_{634}. The current flow through Q_{634} increases, which allows Q_{624} to conduct less, and its collector goes positive. When the collector of Q_{624} goes positive, the bias on Q_{654} increases, resulting in reduced current through series regulator Q_{657}. This means in turn that there is less current through the load, and the output voltage increases. In a similar manner, the series regulator and error amplifier stages compensate for output changes due to ripple.

Note that transistor Q_{614} with its associated circuitry in Fig. 14.19 provides a low-battery warning system. The emitter of Q_{614} is connected to the output of the 10-V series regulator stage and is therefore held constant. In turn, the voltage at the base of Q_{614} is determined by voltage divider R_{695}-R_{696}-R_{697} between the unregulated input voltage and the common negative level of this supply. Quiescently when the batteries are charged, the voltage established at the base of Q_{614} is sufficiently positive to reverse-bias Q_{614}. Therefore, there is no current flowing through B_{699}, the low-battery indicator light. Diode D_{614} is forward biased under these conditions to protect the base-emitter junction of Q_{614} from reverse voltage breakdown.

As the unregulated voltage applied to the 10-V series regulator stage in Fig. 14.19 drops to about 11.5 V, the level at the base of Q_{614} drops sufficiently negative to forward-bias the transistor. In turn, the collector current of Q_{614} flows through B_{699} to indicate that the batteries are low. Operation of the instrument after the low-battery light is energized may not only provide inaccurate indication, but may also result in damage to secondary cells. Since the low-battery indicator stage is operative with all power sources, this stage also provides an indication whenever the unregulated voltage to the series regulator stage drops below approximately 11.5 V. Although continued operation from an external power source with the low-battery light energized will not necessarily damage the instrument, inaccurate measurements will result.

Let us consider the operation of the converter section.

This circuit provides the various output voltages required for instrument operation. Figure 14.18 shows a detailed block diagram for the converter circuit. A schematic diagram is depicted in Fig. 14.21. We observe that the regulated 10-V output from the regulator circuit energizes the converter oscillator stage. Transistors Q_{700} and Q_{710} with transformer T_{701} make up the oscillator stage. Both Q_{700} and Q_{710} have their emitters connected to the positive output of the 10-V regulator, and the bases of these transistors are connected to the common negative output through resistors R_{700} and R_{701} and the base winding of T_{701}. The frequency of oscillation is about 2 kHZ. Since the transformer core is driven into saturation, the output has a square waveform.

Note that the voltage induced into the secondary of T_{701} is rectified by diodes D_{710}-D_{711}-D_{712}-D_{713}-D_{720}-D_{721}-D_{726} in Fig. 14.21. This rectified voltage is then filtered to provide output potentials of +45, +10, −10, −45, and −47.5 V. We observe that the voltage at terminal 16 of T_{701} provides the high potentials required by the crt. Diodes D_{714}-D_{715}-D_{717}-D_{718}-D_{719} and capacitors C_{714}-C_{715}-C_{716}-C_{717}-C_{720} form a voltage multiplier network (pentupler) that supplies about +3350 V to the crt anode. R_{714} and R_{720} are current limiters to protect the rectifiers. Neon bulb B_{714} reduces the input voltage to the multiplier by about 60 V to provide the correct output level (+3350 V). Diode D716 provides rectified voltage for the crt cathode at about −720 V. A separate heater winding is employed, which is elevated to

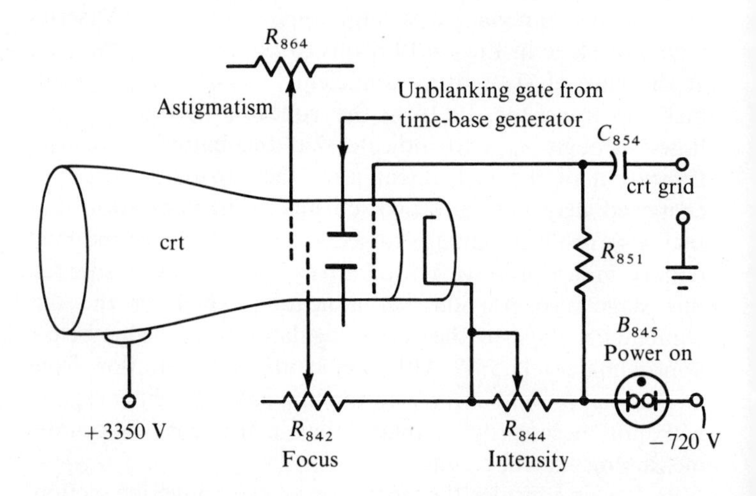

FIG. 14.20 *Crt circuit Schematic.*

-720 V by R_{718}. This prevents arc-over between the crt heater and cathode.

Figure 14.22 shows a schematic diagram for the crt circuit. Positive accelerating voltage for the crt anode (about $+3350$ V) is provided by the converter circuit. Ground return for this supply is through the resistive helix inside the crt to pin 8 of the crt and then to ground through the geometry adjustment R_{861}. Negative accelerating potential is provided by the -720-V supply in the converter circuit. Neon bulb B_{845}, the power-on light, glows when the -720-V potential is applied. The total crt accelerating potential is about 4 kV.

Notice that the geometry adjustment R_{861} in Fig. 14.22 varies the positive level on the deflection-plate isolation shield, thereby controlling the overall geometry of the display. Deflection blanking is utilized to deflect the electron beam off-screen during retrace time and when the sweep is not operating. One deflection-blanking plate has a fixed potential of about $+10$ V, and the other is driven by the unblanking gate pulse from the time-base generator circuit. The potential difference employed has a range from -22 to $+8.4$ V. A z-axis input terminal is also provided, whereby the electron beam can be modulated by means of an external control voltage.

14.11 AMPLIFIER CHARACTERISTICS

High-frequency compensation is required in wideband amplifiers for oscilloscopes. The bandwidth of an amplifier stage can be increased by using a lower value of load resistance, as seen in Fig. 14.23. That is, the stray circuit capacitances have less bypassing effect on high frequencies when the load resistance is reduced. However, this entails a reduction in stage gain, as shown in Fig. 14.23. With all other things being equal, a very large number of stages would have to be provided. Therefore, circuit elaboration is employed to provide extended high-frequency response when a comparatively high value of load resistance is utilized.

One widely used method of high-frequency compensation consists of partial bypassing of a cathode resistor (or emitter resistor), as depicted in Fig. 14.24. For a given value of R_L, a given value of stray capacitance C_S, and a given value of R_K, there is an optimum value for C_K which will provide the flattest response to the highest realizable frequency cutoff

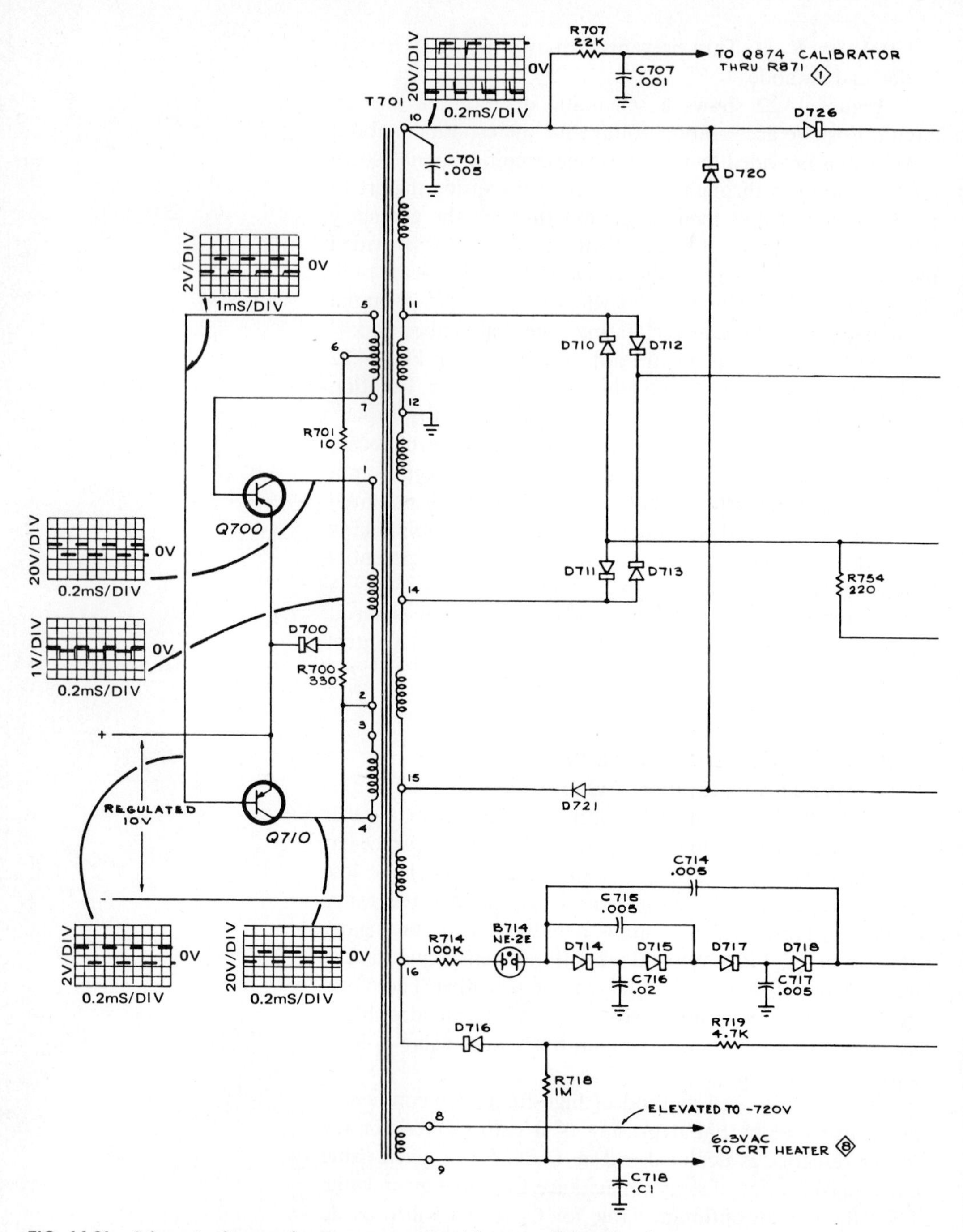

FIG. 14.21 *Schematic diagram for converter section. (Courtesy, Tektronix, Inc.)*

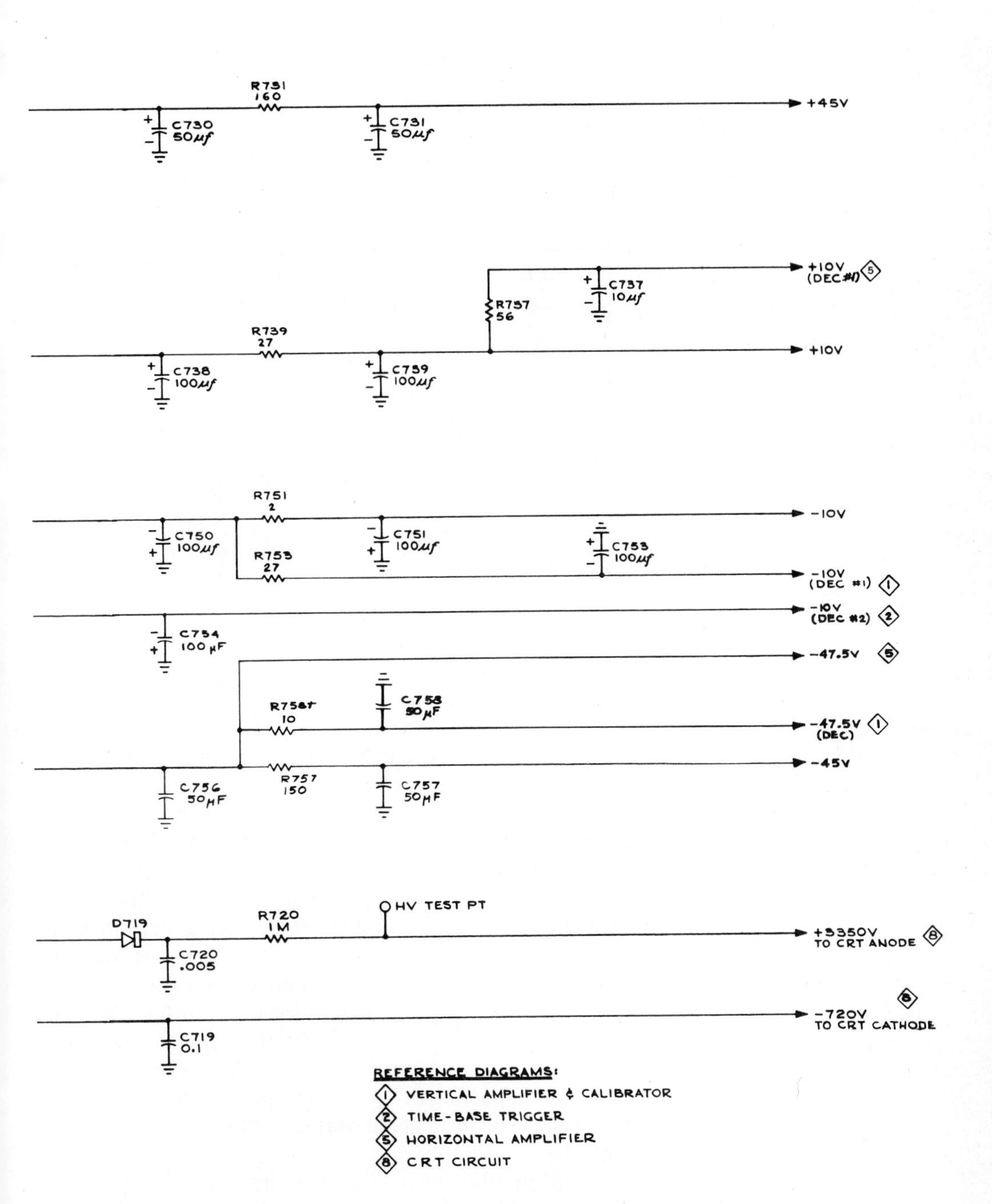
R731
160
+45V
C730
50μf
C731
50μf
+10V
(DEC#1)
C737
10μf
R737
56
R739
27
+10V
C738
100μf
C739
100μf
R751
2
-10V
C750
100μf
C751
100μf
C753
100μf
R753
27
-10V
(DEC #1)
-10V
(DEC #2)
C754
100 μF
-47.5V
R758
10
C758
50μF
-47.5V
(DEC)
-45V
C756
50μF
R757
150
C757
50μF
HV TEST PT
D719
R720
1M
+3350V
TO CRT ANODE
C720
.005
-720V
TO CRT CATHODE
C719
0.1
REFERENCE DIAGRAMS:
1 VERTICAL AMPLIFIER & CALIBRATOR
2 TIME-BASE TRIGGER
5 HORIZONTAL AMPLIFIER
8 CRT CIRCUIT

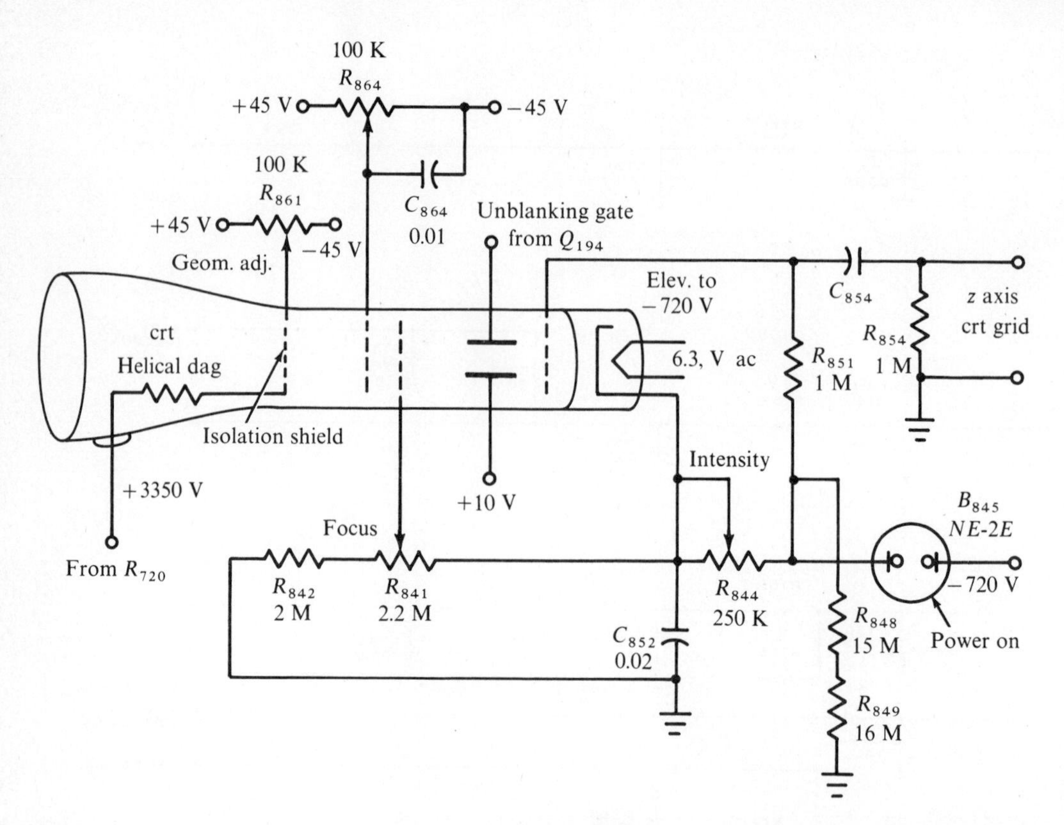

FIG. 14.22 *Schematic diagram for crt circuit.*

point. This method exploits controlled degeneration in the cathode (or emitter) circuit by reducing degeneration at higher frequencies. In turn, the stage gain increases at high frequencies. There is an upper limit, however, which is imposed by the decreasing reactance of C_S as the maximum available gain of the stage is reached.

Another method of high-frequency compensation that is often utilized is shown in Fig. 14.25. This arrangement exploits positive feedback in a push-pull stage; as the frequency increases, the reactances of C_{f1} and C_{f2} decrease, with the result that more signal voltage is fed back. In turn, the stage gain is increased to compensate for the bypassing actions of stray capacitances C_{S1} and C_{S2}. As in the previous example, a practical upper limit is reached due to the progressively decreasing reactances of the stray capacitances with increasing frequency of operation.

Still another method of high-frequency compensation employs peaking coils, as depicted in Fig. 14.26. In the shunt-

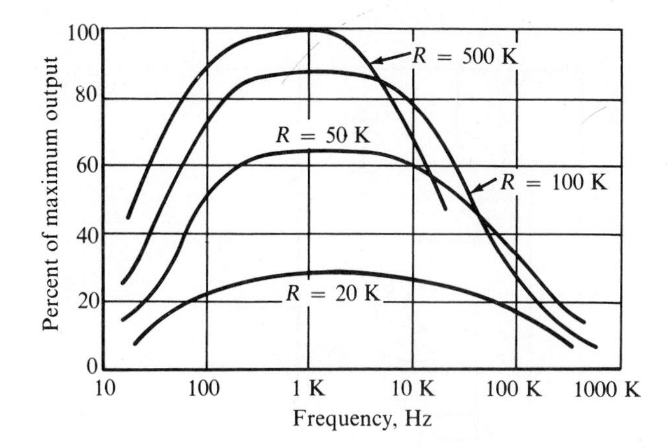

FIG. 14.23 *Comparative gain and bandwidth values for an RC-coupled stage vs. plate or collector load values.*

peaked arrangement, the peaking-coil inductance L is resonated in the vicinity of the high-frequency cutoff point. In turn, the impedance of the load branch rises at high frequencies owing to approach of the parallel-resonant frequency of the peaking coil. This increase in load impedance provides increased stage gain until the operating frequency exceeds the resonant frequency of the peaking coil. In the series-peaked arrangement, also shown in Fig. 14.26, series resonance is exploited to effectively cancel the shunting action of C_S. That is, as the series-resonant frequency of L and C_S is approached, a Q-times voltage magnification develops across C_S, thereby holding up the high-frequency response of the stage. The practical limit is reached when the operating frequency equals the resonant frequency of the peaking circuit.

Figure 14.27 shows a solid-state amplifier configuration

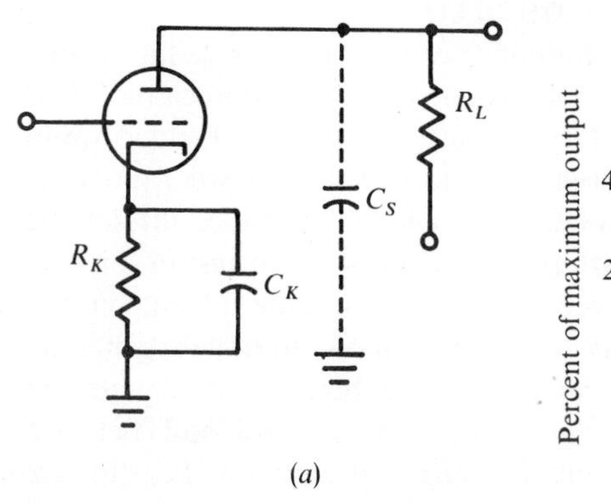

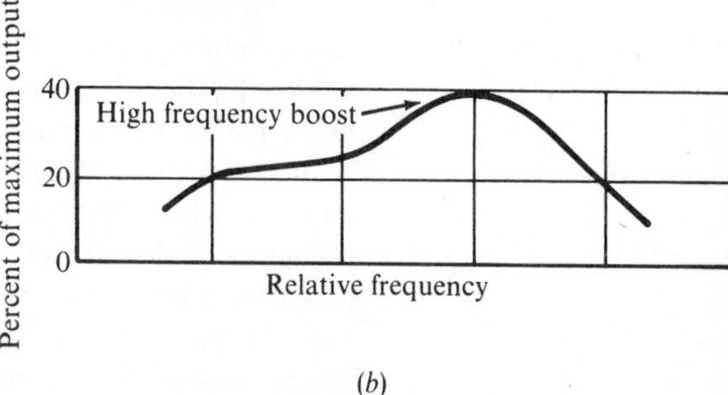

FIG. 14.25 *High-frequency boost. (a) Partial cathode bypassing obtained by suitable choice of value for C_k; (b) typical result of partial cathode bypassing.*

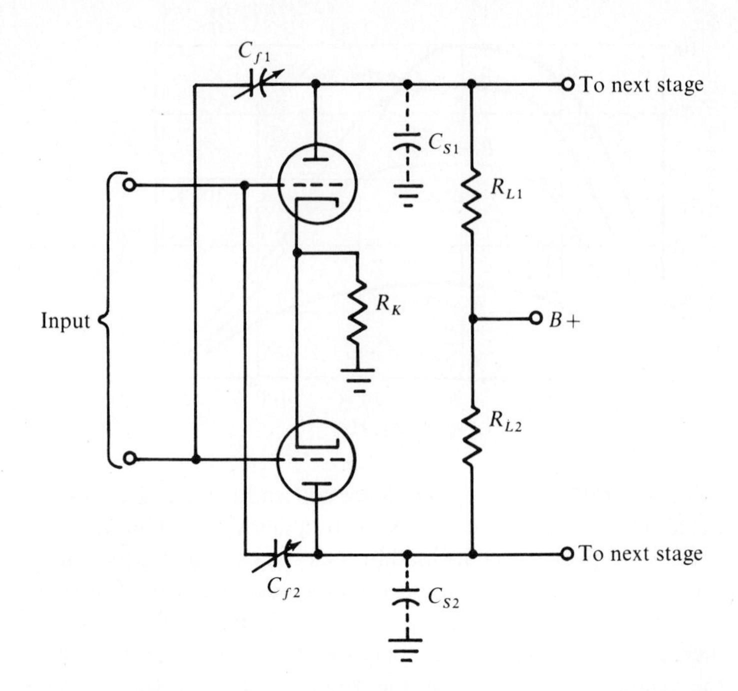

FIG. 14.25 *High-frequency compensation by positive feedback.*

that employs both series and shunt peaking for maximum high-frequency compensation. This method is widely used in service-type instruments, but is not favored in high-performance oscilloscopes because of the comparatively poor transient response. Although greatly extended high-frequency response is provided by the arrangement of Fig. 14.27, the peaking coils tend to ring on fast-rise pulse or square-wave signals, thus distorting the output waveform.

14.12 PROBES FOR OSCILLOSCOPES

Although the vertical amplifier of an oscilloscope has a high input resistance, its input capacitance is appreciable. This results chiefly from the necessity of using a shielded input cable; if a shielded input cable were not employed, the stray fields in the vicinity of the leads would produce intolerable interference in the pattern, particularly when testing in low-level circuits. Therefore, to reduce loading of the circuit under test, a shielded input cable to an oscilloscope is generally provided with a low-capacitance probe such as depicted in Fig. 14.28. This arrangement presents a tradeoff between increased input impedance and signal attenuation. Its principle of operation is the same as that of a frequency-compensated

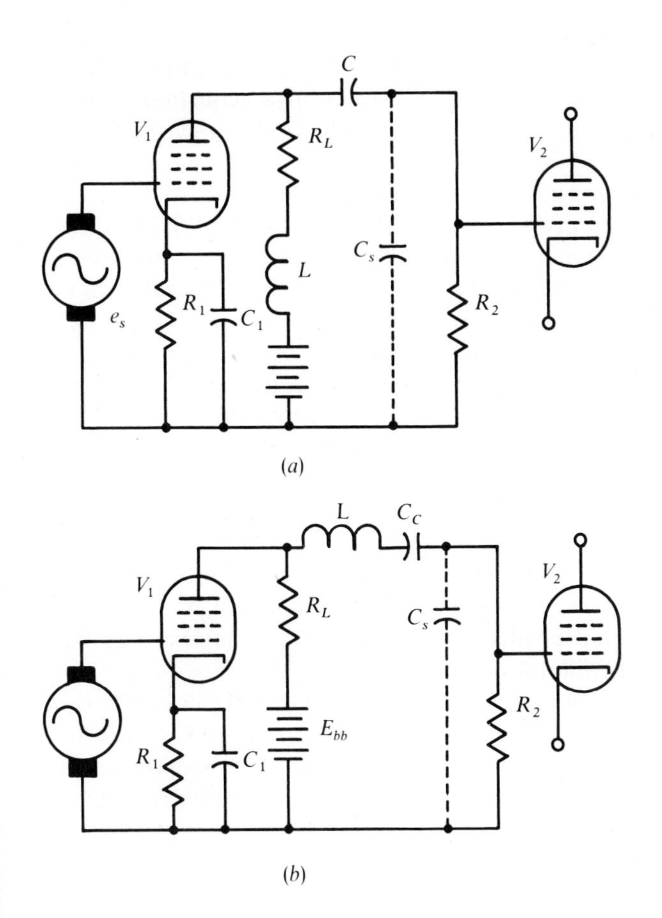

FIG. 14.26 *High-frequency compensation. (a) Impedance-coupled amplifier stage; (b) impedance-coupled amplifier stage, series-peaked.*

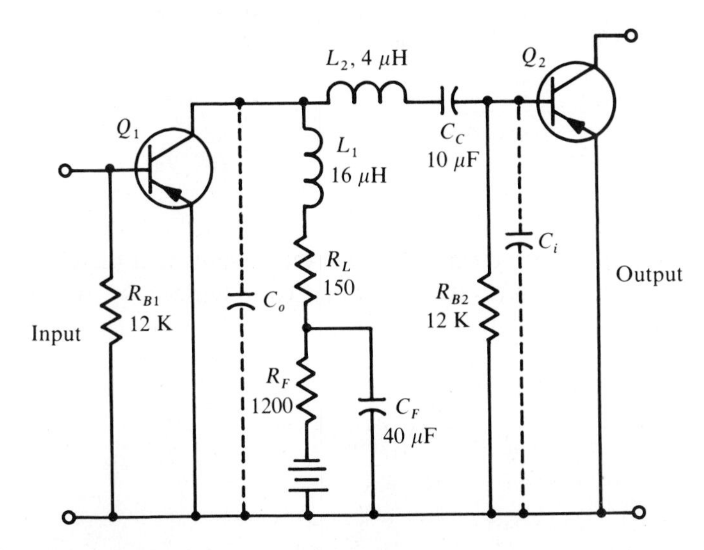

FIG. 14.27 *Typical 4-MHz transistor amplifier.*

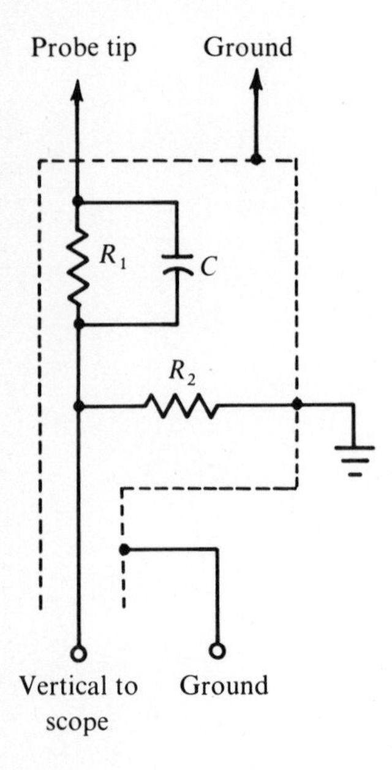

FIG. 14.28 *Low-capacitance probe.*

multiplier or attenuator. In most cases, an attenuation factor of 10:1 is provided, which results in a reduction of input capacitance to one-tenth of its value in the absence of a low-capacitance probe.

To understand the circuit action provided by a low-capacitance probe, let us refer to Fig. 14.29. We shall show that if the time constant (R_1C_1) of the probe is made equal to the time constant (R_2C_2) of the scope input circuit, complex waveforms will be passed without distortion. Also, the input impedance to the probe becomes higher than the input impedance to the scope. As noted above, this transformation is accomplished at the cost of reduced output signal voltage from the probe. We start our analysis by writing Ohm's law for the circuit in Fig. 14.29, and derive the condition for which the voltage division is constant, regardless of frequency. That is, we calculate a parameter relation in which the frequency term cancels out, leaving a purely resistive voltage divider action. In other words, if the circuit is arranged to respond independently of frequency, complex waveform distortion will not occur.

With reference to Fig. 14.29, the parallel circuit R_1C_1 is connected in series with the parallel circuit R_2C_2. In turn, if we let Z_1 equal the impedance of the R_1C_1 section, and let Z_2 equal the impedance of the R_2C_2 section, we can write

$$\frac{e_{out}}{e_{in}} = \frac{iZ_2}{i(Z_1 + Z_2)}$$

We observe that the i variable cancels out. Then, expanding the Z terms, we write

$$\frac{e_{out}}{e_{in}} = \frac{\dfrac{R_2X_{C2}}{R_2 + X_{C2}}}{\dfrac{R_1X_{C1}}{R_1 + X_{C1}} + \dfrac{R_2X_{C2}}{R_2 + X_{C2}}}$$

This equation can be rearranged to show that there is a condition in which the frequency term implied by the X variable cancels out:

$$\frac{e_{out}}{e_{in}} = \frac{\dfrac{R_2/j\omega C_2}{R_2 + 1/j\omega C_2}}{\dfrac{R_1/j\omega C_1}{R_1 + 1/j\omega C_1} + \dfrac{R_2/j\omega C_2}{R_2 + 1/j\omega C_2}}$$

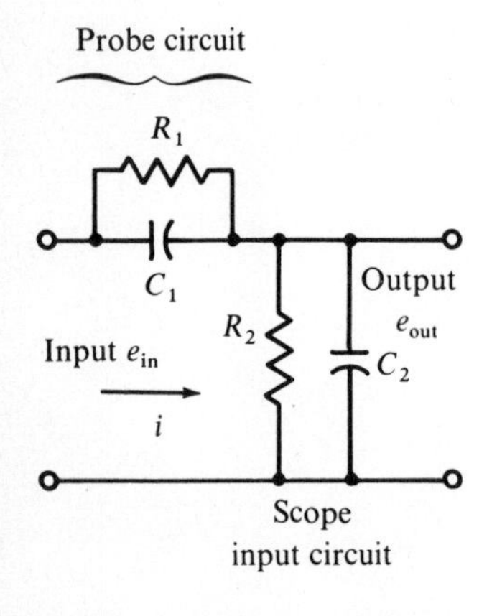

FIG. 14.29 *The probe increases the input impedance and reduces the output voltage.*

In turn, we write

$$\frac{e_{out}}{e_{in}} = \frac{\dfrac{R_2/j\omega C_2}{(j\omega R_2 C_2 + 1)/j\omega C_2}}{\dfrac{R_1/j\omega C_1}{(j\omega R_1 C_1 + 1)/j\omega C_1} + \dfrac{R_2/j\omega C_2}{(j\omega R_2 C_2 + 1)/j\omega C_2}}$$

Canceling the denominator, we write

$$\frac{e_{out}}{e_{in}} = \frac{\dfrac{R_2}{j\omega R_2 C_2 + 1}}{\dfrac{R_1}{j\omega R_1 C_1 + 1} + \dfrac{R_2}{j\omega R_2 C_2 + 1}}$$

Finally, if we stipulate that $R_1 C_1 = R_2 C_2$, we conclude that

$$\frac{e_{out}}{e_{in}} = \frac{R_2}{R_1 + R_2}$$

In other words, when the pair of time constants are made equal, the low-capacitance probe operates as a simple resistive voltage divider (independently of frequency). This circuit action has been obtained at the expense of signal voltage reduction in the ratio of $R_2/(R_1 + R_2)$. Also, since $R_1 C_1 = R_2 C_2$, we perceive that C_1 (the input capacitance to the probe) has also been reduced in the ratio: $C_1 + (R_2 C_2/R_1)$.

As a practical example, let us consider an oscilloscope that has an input impedance comprising 100 pF and 1 M. If we stipulate that a low-capacitance probe shall be constructed to provide a signal voltage attenuation of 10:1, then R_1 will evidently be equal to 9 M. In turn, C_1 will be equal to 100 ÷ 9 pF, or 11.1+ pF.

When testing in circuits operating at frequencies beyond the capability of the vertical amplifier, a demodulator probe such as depicted in Fig. 14.30 often provides a measure of practical utility. This type of probe is a detector arrangement, and is also called a traveling detector. A visualization of its action is shown in Fig. 14.31. The chief disadvantage of a simple demodulator probe is its limited demodulating capability, since only low-frequency envelope waveforms can be reproduced effectively. As tabulated in Fig. 14.30, the input impedance of a demodulator probe is not high, with the result that circuit loading can become a problem.

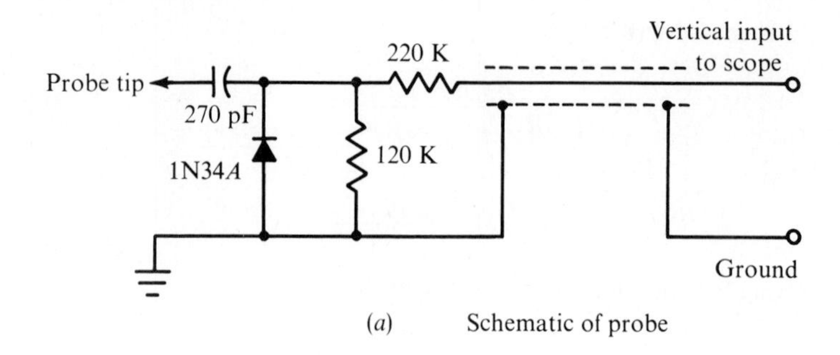

(*a*) Schematic of probe

Frequency response characteristics
rf carrier range: 500 kHz to 250 MHz
Modulated-signal range: 30 to 5000 Hz
Input capacitance (approx.): 2.25 pF
Equivalent input resistance (approx.):

At 500 kHz	25,000 Ω
1 MHz	23,000 Ω
5 MHz	21,000 Ω
10 MHz	18,000 Ω
50 MHz	10,000 Ω
100 MHz	5000 Ω
150 MHz	4500 Ω
200 MHz	2500 Ω

Maximum input:
ac voltage .20 rms V
28 peak V

(*b*) Specifications

FIG. 14.30 *Typical demodulator probe specifications.*

14.13 APPLICATIONS

Among the most basic applications for a triggered-sweep oscilloscope is the measurement of pulse width and pulse rise time. The output waveform from a pulse generator can be checked as depicted in Fig. 14.32. A normal resistive load should be provided for the generator (such as 90 Ω), and the use of a low-capacitance probe with the oscilloscope ensures that capacitive loading will be minimized so that the rise-time measurement will have full accuracy. Figure 14.33 illustrates the results of a typical test. It is evident from the first photo that the pulse width is 20 μs. It is evident from the last photo that the pulse rise time is approximately 0.02 μs.

A triggered-sweep scope is also very useful for checking the condition of a transmission line; artificial delay lines can be checked in the same general manner. With reference to Fig. 14.34, the oscilloscope is connected across the input terminals of the line. A source resistor R_1 is employed, which should have a value equal to the characteristic impedance of the line. The test results can be tabulated as follows:

1. If the line is operating normally and is terminated in its own characteristic resistance, only the pulse applied by the generator is visible in the display.
2. If the line is open-circuited, a reflected pulse of the same polarity as the applied pulse appears in the display. The time from the applied pulse to the reflected pulse corresponds to twice the length of the line.
3. If the line is short-circuited, a reflected pulse of the opposite polarity from the applied pulse appears in the display. The time from the applied pulse to the reflected pulse corresponds to twice the distance between the input and the short-circuit point on the line.

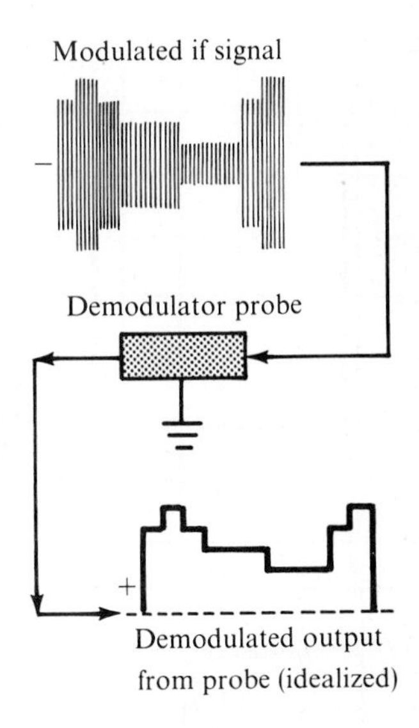

FIG. 14.31 *Visualization of demodulator probe action.*

In some applications, the use of the external trigger facility of an oscilloscope is essential. For example in designing or troubleshooting a chroma-signal delay line, it is helpful to be able to check the signal phase at the input and output of each section in the line. Figure 14.35*a* shows the configuration for a typical delay line. To make this phase display test, we employ the external trigger function of the oscilloscope. That is, the input signal to the delay line is applied to the external trigger input terminals (see Fig. 14.36). In turn, the sweep is initiated by the input signal to the line, regardless of the takeoff point to which the probe is applied. A low-capacitance probe is essential in this example to avoid circuit loading and phase disturbance. It is good practice to use a probe with a comparatively high attenuation factor—such as 100:1—in this type of test work.

In other applications, a dc capability is important because it saves time and provides convenience. Dc response is employed, for example, when making voltage measurements of pulsating dc waveforms, as shown in Fig. 14.37. That is, the signal voltages in a transistor circuit are associated with dc voltage levels, and certain proportions represent normal operation from the standpoint of efficiency, rated distortion, and operating polarities. Remember in pattern evaluation that if an oscilloscope has been calibrated for indication of

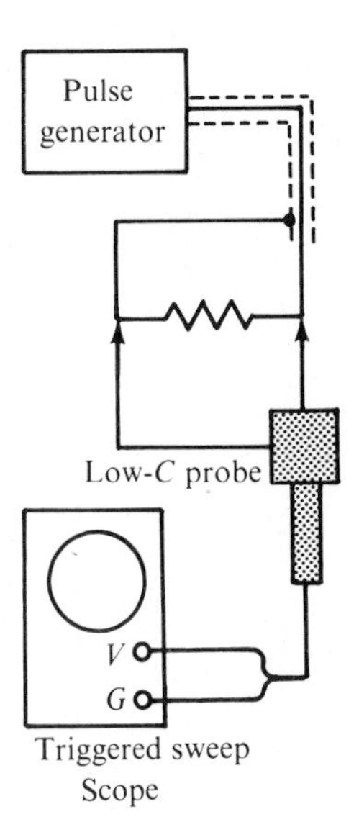

FIG. 14.32 *Measurement of pulse width and rise time.*

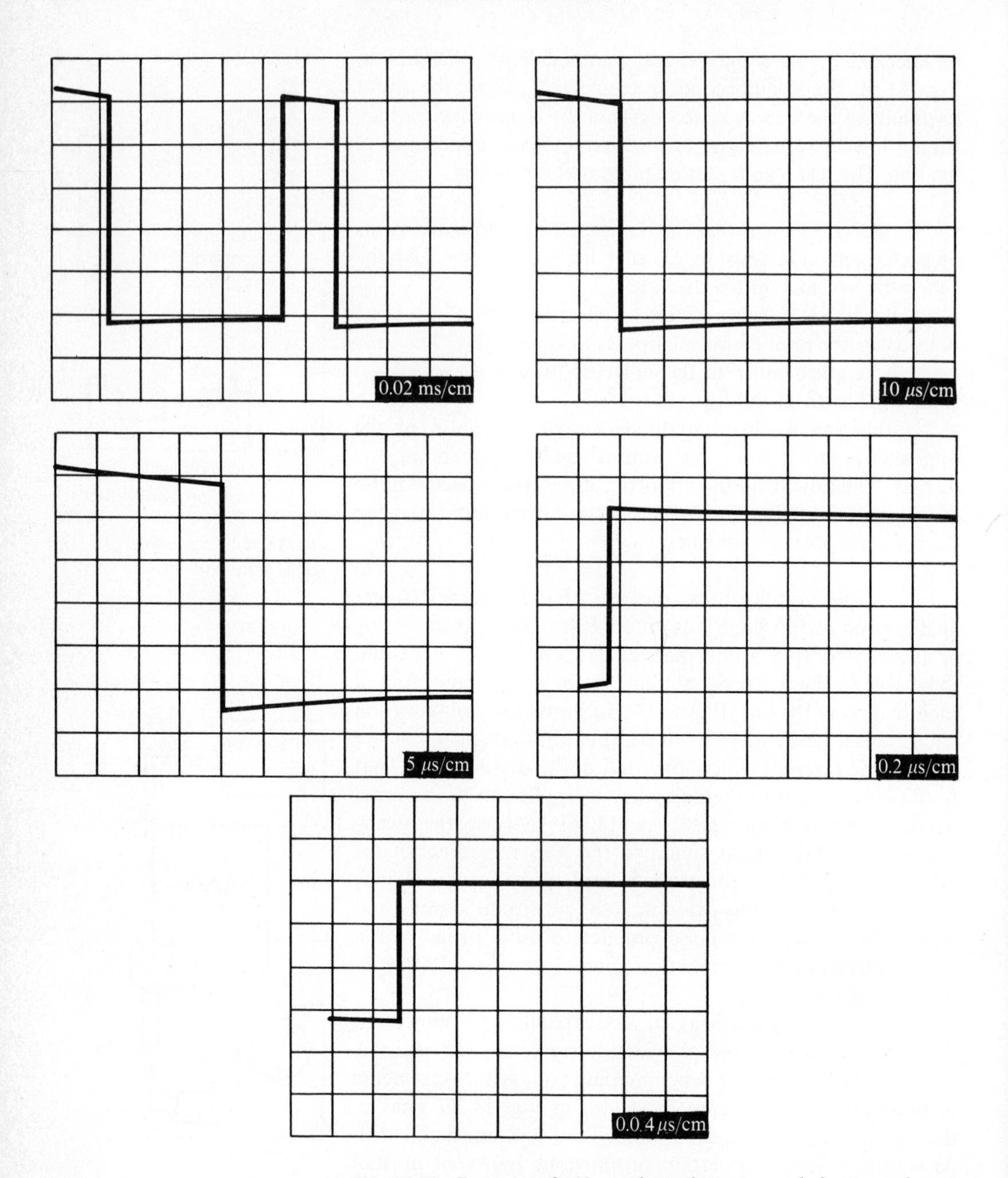

FIG. 14.33 *Expansion of a 20-μs pulse as the sweep speed of a triggered sweep is progressively increased.* (*Courtesy, Howard W. Sams & Co.*)

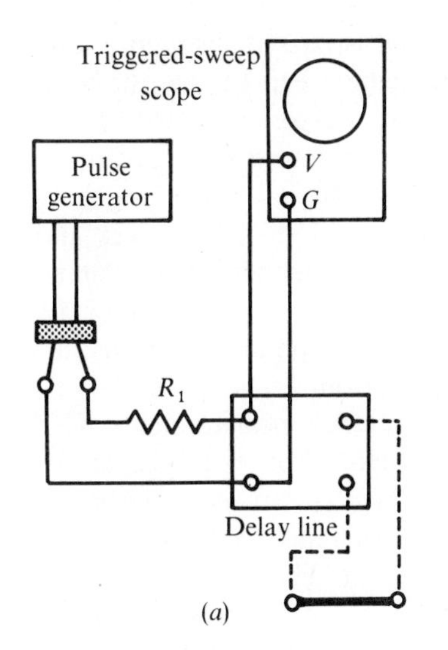

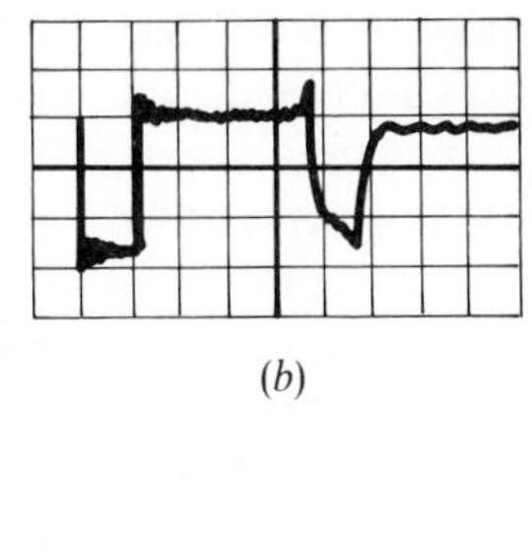

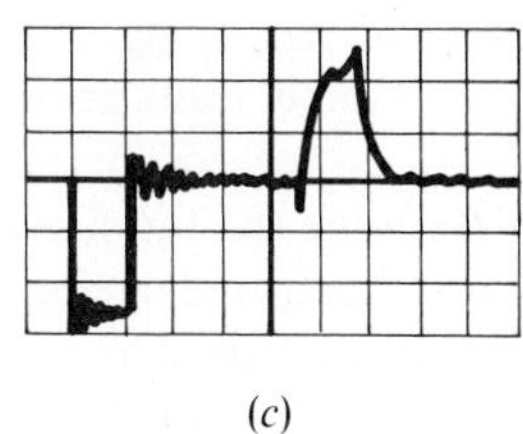

FIG. 14.34 *Evaluation of a transmission line.*

peak-to-peak voltage values, the same calibration holds true for dc voltage values.

There are applications in which it is helpful or even essential to employ the dc trigger function of an oscilloscope. This term is misleading to the beginning student, because the trigger action is not actually based on dc levels, but rather on low-frequency waveform components. When the dc trigger

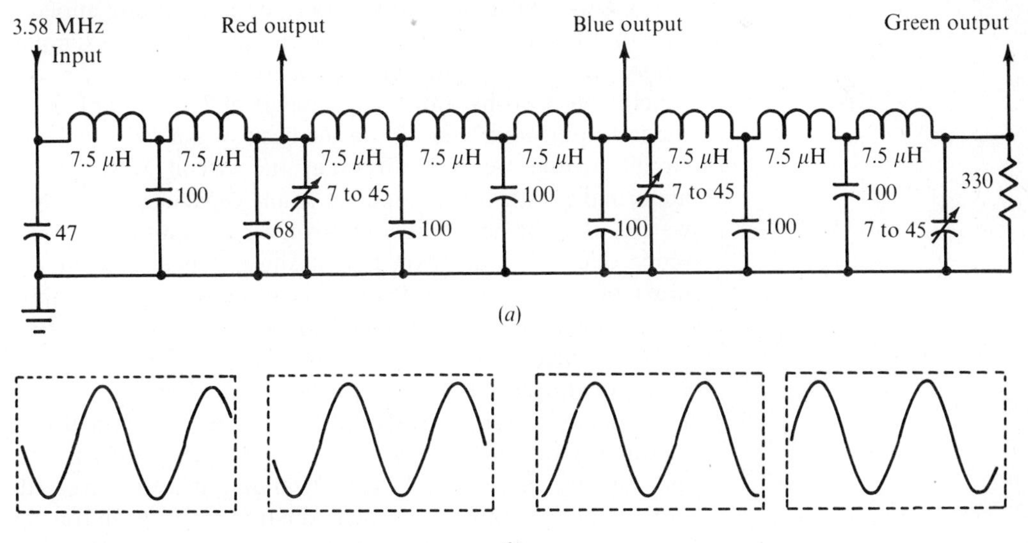

FIG. 14.35 *Checking phase shifts along a delay line. (a) Line configuration; (b) phase-display patterns.*

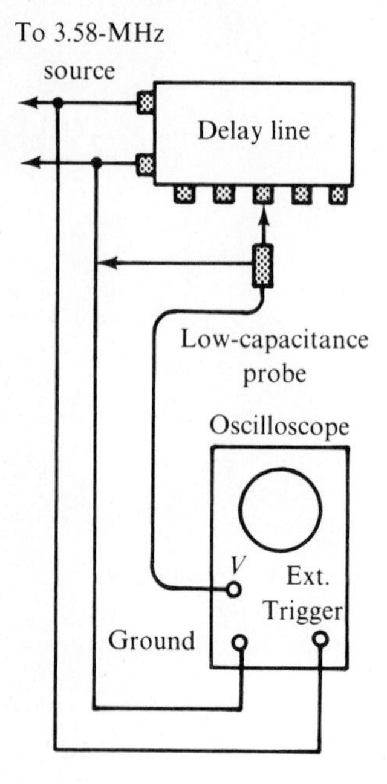

FIG. 14.36 *Oscilloscope connections.*

function is utilized, the input signal is passed through a low-pass filter before it is used to initiate the sweep. In turn, the high-frequency components of the input signal are removed, and the lowest frequency component is enhanced. The advantage of this signal processing is realized when a "noisy" signal is to be displayed. Noise spikes are random occurrences and usually have comparatively high frequency components; therefore, if the noise spikes are rejected, the uniformly repetitive lower frequency signal component remains for reliable triggering of the sweep section.

Some applications entail frequency measurements. This is an aspect of period measurement. For example, let us determine the fundamental frequency contained by the waveform illustrated in Fig. 14.33. With reference to the first photo, the sweep speed is set to 0.02 ms/cm. Therefore, the period of the pulse waveform is a time interval corresponding to 6 cm or 0.120 ms. To calculate the fundamental frequency of the pulse waveform, we take the reciprocal of the period, and we write

$$f_1 = \frac{1}{0.120 \times 10^{-3}} = 8333+ \text{ Hz}$$

To summarize some typical applications for low-capacitance and demodulator probes, we may note:

1. A low-capacitance probe should be used in any application that involves high-impedance circuit tests. For example, before we test a particular circuit, we should consider the general effect of shunting a capacitance of 75 or 100 pF from the test point to ground. If, on the basis of our knowledge of circuit action, we recognize that the circuit will operate abnormally with this amount of shunt capacitance inserted, we abandon the possibility of using a direct input cable to the scope. Next, if it appears that a shunt capacitance in the order of 20 pF will not disturb circuit action objectionably, we choose a 10:1 low-capacitance probe. On the other hand, if the inserted shunt capacitance can be no greater than 2 or 3 pF, we choose a 100:1 low-capacitance probe.
2. A demodulator probe should be used whenever amplitude-modulated waveforms with a carrier frequency higher than the response capability of the vertical amplifier are involved. For example, we must use a demodulator probe when tracing a video signal through the intermediate amplifier of a television

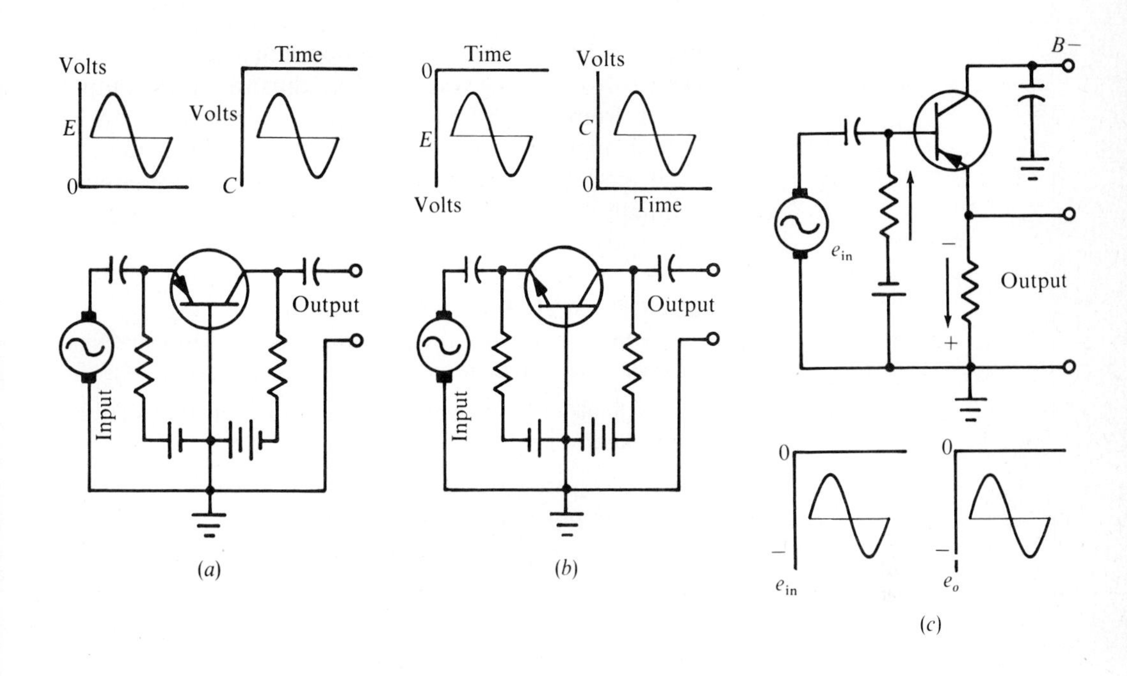

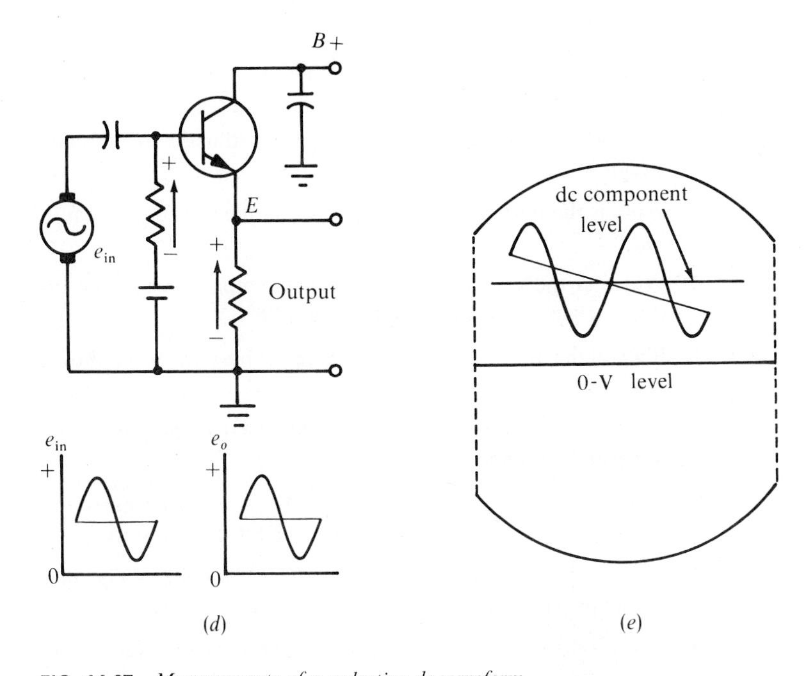

FIG. 14.37 *Measurements of a pulsating dc waveform.*

receiver. If a demodulator probe is used with a dc oscilloscope, we can easily check the percentage modulation in the output waveform from a high-frequency signal generator. The chief precaution to keep in mind in any test situation is the fact that a demodulator probe will load high-impedance circuits substantially.

QUESTIONS AND PROBLEMS

1. Why must high-speed triggered sweep be used to make time measurements on pulses?
2. Why should an oscilloscope have many calibrated steps for the time base?
3. Why is a square-wave signal usually made available at an output connector of a triggered-sweep oscilloscope?
4. What is the function of the time-base generator in the block diagram shown in Fig. 14.4?
5. How many time constants are represented in the waveform in Fig. 14.1 between the 10 percent and the 90 percent points?
6. Explain the operation of the paraphase amplifier (Q_{464} and Q_{467}) in the circuit shown in Fig. 14.6.
7. What determines the gain of the paraphase amplifier in the circuit shown in Fig. 14.6?
8. How is the position of the display established in the oscilloscope circuit in Fig. 14.6?
9. What is the purpose of capacitor C_{506} in the circuit diagram in Fig. 14.6?
10. Where is the vertical amplifier signal sampled to provide internal triggering in the circuit in Fig. 14.6?
11. Explain the purpose of each control in the block diagram shown in Fig. 14.7.
12. In reference to the circuit in Fig. 14.8, what is the purpose of transistors Q_{14} and Q_{24}?
13. In reference to the circuit in Fig. 14.8, what determines the point on the trigger signal at which the crt display starts?
14. Explain the operation of the sweep trigger circuit in Fig. 14.8 in the automatic position of the trigger level control.
15. Explain the operation of the trigger multivibrator stage in Fig. 14.8 for one sweep on the crt.

16. Under what condition will the trigger multivibrator stage in Fig. 14.8 operate as an astable multivibrator?
17. What are the two purposes of the time-base generator circuit?
18. What is the purpose of the unblanking circuit in Fig. 14.11?
19. What is the one function of Q_{199} in the circuit in Fig. 14.11?
20. Why does the sweep generator circuit in Fig. 14.11 use a Miller integrator configuration?
21. How does the control R_{160W} in Fig. 14.11 provide a continuously variable sweep rate?
22. What would be the results if transistor Q_{173} in Fig. 14.11 opened?
23. How is the reset portion of a sawtooth signal produced in the circuit in Fig. 14.11?
24. What is the purpose of the holdoff time circuit in Fig. 14.11?
25. What would be the effect in sweep operation if the holdoff capacitor (C_{180}) in Fig. 14.11 opened?
26. What determines the sweep repetition rate when the trigger level control is turned fully clockwise?
27. What is the purpose of the 5 multiplier control in the circuit shown in Fig. 14.16?
28. What are the correct settings on the oscilloscope shown in Fig. 14.3 to measure a pulse that has a width of 1 μs and a repetition time of 1 kHz?
29. What is the purpose of TK_{621} in Fig. 14.19?
30. The oscilloscope shown in Fig. 14.3 should not be operated after the low-battery light is energized. Why?
31. What is the purpose of the converter section of the oscilloscope depicted in Fig. 14.3?
32. Give three methods of compensation that are used in high-frequency amplifiers.
33. Explain the circuit action of a high-frequency probe.
34. What is the purpose of a high-frequency probe?
35. How is the high-frequency probe adjusted for correct compensation?
36. Why does a low-capacitance probe cause an attenuation of the input signal?

15

OSCILLOGRAPHS AND *XY* RECORDERS

15.1 BASIC DIFFERENCES

There is no sharp dividing line between an oscilloscope and an oscillograph. An oscilloscope provides a display only during the time that a vertical input signal is applied, whereas an oscillograph provides a permanent record of a waveform. A conventional oscilloscope can be converted into an oscillograph by mounting a camera in front of its crt screen. Originally, oscilloscopes employed crt displays, whereas oscillographs utilized ink tracings on moving paper strips. This distinction has been blurred by the development of storage-type cathode-ray tubes, which operate either as conventional display devices or as electronic "memories" that retain a glowing pattern on the screen until the storage circuitry is turned off.

An *XY* recorder is a particular type of oscillograph

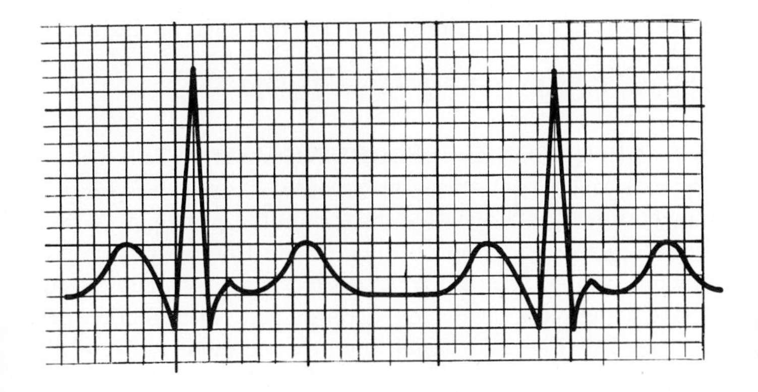

FIG. 15.1 *A typical electrocardiogram.*

which plots inked curves on cartesian graph paper, polar coordinate graph paper, or moving paper strips. Almost everyone has seen electrocardiographs which make *XY* recordings of heart potentials on a moving paper strip. Many of us are somewhat familiar with electroencephalographs which make *XY* recordings of brain potentials. A typical electrocardiogram is shown in Fig. 15.1, and typical electroencephalograms are shown in Fig. 15.2. Again, there is no sharp dividing line between sophisticated mechanical oscil-

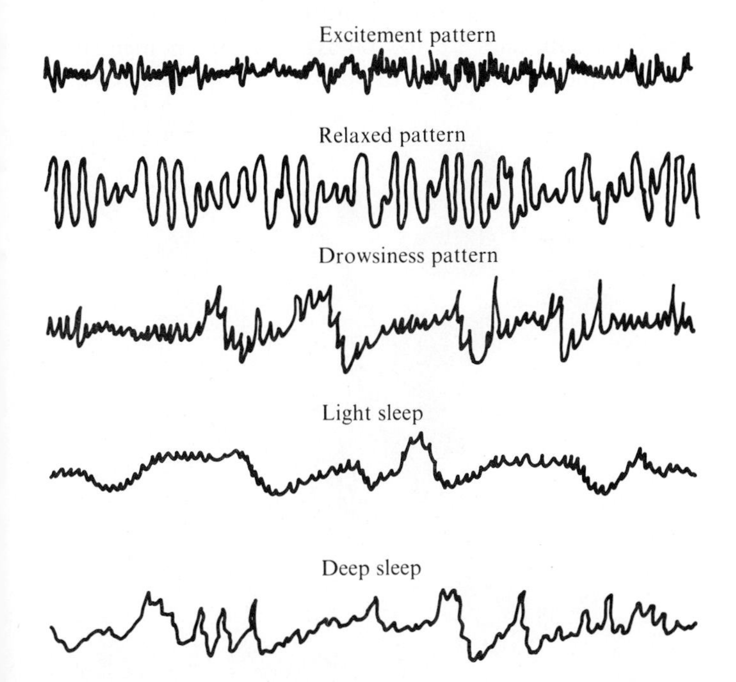

FIG. 15.2 *Typical electroencephalograms.*

lographs and simple recording voltmeters such as illustrated in Fig. 15.3.

15.2 OPERATION OF RECORDING VOLTMETER

The basic magnetic oscillograph employs a movable coil in a magnetic field and operates on the motor principle. A thin stylus is typically driven by the coil and traces a curve on a moving strip of paper. Either sensitized paper may be utilized, or a capillary ink feed may be provided. The multicorder illustrated in Fig. 15.3 uses a high-torque meter movement; the mechanical torque is applied to the moving coil by a taut-band suspension assembly. A synchronous motor and reduction gear system provide accurate travel of the paper strip with respect to time. Sensitized paper is employed, which becomes dark at points where pressure is applied. To avoid friction problems, a vibratory stylus is provided, which makes sequential impressions on the paper every other second. The paper strip moves 12 in./hr.

Simple magnetic oscillographs are useful for monitoring comparatively slow events, such as line-voltage fluctuations over a 24-hr (or longer) period. More elaborate recording voltmeters can operate at comparatively high speed, tracing curves at frequencies up to 100 Hz. To obtain high-speed operation, the mechanism is driven by an amplifier; however,

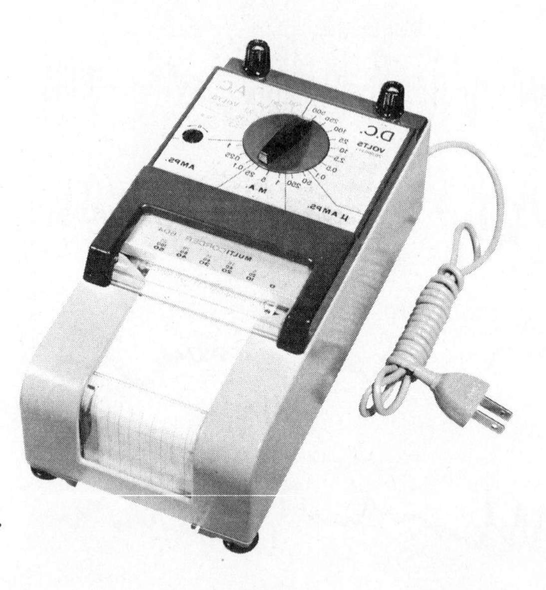

FIG. 15.3 *A multicorder voltage or current curve plotter. (Courtesy, Simpson Electric Co.)*

a limitation on high-frequency response is imposed by the inertia of the pen or stylus. The paper strip travels up to 125 mm/s in this type of oscillograph. The two-channel recorder illustrated in Fig. 15.4 is a direct-writing type of instrument, so-called because it does not utilize a crt and photographic film.

When capillary ink feed is used, the D'Arsonval movement is commonly termed a penmotor. Figure 15.5 depicts a typical arrangement. The mass of the moving-coil system is designed for a minimum practical value in order to obtain maximum speed of response. A flexible tube is provided to feed ink into the pen by capillary action. The sensitivity of a penmotor necessarily decreases as the operating frequency is increased. Therefore, the amplifier in the oscillograph must provide higher gain at high frequencies; for example, more than 10 dB additional gain may be required at the high-frequency limit of the instrument. This limit occurs near the point of mechanical resonance in the moving-coil system.

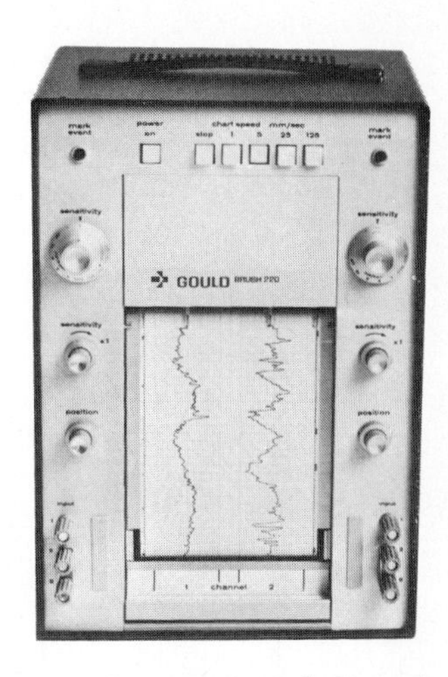

FIG. 15.4 *Brush Mark II recorder. (Courtesy, Brush Instruments, Division of Clevite Corporation)*

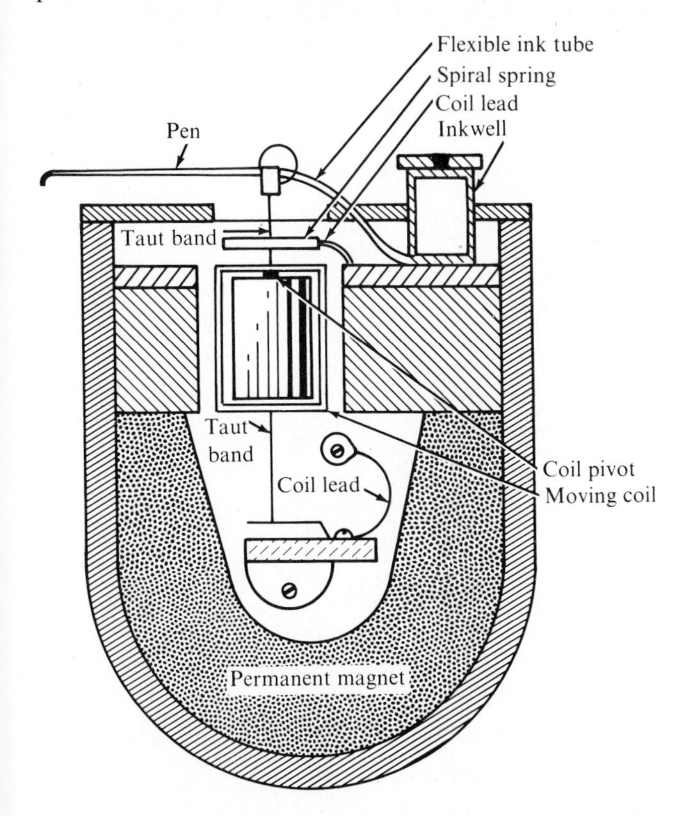

FIG. 15.5 *A D'Arsonval movement for a mechanical oscillograph.*

When operation is required at comparatively high frequencies, the damping of the D'Arsonval movement becomes a basic design consideration. Underdamping results in distortion due to stylus overshoot and "ringing"; overdamping results in reduced speed of response. Critical damping provides maximum speed of response without overshoot. It is possible to operate a D'Arsonval movement at frequencies above mechanical resonance by supplementing the amplifier with a frequency-equalizing network. With a properly designed equalizer, it is possible to operate a movement at frequencies up to 50 percent above resonance. Either a conventional pivot-and-jewel mechanism or a taut-band suspension may be used. Figure 15.6 depicts a taut-band suspension in some detail.

Damping is usually provided in a D'Arsonval movement by eddy current flow in the bobbin that carries the moving coil. This bobbin is fabricated from aluminum and has fairly low resistance. If the bobbin is continuous, a large value of current is induced when a torque is exerted, due to the strong magnetic field in which the bobbin moves. The induced eddy currents develop a countertorque and thereby provide damping. To adjust the amount of damping, holes are punched in the bobbin; these holes increase the resistance in the eddy current flow path and reduce the countertorque that is developed. In most designs, a sufficient number of holes are punched in the

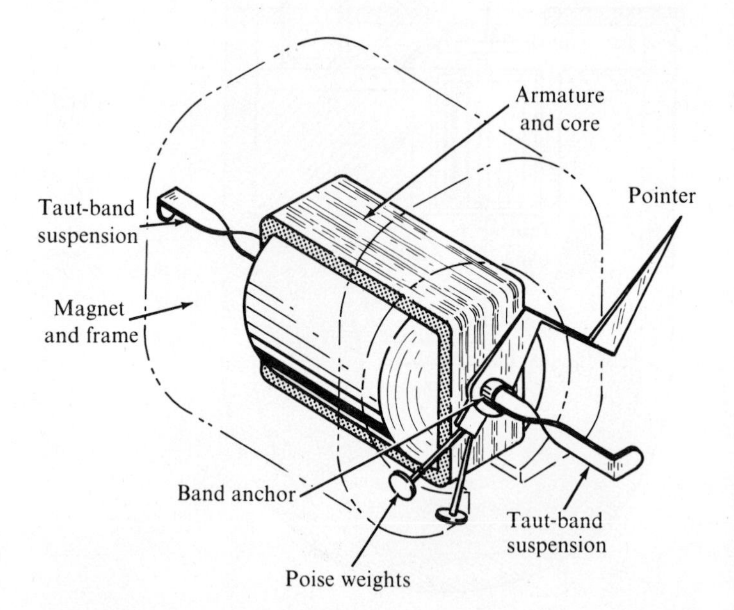

FIG. 15.6 *Hickok panel meter movement using taut-band suspension.*

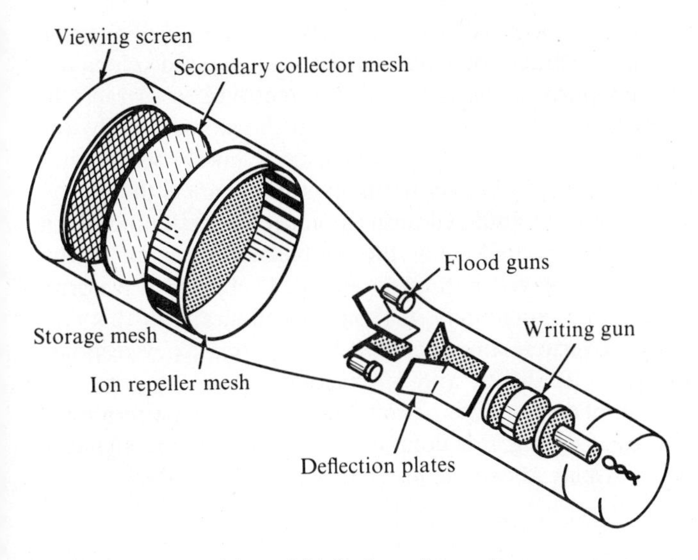

FIG. 15.7 *Construction of the Hughes Memotron tube. (Courtesy, Hughes Aircraft Co.)*

bobbin to provide critical damping when a square-wave voltage is applied to the moving coil.

15.3 STORAGE-TYPE CATHODE-RAY TUBE

Electronic oscillographs are employed to record rapidly occurring events or other high-frequency phenomena. One basic type of electronic oscillograph utilizes a storage-type crt, such as depicted in Fig. 15.7. This device can function either as a conventional crt or as a storage tube. A short distance back of the phosphor screen is a storage assembly consisting of two fine-meshed screens, called the storage mesh and the collector mesh. Next in order are two electron guns, termed flood guns. These guns emit a broad parallel beam of electrons which flood the storage mesh uniformly with low-velocity electrons and thereby charge it to a uniform negative potential.

When the high-energy writing gun beam strikes any part of the storage mesh, secondary emission occurs at the scanned points. In turn, these points lose more electrons than they acquire on the storage mesh. Consequently, local positive charges are established, through which the flood electrons are attracted with increasing velocity, to continue on and to strike the phosphor screen. The screen is operated at an appropriate potential to provide normal fluorescence. Note that each scanned spot continues to glow, as a result of the action of the localized positive charges described above that continue to attract flood electrons at the increased velocity.

The purpose of the collector mesh is to attract the the secondary electrons released from the storage mesh, and thus the scanned traces on the storage mesh acquire the potential of the collector mesh. In this manner, the charge pattern can be maintained until the collector mesh potential is intentionally lowered below the critical storage value. At this point, the displayed pattern is erased. Displayed waveforms can be stored indefinitely. A typical oscillograph of this type employs a vertical amplifier that has flat frequency response from dc to 10 MHz, and a sensitivity of 50 mV/cm. The provision of dc response permits more than one pattern to be stored by inserting a dc component in a subsequent signal to raise its display above the previously recorded pattern.

15.4 OSCILLOGRAPHIC CAMERA EQUIPMENT

As noted previously, oscillograph cameras are widely used in combination with oscilloscopes to provide a permanent record of displayed waveforms. Figure 15.8 illustrates a typical oscillograph camera that can be used with any 5-in. oscilloscope. It is intended primarily for continuous-motion recording, but also provides single-frame recording. The camera in this example is equipped with 400-ft take-up and supply magazines, a controlled selective-speed film-drive

FIG. 15.8 *An oscillograph camera. (Courtesy, Du Mont Corp.)*

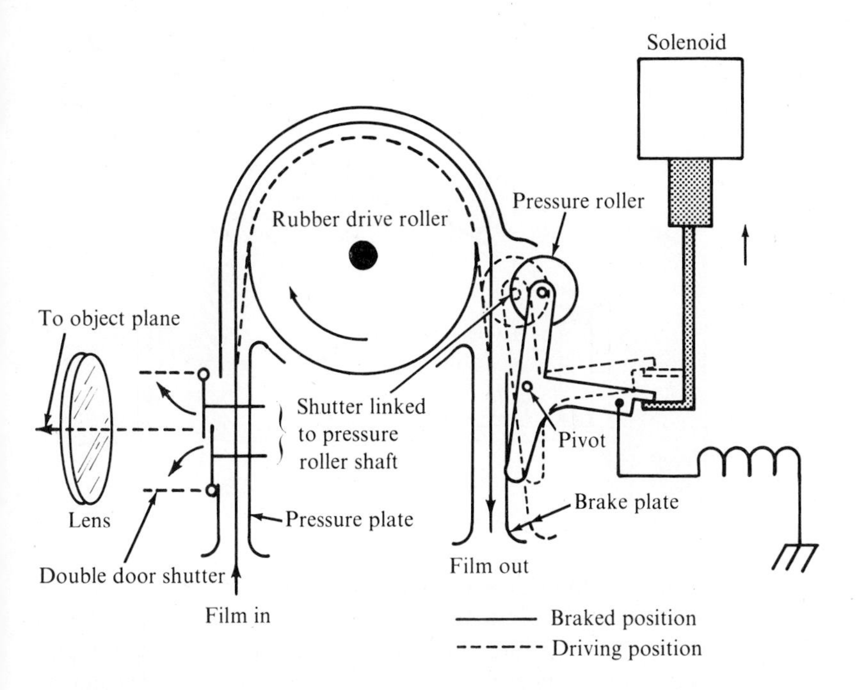

FIG. 15.9 *Mechanical control system, schematic diagram.*

mechanism with an electrical brake, a timing marker, an illuminated data card, and an adjustable floor stand. Either perforated or unperforated 35-mm film or recording paper may be employed.

The film transport system (see Fig. 15.9) provides 16 selectable speeds, ranging from 0.8 to 3600 in./min for full-length recordings. Two additional speeds of 5400 and 10,800 in./min are available for short-length recordings. One end of the viewing housing (see Fig. 15.10) contains an internal expanding type of clamp for attaching the camera to a standard 5-in. crt bezel. At the other end, just above the point of attachment of the camera, is a shuttered and rubber-hooded binocular viewing port to permit continual observation of the crt screen during the recording process. The left side of the housing has an access door for the data card and lens.

Another widely used oscillograph system employs a Polaroid camera for rapid processing of single-frame patterns. A similar arrangement employs roll film, sheet film, or a film pack. Figure 15.11 shows an exploded view of this type of oscillograph camera. The major components consist of the housing, mounting plate, data-recording facility, sliding adapters and film holders, and the optical system. The housing

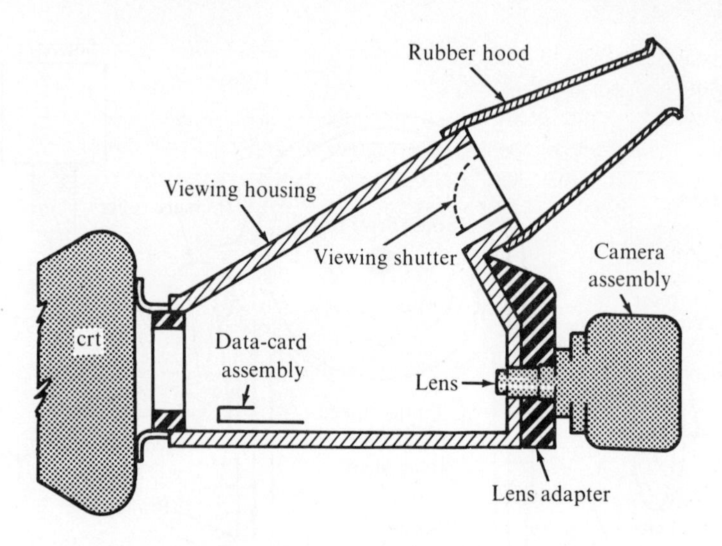

FIG. 15.10 *Viewing housing.*

is of aluminum die-cast construction that houses the optical and data-record systems. The right side of the housing contains the lens access door through which the lens, iris diaphragm, and shutter may be adjusted. A threading insert for attachment to a tripod is included in the bottom of the housing.

The mounting plate in this example is an aluminum die-casting, grooved to accommodate the sliding adapter. It incorporates the lens tube. Three equally spaced detents are provided for use in positioning the film holder for recording three equally spaced events on a single frame of film. The data-recording facility contains two lamps, a pushbutton switch, a data-recording surface, and a dry cell located at the top of the camera housing just above the eye shield. A window in the data-record assembly is fitted with a spring-return door which is covered on its inner side with a white plastic surface. A ground-glass light-diffusing plate covers the opening under the door.

The sliding adapters in this example are of die-cast aluminum and slide in the grooves provided in the mounting plate. To provide detent action, a spring-loaded ball bearing is incorporated on one side to engage in the three equally spaced depressions. Figure 15.12 shows the plan of the optical system. All air-to-glass surfaces of the lens elements are coated to minimize reflections. The dichroic mirror is of the interference type, which is more efficient than a partial-reflecting metallic-coated type of mirror. An interference mirror consists of a number of transparent dielectric films of precisely controlled thickness, deposited on flat glass. In turn, a chromatic selection occurs, in which a certain portion of the spectrum is reflected, while the remainder is transmitted.

Because of its selective transmission and reflection characteristics, this beam-splitting device is termed a dichroic mirror. The descriptive term *dichroic beam splitter* is also utilized. The device in this example is designed to reflect actinic light (blue) and to transmit yellow light when the rays are incident at 45°. Thus, simultaneous direct binocular viewing and recording of the crt screen pattern with the data-recording area is provided. The camera is equipped with a special $f/1.5$ lens, enabling the recording of very high writing rates (up to 200 in./μs with Ektachrome Super XX film). The unit designed for use with a Polaroid camera pack is provided with either an $f/1.9$ or an $f/2.8$ lens.

FIG. 15.11 *Exploded view of major components of the Du Mont oscilloscope camera assembly.*

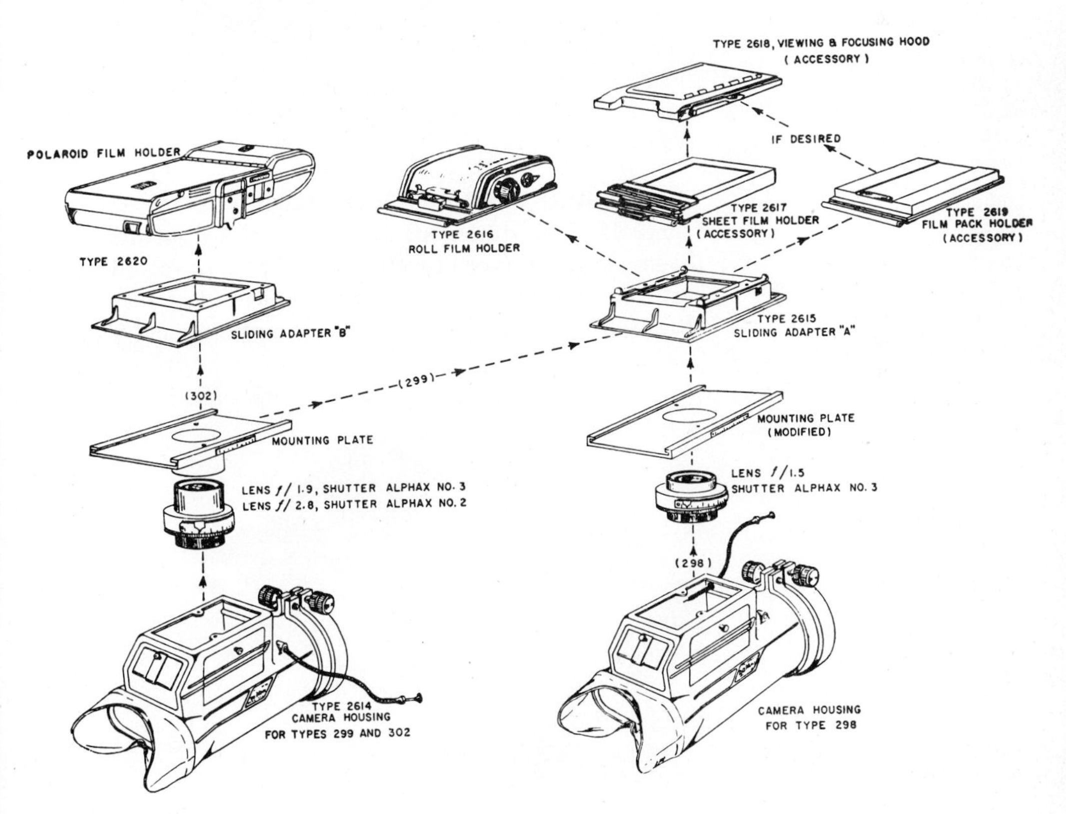

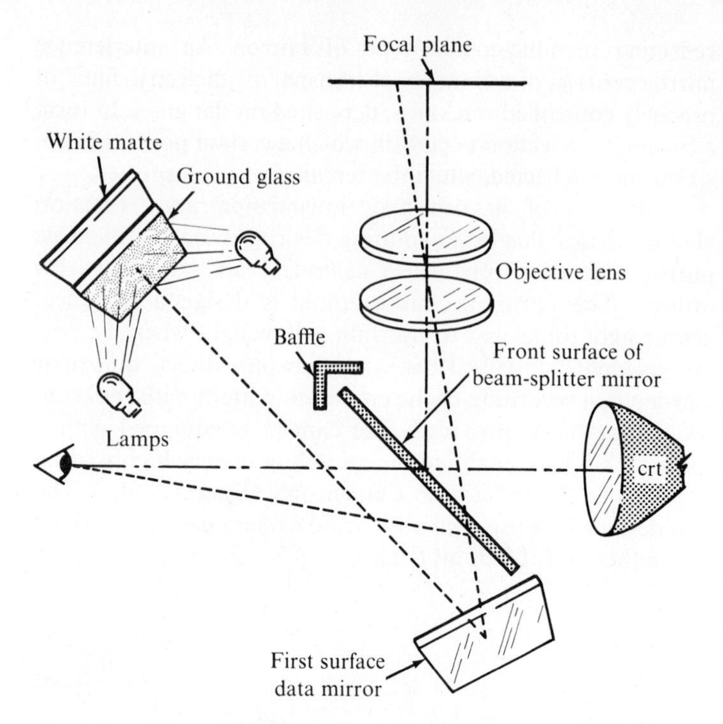

FIG. 15.12 *Plan of optical system of the Du Mont oscilloscope camera.*

15.5 SERVO-TYPE *XY* RECORDER

Figure 15.13 illustrates the servo type of XY recorder. Like the moving-strip type of instrument, the servo recorder plots cartesian coordinate graphs from applied electrical signals. However, the graph paper is fixed in position, and a pair of

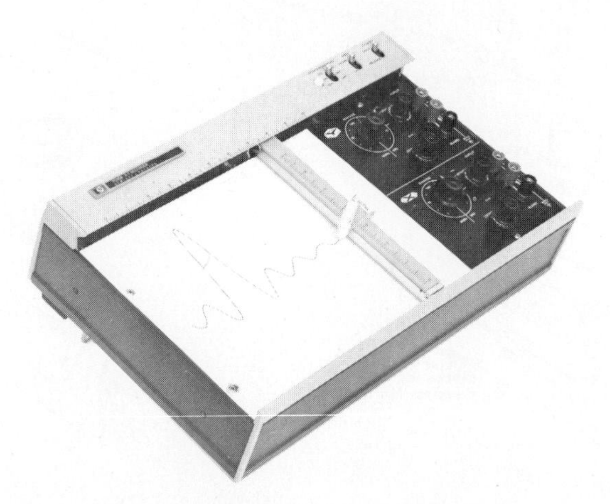

FIG. 15.13 *Appearance of a servo-type XY recorder. (Courtesy, Hewlett-Packard)*

servomechanisms are provided to move the pen and the carriage arm. Each servomechanism is self-balancing, and each is independent of the other. A block diagram is shown in Fig. 15.14. The basic dc voltage range of the servosystems is 1 mV/in. The balancing circuits can be operated with higher voltages when suitable resistors are switched into them. Each range step can also be made continuously variable by switching a potentiometer into the circuit. Thereby, the pen may be driven to full scale by an arbitrary value of voltage.

With reference to Fig. 15.14, the input signal, after passing through an attenuator, is applied to the balance circuit, where it is canceled by an internally supplied opposing

FIG. 15.14 *Block diagram for servo-type XY recorder.*

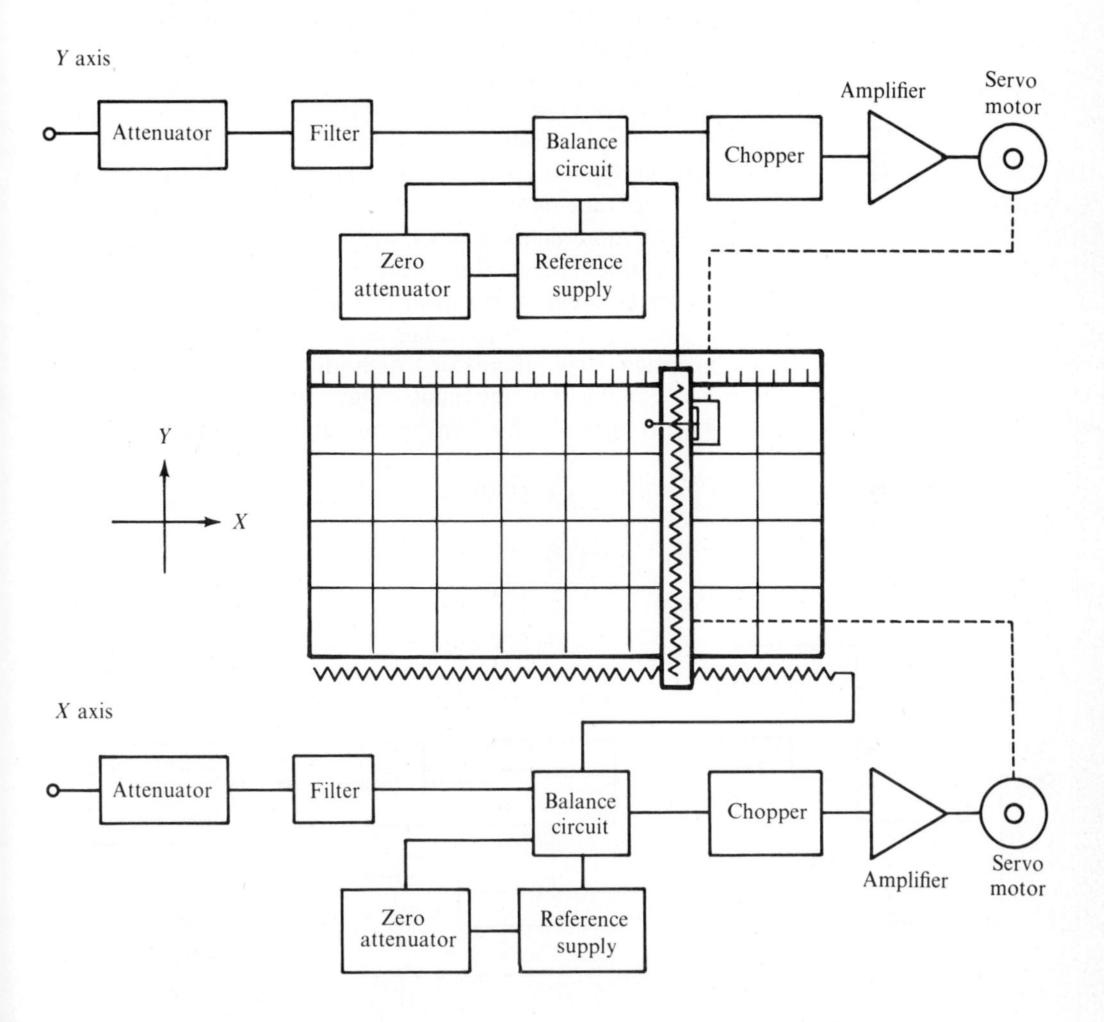

voltage. In the balanced condition, there is no error signal output from the balance circuit, and the servosystem is at a null. On the other hand, when the input signal changes in value, an unbalanced condition exists. The resulting error signal is applied to a photoconductive chopper which converts the dc value into a corresponding 60-Hz voltage, which is amplified and applied to the servomotor. Inasmuch as the motor and the rebalance potentiometer are mechanically coupled, the balance voltage will be changed until the input signal is canceled. In other words, the servosystem maintains a null as the value of the input voltage is varied.

The pen assembly in the recorder consists of a drum-type reservoir resting in a pivot mount which moves along the carriage arm. A rigid capillary tube line leads from the reservoir to the pen point. The pen can be raised and lowered by means of a manually operated lever. Because of the capillary feed, the pen mechanism operates equally well in either a vertical or a horizontal plane. Thus, this type of recorder may be either rack-mounted or table-mounted.

A more detailed block diagram is shown in Fig. 15.15, and a simplified selector circuit is depicted in Fig. 15.16. The schematic diagram is shown in Fig. 15.17. With reference to Fig. 15.16, the input terminals for each axis are connected to a precision step attenuator. With the selector switch in its 1 mV/in. position, the input voltage is applied directly to the balance circuit without any attenuation. The attenuator is also

FIG. 15.15 *Detailed block diagram of an XY recorder.*

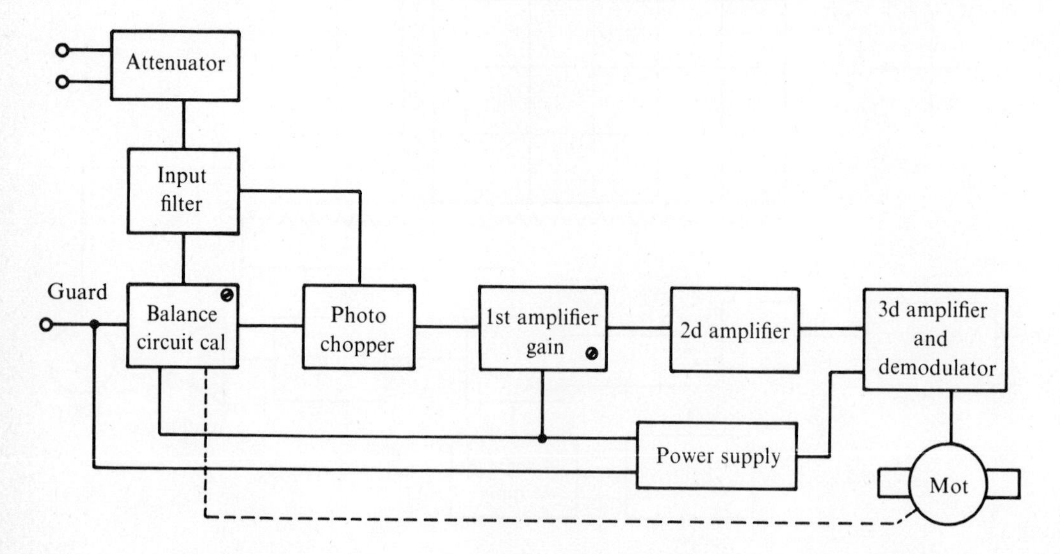

disconnected as a shunt from the input terminals in this position, allowing potentiometric operation. In other words, this provides essentially infinite input impedance when the recorder is at null. The full-scale balanced voltage is always 7 mV for the Y axis and 10 mV for the X axis.

The 10-position range selector has five calibrated positions; the remaining five positions provide variable adjustment by the operator. In an uncalibrated position, a multiturn high-resolution potentiometer R_{107} is inserted in place of the fixed 10-K resistor R_{106} at the base of the attenuator. This potentiometer can change the sensitivity of the recorder over a range of 10:1. Next, note that the input filter consists of three RC sections. It provides low-pass action with a minimum of 20 dB attenuation at 60 Hz and a cutoff characteristic of 18 dB per octave above 60 Hz to minimize the effects of extraneous noise. (Each octave represents a doubling in frequency.) Diodes CR_{101} and CR_{102} are protective devices for minimizing the possibility of component damage due to excessive input voltage.

After passing through the input attenuator, the input signal is applied to the balance circuit, where it is opposed by a dc cancellation voltage from the internal reference supply. The difference between the input voltage and the reference supply voltage is called the error signal, and is applied to the servo amplifier which drives the servomotor. The rebalance potentiometer, being mechanically coupled to the servomotor, is driven in a direction to produce a voltage which cancels the error signal produced by the input voltage. The zero control R_{114} introduces into the balance circuit a controlled error signal, and thus provides the means for locating the origin of coordinates anywhere within the limits of the graph.

The phase lead network C_{105}-R_{118} draws a charging current whenever a change in output occurs, thus increasing the rate of appearance of the balance voltage. This phase advance in the slowly varying error signal causes an "anticipatory" approach to the balance point, thus providing desirable damping characteristics. C_{107} serves as a low-pass filter to attenuate slide-wire noise. The reference voltage for each axis is independently derived from a zener-controlled dc power supply of nominally 9 V. Note that the error signal, or difference between the input signal and the rebalance voltage, is converted into 60 Hz ac by the chopper. A photoconductive chopper is employed, in which photocells V_{101} and V_{102} are

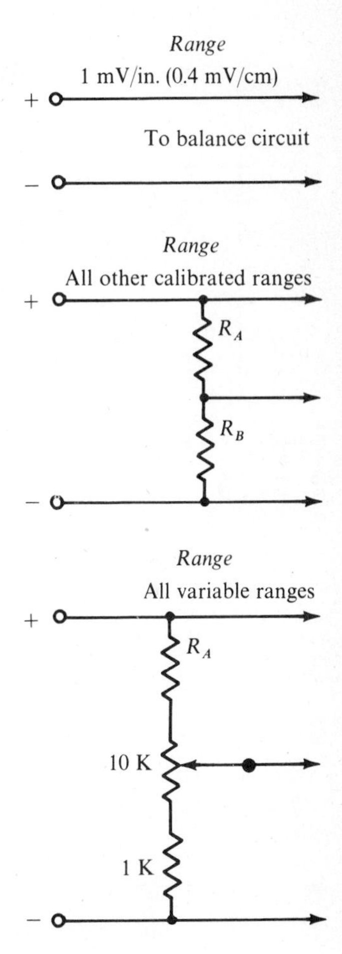

FIG. 15.16 *Simplified selector circuit for an XY recorder.*

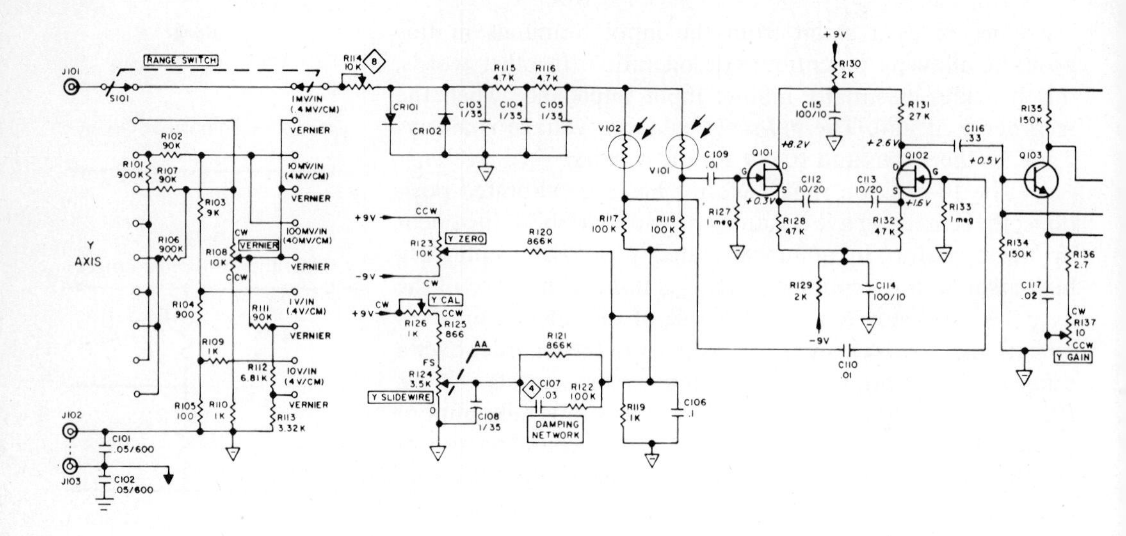

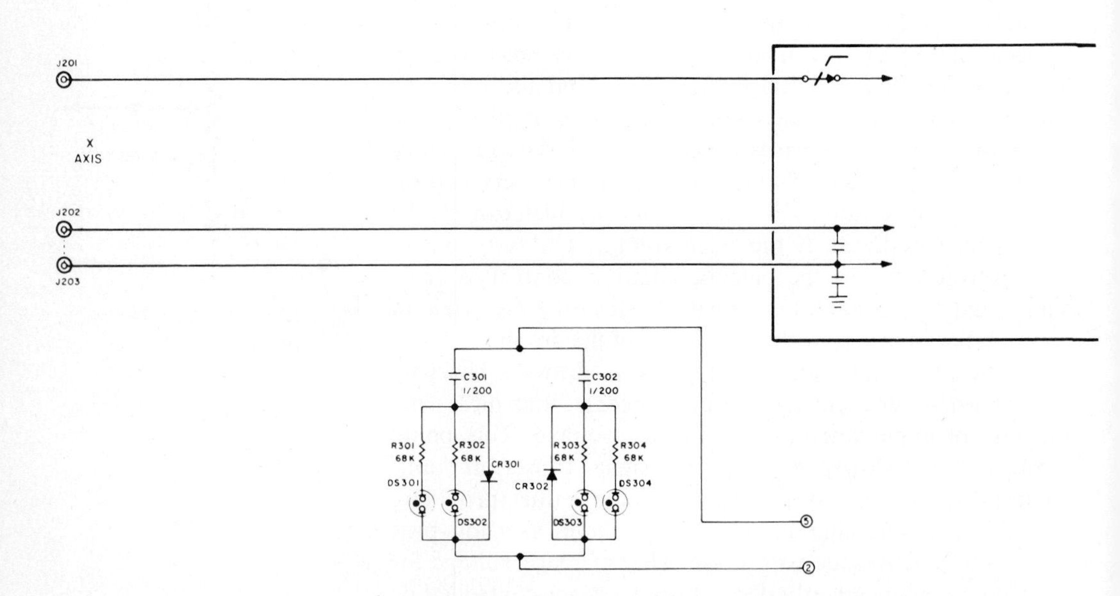

FIG. 15.17 *Schematic diagram for servo-type XY recorder.* (*Courtesy, Hewlett-Packard*)

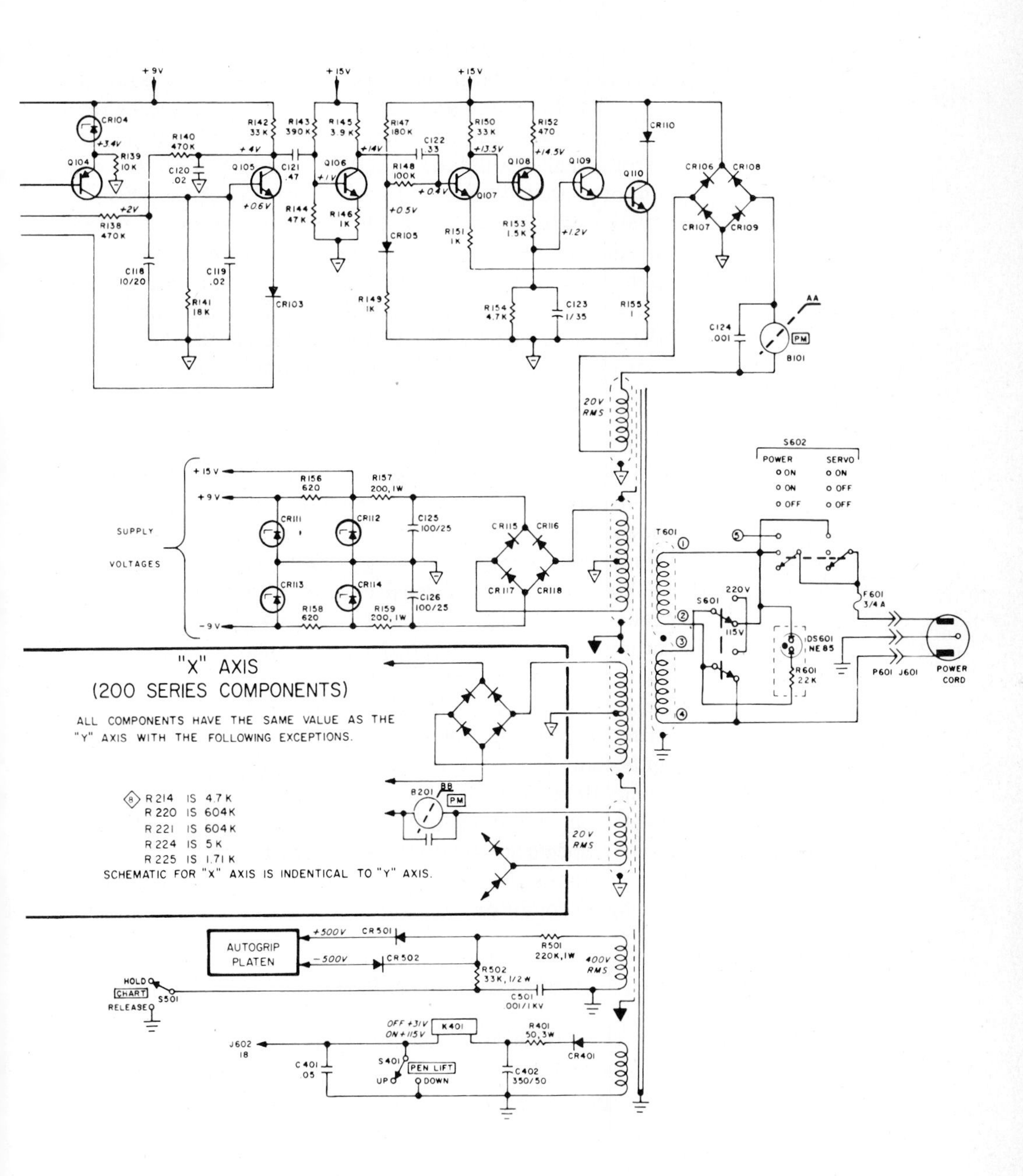
+9V
+15V
+15V
CR104
+3.4V
Q104
R139 10K
R140 470K
C120 .02
+4V
R142 33K
R143 390K
Q105
C121 .47
R145 3.9K
+14V
Q106
+1V
R147 180K
C122 .33
R148 100K
+0.4V
R150 33K
+13.5V
Q107
R152 470
+14.5V
Q108
Q109
Q110
CR110
CR106
CR108
CR107
CR109
+2V
R138 470K
+0.6V
R144 47K
R146 1K
+0.5V
CR105
R151 1K
R153 1.5K
+1.2V
C118 10/20
C119 .02
R141 18K
CR103
R149 1K
R154 4.7K
C123 1/35
R155 1
C124 .001
AA
PM
B101
20V RMS
S602
POWER
SERVO
O ON
O ON
O ON
O OFF
O OFF
O OFF
+15V
R156 620
R157 200, 1W
+9V
SUPPLY VOLTAGES
CR111
CR112
C125 100/25
CR115
CR116
CR117
CR118
CR113
CR114
C126 100/25
R158 620
R159 200, 1W
-9V
T601
220V
S601
115V
F601 3/4 A
DS601 NE 85
R601 22K
P601 J601
POWER CORD
"X" AXIS
(200 SERIES COMPONENTS)
ALL COMPONENTS HAVE THE SAME VALUE AS THE "Y" AXIS WITH THE FOLLOWING EXCEPTIONS.
R 214 IS 4.7 K
R 220 IS 604K
R 221 IS 604K
R 224 IS 5 K
R 225 IS 1.71 K
SCHEMATIC FOR "X" AXIS IS INDENTICAL TO "Y" AXIS.
B201
BB
PM
20V RMS
AUTOGRIP PLATEN
+500V
CR501
-500V
CR502
R501 220K, 1W
400V RMS
R502 33K, 1/2W
HOLD
CHART
RELEASE
S501
C501 .001/1KV
OFF +31V
ON +115V
K401
R401 50, 3W
J602 18
C401 .05
S401
PEN LIFT
UP
DOWN
C402 350/50
CR401

alternately turned on and off by neon bulbs DS_2 and DS_4. The bulbs are energized in synchronism with the power line.

The foregoing switching action of the photocells results in an ac error output signal that is proportional to the dc error signal in amplitude, and that is either in phase or 180° out of phase with respect to the power-line phase. This phase relationship is dependent upon the polarity of the dc error signal. The direction of rotation of the servomotor is determined by the relative phase excitation of the winding in the servo amplifier's output demodulator stage and the phase of the dc error signal. This phase sensing causes the motor to drive potentiometer R_{115} in the direction of balance.

15.6 APPLICATIONS

XY recorders are used instead of conventional voltmeters or ammeters when a permanent record of the measured values is desired. An *XY* recorder also serves as a convenient monitor, which can operate over long periods without an attendant. Public utilities make extensive use of *XY* recorders to monitor line voltages and power demands. An *XY* recording is essential in applications that require detailed evaluation of the measured values, as in medical electrocardiograms and electroencephalograms. The same utility is provided by *XY* recordings in criminal investigations during lie-detector tests. Another important field of application is in telemetry operations, in which these devices provide a permanent record of incoming data from a remote transmitter. In the design laboratory, some projects are greatly facilitated by the ability to maintain a continuous record of device or system performance under changing conditions of operation.

QUESTIONS AND PROBLEMS

1. What is the distinction between an oscilloscope and an oscillograph?
2. What are the basic components of a magnetic oscillograph?
3. What factors limit the high-frequency response of a magnetic recorder?
4. What is the purpose of a penmotor in a magnetic recorder?
5. Explain the operation of the taut-band suspension meter movement in a magnetic recorder.

6. How is damping usually provided in a D'Arsonval meter movement?
7. Explain the operation of the memotron cathode-ray tube.
8. What are the advantages of the oscilloscope-record systems over the memotron cathode-ray tube?
9. When is the Polaroid camera system most useful as a memory system for cathode-ray tube waveforms?
10. Explain the operation of the balance circuit in the *XY* recorder shown in Fig. 15.14.
11. How is a high input impedance maintained on all ranges of the *XY* recorder depicted in Fig. 15.16?
12. What is the purpose of the input filter circuits in each of the amplifiers?
13. What is the purpose of the error signal that is applied to the servo amplifier?
14. What is the function of the zero control, and how does the control perform this function?
15. What is the function of components C_{105} and R_{118} in the circuit diagram depicted in Fig. 15.7?
16. What is the purpose of capacitor C_{107} in the circuit diagram depicted in Fig. 15.7?
17. What is the frequency of the photoconductive chopper circuit depicted in the circuit diagram in Fig. 15.7?
18. What determines the direction of the rotation of the servomotor in the *XY* recorder?

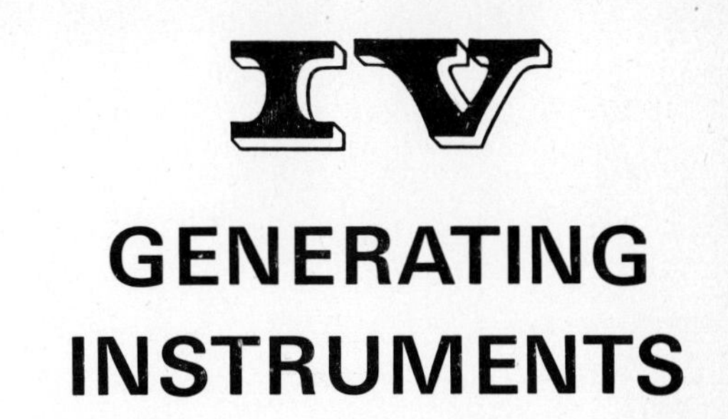

IV GENERATING INSTRUMENTS

AUDIO OSCILLATORS

16.1 BASIC REQUIREMENTS

Basically, an oscillator is an amplifier which derives its input from its own output. This basic principle of oscillator operation is depicted in Fig. 16.1*b* and *c*. The amplifier in (*b*) has a gain of 10. It produces an output of 10 V when 1 V is applied to its input terminal from the generator through the switch in position 1. It is important to note that the input and output voltages of the amplifier have the same polarity (are in phase). In (*c*), switch contact 2 has been connected to the load resistor. Thus, when the amplifier input switch is thrown from position 1 to position 2, the external input voltage indicated in (*b*) is replaced by its exact duplicate in (*c*)—obtained, however, from a tap on the output load resistor of the amplifier.

Note that the new input voltage employed in Fig. 16.1*c* is identical to the original external input signal in both amplitude and phase. Therefore, the external input signal is no longer needed, and the amplifier continues to produce an output voltage as long as the feedback path from output to input is maintained. Under these conditions, the amplifier is said to *oscillate*. In practical oscillators, the application of collector (or plate) voltage causes a rising surge of current in the configuration which is amplified and fed back to the input of the oscillator. Thus the arrangement is energized into the oscillatory state which is self-sustaining.

We recognize that an amplifier performs as an oscillator, provided that a portion of its amplified output voltage is fed back with a specified *amplitude* and *phase*. These are the two basic requirements for any type of oscillator used in an audio signal source. With reference to Fig. 16.2, the power output is divided as shown. A portion (P_{load}) is supplied to the load, with the remainder ($P_{feedback}$) going to the feedback network. The feedback power is equal to the input power (P_{in}) plus the loss incurred in the feedback network. For example, if the power gain of a transistor amplifier is 20, and the input power level is 1 mW, the output power is 20 mW. If the loss incurred in the feedback network is 10 dBm, the feedback power must be 11 mW to sustain oscillation.

An audio oscillator is an instrument that supplies a sinusoidal output voltage over the range of audible frequencies. In its generic sense, an audio oscillator is often a source of supersonic sinusoidal voltage, also. Thus the highest range of

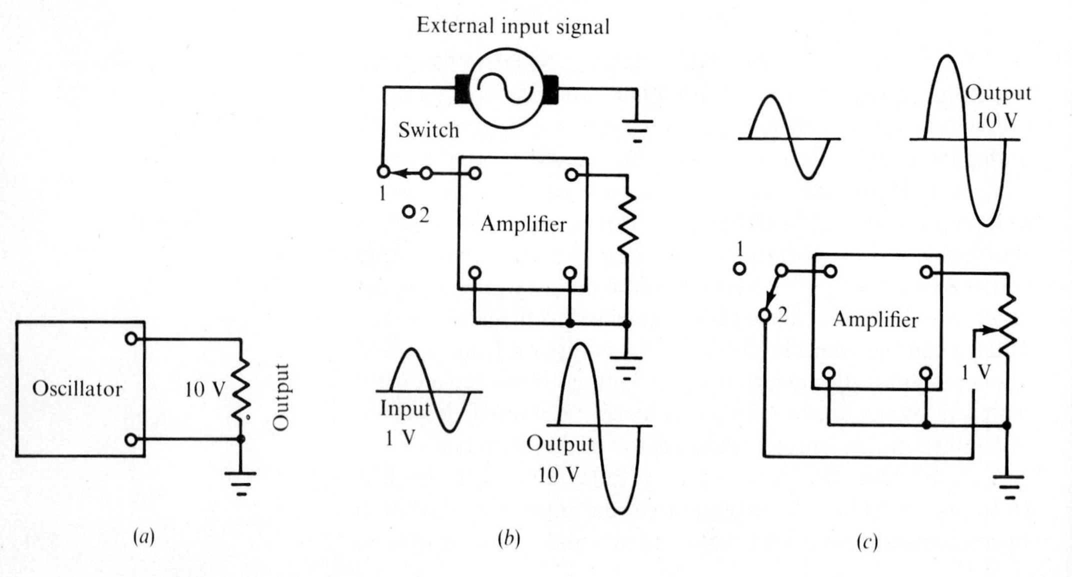

FIG. 16.1 *Operation of an oscillator. (a) No external signal input; (b) amplifier operating normally; (c) amplifier operating as an oscillator.*

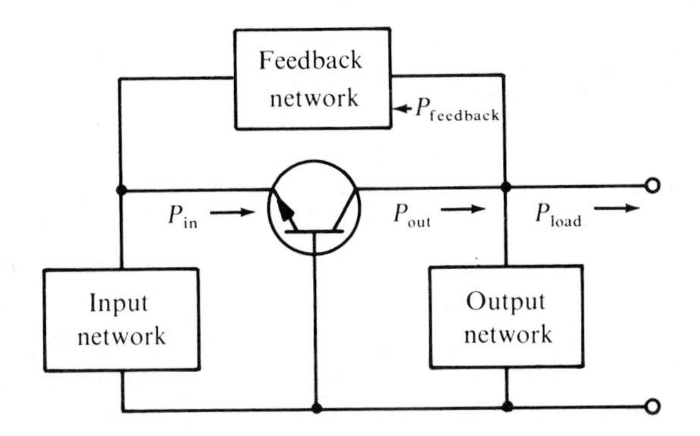

FIG. 16.2 *Transistor oscillator showing the important networks in block form.*

an audio oscillator may overlap the lowest range of a radio-frequency instrument. Although the distinction is not sharp, some writers classify sine-wave voltage sources into oscillator- and generator-type instruments. In this system of classification, an oscillator provides a controllable output voltage level which is not accurately calibrated, and may be uncalibrated. On the other hand, a generator provides a level meter and an accurately designed low-impedance step attenuator, whereby the operator is enabled to apply precise values of signal voltage to the circuit or device under test.

Distinction between oscillators and generators may also be made on the basis of accuracy in frequency calibration, operating stability, and percentage distortion (waveform purity). When thus classified, generating instruments are characterized by high accuracy ratings and laboratory-type performance. Both classes of instruments can be further subdivided into beat-frequency or *RC* types. We shall find that although beat-frequency instruments have certain advantages, these are largely outweighed by various disadvantages. For this reason, nearly all modern audio oscillators are of the *RC* type.

16.2 WIEN-BRIDGE *RC* OSCILLATOR

The basic Wien-bridge configuration is depicted in Fig. 16.3. Note that the circuit is frequency selective in that the impedance of the series *RC* arm decreases as the frequency increases, whereas the impedance of the parallel *RC* arm increases as the frequency increases. The reader may observe from the relations in Fig. 16.2 that the bridge circuit in Fig. 16.4 will be balanced when

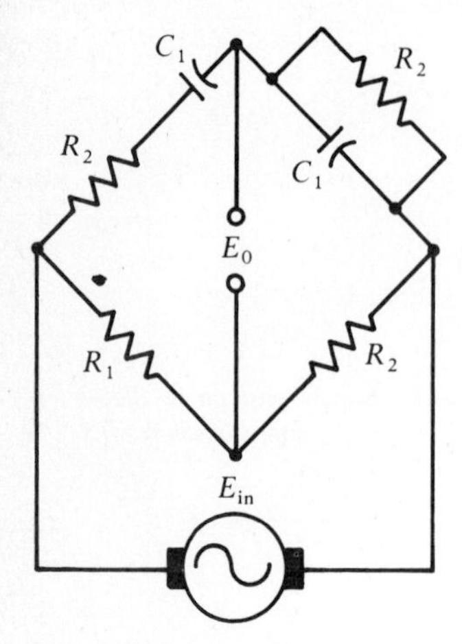

FIG. 16.3 *Wien-bridge configuration.*

$$R_2 = X_{C_1} \tag{16.1}$$

If Formula (16.1) is rearranged we obtain

$$f_o = \frac{1}{2R_2C_1} \quad \text{when } R_1 = R_2 \text{ and } C_1 = C_2 \tag{16.2}$$

In other words, $E_o = 0$ at frequency f_o. This characteristic is exploited in the Wien-bridge oscillator circuit depicted in Fig. 16.5. V_1 operates as an oscillator tube, and V_2 provides phase inversion. Both tubes provide amplification to overcome circuit losses. Note that if the Wien-bridge section were not present, the circuit would operate as a conventional multivibrator and generate a square-wave output. However, the Wien bridge rejects feedback frequencies other than f_o as given by Formula (16.2). When all the component frequencies except one are removed from a square wave, a sine wave remains. Therefore, the output from the oscillator circuit is sinusoidal.

A more detailed analysis of the oscillator configuration in Fig. 16.5 will show that only one frequency can feed back in aiding phase because of the degeneration and phase shift that occur at any frequency other than f_o. In other words, the oscillator circuit develops cathode degeneration at frequencies other than f_o. Moreover, the feedback voltage shifts out of phase with the reference oscillatory phase at any frequency higher or lower than f_o. To summarize briefly, oscillation can take place only at the frequency which permits the signal voltage across R_2 (input signal to V_1) to be in phase with the V_2 output voltage, and also provided that the positive feedback voltage exceeds the negative feedback voltage and cancels out the circuit losses.

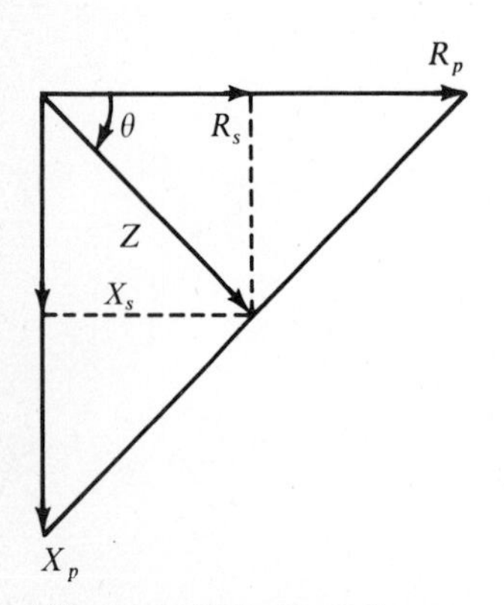

FIG. 16.4 *Equivalent RC series and parallel relations.*

Signal voltages at frequencies other than f_o in Fig. 16.5 cause a phase shift between the output of V_2 and the input of V_1. These voltages become attenuated by the comparatively large degenerative action in the circuit, and in turn have insufficient amplitude to produce oscillation. A degenerative feedback voltage is provided by the voltage divider consisting of R_3 and lamp R_4. Since there is zero phase shift across the voltage divider, and since these resistance values are virtually constant at various frequencies, the amplitude of the negative feedback voltage is constant for any frequency that may be present in the output circuit of V_2. The basic relations are seen in Fig. 16.6, which depicts the constancy of the negative feedback voltage 1 vs. frequency.

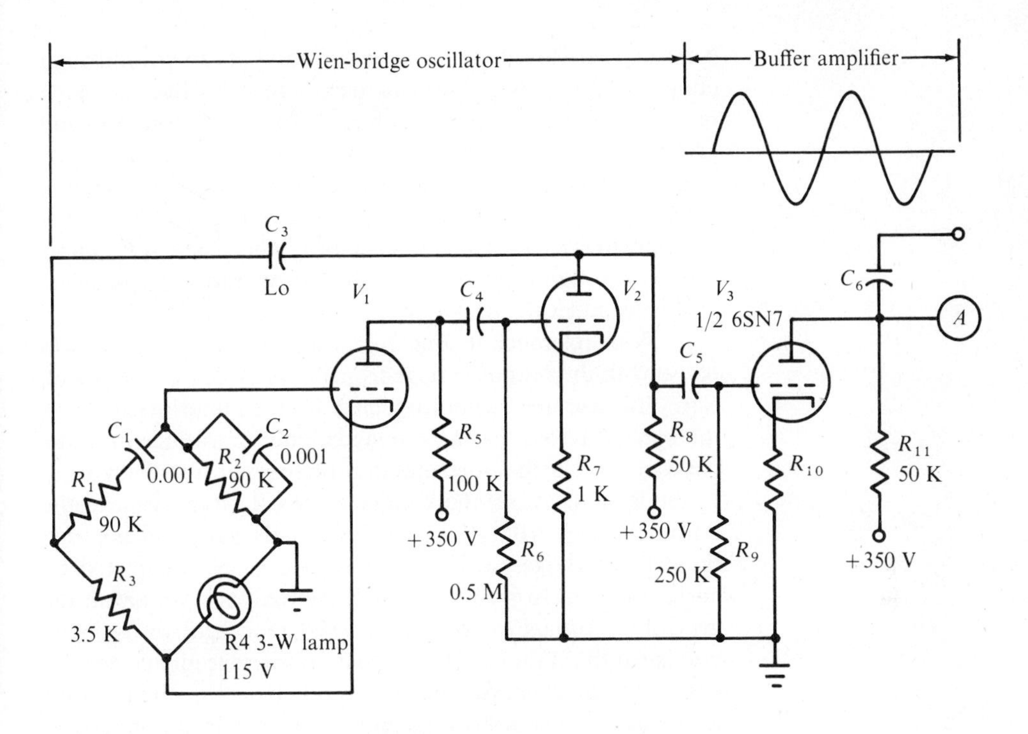

FIG. 16.5 *Diagram showing a Wien-bridge oscillator.*

The positive feedback voltage 2 in Fig. 16.6 is provided by the voltage divider R_1-C_1-R_2-C_2 in Fig. 16.5. At a very high frequency, the reactances of the capacitors are almost zero. In turn, R_2 is shunted by a very low reactance, bring the voltage between the grid of V_1 and ground to almost zero.

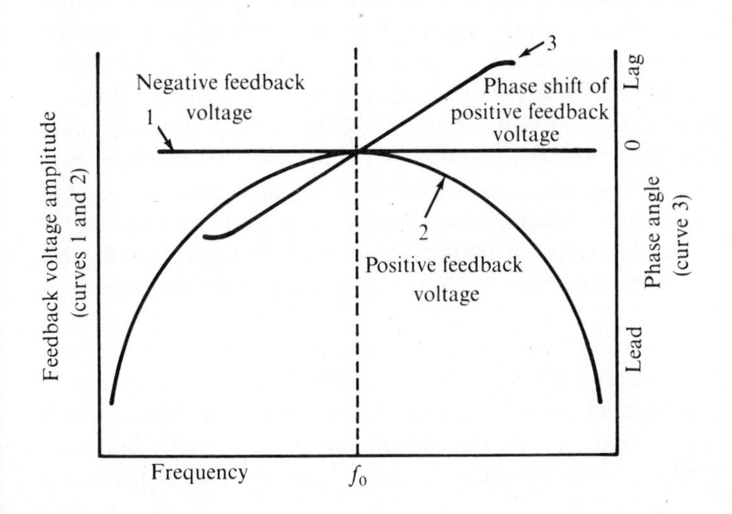

FIG. 16.6 *Variation of feedback voltage vs. frequency for a Wien-bridge circuit.*

Next, at a very low frequency, the current that can flow through either C_2 or R_2 is reduced to nearly zero by the very high reactance of C_1. In turn, the voltage between the grid of V_1 and ground is brought almost to zero. At some intermediate frequency, the positive feedback voltage rises to a maximum value, as seen on curve 2 in Fig. 16.6. Although the curve is comparatively flat in the vicinity of f_o, the phase shift varies rapidly through f_o, with the result that operation is stabilized at this frequency.

With reference to Fig. 16.5, the voltage across R_2 is in phase with the output voltage from V_2, provided that $R_1C_1 = R_2C_2$. In case the output frequency from V_2 tends to increase, the voltage across R_2 starts to lag behind the voltage from the plate of V_2. On the other hand, if the output frequency from V_2 tends to decrease, the voltage across R_2 starts to lead the output voltage from V_2. The phase-angle variation between these two voltages with respect to frequency is depicted by curve 3 in Fig. 16.6. At f_o, the positive feedback voltage at the grid of V_1 equals (or barely exceeds) the negative feedback voltage at the cathode. This in-phase positive feedback voltage sustains oscillation. At frequencies other than f_o, the negative feedback voltage is greater, and oscillation is suppressed at such frequencies.

Note that lamp R_4 in Fig. 16.5 is employed as a cathode resistor for V_1, and serves to stabilize the amplitude of the generated sine wave. If the amplitude of oscillation tends to increase, the current through the lamp filament also tends to increase; in turn, the filament becomes hotter and its resistance increases. A greater negative feedback voltage is accordingly developed across the increased resistance. The resulting increase in degeneration holds the output signal voltage at practically constant amplitude. If the output from V_1 were quite substantial, the output waveform would no longer be a true sine wave; a good waveform is obtained when the output amplitude is comparatively limited. Since the lamp is chosen for a suitable value of filament resistance, this component also serves to minimize harmonic distortion of the output waveform.

Because the output amplitude from the Wien bridge oscillator has a comparatively low level, an amplifier is usually provided, as depicted in Fig. 16.5. Tube V_3 not only increases the level of the output signal, but also provides buffering action between the oscillator section and the load or utilization circuit. A change in load does not affect the oscillator frequency. Note that the output impedance at point A is comparatively high. In most applications, it is desirable to employ a low-impedance attenuator in the output circuit. Therefore, point A typically drives a cathode follower stage; the cathode follower operates as an electronic impedance transformer and provides substantial power gain.

The circuit in Fig. 16.7 is a modification of the standard Wien bridge oscillator. The principal difference is that silicon diodes are used as controllable resistance elements to make the oscillator frequency vary in proportion to a control voltage.

The control diodes have a forward conduction characteristic such that their dynamic resistance is inversely proportional to current over a wide range. It varies from about 1500 Ω at 30 μA to 160 Ω at 300 μA. By using a large resistance (R_1), between the diodes and a control source, diode resistance is made inversely proportional to control voltage. It also serves to isolate the oscillator from the control source.

For the type of bridge chosen (made up of the components with the dashed lines plus resistors R_6 and R_7), oscillation frequency is $1/(2\,RC)$.

By using resistance elements that vary inversely proportional to a control voltage, the oscillator frequency is made to vary directly proportional to this voltage. A gain of 3 is required for oscillation and is provided by adjustment of feedback resistors R_6 and R_7.

The three-stage oscillator amplifier need only provide high open-loop gain with no phase shift. The two stages of output amplification provide a 1 V rms signal at 1000 Ω. Automatic gain control to compensate for bridge dynamic variations is accomplished by controlling the resistance of the shunt diode C_{R_3}.

With this circuit, 5 percent linearity of frequency control from 10 kHz to 70 kHz is produced by a control voltage of 2.5 V to 27 V, respectively. Frequency range is determined by bridge capacitors C_1 and C_2 and any required frequency in the audio or low rf range may be generated, consistent with amplifier requirements and diode capabilities.

16.3 BRIDGED-T OSCILLATOR

A bridged-T network can be compared with a Wien bridge in that it is frequency selective. Figure 16.8 depicts the basic

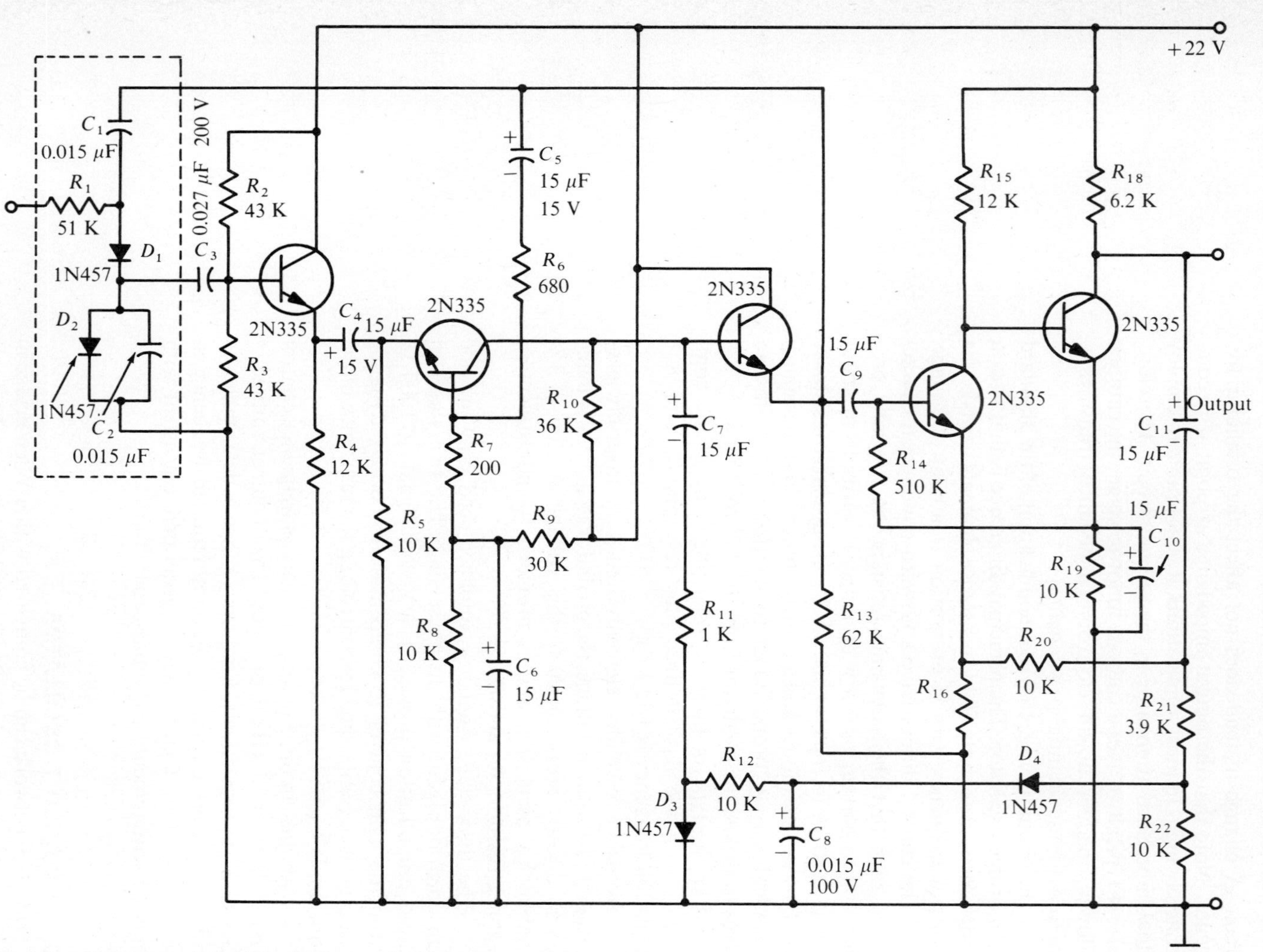

FIG. 16.7 *Voltage-controlled Wien-bridge oscillator circuit.*

bridged-T circuit. An input signal branches into R_1 and into C_1. From C_1 the signal branches into R_2 and C_2. From C_2, the signal combines with the output from R_1. Note that the signal through R_1 appears at the output without any phase shift. On the other hand, the signal through C_1 produces a leading voltage drop across R_2. An additional lead is imposed on the signal by flow through C_2 into the output circuit. Thus the combined output signals from R_1 and C_2 form a resultant voltage with an amplitude that depends on the phase difference between the two signals. At a certain frequency, the phase shift is almost 180°, and this is called the null frequency of the network; the output voltage is nearly zero at this null frequency.

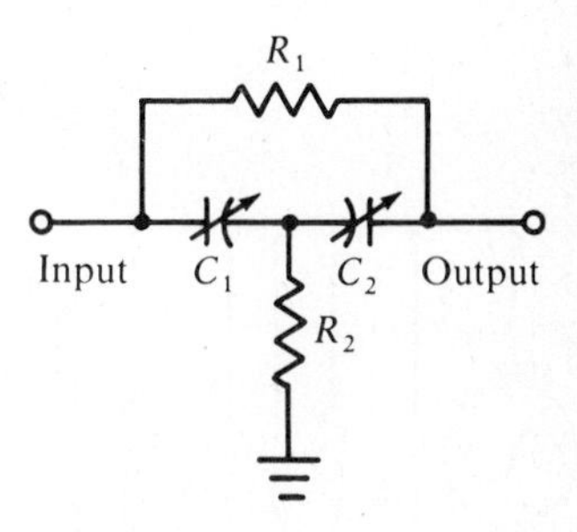

FIG. 16.8 *Basic bridged-T-circuit.*

When R_1 is much larger than R_2, and $C_1 = C_2$ in Fig. 16.8, the null frequency is given by the approximate formula:

$$f_o = \frac{1}{2C_1R_1R_2} \qquad (16.3)$$

16.4 PHASE-SHIFT OSCILLATOR

When an audio oscillator is designed to operate over a comparatively limited frequency range, or on a specified frequency (as in the modulator section of a radio-frequency generator), simplified types of RC networks are commonly employed. For example, Fig. 16.9 depicts a phase-shift oscillator circuit. We observe that this arrangement dispenses with negative feedback and nonlinear resistance stabilization. Note that the signal voltage is fed back from drain to gate via three RC sections, which reverse the signal phase at a frequency that

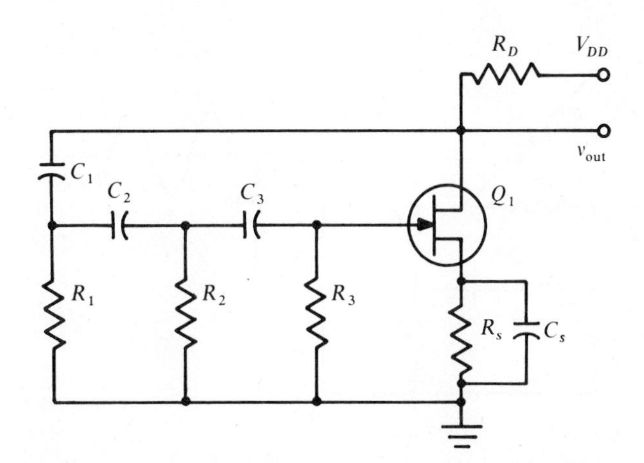

FIG. 16.9 *Phase-shift oscillator using a field-effect transistor as an amplifier.*

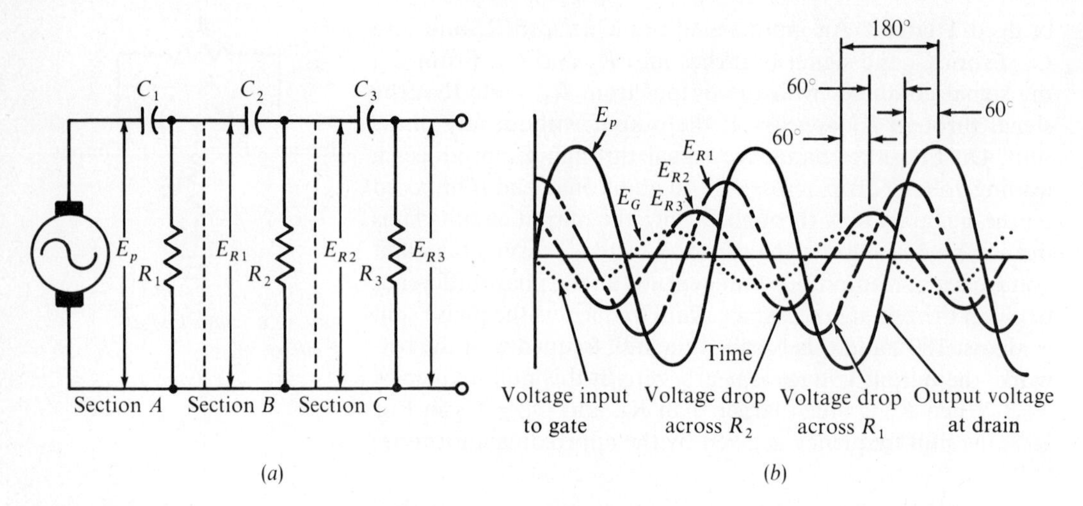

FIG. 16.10 *Relative phases of waveforms in the phase-shifting network. (a) Simplified phase shifter; (b) phase relations.*

depends on the values of R_1, R_2, R_3, C_1, C_2, and C_3. This circuit action takes place as depicted in Fig. 16.10.

The relative signal phases shown in Fig. 16.10 are the same at any frequency of operation. In other words, a total phase shift of 180° is required to produce oscillation in the configuration of Fig. 16.11. To obtain a range of operating frequencies, R_1, R_2, and R_3 are made variable, and are ganged on a single shaft. The chief disadvantage of this arrangement is that the amplitude of oscillation decreases as the resistance values are reduced. Accordingly, only a limited range of frequency adjustment is realizable in practice. Additional ranges may be provided by switching suitable values of C_1, C_2, and C_3 into the circuit.

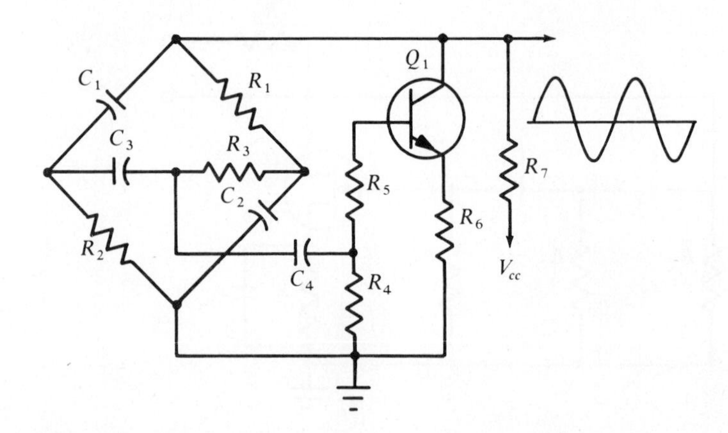

FIG. 16.11 *Bridge-type phase-shift oscillator.*

16.5 BRIDGE-TYPE PHASE-SHIFT OSCILLATOR

Another RC oscillator configuration that may be employed for operation at a single frequency is termed the bridge-type phase-shift network. This is a hybrid configuration that employs reactance in both the arms and the bridge lead, as shown in Fig. 16.12. Positive feedback is obtained by coupling the collector of Q_1 back to the base via the phase-shifting network. This network reduces the signal from plate to ground by an amount that is almost equal to the current gain (beta) of the transistor. Because the signal is shifted in phase by 180° at a particular frequency which depends upon the R and C values in the circuit, oscillation takes place at this frequency.

To obtain good waveform, the amplitude of oscillation is limited in the circuit of Fig. 16.12 by the proportions of voltage divider $R_4 - R_5$, and by the negative feedback provided by unbypassed emitter resistor R_6. An advantage of this arrangement is its frequency stability. On the other hand, the circuit is not well adapted to operation over a range of frequencies because of the multiplicity of components that are in the frequency-determining network, and also because of the difficulties that would be encountered in obtaining proper tracking. However, for single-frequency operation, as in the modulator section of an rf generator, the bridge-type phase-shift oscillator finds practical application.

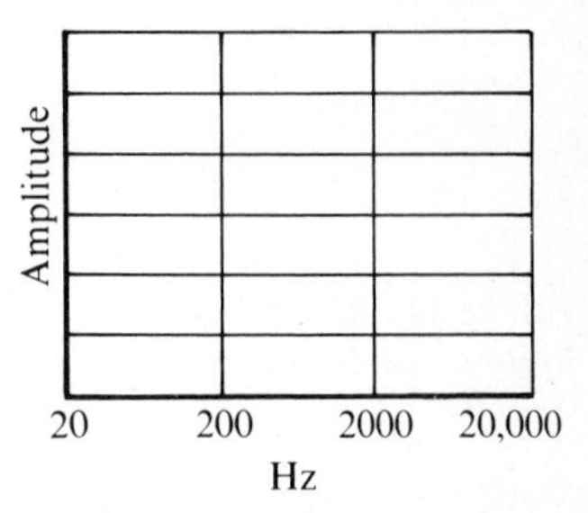

FIG. 16.12 *Display of amplitude vs. logarithmic frequency.*

16.6 BEAT-FREQUENCY OSCILLATORS

As noted previously, beat-frequency audio oscillators are much less common than before RC oscillators attained their present state of development. One of the advantages of a beat-frequency oscillator is its wide frequency range on a given switch position. A typical instrument covers a range of 20 Hz to 20 kHz on its first switch position. The output frequency is logarithmic with respect to dial rotation; that is, the frequency variation is much more rapid in the vicinity of 20 kHz than in the vicinity of 20 Hz. The meaning of a logarithmic frequency variation is seen in Fig. 16.13. If the output from this type of generator is displayed on a scope screen, the presentation can be compared to a plot on semilog graph paper.

Beat-frequency oscillators have the disadvantage that a comparatively large number of tubes (or transistors) and associated components are required. We also find that pulling

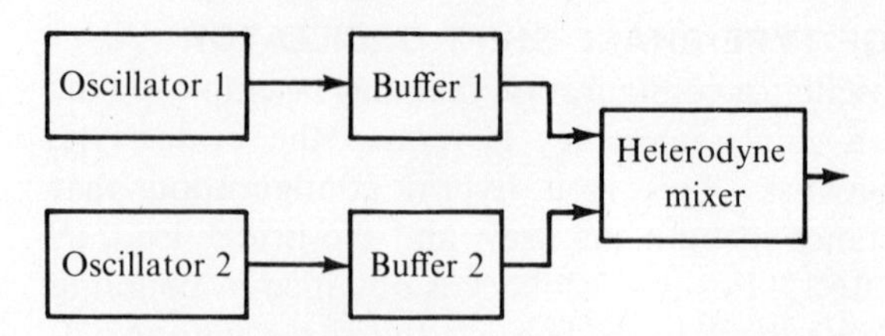

FIG. 16.13 *Basic plan of a beat-frequency oscillator.*

and locking (interaction) of the beating oscillators can be eliminated only by using buffer sections between each oscillator and the heterodyne mixer (see Fig. 16.13). Harmonics and resulting spurious beats impose design problems that can be minimized but not completely overcome. The inherent frequency stability of a beat-frequency oscillator is considerably poorer than that of a Wien-bridge oscillator. For these reasons, beat-frequency audio oscillators are now used only in specialized applications.

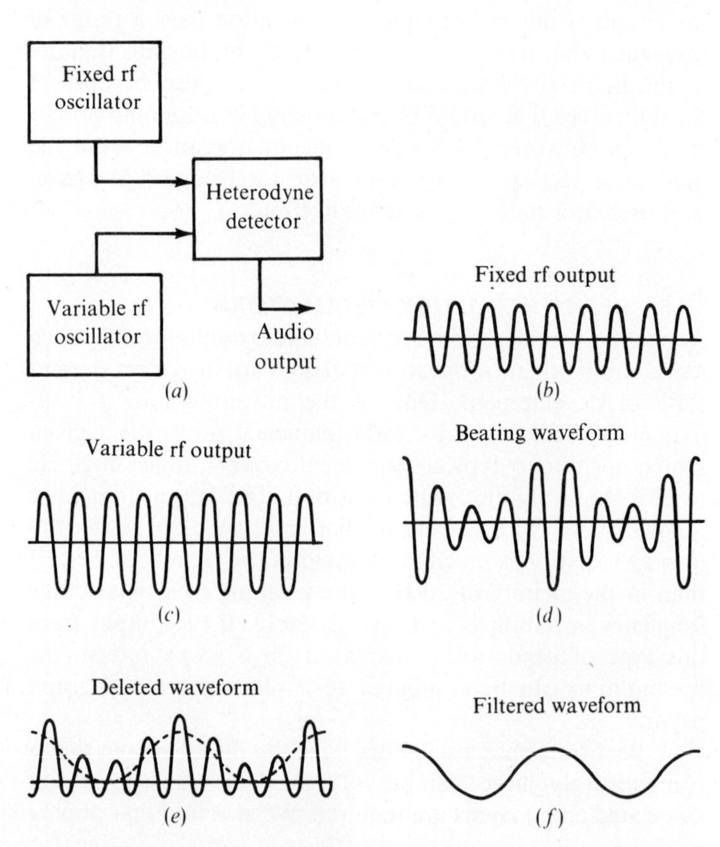

FIG. 16.14 *Action of a heterodyne detector.*

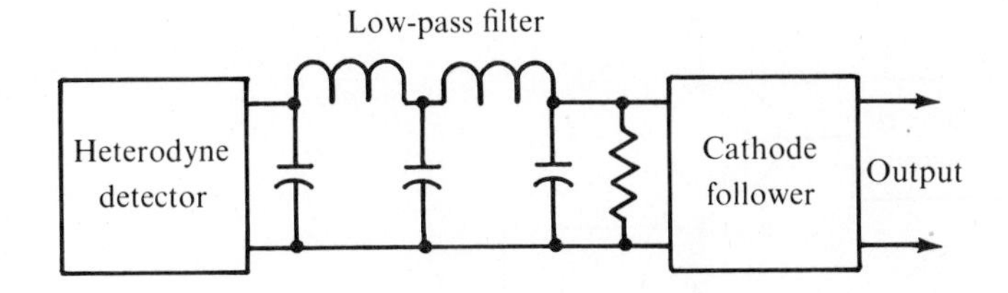

FIG. 16.15 *Diagram showing that lowpass filter removes all but the difference frequency.*

Figure 16.14 depicts the principle of beat-frequency generation. Note that the audio frequency is developed as the difference between two radio frequencies. A beat waveform is produced by mixing the two radio frequencies. Although the average value of the beat waveform is zero, it is next heterodyned through a nonlinear resistance, and its average value is then the difference or audio frequency. By passage through a low-pass filter, as shown in Fig. 16.15 the rf pulse components are removed, and only the difference or audio frequency appears in the filtered output. Note in Fig. 16.14 that the audio-frequency output is made variable in frequency by providing a variable rf oscillator to beat against a fixed rf oscillator.

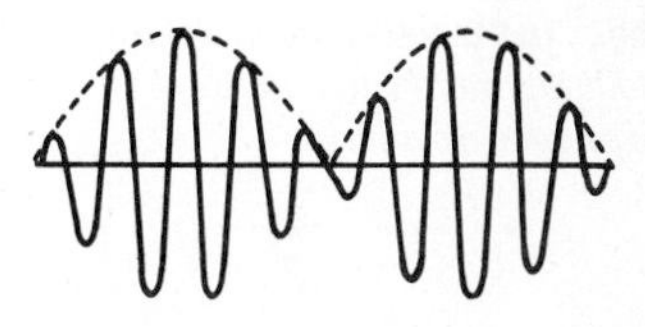

FIG. 16.16 *Diagram showing that beat envelope is a half-sine wave when the two beating frequencies have the same amplitude.*

It is evident that both the fixed and variable rf oscillators must have a pure sine waveform if spurious beat outputs are to be avoided. That is, harmonics will beat together in the heterodyne mixer, just as fundamentals beat together. If harmonics are present, the result is a contaminated audio output. Therefore, the design of beat-frequency oscillators is concerned with oscillator arrangements that provide as nearly a pure sine waveform as possible. Good audio output waveform also requires a suitable relation between the amplitudes of the beating frequencies. For example, Fig. 16.17 shows that the beat envelope is not sinusoidal when the beating frequencies have the same amplitude. That is, the envelope has the shape of half-sine waves in this mode of operation. However, if one of the beating frequencies has a much greater amplitude than the other, the beat envelope then approximates a sine waveform, as shown in Fig. 16.17.

FIG. 16.17 *Diagram showing that beat envelope approximates a sine wave when one of the two beating frequencies has a greater amplitude than the other.*

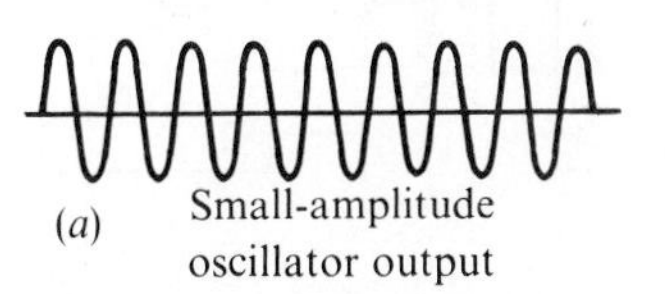

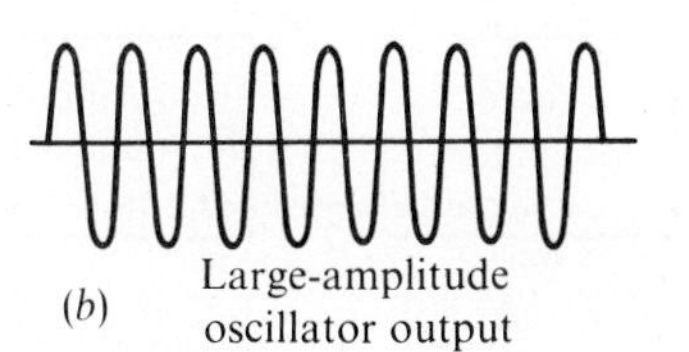

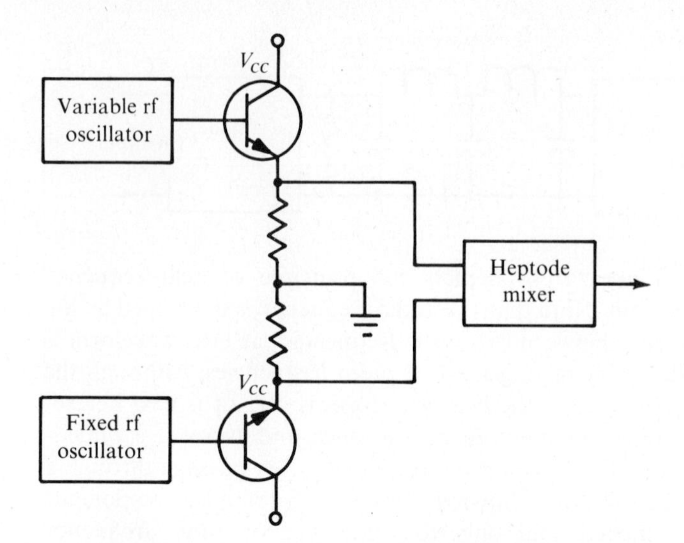

FIG. 16.18 *Addition of emitter-follower buffers to improve oscillator isolation.*

When comparatively low audio-frequency output is desired, additional buffering is employed, as depicted in Fig. 16.18. The emitter followers minimize interaction between the beating oscillators, so that the oscillators have little tendency to pull and lock when they are operated at nearly the same frequency. The best beat-frequency oscillators have a lower frequency limit of about 20 Hz. By way of comparison, *RC* oscillators can be designed to have a lower frequency limit of 0.01 Hz. The chief advantage of the beat-frequency type of audio oscillator is found in applications which require complete coverage of the audio-frequency spectrum in one turn of the tuning dial.

16.7 APPLICATIONS

One of the most basic applications for an audio oscillator is measurement of the frequency response of an audio amplifier, as depicted in Fig. 16.19. The amplifier should be terminated

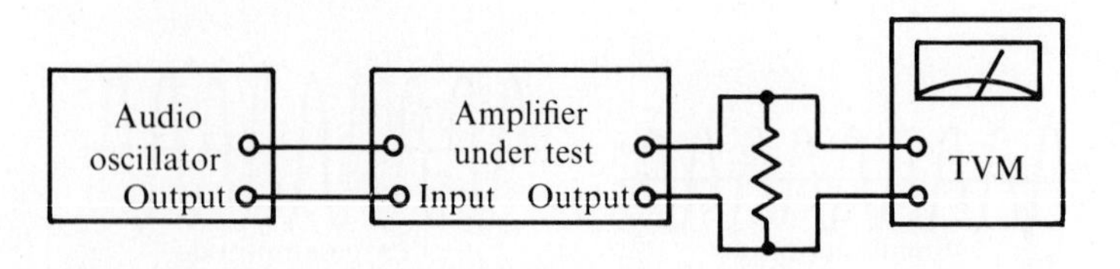

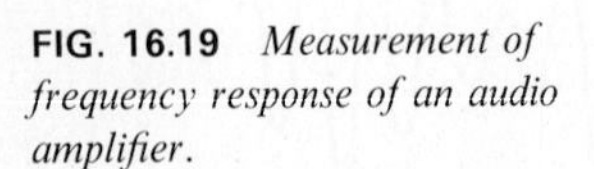

FIG. 16.19 *Measurement of frequency response of an audio amplifier.*

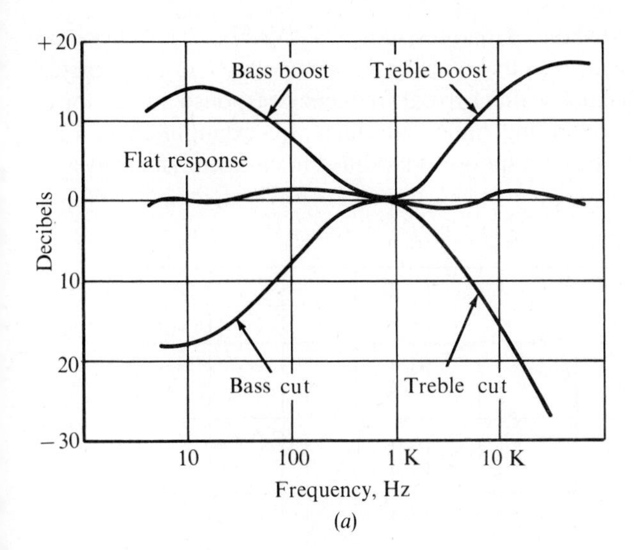

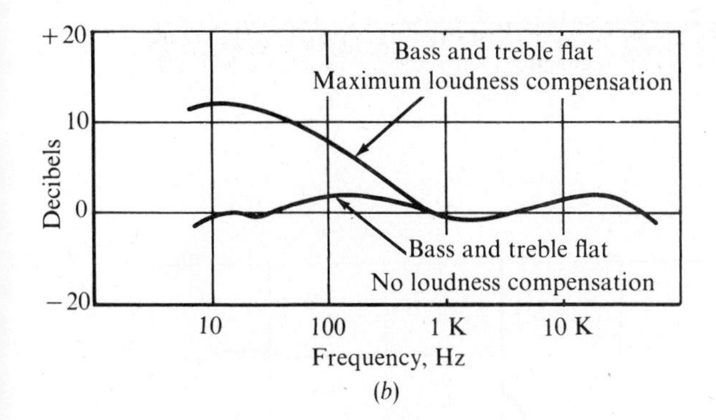

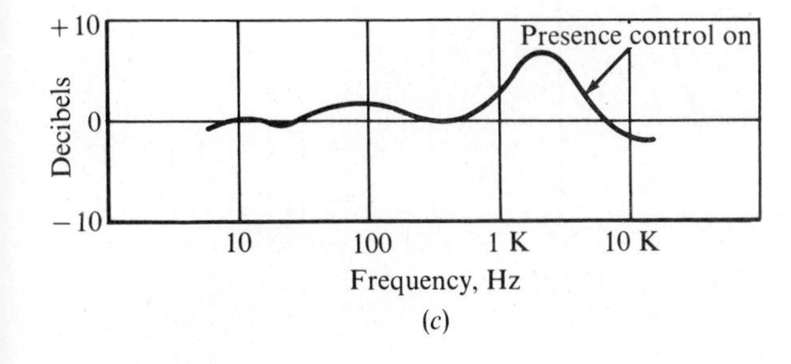

FIG. 16.20 *Typical audio frequency amplifier response curves. (a) Flat response modified by setting of tone controls; (b) flat response modified by loudness compensation; (c) basic response modified by presence control.*

in its rated value of load resistance; a TVM or VOM serves as a convenient output indicator. It is customary to use a meter with a decibel scale. Typical frequency-response curves for a normally operating audio amplifier are exemplified in Fig. 16.20. The basic response is modified in characteristic ways by

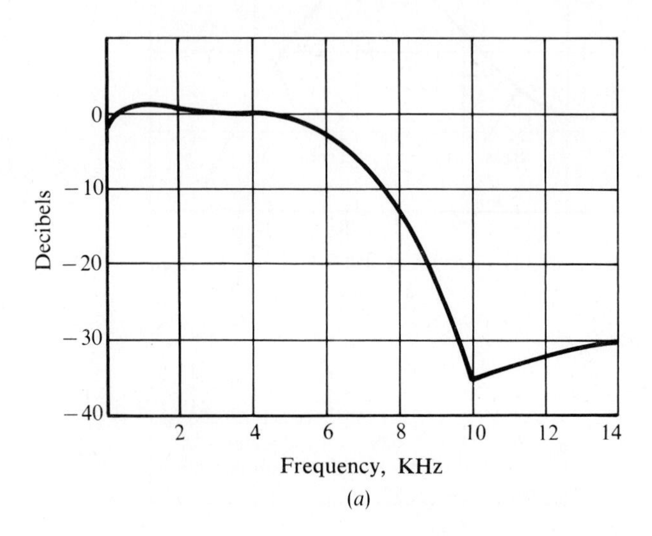

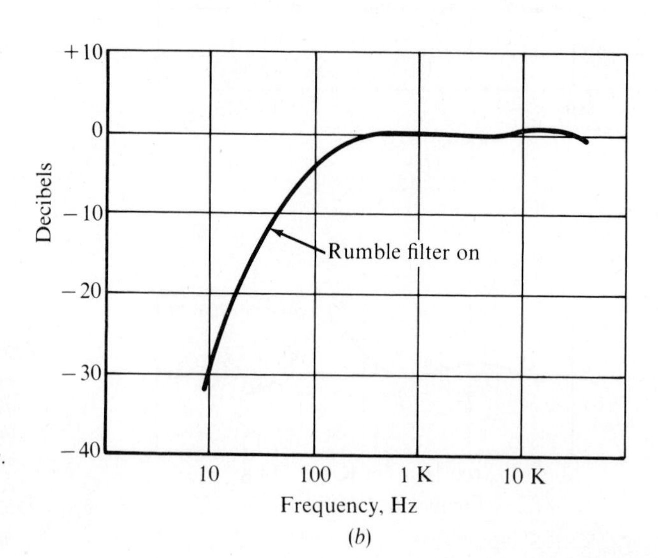

FIG. 16.21 *Effect of audio filters. (a) Basic response modified by a scratch filter; (b) basic response modified by a rumble filter.*

bass boost, treble boost, loudness, and presence controls. A preamplifier with a phonograph input often provides scratch and rumble filters, also. The effect of these filter responses on the basic audio-frequency response curve is shown in Fig. 16.21. Note that equalization networks used in record player and

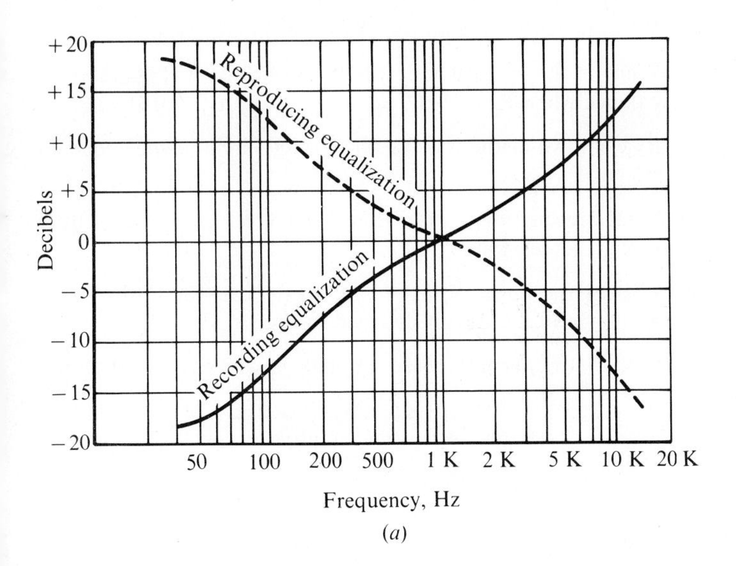

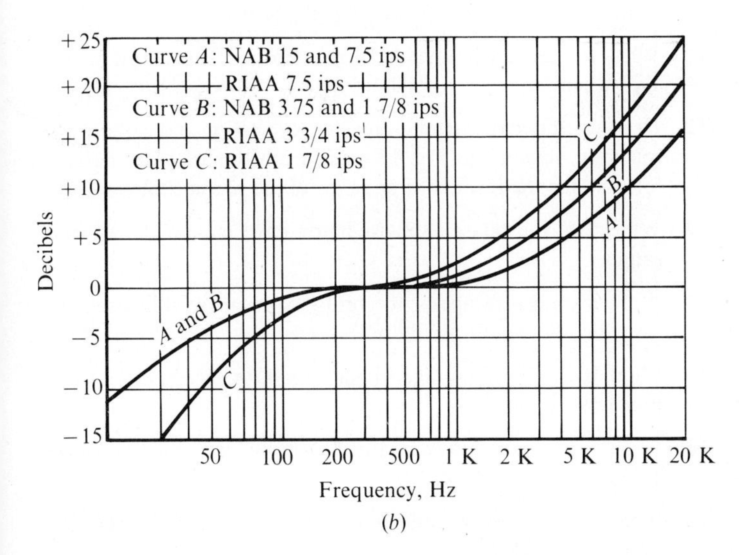

FIG. 16.22 *Effect of equalization networks on audio-amplifier frequency response. (a) Standard RIAA recording and playback characteristics for phonograph records; (b) standard playback equalization characteristics for tape recorders.*

tape player amplifiers also modify the basic frequency response of an audio amplifier as shown in Fig. 16.22.

QUESTIONS AND PROBLEMS

1. What is the distinction between an audio oscillator and an audio generator?
2. Why are nearly all audio oscillators of the *RC* type?
3. What is the frequency at which the bridge in Fig. 16.3 will balance when $R_1 = 2$ kΩ, $R_2 = 3$ kΩ, and $C_1 = 0.01$ μF?
4. By what two methods is the Wien-bridge circuit made frequency selective?
5. What prevents the output from a Wien-bridge oscillator from being a square wave?
6. What is the purpose of the lamp in the cathode of the Wien-bridge oscillator circuit depicted in Fig. 16.5?
7. What is the purpose of the amplifier (V_3) in the circuit shown in Fig. 16.5?
8. How is harmonic distortion minimized in a Wien-bridge oscillator?
9. Why is a cathode follower usually used as the output stage of an audio oscillator?
10. In the circuit shown in Fig. 16.7, what is the function of each of the following components: C_1, R_9, and R_{22}?
11. What is the null frequency of the bridged-T network in Fig. 16.8 when $C_1 = C_2 = 0.001$ μF, $R_1 = 100$ K, and $R_2 = 10$ K?
12. Draw an impedance triangle for one *RC* section of the frequency-selection network in Fig. 16.9 and from this diagram derive a formula for the resonant frequency of the phase-shift oscillator.
13. Why does the phase-shift oscillator have a limited frequency range?
14. What is an advantage of the bridge-type phase-shift oscillator over the phase-shift oscillator?
15. What is the disadvantage of the oscillator circuit depicted in Fig. 16.9?
16. What is the advantage of the beat-frequency oscillator over an *RC* type?
17. What are the disadvantages of the beat-frequency oscillator over an *RC* type?

18. What is the principle of operation of a beat-frequency oscillator?

19. Why must the outputs from the oscillators in a beat-frequency oscillator be low-distortion sine waveforms?

20. What is the purpose of the emitter-follower amplifiers in the circuit depicted in Fig. 16.18?

17

RF SIGNAL GENERATORS

17.1 BASIC REQUIREMENTS

Various types of rf signal generators are in general use. The amplitude-modulated (am) signal generator is found in almost every shop and laboratory. It is used to align radio receivers, to measure resonant frequencies, to calibrate auxiliary equipment in terms of frequency, to make signal-substitution tests, and for many other purposes. In the strict sense of the term, a generator provides a known output level which is determined by means of an output meter and a calibrated step attenuator. This type of instrument is required to measure the sensitivity of a radio receiver, for example. On the other hand, an rf oscillator provides an output signal that has an unknown level, although its level can be changed by means of an uncalibrated attenuator. Students frequently make no distinction between oscillators and generators.

Marker generators are a class of comparatively accurate rf oscillators. They generally provide frequency-calibration facilities (usually quartz crystals), although few of them have output meters and calibrated attenuators. In this chapter, we shall consider marker generators simply from the viewpoint of RF generating equipment. In an attempt to reduce confusion in terminology, manufacturers of complete and highly accurate sine-wave signal sources have applied the name *standard signal generator* to this class of equipment. Although an rf test oscillator might be called a signal generator, it is never called a standard signal generator.

17.2 BASIC CONSIDERATIONS

In the strict sense of the term, a signal generator is a source of sine-wave voltage with an appreciable range of frequency and amplitude, both of which are known to a high degree of accuracy. An rf signal generator provides a range of frequencies within the limits of 10 kHz and 10,000 MHz. The output amplitude is usually adjustable to less than 1 μV and up to 0.1 V or more. Figure 17.1 illustrates an intermediate-type signal generator, representative of instruments between service-type generators and standard-signal generators. This instrument has a frequency range from 100 kHz to 9.5 MHz. Its maximum output voltage is 0.1 V rms in 50 Ω. An output meter is provided, with a calibrated five-step attenuator; a continuous attenuator is also provided to fill in between the steps of the coarse attenuator. This continuous attenuator also has a calibrated output, indicated by the output meter. Internal amplitude modulation of 400 Hz, with adjustable percentage from zero to 50 percent, is also provided.

FIG. 17.1 *An intermediate-type signal generator.*

The plan of a basic signal generator is depicted in Fig. 17.2. It comprises an rf oscillator with a calibrated tuning capacitor, an output (or carrier level) meter, and a calibrated attenuator. Shield compartments are provided to prevent radiation of rf energy into surrounding space. Note that in this example the components are housed in a double-shielded case. Double shielding minimizes the value of rf leakage; however, it is essential to ground both shield boxes at the same point. If each box is grounded at an arbitrary point, circulating rf ground currents are established which increase the leakage and radiation of rf energy into surrounding space. Since the attenuator supplies energy to the output cable, designers often

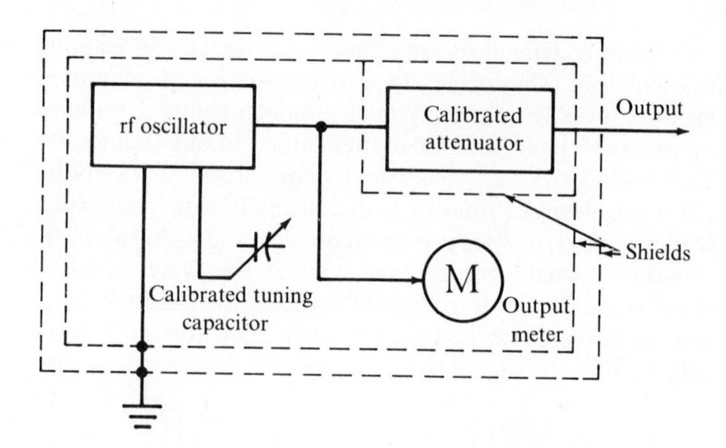

FIG. 17.2 *Plan of a basic signal generator.*

provide additional shielding for the attenuator section. Details are reserved for subsequent discussion.

The sine-wave oscillators employed in basic generators are tunable over various frequency bands. The generator illustrated in Fig. 17.1 provides four frequency bands, with ranges from 100 to 290 kHz, 280 to 1000 kHz, 0.95 to 3.1 MHz, and 2.9 to 9.5 MHz. The practical range for an *LC* oscillator is about 3:1 as exemplified by the foregoing example. Fixed inductors are commonly employed, and tuning is obtained by means of variable shunt capacitance. A 3:1 tuning ratio is the practical maximum in order to avoid an objectionably small L/C ratio; a fairly high L/C ratio is required to obtain good oscillator characteristics. Band switching is used to select suitable coil sizes for each frequency band.

Since a good sine-wave output is desired (minimum harmonic distortion), oscillator circuits used in signal generators are designed primarily to provide high purity of output waveform. Generators that cover the frequency ranges up to 250 MHz usually employ some version of the Hartley or Colpitts oscillator circuits; the most basic configurations are depicted in Fig. 17.3. Frequency stability is also a prime design factor, and the accuracy rating of a signal generator depends directly upon the frequency stability of its oscillator. The same basic design problems apply also to oscillator circuits in transistorized generators. Figure 17.4 shows a basic Hartley oscillator circuit with a junction transistor. To obtain a good sine waveform, the resonant circuit in any oscillator arrangement must have a high Q value.

To realize high Q operation, it is not sufficient that the oscillator coil have low rf resistance; it is also necessary that the tank circuit be lightly loaded, because heavy loading is equivalent to increasing the rf resistance of the circuit. (Details are reserved for subsequent discussion.) Moreover, the coil must have reasonably low distributed capacitance, so that the L/C ratio is not unduly small at the low-frequency end of the band. An oscillator tank circuit has an equivalent series-resonant configuration. In a series RLC circuit, the Q value is related to the L/C ratio by the following formula:

$$Q = \frac{LC}{R} \tag{17.1}$$

Frequency stability depends upon the constancy of the oscillator supply voltages, and this is why regulated power supplies are commonly employed in signal generators. Frequency stability also depends upon temperature compensation. Although warm-up drift is unavoidable in a tube-type generator, a well-designed instrument soon attains a stable operating state. Thermal stability is maintained chiefly by means of temperature-compensating capacitors which are inserted at strategic points in the circuitry. In the case of transistor oscillators, additional means are employed to obtain stable operation over a specified range of temperature.

To some extent, frequency stability is also dependent upon the method of biasing that is utilized. Thus an oscillator

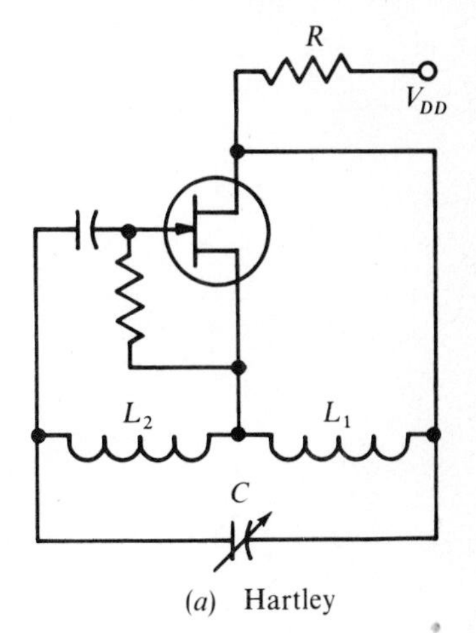

(a) Hartley

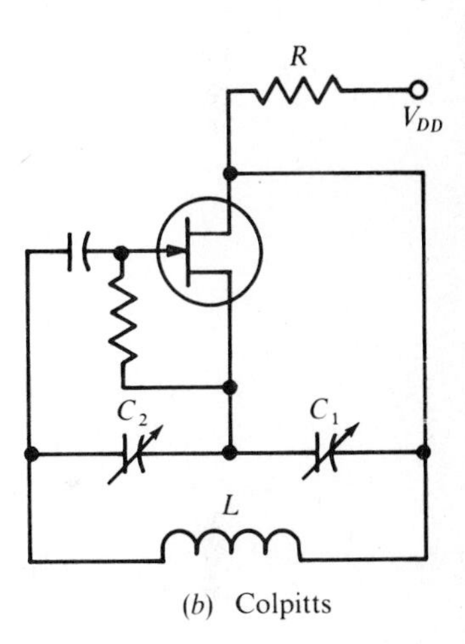

(b) Colpitts

FIG. 17.3 *Basic oscillator circuits.*

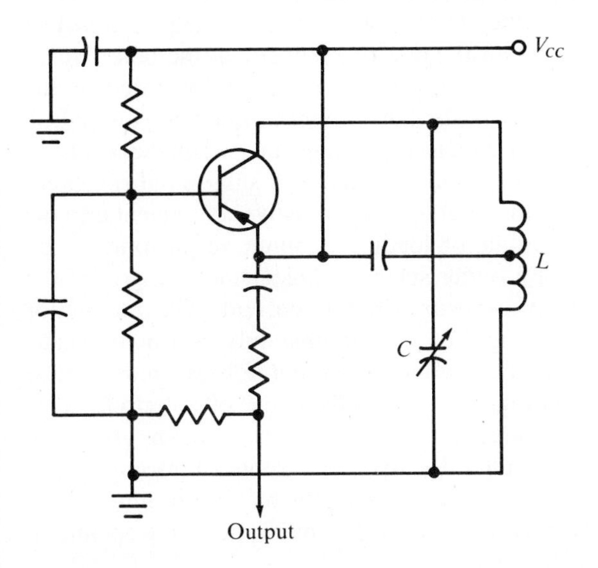

FIG. 17.4 *Basic transistorized Hartley oscillator for an rf generator.*

can be operated with fixed bias, cathode bias, or signal-developed bias. Signal-developed bias, as exemplified in Fig. 17.3, provides maximum frequency stability, particularly if the grid-leak resistance is made as high as practical while still permitting reliable operation. Note in passing that signal-developed bias also provides some degree of amplitude regulation. For example, as the generated voltage tends to change with a change in the L/C ratio, the change is opposed somewhat by the accompanying shift in bias voltage. Most rf signal generators employ signal-developed bias for this reason.

17.3 WAVEFORM OPTIMIZATION

It follows from previous discussion that the oscillator in a signal generator does not operate in Class A. The plate current flows in surges or dc pulses, but it does not flow over the entire operating cycle. The output waveform would be highly non-sinusoidal if it were not for the flywheel action of the plate tank. Let us consider this circuit action. With reference to Fig. 17.5, the plate tank consists of inductor L_2 and capacitor C_{10}. When a pulse of plate current flows through L_2 from the rf oscillator, the tank circuit is shock excited, as depicted in Fig. 17.6. This is a basic damped sine waveform. Since a pulse of plate current flows at the beginning of each period, a good sine waveform is maintained.

Note that the amount of decay over one period in Fig. 17.6 will depend on the effective Q value of the tank circuit. The output waveform can be optimized by reducing the effective resistance of the tank circuit as much as is practical. The chief consideration is the loading imposed on the oscillator circuit by the output circuit. That is, a substantial rf current demand (heavy loading) reflects a resistive component into the plate circuit. Oscillator loading is minimized in practice by employment of a buffer stage or stages, since the grid of the buffer triode draws very little rf current. The modulator section of V_2 in Fig. 17.5 does double duty as a buffer stage.

Another buffer configuration that utilizes a tetrode tube is depicted in Fig. 17.7. The advantage of a tetrode is its comparatively large power output for a given grid-drive voltage. Since a buffer does not drive a tuned tank circuit, but works instead into a resistive load, the buffer tube must necessarily be operated in Class A. If a buffer does not operate in

Class A, the output waveform will be distorted accordingly; it would consist of half-sine waves if the buffer tube were biased to cutoff for Class B operation. A buffer tube is basically a current amplifier, or electronic impedance transformer; in Fig. 17.5 it operates in a cathode-follower circuit.

The buffer tube in Fig. 17.7 does not operate as a cathode follower, but as a conventional amplifier with a low-value plate load. The effective plate load value is about 200 Ω. This low value is necessary because the operating frequency may be as high as 10 MHz in this example. A 6AV5 tube has a trans-

FIG. 17.5 *Configuration of a simple rf signal generator. (Courtesy, Knight Electronics)*

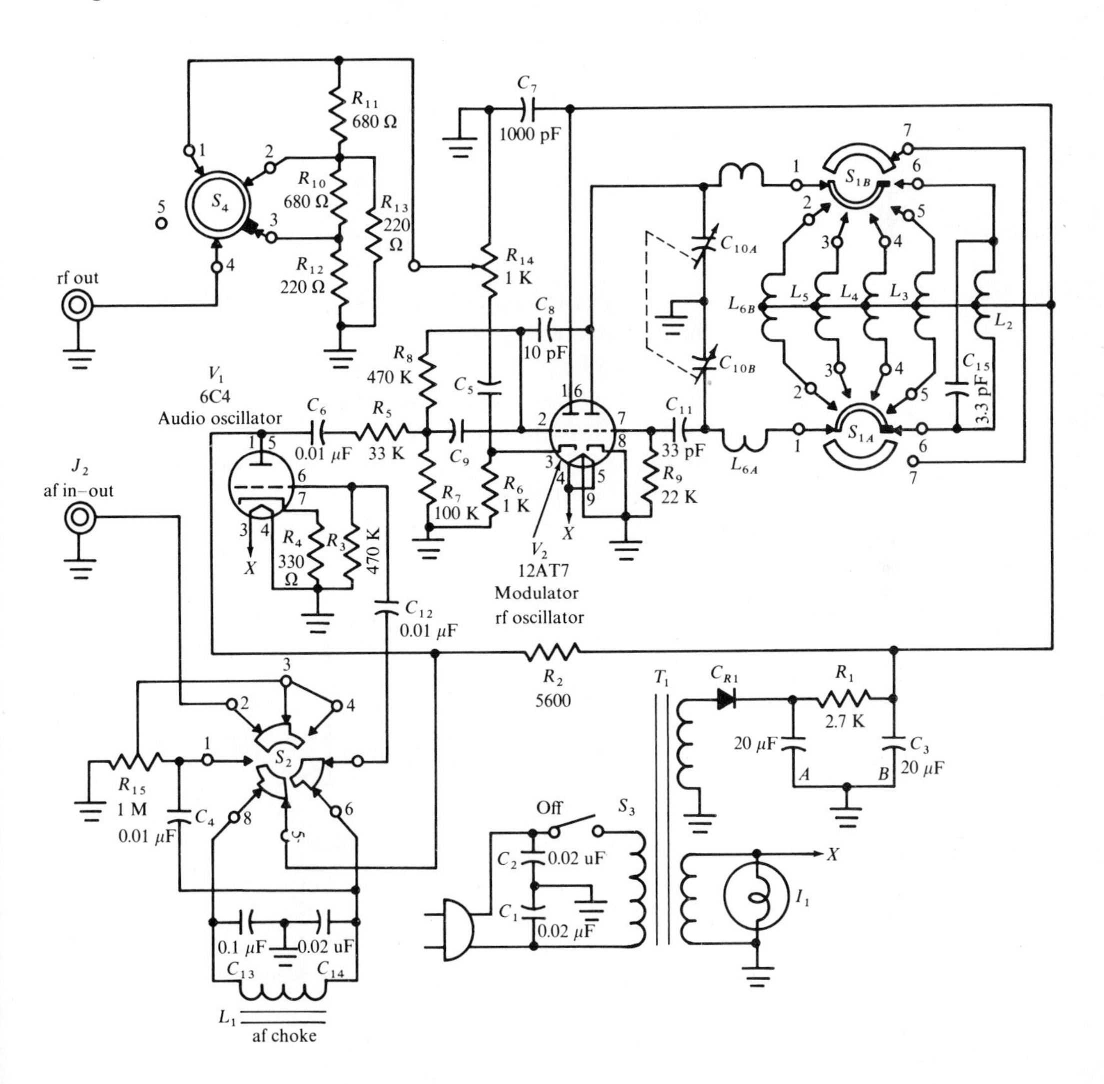

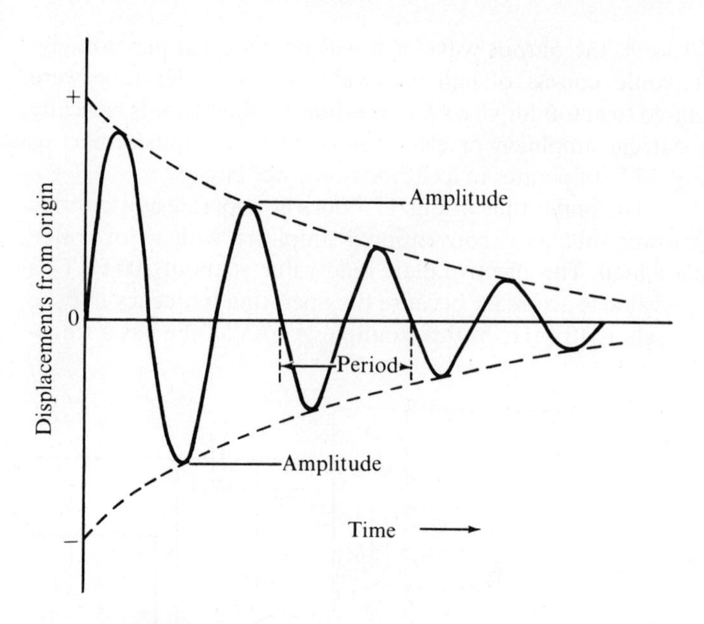

FIG. 17.6 *Waveform of damped oscillations.*

conductance value of about 5900 μS; thus the voltage gain of the buffer stage is approximately unity. However, the current gain is very large, because the grid draws practically no current, while the output load of 200 Ω draws a heavy current. The result is that although very little rf power demand is placed on the oscillator, the generator can supply appreciable rf power for testing devices or circuits.

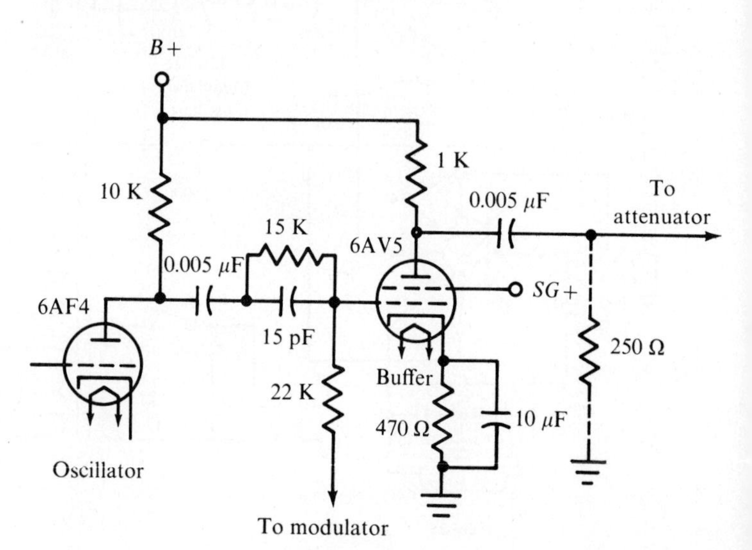

FIG. 17.7 *Buffer stage to minimize loading of the oscillator.*

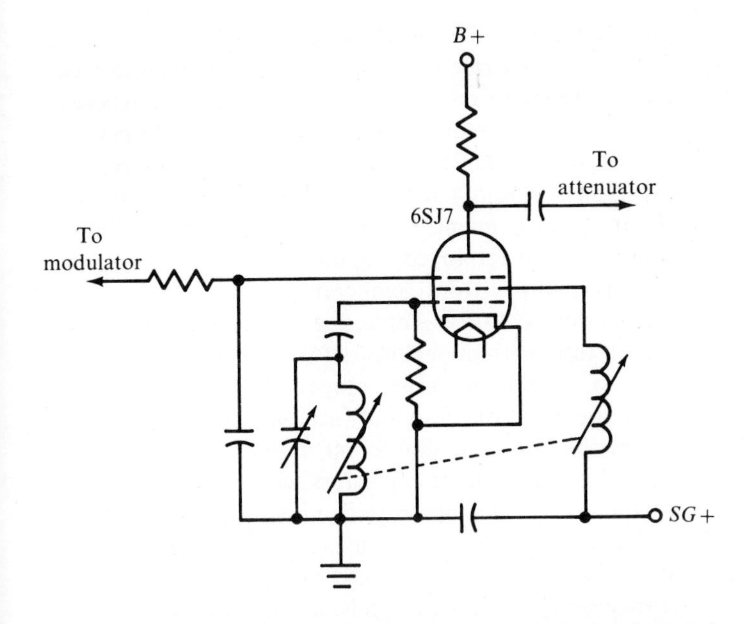

FIG. 17.8 *Electron-coupled oscillator circuit.*

Another method of oscillator isolation employed in service-type signal generators exploits the characteristics of the electron-coupled oscillator circuit depicted in Fig. 17.8. The screen grid, control grid, and cathode of the tube operate as a triode in the oscillator circuit. Consequently, the electron stream becomes stronger and weaker as the oscillator waveform rises and falls. Since a substantial portion of the electron stream passes through the screen grid, a plate current flows, and this plate current has the same waveform as in the oscillator section. If the tank circuit has a high Q value, this waveform will approximate a true sine wave.

Note that the grid and screen-grid coils are coupled in Fig. 17.8. Sufficient coupling must be provided to sustain oscillation over the associated band of operating frequencies. On the other hand, excessive feedback must be avoided or the oscillatory waveform will be impaired, as exemplified in Fig. 17.9. This type of distortion is caused by excessive grid current that flows when the grid is driven considerably positive. In turn, grid current flow "robs" the space current of electrons that would otherwise travel to the plate. Heavy loading is also imposed on the plate coil during the time that energy is being extracted by the grid coil to drive the grid into heavy grid current flow. Under this condition of operation the output waveform from the oscillator becomes distorted.

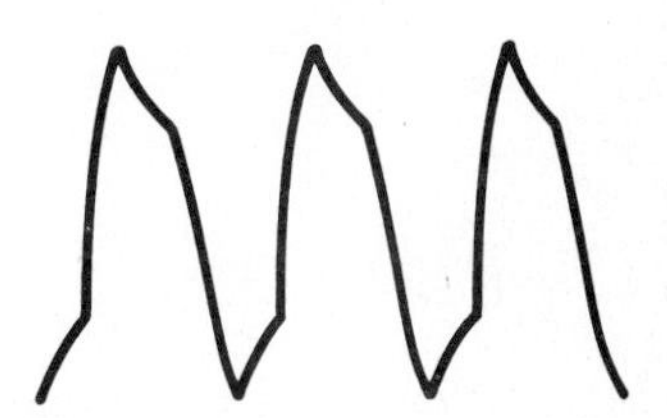

FIG. 17.9 *Distorted output waveform caused by excessive feedback.*

17.4 FREQUENCY STABILIZATION

Signal generators are rated for calibration accuracy. A service-type generator may be rated for ±2 percent accuracy, whereas a laboratory-type generator may be rated for ±0.25 percent accuracy. It follows that the designer must observe electrical and mechanical stability requirements in order to attain a given accuracy assignment. From the mechanical standpoint, rigid construction is employed, and the effects of expansion and contraction resulting from temperature changes are taken into consideration. For example, the design of an inductor is oriented toward maintenance of a constant inductance value over a wide temperature range. The same basic consideration applies to the design of variable capacitors.

From the electrical standpoint, frequency stabilization is optimized by use of fixed capacitors with suitable temperature coefficients. A temperature-compensating capacitor is mounted at a point where it provides optimum compensation during both warm-up and sustained operation. The temperature distribution during warm-up is not the same as in steady operation; moreover, since the distribution is changing, it is comparatively difficult to realize high stability during warm-up. Therefore, most lab-type generators, both tube-type and solid-state, are rated for accuracy following a specified warm-up period.

Frequency stabilization also refers to calibration accuracy under various loads. To illustrate, we might connect the output cable from a generator across an inductor or across a capacitor. The load may vary in current demand over a wide range. A good generator is designed so that the load condition has practically no effect on the oscillator frequency. The chief requirement in this respect is efficient buffering between the oscillator and the output section, so that the oscillator circuit is virtually isolated from the load circuit. A screen-grid tube provides better buffer action than a triode, because changes in plate voltage have less effect on grid circuit parameters. In other words, the Miller effect is minimized in a screen-grid tube.

17.5 MODULATION OF THE rf SIGNAL

Conventional rf signal generators provide a choice of continuous-wave (cw) or am output. Although an oscillator can be modulated directly, it is preferable to modulate the buffer stage, as exemplified in Fig. 17.5. Details are explained subsequently. Figure 17.10 shows an unmodulated rf carrier, and a 100 percent modulated wave. Note that the peak voltage of the 100 percent modulated signal is double the peak voltage of the unmodulated signal. The modulated signal rises to its peak once each period, and falls to zero once each period. In most cases, an am generator is operated at less than 100 percent modulation. Figure 17.11 shows a 50 percent modulated wave. Some generators do not provide an adjustable percentage of modulation, and can be operated only at 30 percent modulated output.

It is practical to modulate an oscillator directly at 30 percent modulation but not at high percentages of modulation. The reason is that the plate voltage of an oscillator changes during the modulation process, and in turn, the oscillating

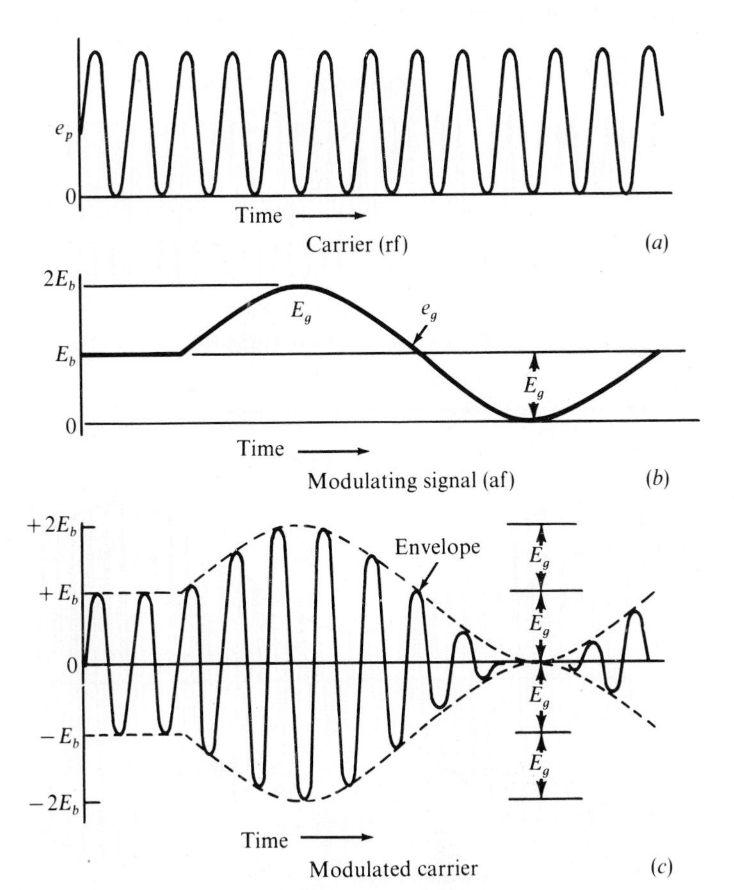

FIG. 17.10 *Modulation of an rf carrier.*

frequency tends to vary. This frequency shift over the modulation cycle is called incidental frequency modulation (incidental fm). When high percentages of modulation are desired, it is necessary to modulate a buffer stage in order to minimize incidental fm. It is obvious that a generator with appreciable incidental fm necessarily has a low accuracy rating and poor operating stability. Lab-type generators have extremely low incidental fm, such as 1 part per million of the carrier frequency plus 100 Hz at 1 kHz and 50 percent modulation.

Sine-wave modulation is almost universal in conventional rf signal generators. Many generators provide 400-Hz internal modulation, with provisions for feeding in an external modulating signal, if desired. The more elaborate generators contain built-in audio oscillators to modulate the rf carrier over an appreciable range of audio frequencies. As seen in Fig. 17.12, the output meter may be used to measure the

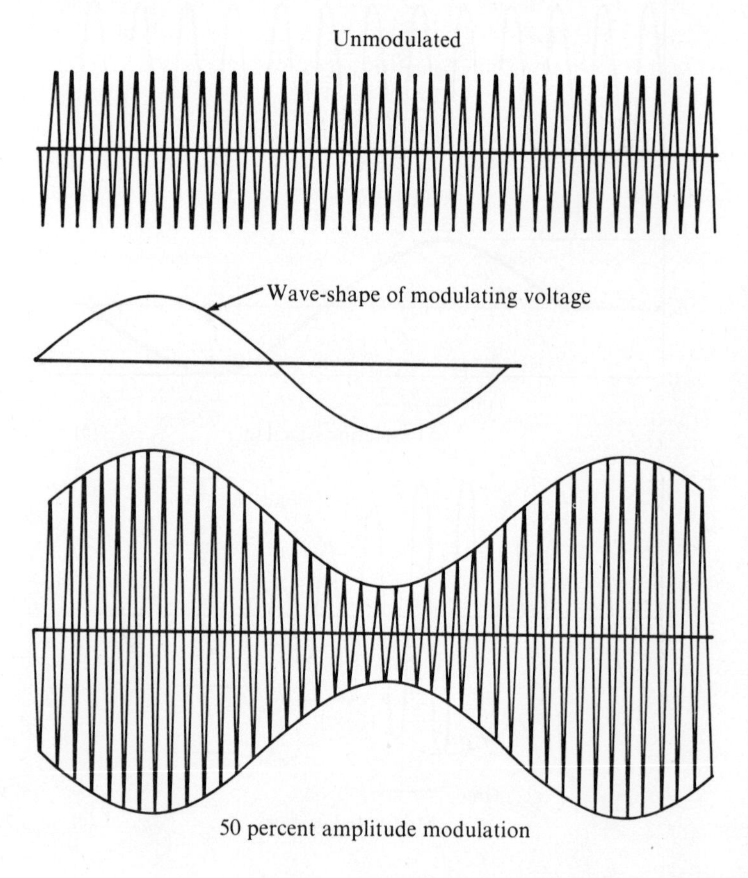

FIG. 17.11 *A 50 percent modulated waveform.*

percentage of modulation, as well as to measure the carrier level. The audio oscillator in this example operates at 400 Hz. When the meter is switched to indicate the modulating voltage, it is energized by a half bridge and responds to the average value of half-sine waves. On the other hand, when the meter is switched to indicate the carrier level, it is energized by a rectifier diode and charging capacitor; thus the meter responds to the peak value of the rf carrier. The meter scales are calibrated in terms of rms volts.

17.6 ATTENUATORS FOR SIGNAL GENERATORS

All rf signal generators and test oscillators are provided with some type of output attenuator, whereby the generated signal can be adjusted to a desired level. An uncalibrated attenuator may consist merely of a potentiometer; greater attenuation is obtained with two potentiometers connected in series, as depicted in Fig. 17.13. To attenuate the generated signal down to a very low level, rf leakage must be prevented. Accordingly, the attenuator section is usually enclosed in an individual shield compartment, as depicted in Fig. 17.2. The line cord to the power supply is another source of rf leakage. Therefore, the line conductors are bypassed to the case of the instrument, as exemplified in Fig. 17.5. Better suppression of leakage current is provided by the addition of rf chokes in the line, as seen in Fig. 17.12.

Calibrated attenuators are basically resistive ladders, as shown in Fig. 17.12. The output level is adjustable in 10:1 steps. Five steps are provided, with a level of 100,000 μV on the highest step and a level of 10 μV on the lowest step. The output can be reduced down to 5 μV, if desired, by reduction of the carrier level with the fine attenuator. Calibration of the output below 5 μV is impractical in this example, because of residual rf leakage. Calibrated attenuators have low-resistance branches to minimize the effect of stray capacitances in the attenuator network. Otherwise, the calibration accuracy would be impaired at the higher frequencies of operation. Low-resistance attenuators also minimize calibration errors that are due to loading effects of the device or circuit under test.

A high-quality step attenuator has a typical calibration accuracy of ± 1 percent. At a frequency of 100 MHz, 5 pF of capacitance, for example, has a reactance of approximately

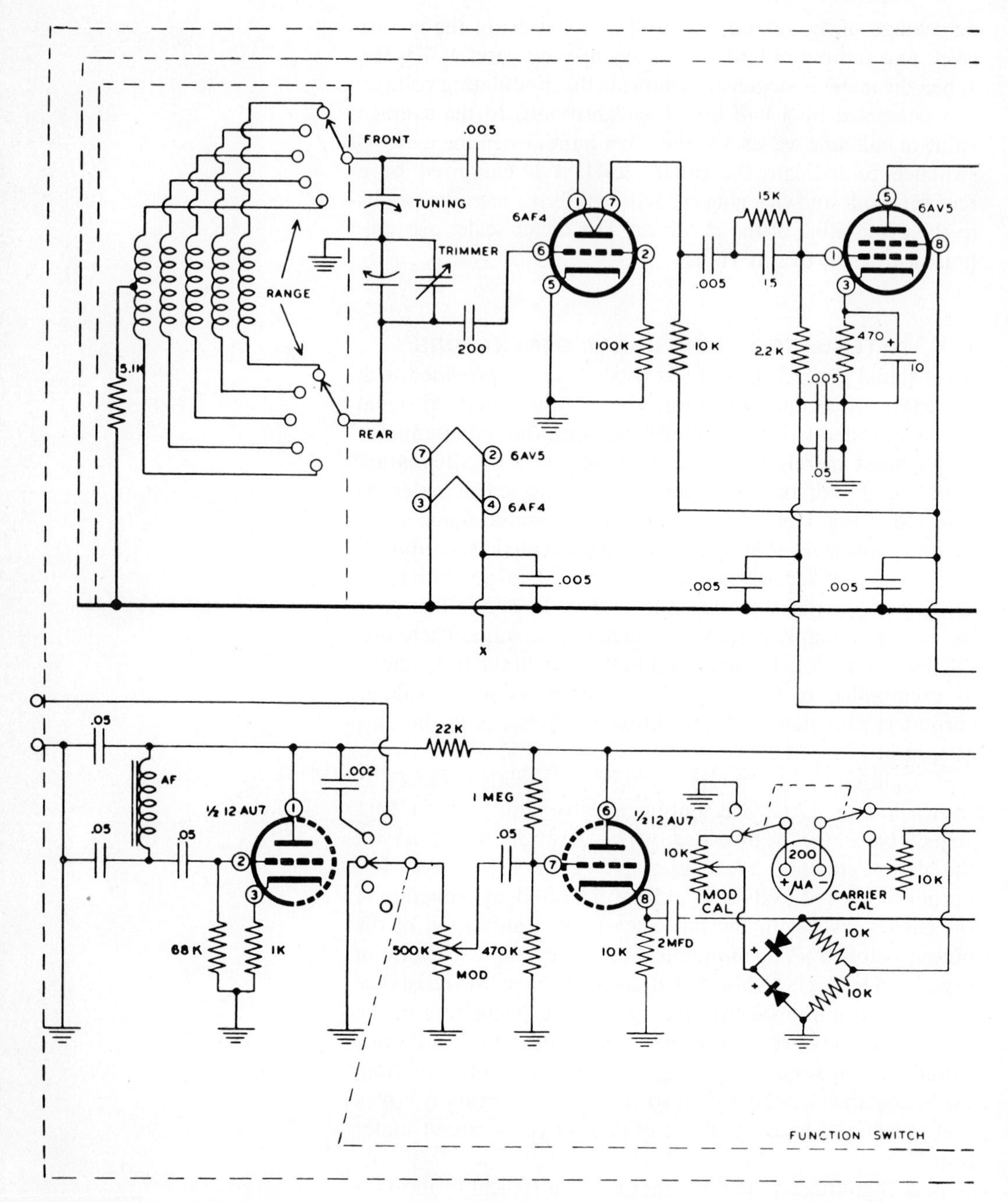

FIG. 17.12 *Configuration of an intermediate-type signal generator.*

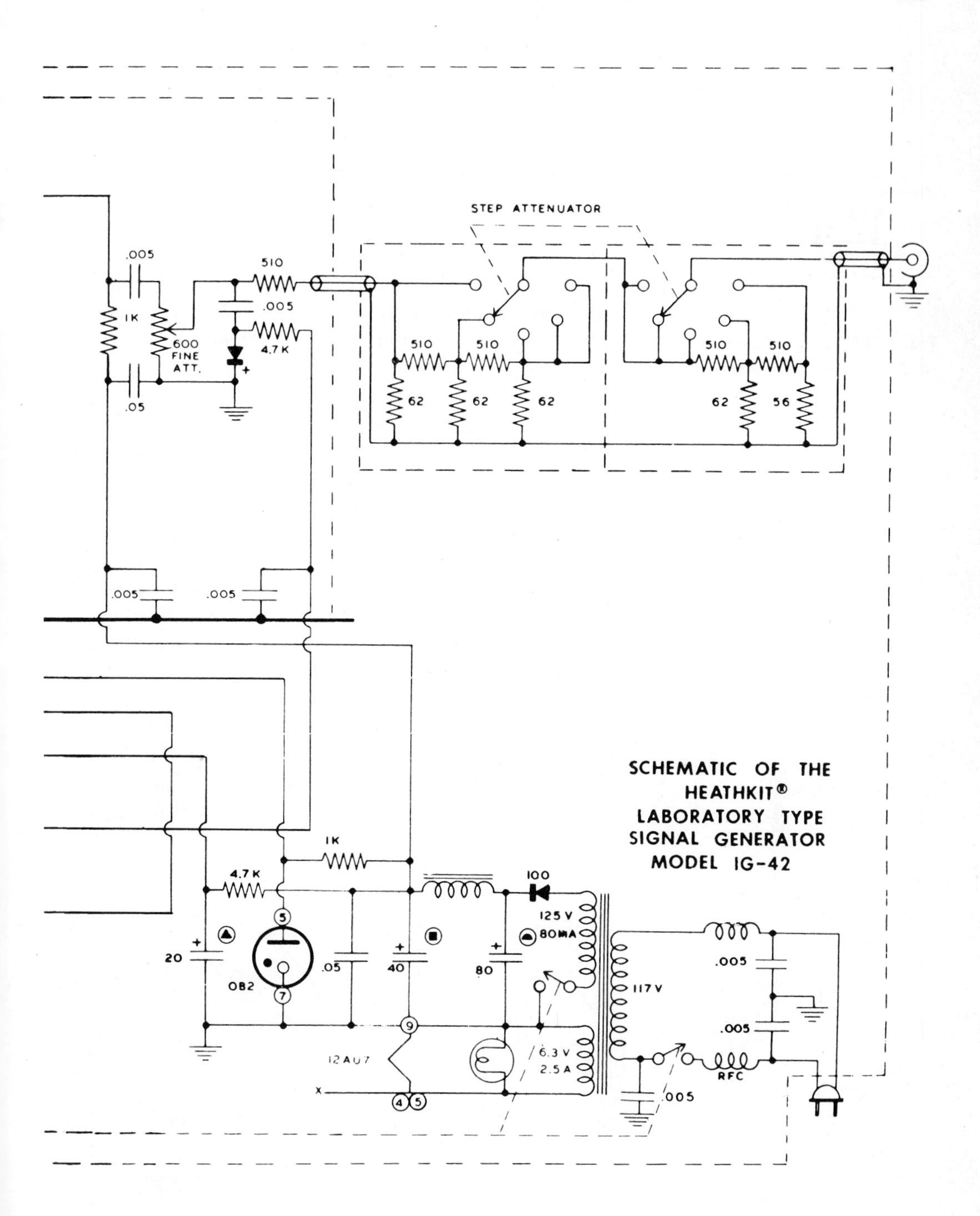
STEP ATTENUATOR
.005
1K
600
FINE
ATT.
.05
510
.005
4.7K
510
510
510
510
62
62
62
62
56
.005
.005
1K
4.7K
100
125 V
80MA
20
OB2
.05
40
80
117 V
.005
.005
RFC
12AU7
6.3 V
2.5 A
.005
SCHEMATIC OF THE
HEATHKIT®
LABORATORY TYPE
SIGNAL GENERATOR
MODEL IG-42

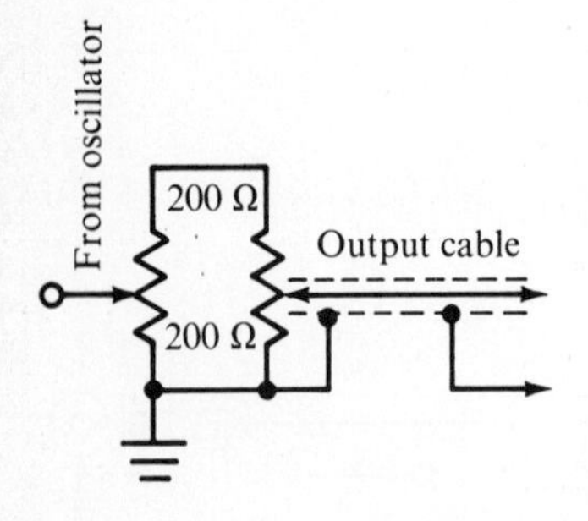

FIG. 17.13 *Two potentiometers used as an uncalibrated attenuator.*

300 Ω. Therefore, unless the attenuator resistance is a small fraction of 300 Ω, its calibration accuracy will be poor at this frequency. In this situation, the designer might choose an attenuator resistance of 20 Ω. With reference to Fig. 17.12, the maximum operating frequency is 10 MHz. In turn, a capacitance of 5 pF will have 10 times as much reactance as in the previous example, or approximately 3000 Ω. Therefore, the attenuator resistance can be chosen somewhat greater without objectionable impairment of accuracy. The resistance of this ladder attenuator is approximately 50 Ω on each step. A coaxial output cable is employed, terminated in approximately 50 Ω.

In the example of Fig. 17.12, the step attenuator employs the construction depicted in Fig. 17.14. A shield plate is placed between the series and parallel banks of resistors to minimize capacitive coupling between sections, and the entire attenuator is housed in a metal case. When greater calibration accuracy is required at comparatively high operating frequencies, the step attenuator is designed with individual shield compartments, as exemplified in Fig. 17.15. Note also that the resistive ladder employs a common ground point, which minimizes circulating rf ground currents. Precision attenuator design is a specialized branch of engineering, and interested students are referred to engineering handbooks for detailed discussion.

The output cable of a signal generator is an integral part of the attenuator system, and places a certain impedance across the ladder network. A shielded (coaxial) cable is necessarily employed in order to confine the rf energy and prevent its radiation into surrounding space. To maintain

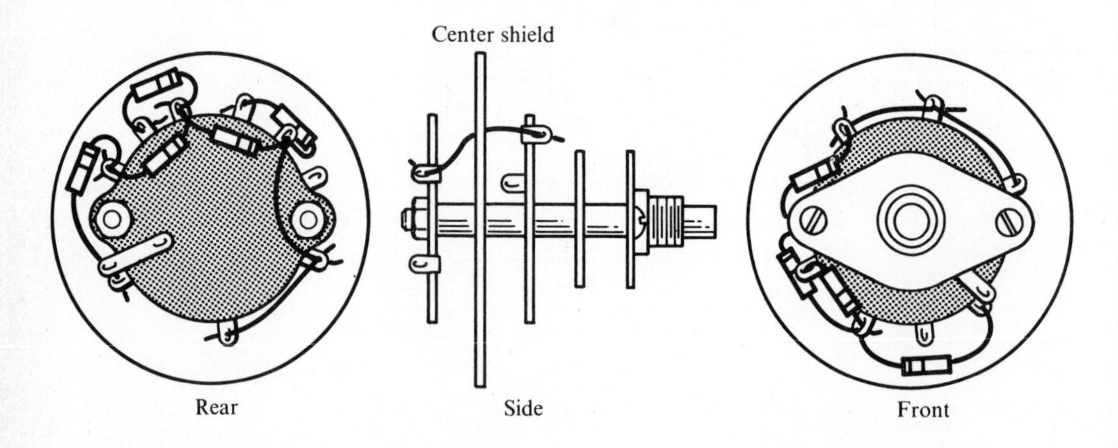

FIG. 17.14 *Construction of a simple step attenuator.* (*Courtesy, Heath Co.*)

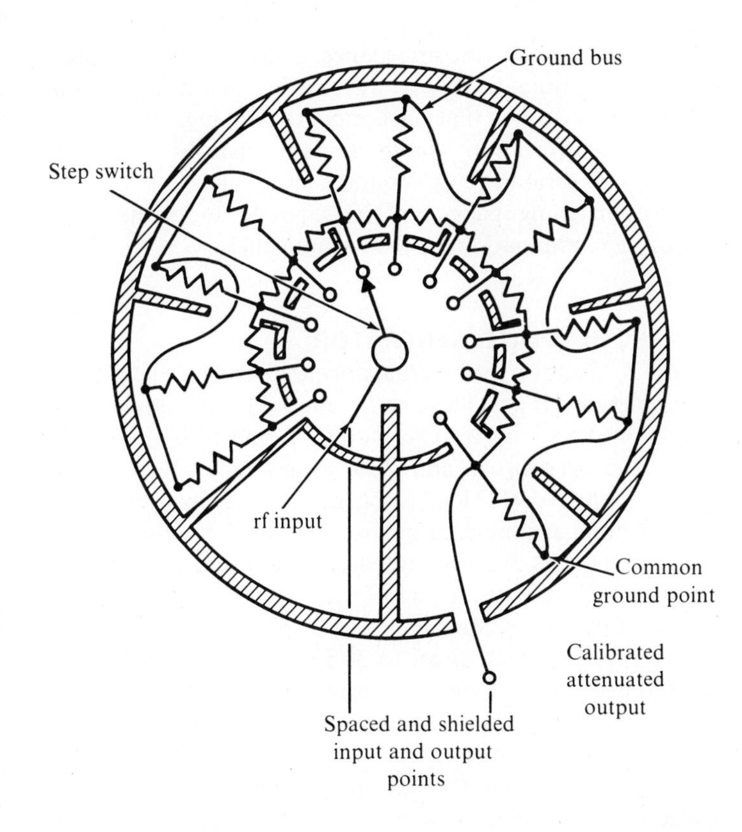

FIG. 17.15 *A step attenuator with individual shield compartments.*

accurate calibration of the attenuator, the output cable must be resistive at all operating frequencies; its input end must not present a reactive component. This requirement means that the cable must be terminated in its characteristic impedance, as shown in Fig. 17.16. Thus, if a cable with a characteristic impedance of 50 Ω is terminated with a 50-Ω resistor, the input end of the cable will look like a 50-Ω resistor.

Next, if a cable is unterminated, its input end will look like a capacitor at comparatively low frequencies. As the operating frequency is increased, the input end of the cable will look like an open circuit at a certain frequency that depends on the length of the cable. At still higher frequencies, it will look like an inductor. If the operating frequency is further increased, a frequency will be found at which the input end of the cable looks like a short circuit. These reactive parameters are caused by reflections of energy and standing waves in the coaxial cable. Standing waves are eliminated by terminating a

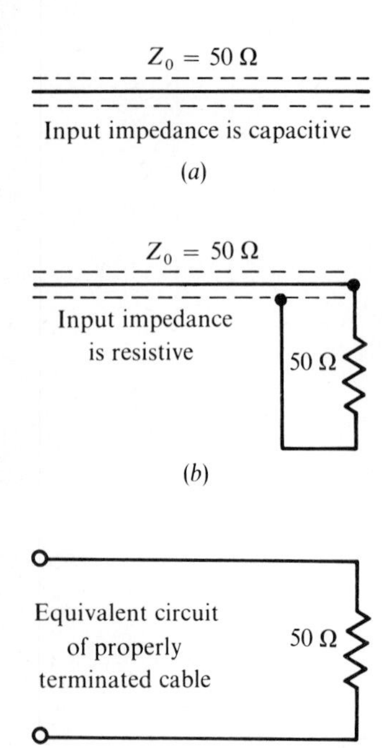

FIG. 17.16 *Correct value of resistor eliminates standing waves.*

cable in its characteristic impedance. In Fig. 17.16*b*, the cable is correctly terminated. Since we usually apply the rf output to circuits or devices that have moderate or high impedance values, the utilization circuit seldom disturbs the cable termination seriously. Of course, in special applications, resistive matching pads may be required to maintain an appropriate termination for the output cable.

17.7 SOLID-STATE GENERATOR

Figure 17.17 shows the external appearance of a typical solid-state generator. It provides a wide range of output frequencies, from 5 Hz to 1.2 MHz. The output waveform is sinusoidal, and has an open-circuit amplitude range from 5 V rms down to 150 μV rms. A 600-Ω load is generally utilized, which reduces the maximum available output voltage to 2.5 V. Distortion is less than 1 percent over the entire frequency range, and does not exceed 0.1 percent from 200 Hz to 100 kHz. As explained subsequently, an optional function is provided that extends the 0.1 percent rating down to 30 Hz. The oscillator can be synchronized by an external source, and will remain locked over a frequency range of ± 5 percent. A synchronizing voltage of 5 V is required for optimum locking action.

With reference to the block diagram in Fig. 17.18, a Wien bridge oscillator arrangement is employed. A schematic diagram is shown in Fig. 17.19. The output from the amplifier is attenuated by a 600-Ω variable attenuator, followed by a step attenuator. An overall loop gain of at least unity is a requirement for oscillation. This gain figure is provided by positive feedback combined with negative feedback. We observe in Fig. 17.19 that the oscillator bridge comprises a frequency-selective network and a negative-feedback network. That is, positive feedback is developed through the frequency-determining network, C_{1A}, R_8, C_{1B}, and R_{16}. At the frequency that makes the phase of positive feedback 0^0, $X_c = R$, and the maximum ratio of output voltage is supplied to the amplifier. (See Fig. 17.20.)

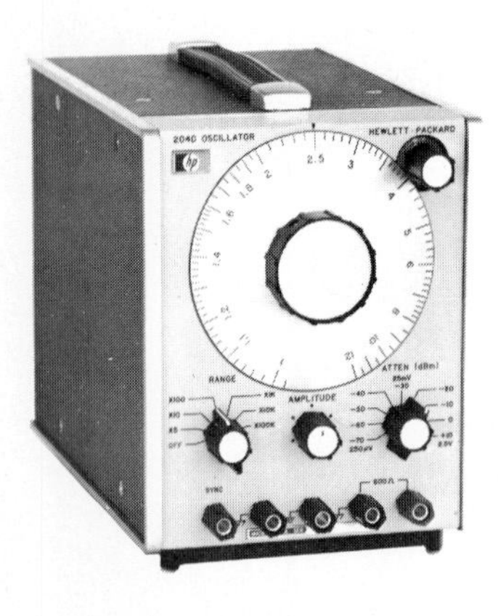

FIG. 17.17 *A wide-range solid-state oscillator.* (*Courtesy, Hewlett-Packard*)

The characteristics of the Wien bridge are such that the output voltage to the + input of the amplifier at f_o is one-third the amplitude of the positive-feedback voltage. Therefore, to maintain unity gain and oscillation, the negative-feedback network in Fig. 17.19 (R_{28}, R_{24}, and agc) is designed with a divider ratio of 2:1, in order to provide an amplifier gain of three times. The output voltage from the Wien bridge to the input of the amplifier is not always one-third of the positive-feedback voltage at all operating frequencies, nor is the amplifier gain constant at all operating frequencies. One technique used for maintaining unity gain in the oscillator circuit at all operating frequencies is to have a dynamic resistance, variable with changes in gain, in the negative-feedback network. This is accomplished by means of the peak comparator and agc circuits.

We observe that the peak comparator in Fig. 17.19 compares the negative peak of the oscillator output to a 7.2-V reference. If the output varies above or below the reference value, a difference voltage will be supplied to the agc circuit. In turn, the "dynamic resistance" of the agc circuit is a field-effect transistor in which the gate is controlled by the difference signal from the comparator. Thus the output from the oscillator amplifier is held at 7.2 V peak amplitude.

When the oscillator is first turned on, the agc system provides an amplifier gain greater than three times. Noise in the amplifier is greatly amplified, and the frequency-selective network in the Wien bridge selects the noise frequency to which it is tuned. This selected noise frequency provides

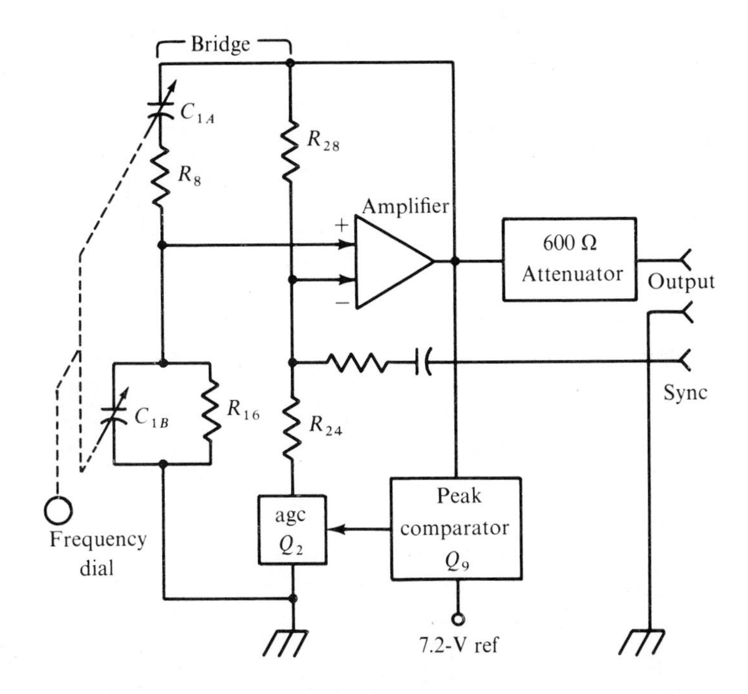

FIG. 17.18 *Model 204C block diagram.*

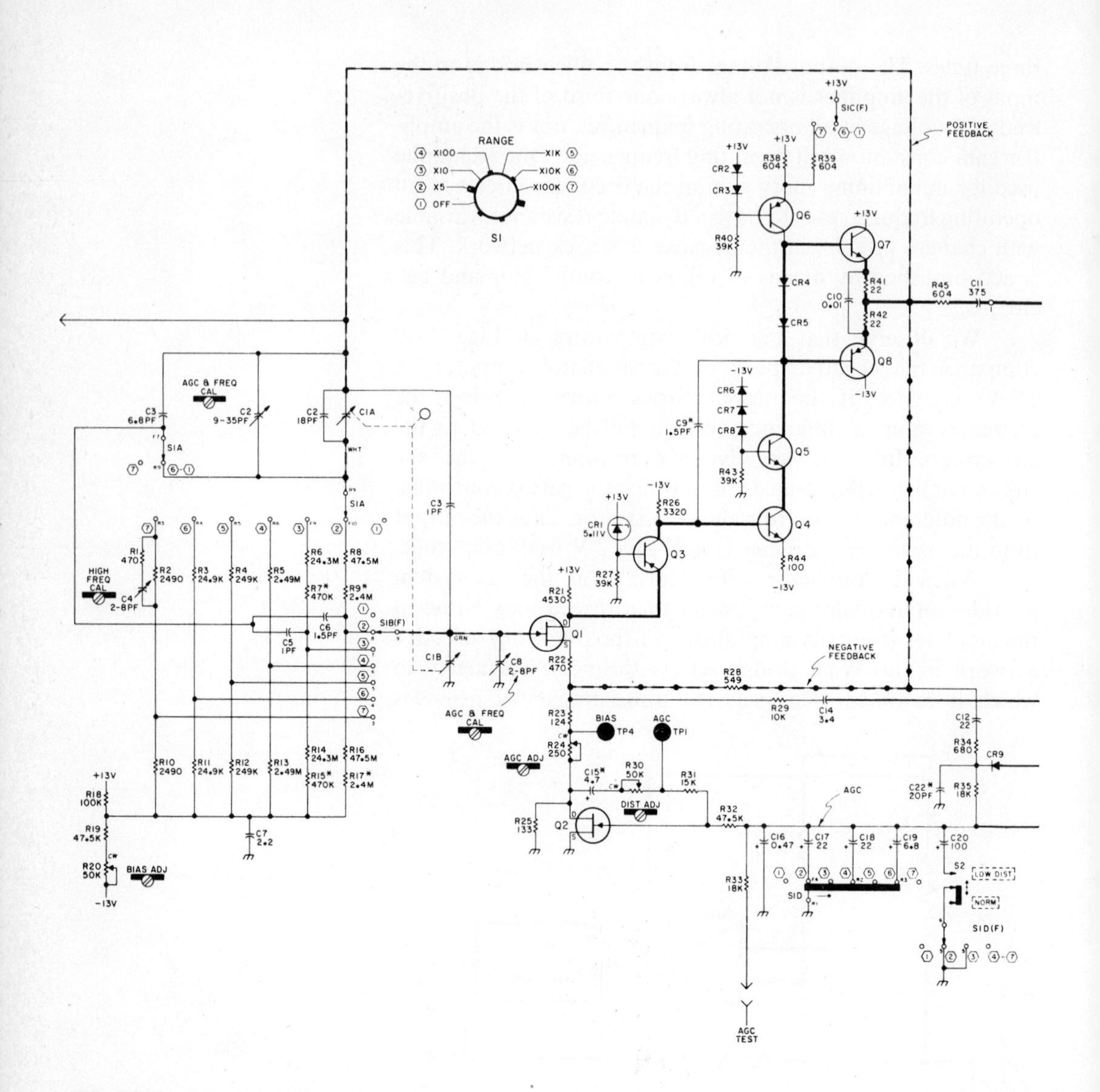

FIG. 17.19 *Schematic diagram of a wide-range oscillator.* (*Courtesy, Hewlett-Packard*)

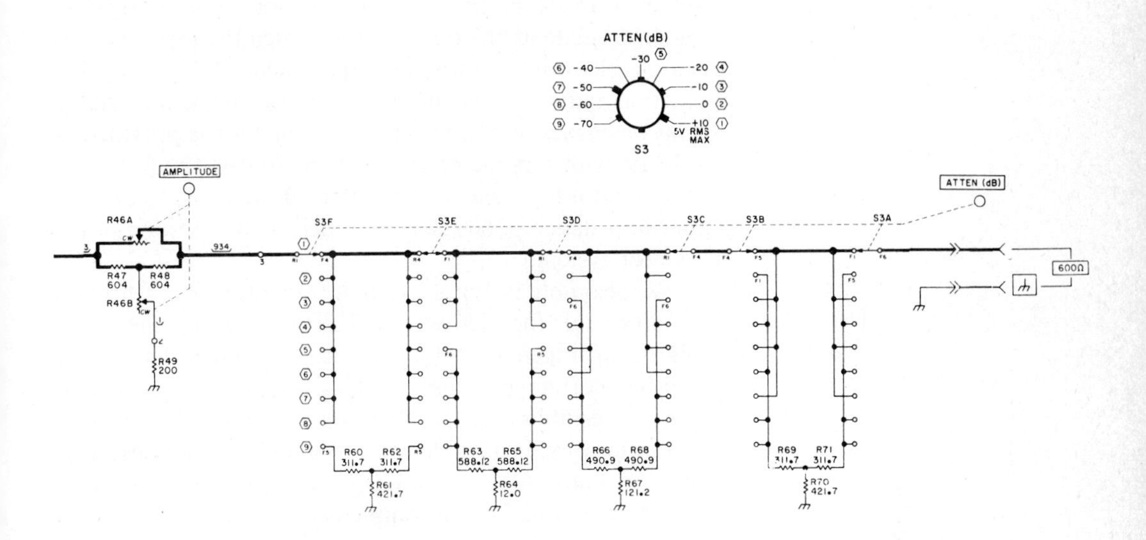
ATTEN (dB)
S3
AMPLITUDE
ATTEN (dB)
R46A
R47 604
R48 604
R46B
R49 200
S3F
S3E
S3D
S3C
S3B
S3A
R60 311.7
R62 311.7
R61 421.7
R63 588.12
R65 588.12
R64 12.0
R66 490.9
R68 490.9
R67 121.2
R69 311.7
R71 311.7
R70 421.7
600Ω

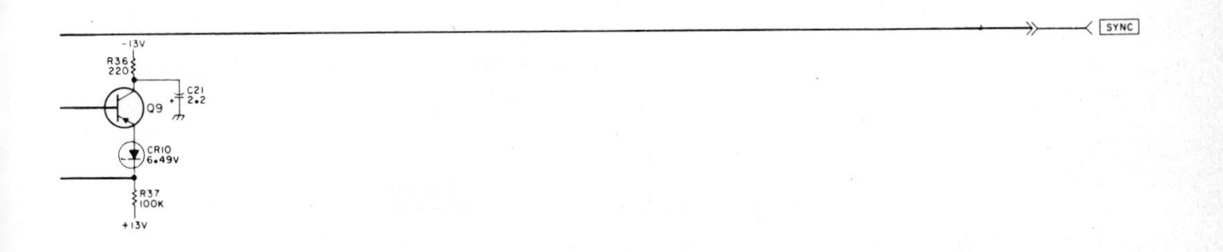
SYNC
-13V
R36 220
Q9
C21 2.2
CR10 6.49V
R37 100K
+13V

positive feedback at the amplifier section, and the system starts to oscillate at the frequency to which the instrument is tuned. As the output amplitude approaches 7.2 V peak, the agc action reduces the amplifier to three times, and stable oscillation ensues. A 600-Ω variable attenuator is provided in Fig. 17.19, with a range of greater than 10 dB. The following step attenuator has a range of 80 dB in 10-dB steps. A constant output impedance of 600 Ω is provided on any setting of the attenuator system.

We also note in Fig. 17.19 that transistors A_{1Q1} through A_{1Q8} comprise the active amplifier devices in the basic oscillator arrangement. A_{1Q1} is an N-channel fet. A_{1CR1} establishes correct dc bias for A_{1Q3}. Diodes A_{1CR6}, A_{1CR7}, and A_{1CR8} establish proper bias for A_{1Q5}. Capacitor A_{1C9} is chosen to provide a stable roll-off at high frequencies. A_{1Q6} is a current source for A_{1Q4} and A_{1Q5}. A_{1CR4} and A_{1CR5} provide appropriate biasing for the complementary output transistors A_{1Q7} and A_{1Q8}.

Note that the positive-feedback arm of the Wien bridge in Fig. 17.19 consists of tuning capacitors A_{1C1A} and A_{1C1B}, and the range-switching resistors A_{1R1} through A_{1R17}. The negative-feedback arm of the Wien bridge depends upon the ratio of the impedance of A_{1R28} to the total impedance of A_{1R23}, A_{1R24}, A_{1R25}, and A_{1Q2}. A_{1R25} reduces the effect of

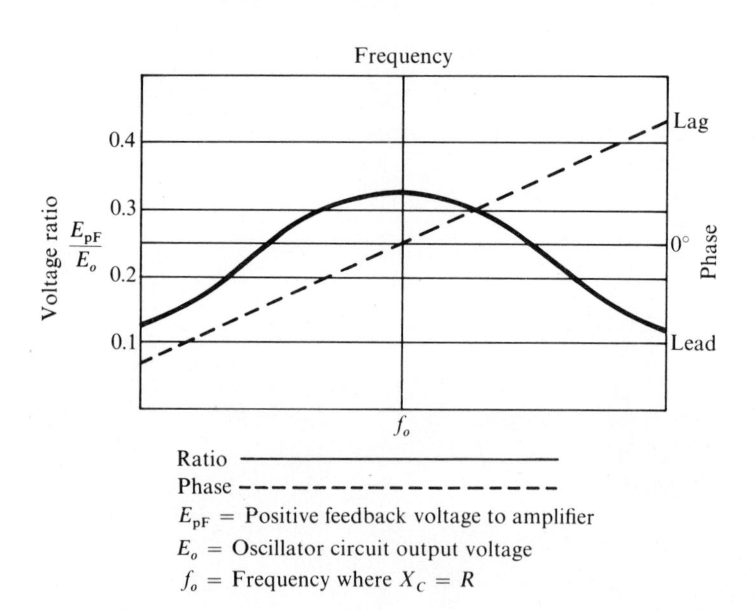

FIG. 17.20 RC *network characteristics.*

the fet A_{1Q2} and thereby increases operating stability. A_{1Q2} provides agc for this amplifier by means of impedance variation to maintain optimum feedback action.

Next, the conduction of the fet A_{1Q2} in Fig. 17.19 is controlled by the peak-detector circuit associated with A_{1Q9}. We recognize that A_{1Q9} conducts during the most negative portion of each half cycle, thereby developing a negative charge in A_{1C16} and its parallel-connected capacitors. As the output amplitude of the amplifier increases, A_{1Q9} conducts more current and A_{1Q16} becomes more negatively charged. This, in turn, makes the fet input voltage more negative, and increases its impedance to reduce the amount of negative feedback and thereby reducing the output voltage from the amplifier. At low operating frequencies, a longer agc time constant is required for minimum distortion. This is provided by C_{20} with switch S_2.

With reference to Fig. 17.21, the power supply is arranged to operate from either 115 or 230 V ac. It provides output voltages of +13 and −13 V. We observe that A_{2CR7} serves as a reference for the positive supply; the negative supply follows the positive supply. Current limiting is provided by A_{2R1} and A_{2R8}, respectively. Transistors A_{2Q2} and A_{2Q5} serve as current sources for the amplifier transistors A_{2Q3} and A_{2Q4}.

17.8 GRID-DIP METERS

A grid-dip meter is a miniaturized type of rf test oscillator with its tank coil mounted externally. It is chiefly used to measure resonant frequencies of, for example, an antenna tuning coil, a dipole antenna, or a transmission line stub. A typical grid-dip meter is illustrated in Fig. 17.22, and a circuit arrangement is depicted in Fig. 17.23. Note that a small current meter is connected in series with the grid-leak resistance. In turn, the meter indicates the grid current flow, and will indicate a lower value when power is demanded from the tank circuit.

If the tank coil of the instrument is coupled to an *LC* circuit, the reading of the grid-dip meter will decrease as the tank coil is tuned to the resonant frequency of the *LC* circuit. A grid-dip meter is used to test deenergized circuits in a manner analogous to an ohmmeter. It can also be used as a test oscillator, although it is less satisfactory for this purpose than the instruments discussed previously. In addition, it

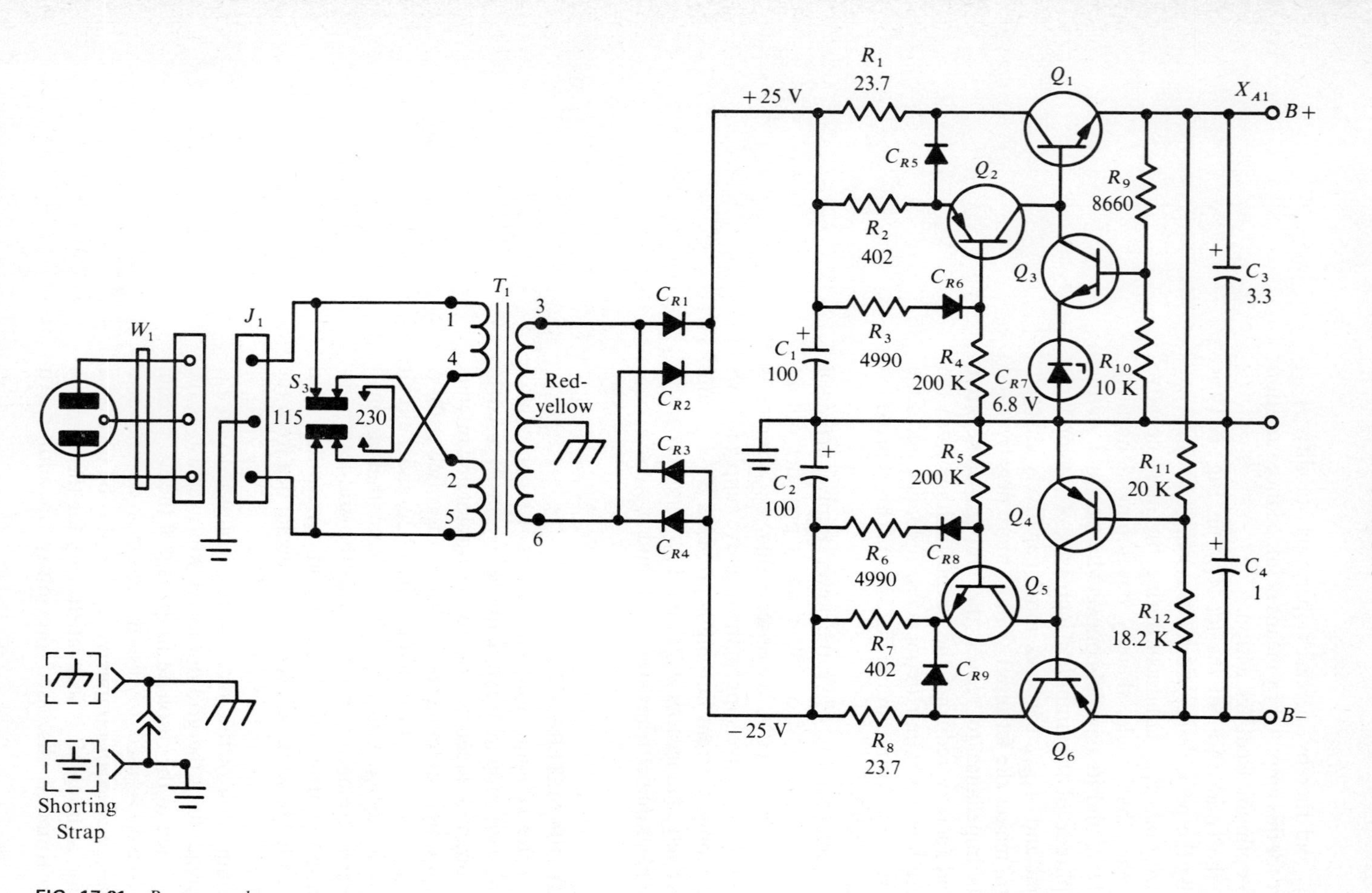

FIG. 17.21 *Power supply configuration.*

can be used to measure inductance values by measuring the frequency at which the inductance resonates with a known value of capacitance. Note that the oscillator is unmodulated. Potentiometer R_2 in Fig. 17.23*b* provides control of plate voltage so that the pointer can be brought to full scale on the meter.

Best accuracy is obtained when comparatively loose coupling is used between the instrument and the coil or device under test. If the grid-dip meter is tuned to resonance with loose coupling, the dip is small, but the detuning of the tank circuit by coupled reactance is thereby minimized. Note that a grid-dip meter can also be used as an indicating-type absorption wavemeter by reducing the plate voltage to zero. If the tank is brought into the field of an operating transmitter, for example, the rf signal picked up by the tank coil is rectified in the grid-cathode section of the tube, and the rectified current deflects the meter. In this application the grid and cathode of the triode operate as a simple diode.

The instrument depicted in Fig. 17.23 operates over a frequency range of from 1.6 to 190 MHz. Six plug-in coils are provided to accommodate this frequency range in six bands. In general, a grid-dip meter lacks the accuracy of a well-designed rf test oscillator or signal generator. However, in

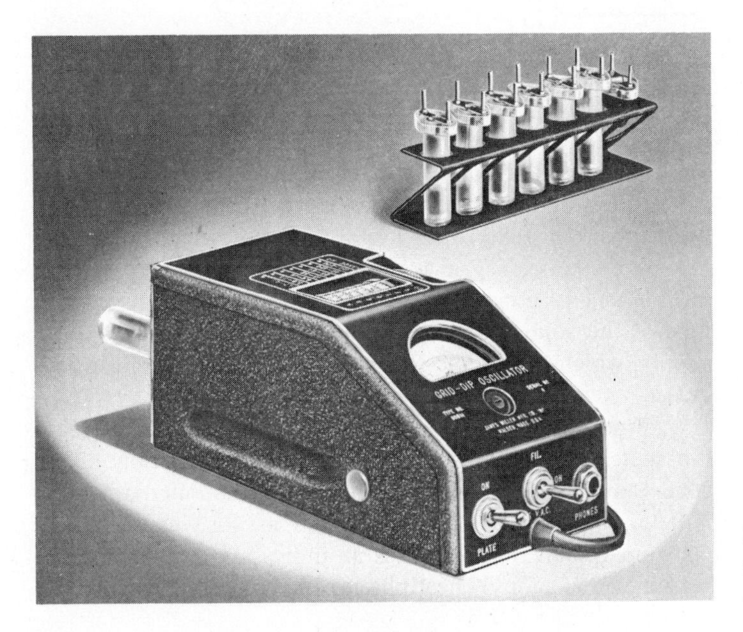

FIG. 17.22 *Appearance of a typical grid-dip meter. (Courtesy, Millen Inc.)*

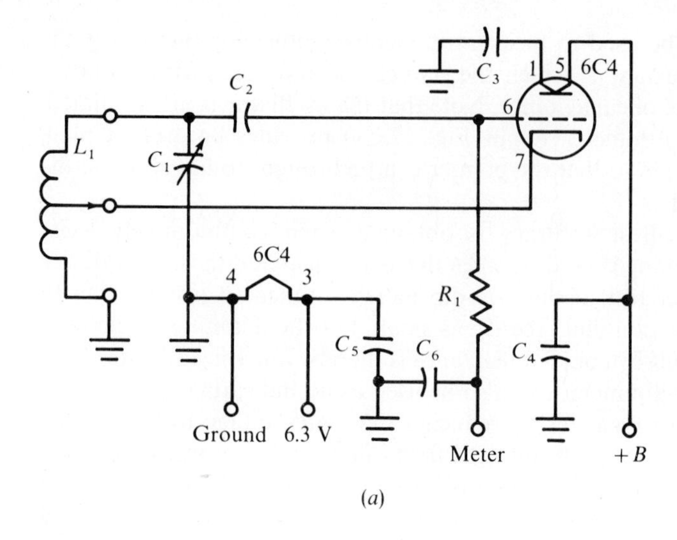

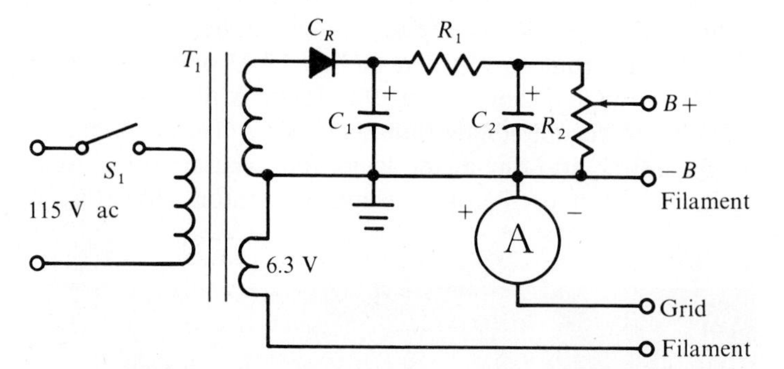

FIG. 17.23 *Grid-dip meter. (a) Oscillator section; (b) power supply and indicating meter.*

applications where high accuracy is not essential, the grid-dip meter often simplifies test procedures considerably. There has been a marked trend to the use of solid-state circuitry in grid-dip meters. Since a grid is absent in this type of circuit, the term *dip meter* is commonly employed. An important advantage of the solid-state dip meter is its self-contained power supply, which makes the instrument independent of a power outlet.

Figure 17.24 shows the circuit for a typical transistor dip meter. The 1N34A diode provides rectification of the generated rf voltage for indication by the microammeter. Note that a phone jack is provided in this version of the instrument. When operated as an indicating-type absorption wavemeter, a pair of headphones may be plugged into the

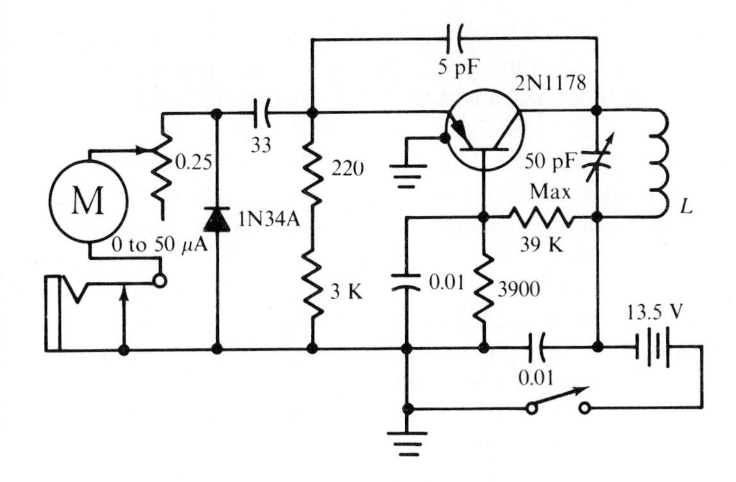

FIG. 17.24 *Transistor dip meter.*

jack in order to check the modulation of the signal to which the instrument is tuned. This feature facilitates signal identification in some situations. Some dip meters employ tunnel diodes, as exemplified in Fig. 17.25. A semiconductor is provided for rf rectification, and instrument operation is basically the same as for the transistor type of dip meter.

17.9 APPLICATIONS

A simple and useful application for a signal generator entails signal substitution for a dead local oscillator in a radio or

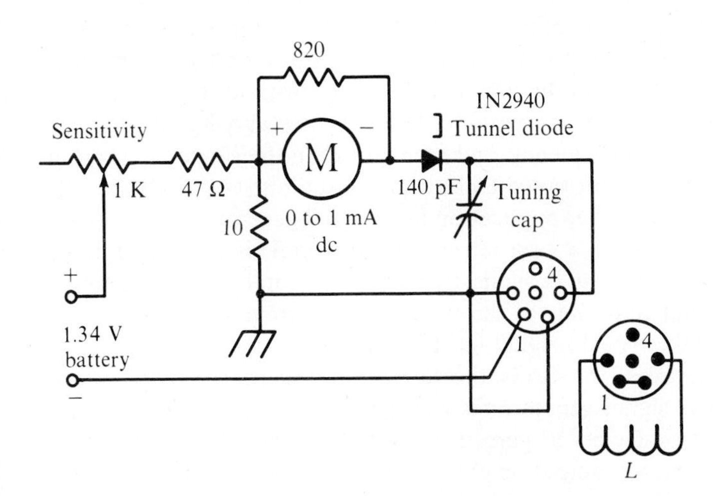

FIG. 17.25 *Tunnel-diode dip meter.*

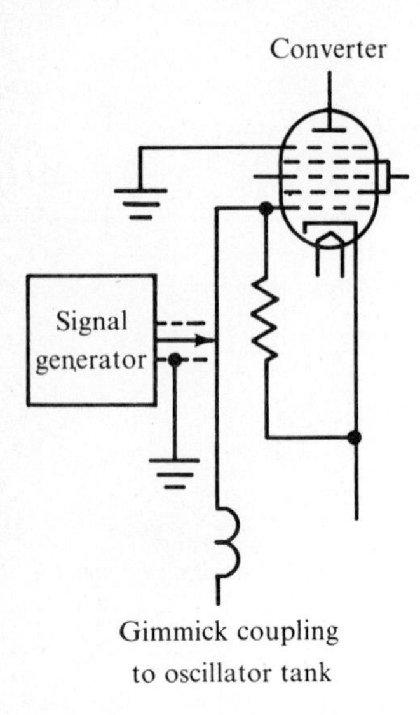

FIG. 17.26 *Substituting signal generator for local oscillator signal.*

television receiver. With reference to Fig. 17.26, the generator output cable is connected to the oscillator input lead of the converter tube (or transistor). Using maximum output amplitude from the generator, and unmodulated output, the generator is tuned to the appropriate oscillator frequency. In turn, the converter becomes operative as if the local oscillator were working, thereby facilitating preliminary troubleshooting.

Figure 17.27 shows how a signal generator is applied to determine the frequency response of an rf or if stage and its Q value. The generator signal is coupled to the grid of the input tube through a blocking capacitor to avoid drain-off of grid-bias voltage. A VTVM is connected through an rf probe to the plate of the output tube, and the primary and secondary of the output transformer are damped by shunting with 200-Ω resistors as shown in the diagram. The frequency response of the stage is plotted as a function of output voltage vs. frequency. To determine the Q value of the stage, the frequency difference between the 70.7-percent-of-maximum response points is measured. The quotient of the center frequency and the foregoing frequency difference is approximately equal to the Q value of the stage.

In many cases, the tuned circuits in a radio or television receiver are peak aligned. This is a procedure in which the resonant frequency of a circuit is checked without determination of its bandwidth. Figure 17.28 shows a test arrangement for alignment of a bandpass amplifier in a television receiver. The output from the signal generator is coupled to the grid of the bandpass amplifier tube through a coupling capacitor. A VTVM is commonly used as an indicator, and is connected at the output of the bandpass amplifier with an rf probe. Peak-alignment procedure entails setting the signal generator to a specified frequency, and then adjusting the slug or trimmer of the pertinent tuned coil for maximum output indication.

Another application for a signal generator is to measure the sensitivity of a television receiver. This is defined as the number of microvolts that must be applied to the antenna input terminals of the tv receiver in order to develop a 1-V peak-to-peak signal at the output of the video detector. Figure 17.29 shows the test arrangement that is employed. The signal generator is tuned to the same channel as the tv receiver, and 30 percent amplitude modulation is employed. Then, the output level from the signal generator is advanced

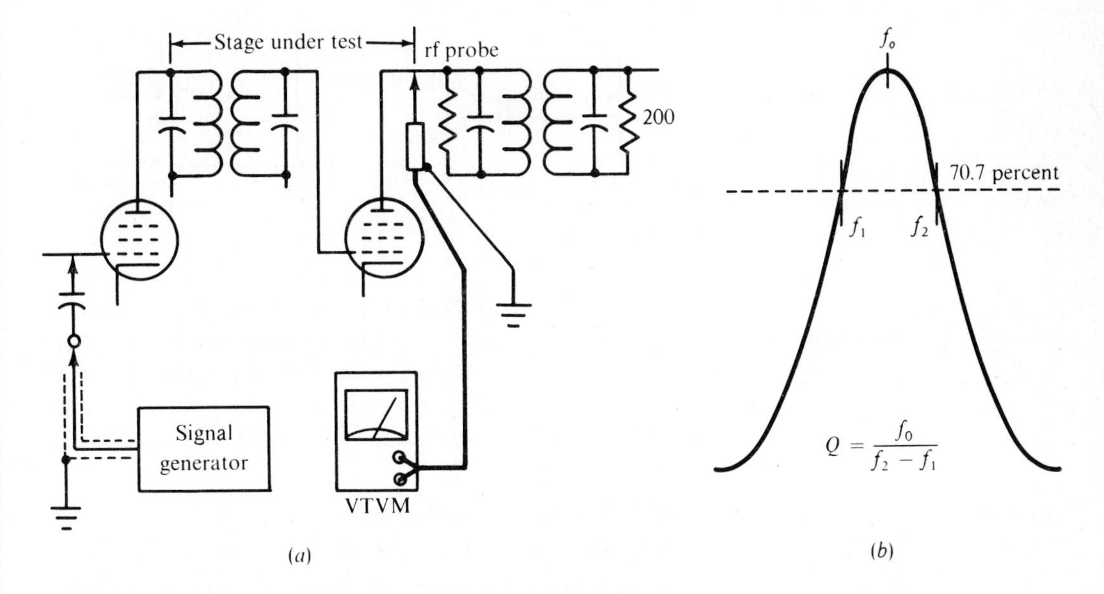

FIG. 17.27 *Frequency-response determination. (a) Test arrangement; (b) frequency-response curve, showing bandwidth and Q relation.*

until the voltmeter indicates 1 V p-p. In turn, the receiver sensitivity is equal to the number of microvolts supplied by the signal generator. Since a tv receiver has 12 operating channels, a complete test requires a sensitivity measurement on each channel. In general, we find maximum sensitivity on Channel 2, and minimum sensitivity on Channel 13.

One of the finer points to be considered in the test arrangement of Fig. 17.29 is that of impedance matching. A typical signal generator provides an output impedance of 75 Ω. Some tv receivers also have an input impedance of 75 Ω, in which case, the generator can be used to energize the receiver directly. However, most tv receivers have an input impedance of 300 Ω. Since the receiver sensitivity tends to change as the source impedance of the signal changes, it is good practice to provide a matching pad in this situation, as

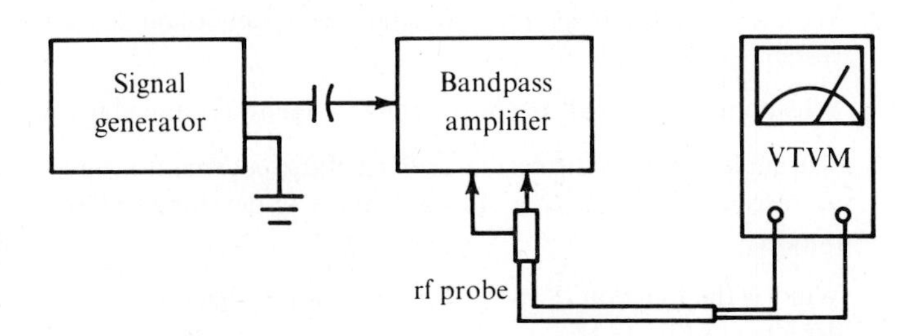

FIG. 17.28 *Test arrangement for bandpass amplifier alignment.*

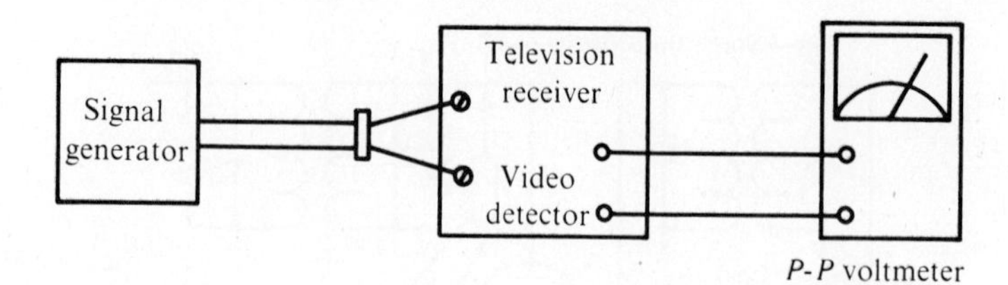

FIG. 17.29 *Measurement of television receiver sensitivity.*

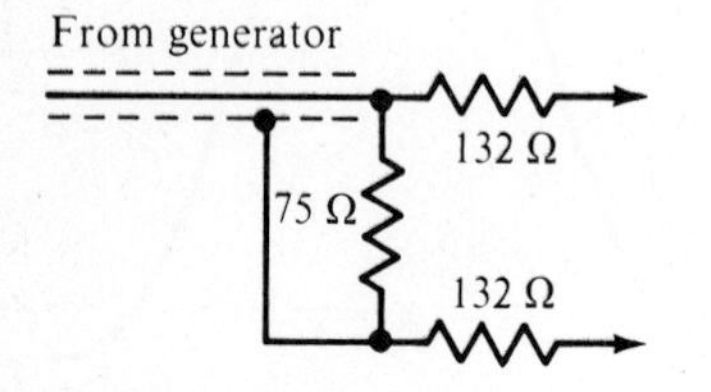

FIG. 17.30 *Matching pad for driving a 300-Ω receiver.*

depicted in Fig. 17.30. Note that when a pad is employed, a a voltage divider action is entailed with respect to the signal level. In this example, the generator signal is reduced to 0.53 of the value indicated by the generator, and a corresponding correction must be made when evaluating receiver sensitivity.

QUESTIONS AND PROBLEMS

1. What are some of the applications of rf generators in electronics?
2. Explain the terminology *standard signal generator* as it is applied to sine-wave signal sources.
3. What is the purpose of the shields in the plan of the basic signal generator depicted in Fig. 17.2?
4. Why is additional shielding sometimes provided in the attenuator section of an rf signal generator?
5. What is a practical range of an *LC* oscillator in an rf signal generator?
6. Why are the circuits in signal generators designed primarily to provide high purity of output waveform?
7. What are two important design factors for a signal generator?
8. What are two necessary requirements for an oscillator coil so that it can have a high *Q* value?
9. Why are regulated power supplies employed in signal generators?
10. Why is thermal stability necessary and how is it maintained in an rf signal generator?
11. Why do most rf signal generators employ signal-developed bias?
12. The current flow in the oscillator of an rf signal generator flows in surges. Explain how a sine waveform is developed at the output.
13. What is the function of the buffer stage in the signal generator depicted in Fig. 17.5?

14. Why does the buffer stage depicted in Fig. 17.7 operate in Class *A*?
15. Why is the plate load of the buffer stage in Fig. 17.7 such a low value?
16. What is the most important characteristic of the electron-coupled oscillator?
17. How is the accuracy of a signal rated?
18. How is frequency stabilization optimized in a signal generator from an electrical standpoint?
19. Why are most lab-type generators rated for accuracy following a specified warm-up period?
20. How are generators designed to prevent frequency shifts due to load conditions?
21. Why do screen-grid tubes provide better buffering than triode tubes?
22. Why isn't it practical to directly modulate an oscillator for percentages greater than 30 percent?
23. How are high percentages of modulation developed in signal generators?
24. What are the functions of the output meter in the signal generator depicted in Fig. 17.12?
25. What is the purpose of the rf chokes in the lines in the circuit shown in Fig. 17.12?
26. What is the purpose of the shielded plate between the series and parallel banks of resistors in Fig. 17.12?
27. Of what importance is the output cable of a signal generator to the operation of the attenuator?
28. Why must the output cable of a signal generator be terminated in its characteristic impedance?
29. Discuss the operation of a basic grid-dip meter that uses a triode vacuum tube as the oscillator.
30. What are four applications of grid-dip meters?
31. How is a grid-dip meter circuit operated when it is used as an indicating-type absorption wavemeter?
32. What is the purpose of the phone jack in the circuit shown in Fig. 17.20?

18

SWEEP-FREQUENCY GENERATORS

18.1 BASIC REQUIREMENTS

Sweep-frequency (or sweep) generators are frequency-modulated (fm) generators that are used to automatically plot frequency-response curves on the screen of an oscilloscope. An rf carrier is frequency modulated at a 60-Hz rate in most sweep generators. Figure 18.1 shows the distinction between a cw wave and an fm wave. Note that the sweep signal has a constant amplitude, and that its instantaneous frequency frequency changes continually. That is, the center frequency of the waveform deviates periodically between an upper and a lower frequency limit. Let us consider some typical examples.

The ratio detector in the sound section of a television receiver has a center frequency of 4.5 MHz. To display the frequency-response curve of this receiver section, the center frequency of the sweep generator would be set to 4.5 MHz. A deviation (or sweep width) of 75 or 100 kHz would be used in this situation, since a ratio detector employs a comparatively narrow-band circuit. Next, if we wish to display the frequency-response curve of an rf tuner, the center frequency of the sweep generator might be set to 214 MHz, with a sweep width of 10 or 15 MHz, since an rf tuner in a television receiver employs comparatively large bandwidth.

18.2 BASIC SWEEP GENERATOR CHARACTERISTICS

A typical sweep generator used in television service procedures has a range of center frequencies from 40 kHz to 214 MHz. Continuous coverage is not provided, however, since the sweep generator can be switched to each of the 12 tv channels, and to a lower frequency band that is tunable from 40 kHz to 50 MHz. Sweep generators that are designed for

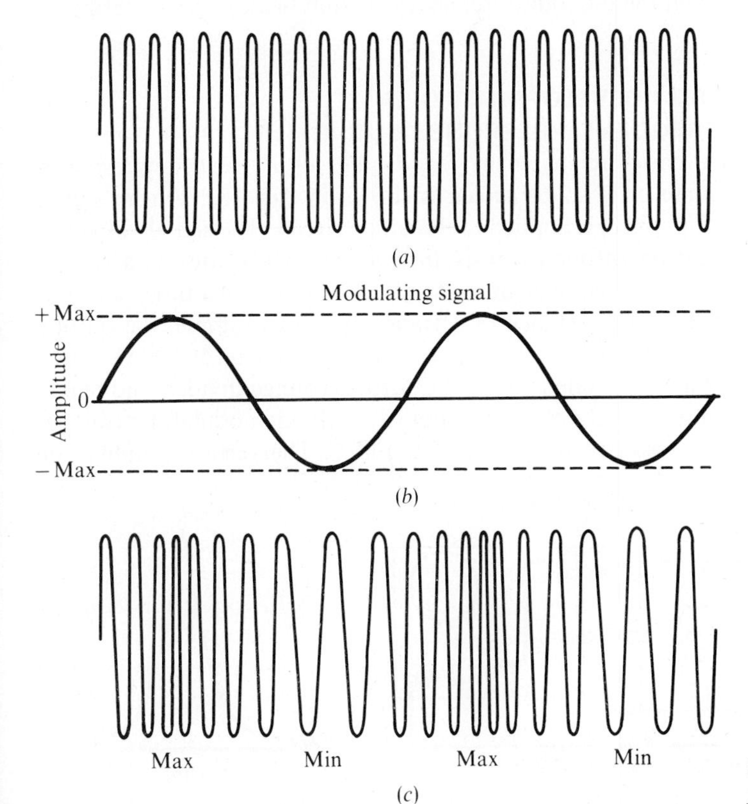

FIG. 18.1 *Development of a frequency-modulated signal.*

both tv and fm service also provide a band from 88 to 108 MHz, which is continuously tunable. The uhf sweep generators are designed as separate instruments, because uhf circuitry employs cavity resonators instead of lumped *LC* tanks. Most sweep generators have a maximum output of approximately 0.1 V rms. Simple attenuators are provided to adjust the output level. Sweep generators ordinarily are not provided with output meters and calibrated attenuators.

Unless a sweep generator has an output with uniform amplitude (flat output) over the swept band, the displayed response curve will be distorted accordingly. Designers of sweep generators thus employ oscillator circuitry that provides essentially flat output. It is also desirable that the output have good waveform with a minimum harmonic content. When beat-frequency oscillators are employed, extensive buffering must be utilized to maintain good waveform at low frequencies. In addition, sum and feedthrough frequencies must be removed from the difference frequency by suitable low-pass filtering.

18.3 DEVELOPMENT OF A FREQUENCY-RESPONSE CURVE

A basic arrangement for display of a frequency-response curve on a scope screen is shown in Fig. 18.2. An fm signal (Fig. 18.1*c*) is applied to the tuned circuit. The signal voltage is applied through a resistance *R*, which operates as a source resistance equivalent to the plate resistance of a tube, or to the collector resistance of a transistor. For example, the output resistance of a common-emitter stage may be 50,000 Ω. Thus, *R* and the tuned circuit *LC* form a voltage divider, and maximum signal voltage is applied to the demodulator probe at the resonant frequency of *L* and *C*. The vertical amplifier of

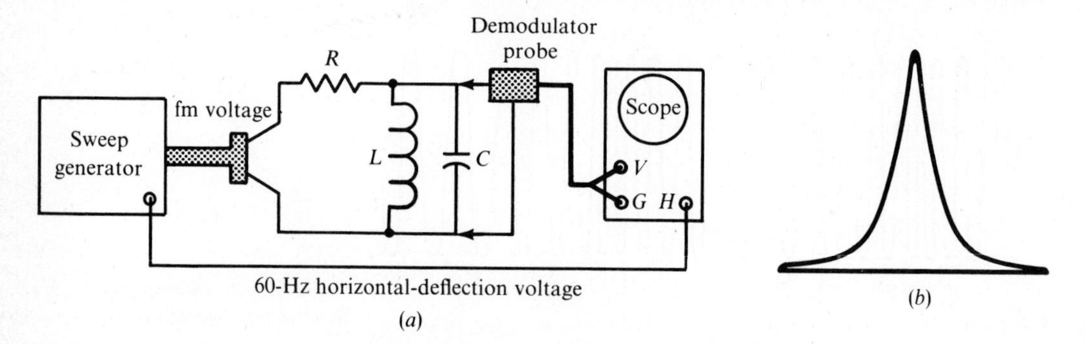

FIG. 18.2 *Display for a frequency-response curve. (a) Arrangement of test instruments; (b) screen pattern.*

the scope is then energized by a changing voltage that is determined by the frequency-response characteristic of the tuned circuit.

Note in Fig. 18.2*a* that the horizontal amplifier of the scope is energized by a 60-Hz sine-wave voltage supplied by the sweep generator. Although 60-Hz sawtooth deflection can be utilized, the test procedure is simplified by use of 60-Hz sine-wave deflection. The reason is evident from the relations depicted in Fig. 18.1. That is, the sweep-signal deviation is produced by a 60-Hz sine-wave voltage. Accordingly, if the scope is deflected by a 60-Hz sine-wave voltage, the resulting pattern is automatically synchronized. Moreover, sine-wave deflection provides horizontal deflection intervals that are proportional to frequency. On the other hand, if we employ 60-Hz sawtooth deflection, the horizontal baseline intervals will not be entirely proportional to frequency. Designers commonly employ a 60-Hz frequency-modulating voltage in sweep generators for reasons of economy.

We observe also in Fig. 18.2*b* that the pattern includes a baseline trace under the response curve. This horizontal trace is obtained by disabling the sweep oscillator for one-half of each modulating cycle shown in Fig. 18.1. To understand this circuit action, let us first consider the pattern that is displayed when the scope is operated on 60-Hz sawtooth deflection. In such case the response curve is displayed twice, in mirror-image form, as illustrated in Fig. 18.3; the second image is split, and part of it is contained in the retrace in this example. The reason for this form of display is seen in Fig. 18.1, where the circuit under test is swept twice each $\frac{1}{60}$ s. The sweep signal proceeds from its low-frequency limit to its high-frequency limit, and then back to its low-frequency limit.

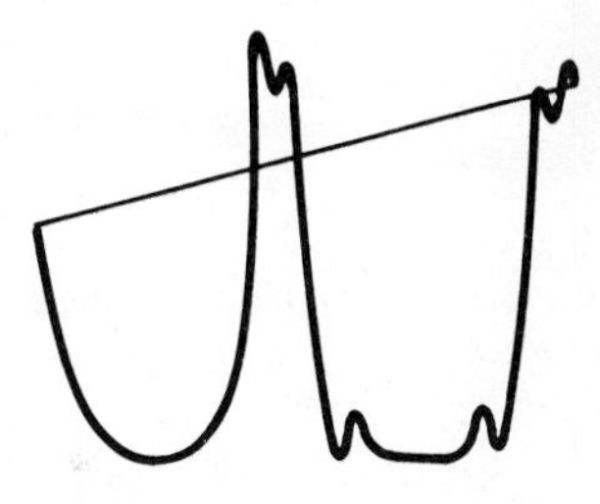

FIG. 18.3 *A response curve displayed on a 60-Hz sawtooth deflection.*

Next, if we employ 60-Hz sine-wave deflection, the response curve is displayed twice, as before; however, because of the reversal of the deflection voltage each $\frac{1}{120}$ s, the two curves are superimposed as seen in Fig. 18.4. This form of display is quite usable; however, perfect layover of the two curves is seldom obtained, with the result that there is usually a slight double-image effect which the operator may find confusing. Also, when a wide-band circuit is being swept, the available deviation in the sweep generator may be insufficient to display the curve down to its zero level. For these reasons, it is customary to blank the return trace and thereby develop the zero-volt level in the pattern as a baseline.

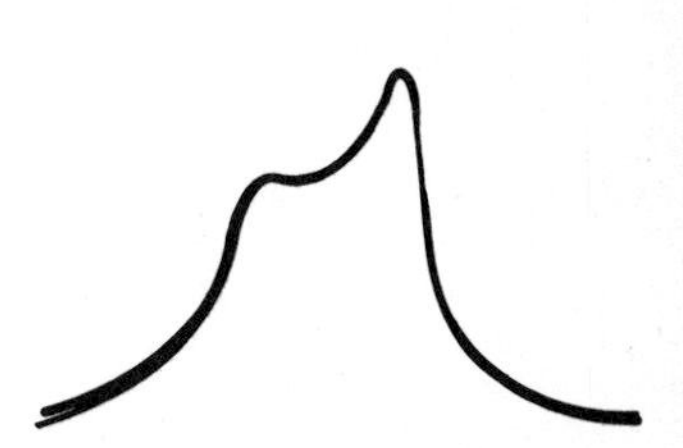

FIG. 18.4 *A response curve displayed on a 60-Hz sine-wave deflection.*

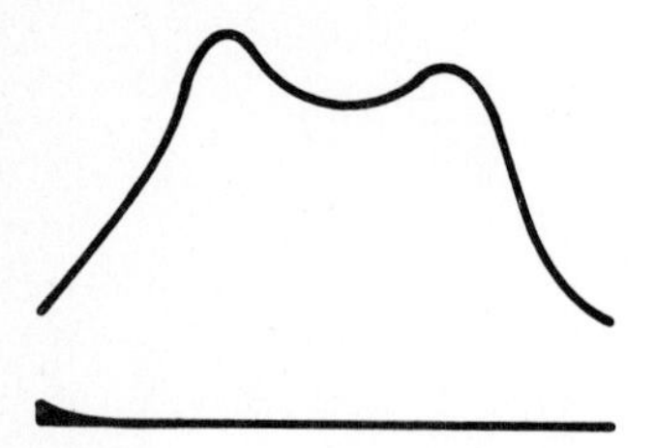

FIG. 18.5 *An rf response curve.*

When a wideband circuit is swept, the ends of the skirts may be clipped, as illustrated in Fig. 18.5, due to limited deviation. Nevertheless, when the second curve, or retrace, is blanked to form a zero-volt reference line, the pattern is practically as useful as if the skirts were completed. That is, the amplitude of the waveform is evident, which permits the operator to measure stage gains, if desired. We also find that a zero-volt reference line is very useful to facilitate the location of 50-percent- and 70-percent-of-maximum amplitude points on the sides of a wideband response curve. That is, bandwidth is measured between the 50-percent-of-maximum points on a tv response curve, for example. If the zero level were absent in Fig. 18.5, it is evident that the 50-percent-of-maximum level would not be clearly apparent.

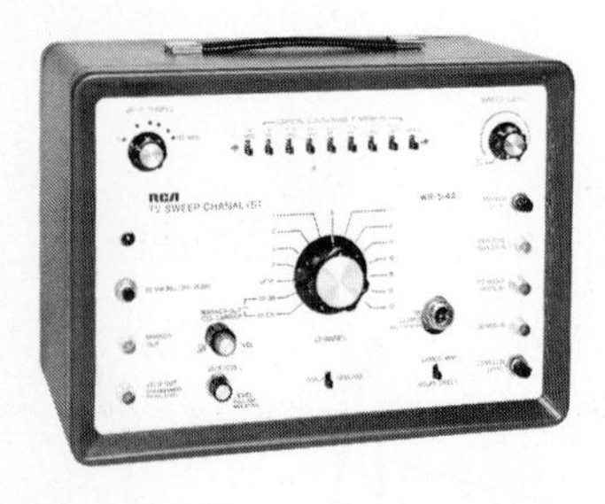

FIG. 18.6 *Appearance of a typical sweep generator. (Courtesy, Radio Corporation of America)*

18.4 SWEEP SIGNAL GENERATION

A sweep generator for tv and fm service is illustrated in Fig. 18.6. This instrument is typical of the group that employs an electromechanical sweep modulator. Figure 18.7 depicts the plan of the frequency-modulator unit. A capacitor with circular concentric "plates" is connected across the oscillator tank coil. The capacitance value is varied at a 60-Hz rate by means of an electromechanical vibrator arrangement. To adjust the deviation (sweep width), a rheostat is provided to set the current value for the moving coil. This arrangement can provide up to 12-MHz sweep width on the vhf tv channels. Greater deviations are impractical because the L/C ratio becomes too small, and the output amplitude is not sufficiently uniform.

Another widely used method of sweep signal generation

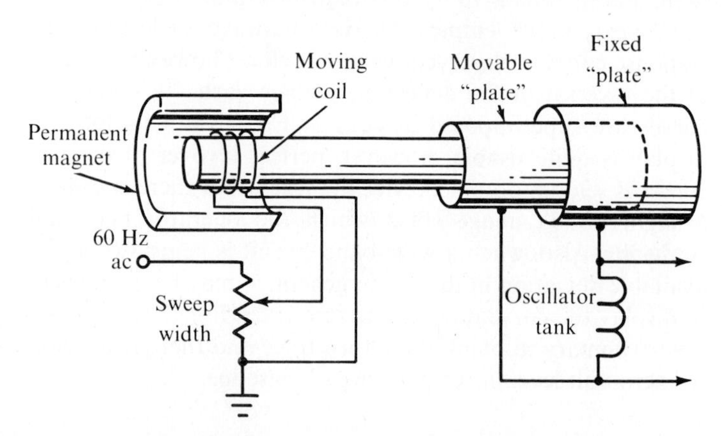

FIG. 18.7 *Plan of an electromechanical fm modulator unit.*

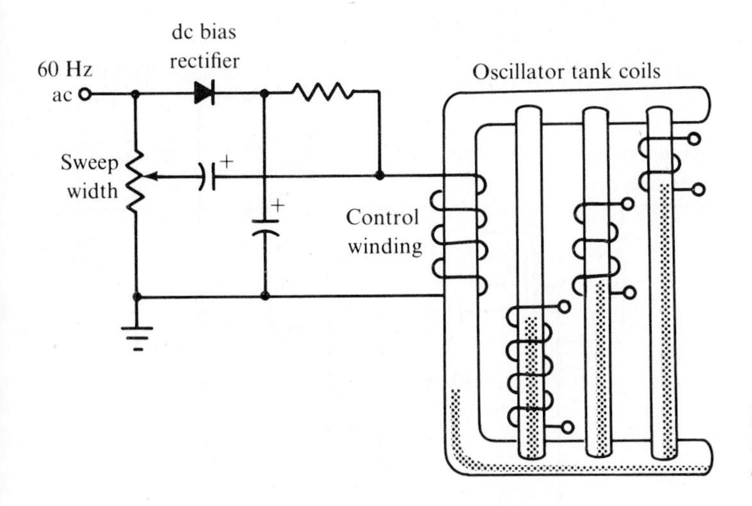

FIG. 18.8 *Current-controlled inductor (saturable reactor) that is another form of fm modulator.*

employs a saturable reactor, as depicted in Fig. 18.8. The tank coils are wound on a ferrite core, and the permeability of the core is varied by a 60-Hz magnetic field from a control winding. We recall that a *B-H* curve has its maximum linearity at medium flux densities. Therefore, a dc bias current is passed through the control winding, in addition to the 60-Hz ac current. Practically linear frequency deviation is thereby obtained. The chief advantage of the saturable-reactor method of sweep signal generation is its absence of moving parts.

Next, let us consider the semiconductor frequency-modulator unit shown in Fig. 18.9. A reverse-biased varicap diode is connected across the grid tank coil in the oscillator circuit. When the bias voltage for the diode is varied at a 60-Hz rate, the oscillator frequency is swept back and forth accordingly. This method has the same advantage as the saturable-reactor method in that no moving parts are employed in the frequency-modulator section. Varicap diodes are used as modulators in numerous fm sweep generators, and are used to some extent in tv sweep generators.

The sweep oscillators discussed above are of the single-ended type. At high operating frequencies, it is advantageous to employ a push-pull oscillator configuration, as depicted in Fig. 18.10. It will be observed that the interelectrode capacitances of the duo-triode are effectively connected in series with respect to the tank. Thus the total capacitance in shunt with the tank is reduced, and a higher *L/C* ratio can be realized. Push-pull operation also has the advantage of cancellation of the second harmonic in the generated waveform. Although

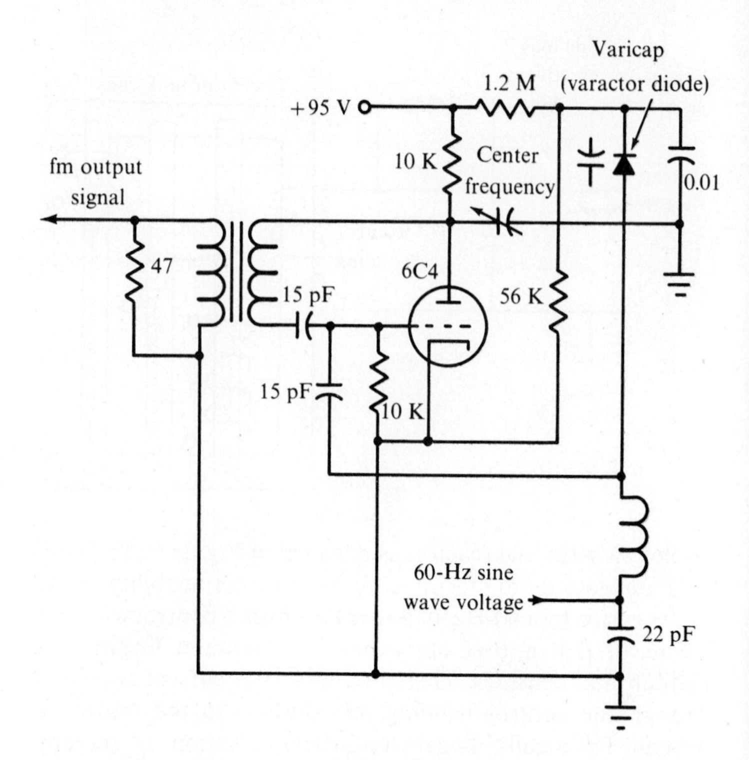

FIG. 18.9 *Semiconductor-diode frequency-modulator arrangement.*

third-harmonic distortion is increased by push-pull operation, the second harmonic is the most troublesome.

Note how the oscillator is disabled for one-half of each modulation period in Fig. 18.10. This retrace-blanking action is accomplished by feeding a 60-Hz square-wave signal to the grids of the oscillator tube. In turn, the duo-triode is biased beyond cutoff for one-half of each modulation period. The 60-Hz square wave is obtained by means of a diode clipper that changes a 60-Hz sine-wave voltage into a semisquare wave. When it is desired to disable the blanking action, the 60-Hz square-wave voltage is short-circuited to ground by means of switch S_2. Thereupon, no bias voltage is supplied to the oscillator tube, and the circuit operates in the conventional manner with signal-developed bias.

18.5 rf OUTPUT CONSIDERATIONS

It is as essential to maintain a constant signal amplitude in the output system as it is to generate the signal. With reference to Fig. 18.10, we observe that the push-pull (double-ended) rf

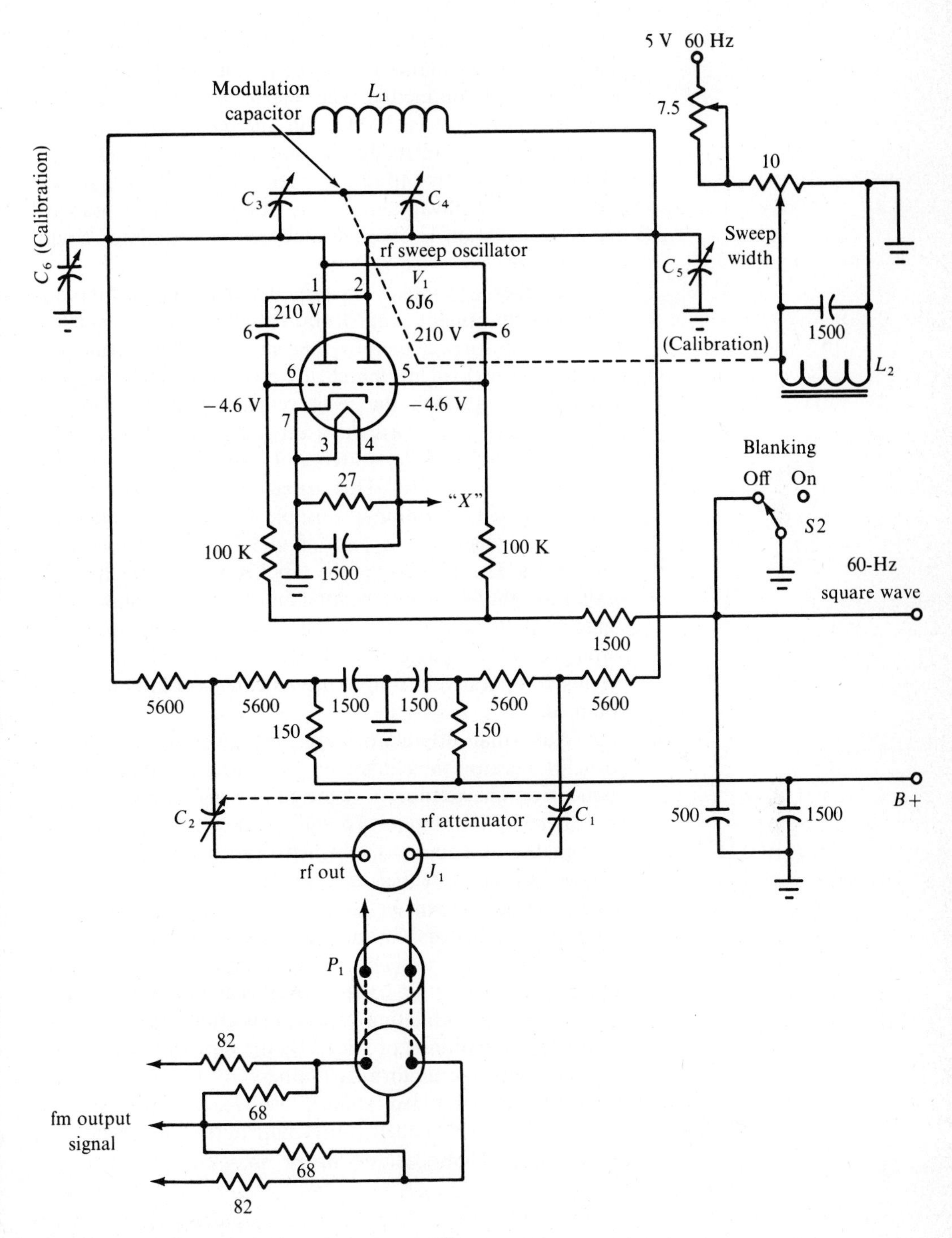

FIG. 18.10 *Configuration of a push-pull sweep oscillator.*

sweep signal is fed into a twin-conductor coaxial cable. The characteristic impedance of the coaxial output cable is 100 Ω (50 Ω for each conductor). The reader may show that if the cable is connected to a 300-Ω load, the resistive network terminates the cable correctly. We observe that the cable conductors are connected to resistive pads at each end, as seen in Fig. 18.10. The advantage of this design is that a practical degree of isolation is obtained between the oscillator circuit and the load.

Another advantage of resistive padding is that any reflections that might occur from the load (due to mismatch) tend to be absorbed, thereby minimizing the impairment of flatness due to load mismatch. In other words, the original signal passes through the resistive pad once, whereas any reflected signal must pass back through the pad once more before it returns to the source end. Note the rf attenuator depicted in Fig. 18.10. This capacitive attenuator is not designed as a conventional variable capacitor. Instead, the attenuator consists of a pair of metal tubes operating as waveguides. The diameter of the tubes is comparatively small, so that the guides operate beyond cutoff, thus attenuating the signal rapidly down the guide. Movable electrodes are employed in the guides to effectively change their length; by this comparatively simple means extensive attenuation is obtained.

Waveguide attenuators are practical only at high frequencies, because their diameter becomes prohibitive at low frequencies. The attenuator depicted in Fig. 18.10 is used only for the vhf signal outputs. To change the center frequency of the signal, tank coil L_1 is switched in the same manner that coils are switched in the tuner section of a tv receiver. The center frequency can be adjusted precisely by means of calibration capacitors C_5 and C_6. Two sweep-width controls appear in the diagram. The 7.5-Ω rheostat is a maintenance adjustment, whereas the 10-Ω rheostat is an operating control. A twin coaxial modulating capacitor is employed. Feedback to sustain oscillation is provided by the 6-pF capacitors.

To check the uniformity (flatness) of the output from a rf sweep generator, a demodulator probe and oscilloscope are used. When double-ended (push-pull) output is provided, a double-ended demodulator probe is used, as shown in Fig. 18.11.

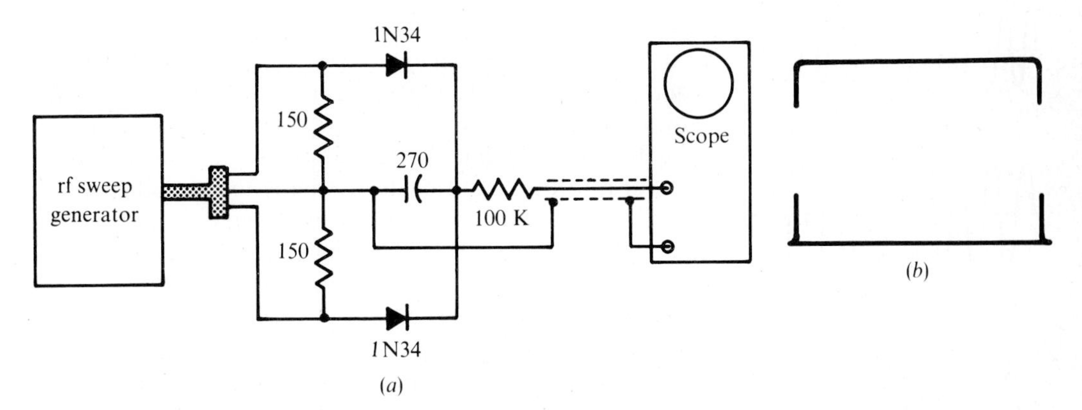

FIG. 18.11 *Checking an rf sweep signal. (a) Double-ended demodulator probe used to check the rf sweep signal; (b) pattern displayed with a flat output.*

18.6 VIDEO-FREQUENCY SWEEP SIGNALS

It is quite possible to generate sweep signals at if frequencies in the same manner as rf signals. However, some sweep generators employ the beat-frequency method to generate if sweep signals. In the case of video-frequency sweep signals, beat-frequency oscillators are mandatory, because of the extremely large change in instantaneous frequency that is required. As noted previously, a typical video-frequency sweep signal may have a 40-kHz low-frequency limit and a 5-MHz upper frequency limit. It is obviously impossible to generate this form of sweep signal by means of a simple variably tuned tank. Instead, the beat method must be employed.

The basic plan of a beat-frequency arrangement for a sweep generator is depicted in Fig. 18.12. The fixed rf oscillator might operate at 100 MHz, while the variable rf oscillator is deviated from 100 to 105 MHz. In turn, the heterodyne detector develops the difference frequency between the two input signals. This difference signal swings nominally from zero frequency to 5 MHz. In practice, however, there is no output below approximately 35 kHz because the beat oscillators have residual coupling that causes pulling below 40 kHz, and locking below 35 kHz. Of course, a lower frequency sweep output can be obtained in a specialized generator by providing extensive buffering between the oscillators and the heterodyne detector.

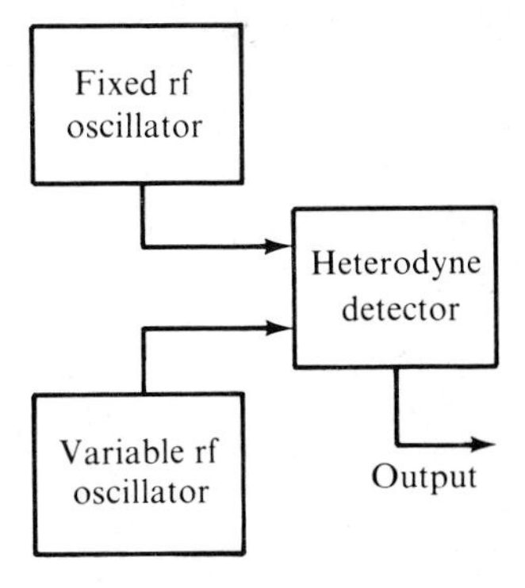

FIG. 18.12 *Basic plan of a beat-frequency sweep oscillator.*

Figure 18.13 shows the action of a mixer, or heterodyne detector. Rectifier action develops one polarity of the beat waveform, and suppresses the other polarity. In turn, the

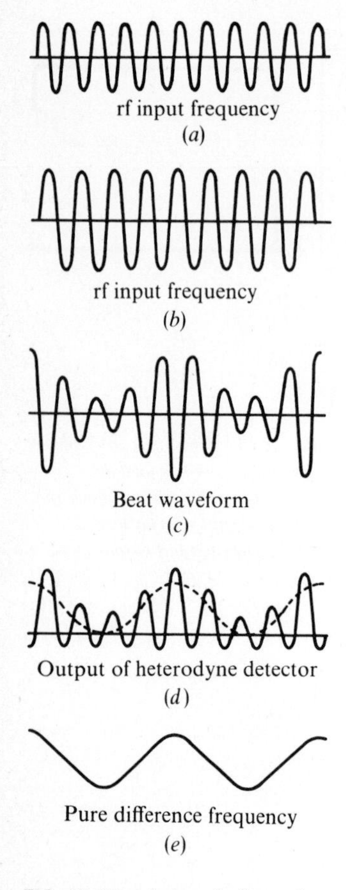

FIG. 18.13 *Action of a heterodyne detector or mixer.*

output from the heterodyne detector has an average value that provides the frequency of the envelope in the beat waveform. Finally, the output waveform is fed to a low-pass filter which removes the high-frequency components associated with the rectified pulses, and supplies the pure difference waveform to the circuit or device under test. In other words, heterodyne detection develops a sum frequency as well as a difference frequency. Feedthrough frequencies from the oscillators are also removed by the low-pass filter (see Fig. 18.14).

It is very desirable to employ oscillator configurations in Fig. 18.12 that generate as pure a sine waveform as is practical. The reason for this design feature is that harmonics will cross-beat in the heterodyne detector, just as the fundamentals cross-beat with each other. If the oscillators have significant harmonic content, the output signal from the heterodyne detector will be contaminated with spurious frequencies, and its flatness will be impaired to some extent. Therefore, high Q tank circuits are employed, with light loading on the oscillators. Even if the oscillators generate a perfect sine waveform, the output signal from the heterodyne detector may not have a good sine waveshape. Let us see why this is so.

With reference to Fig. 18.15, we find that when two frequencies with the same amplitude beat through an ideal heterodyne detector, the output waveform has an envelope with the shape of half-sine waves. In turn, the average value of rectified waveform does not have a sinusoidal waveshape, but follows the contour of the beat envelope. To minimize this source of waveform distortion, a sweep generator applies signals with unequal amplitudes to the heterodyne detector, as depicted in Fig. 18.16. That is, if one signal has substantially greater amplitude than the other, the beat output waveform will be more nearly sinusoidal.

18.7 COMPLETE TV-FM SWEEP GENERATOR

The configuration for a complete tv-fm sweep generator is shown in Fig. 18.17. This instrument employs a push-pull oscillator arrangement with switched tank coils to accommodate the tv vhf channels. One of these coils is also employed on the beat-frequency function. That is, the output of V_1 beats with the output of V_2A to cover the range from 40 kHz to 50 MHz. V_3 operates as a heterodyne mixer, and V_4 provides

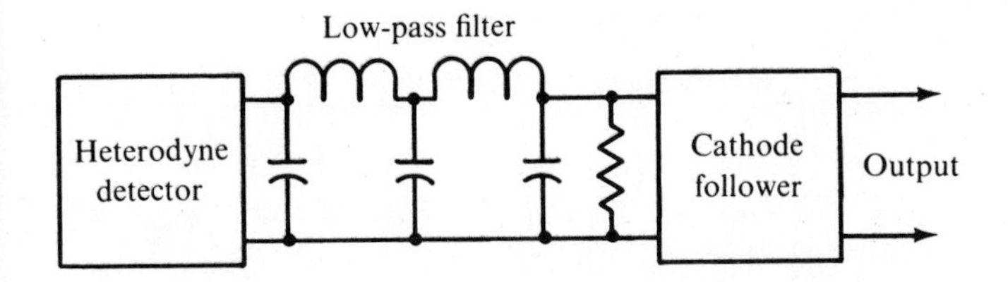

FIG. 18.14 *Diagram showing low-pass filter that removes all mixer products except the difference frequency.*

cathode-follower isolation of the output system. V_2B is a clipper; it forms a semisquare wave from the 60-Hz sine-wave source for blanking the retrace in the displayed pattern.

It is evident that the phase of the blanking voltage must be the same as that of the fm modulating voltage; otherwise, the pattern will be clipped more or less at one end, and the curve will not "fit" properly on the baseline. Figure 18.5 illustrates the proper phase relations. Since the modulating and blanking circuits entail both inductance and capacitance, a phase adjustment must be provided to bring the operation of these circuits into proper phase relation. This adjustment is made by means of R_{39}, which operates in combination with C_{34} in a phase-shifting configuration. Note that this instrument does not provide a 60-Hz source for horizontal deflection of the associated scope. In other words, the scope must have a built-in 60-Hz sine-wave source for horizontal deflection during visual alignment procedures.

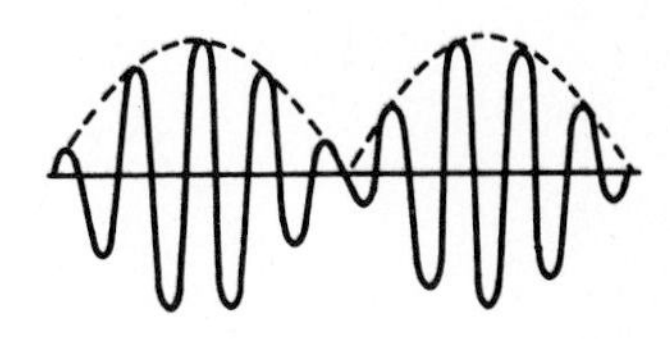

FIG. 18.15 *Diagram showing that beat envelope is a half-sine wave when the two frequencies have the same amplitude.*

Note also that J_2 in Fig. 18.17 provides a sample of the sweep signal for use by auxiliary equipment. The chief application for this sampled signal is in the operation of auxiliary frequency-marking devices. It is evident that a frequency-response curve does not provide full information unless its center frequency can be measured and its bandwidth determined. This requirement entails some form of irregularity which is introduced into the pattern at a point corresponding to a particular frequency. We shall see how this is accomplished in the following section.

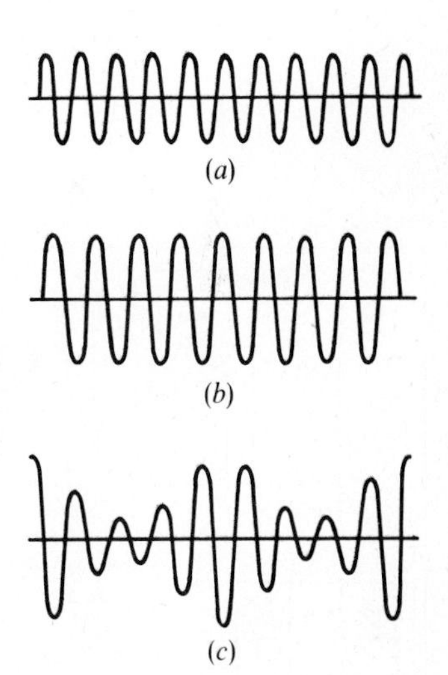

FIG. 18.16 *Diagram showing that beat envelope approximates a sine wave if one of the two beating frequencies has a greater amplitude than the other.*

18.8 MARKING THE RESPONSE CURVE

Beat markers, also called pips or birdies, are the most common form of frequency indication in visual-alignment procedures. Figure 18.18 illustrates a beat marker. The marker is produced by loosely coupling the output from a marker generator into the input of the receiver or circuit under test. In a typical situation, we merely place the output leads from the marker

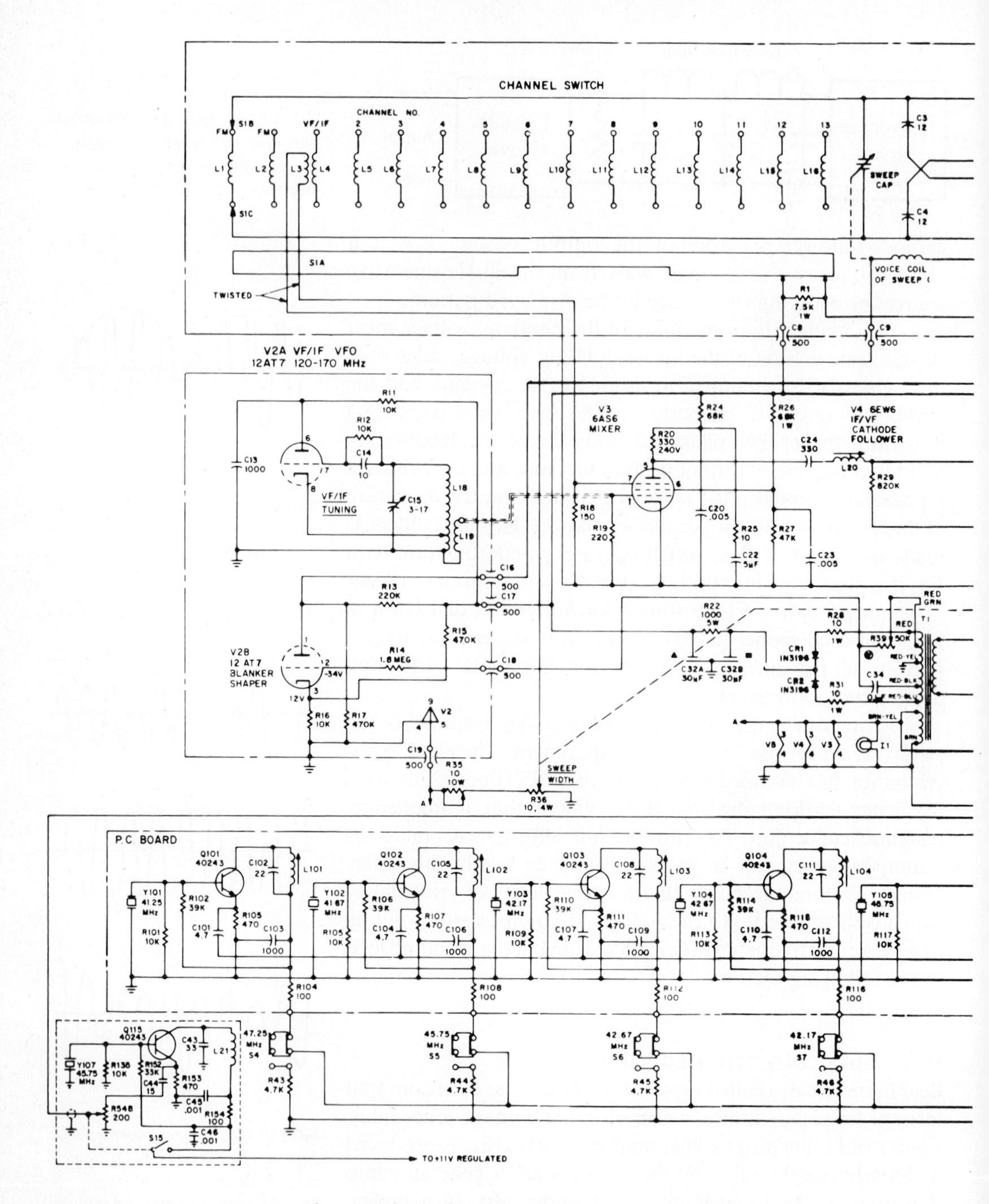

FIG. 18.17 *Circuit arrangement for a complete tv-fm sweep generator.*

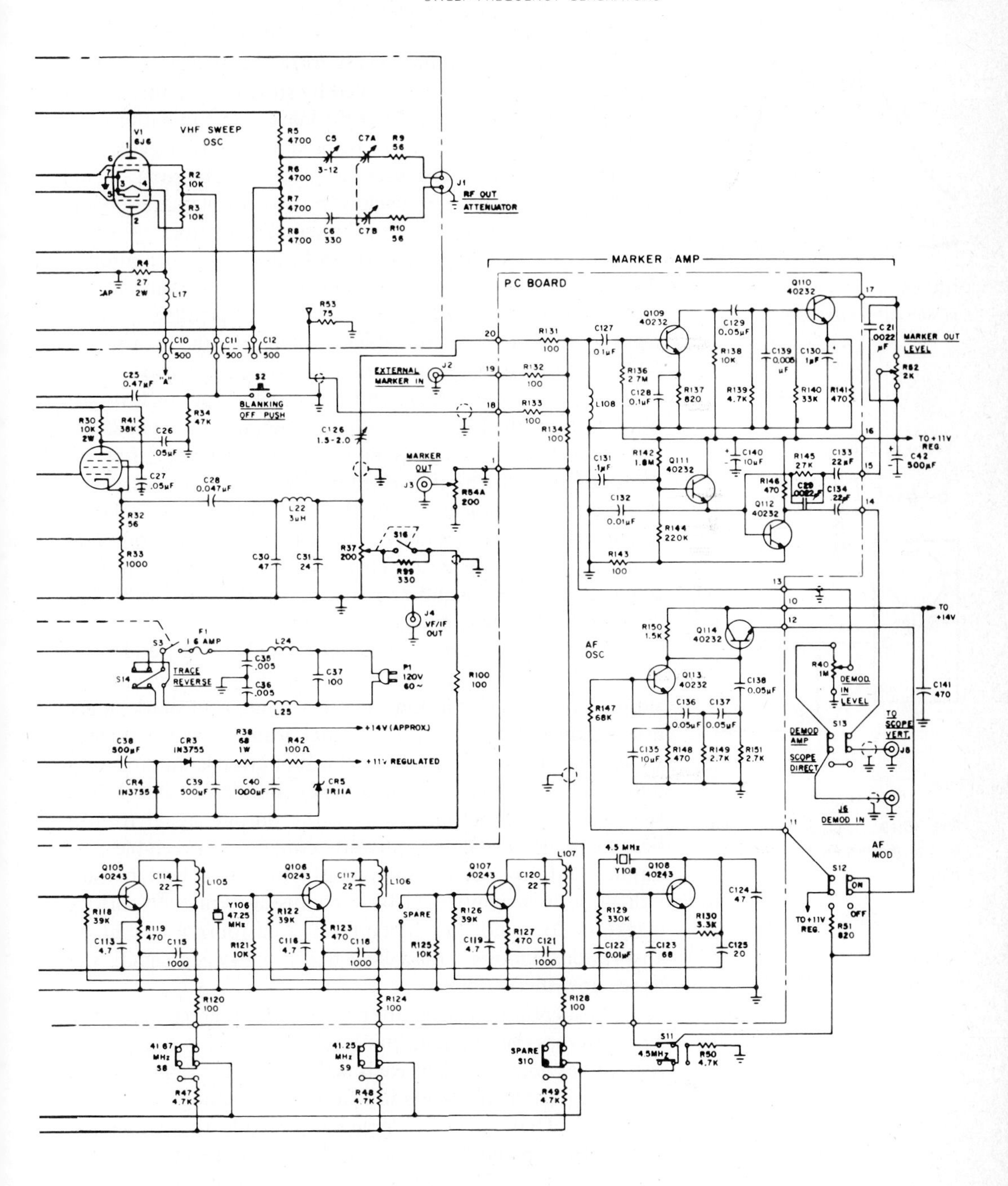

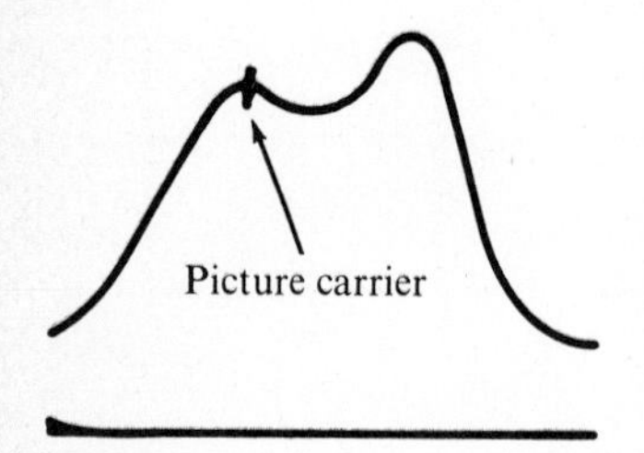

FIG. 18.18 *A beat marker at the picture carrier frequency.*

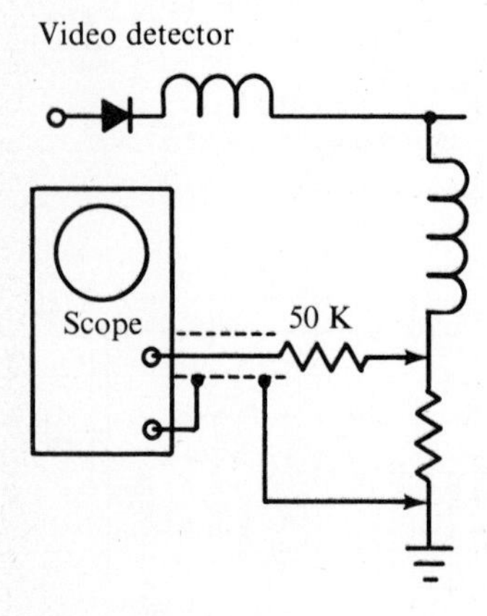

FIG. 18.19 *An isolating resistor used to sharpen the beat marker indication.*

generator near the output leads from the sweep generator. Sufficient signal energy is coupled by stray capacitance in this example to produce a visible beat marker in the pattern. An ordinary signal generator can be used as a marker generator, although precise procedures require a calibrated signal generator; calibration is usually provided by quartz-crystal frequency standards built into the signal generator.

An essential consideration in beat-marker formation is the provision of appropriate filter action. For example, the scope input circuit in Fig. 18.19 comprises a so-called isolating resistor in series with the shielded cable to the scope. The 50-K resistor operates in combination with the capacitance of the coaxial cable to form an integrating circuit, or low-pass filter. In turn, the higher beat frequencies are suppressed, and only the lower frequencies in the vicinity of zero beat are reproduced in the marker display. The result is a sharp and definite marker indication. On the other hand, if the isolating resistor is omitted, the marker indication becomes comparatively broad and indefinite. This difficulty is aggravated if the scope also has wideband response and if the input cable has comparatively low capacitance.

Beat markers are satisfactory for indicating frequencies along a response curve that is produced by a signal from a simple sweep oscillator. On the other hand, beat markers are not practical for use with a signal from a beat-frequency sweep oscillator. The reason for this limitation is that it is very difficult to design a beat-frequency oscillator that has a sufficiently pure output. That is, the output signal from a beat-frequency oscillator generally contains appreciable harmonics and cross-beat components. In turn, a beat marker indication is accompanied by a confusing array of spurious markers. To avoid this difficulty, it is customary to employ absorption markers with a beat-frequency sweep signal. Let us see how this is done.

Absorption markers are produced by high Q traps connected across the sweep output line, as depicted in Fig. 18.20*a*. A terminal is provided at each coil so that the operator can identify a particular marker indication by touching the various terminals. That is, body capacitance tends to detune the associated trap and causes the marker to move slightly on the curve. Figure 18.20*b* shows the external appearance of an absorption marker box, and a response curve with two absorption markers is illustrated in Fig. 18.20*c*. Spurious markers are not produced because an absorption marker box

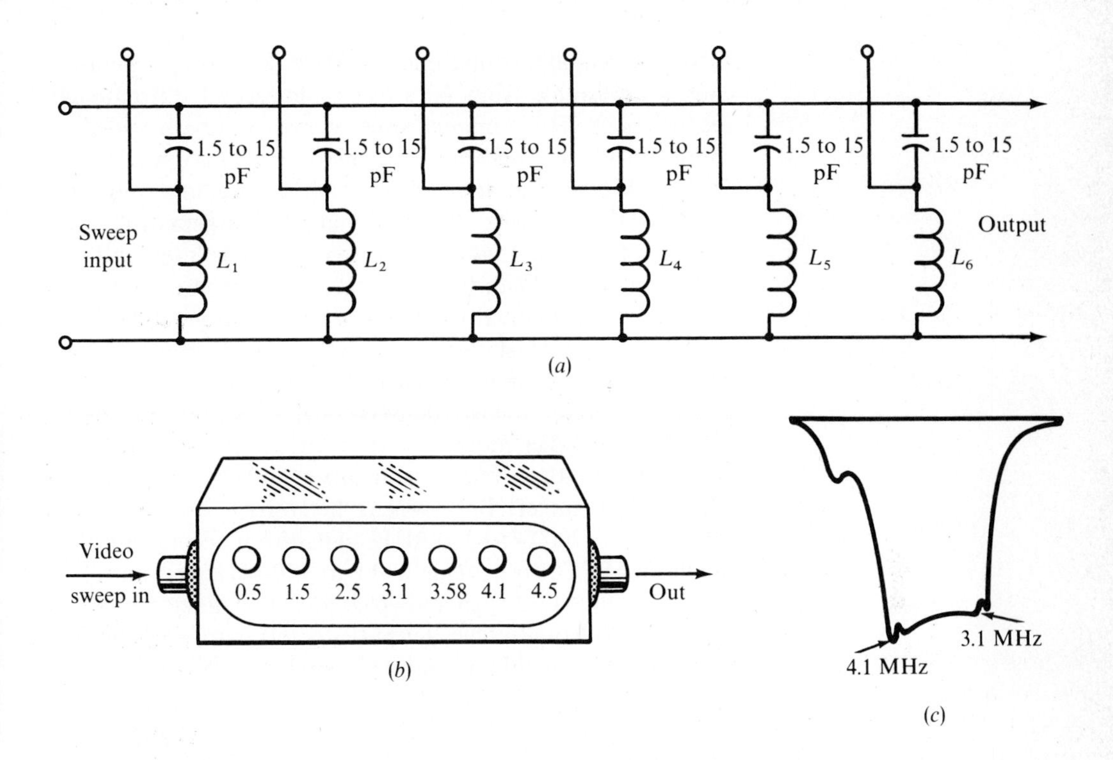

FIG. 18.20 *Absorption marker system. (a) Circuit for an absorption marker system; (b) view of a typical marker box; (c) absorption markers on a response curve.*

is a passive device; it does not heterodyne with the spurious frequencies that are present in the beat-frequency signal.

18.9 SOLID-STATE FM GENERATOR

Next, let us consider the features of a solid-state fm generator. A typical instrument is illustrated in Fig. 18.21. It is an rf signal generator with provisions for either amplitude or frequency modulation. Frequency range is from 0.1 to 110 MHz, with a choice of three modes of frequency modulation. In the Full mode, the entire 0.1 to 11 MHz band, or the entire 1.0- to 110-MHz band is swept. However, when set to the Video mode, the sweep starts at the lower band limit and extends up to the frequency to which the front-panel frequency scale is set. Finally, the Symmetrical mode provides a swept output of adjustable width above and below a center frequency to which the front-panel frequency scale is set.

Output-level and vernier controls are provided for adjusting the calibrated rf output over a range from +20 to −110 dBm (2.223 V to 1 μV) into 50 Ω. An internal 5-MHz quartz-crystal oscillator permits frequency calibration to

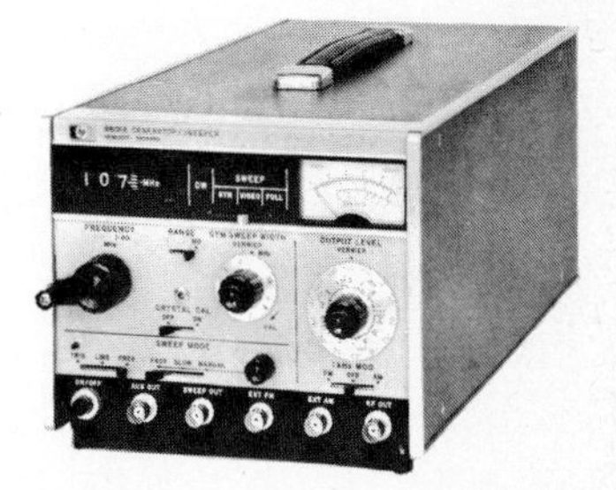

FIG. 18.21 *A typical solid-state fm generator. (Courtesy, Hewlett-Packard)*

± 0.01 percent at any multiple of 5 MHz. The sweep repetition rate is adjustable from 60 sweeps per second down to 60 seconds per sweep; a manual sweep control is also provided. A simplified block diagram of the instrument is shown in Fig. 18.22. Observe that four major sections are involved: the primary signal path, the automatic-leveling-control (alc) feedback loop, the frequency-control feedback loop, and the sweep generator. With this preliminary survey in mind, let us consider the individual functions in somewhat greater detail.

Figure 18.23 shows a detailed block diagram for the generator. The primary signal path starts with a 100-MHz crystal oscillator. This 100-MHz signal is doubled by a 200-MHz amplifier/modulator. In turn, the 200-MHz output is applied to (1) the primary signal mixer, and (2) the frequency-control loop mixer. Both mixers heterodyne the 200-MHz signal with a 200.1- to 310-MHz signal from a voltage-tuned oscillator (VTO) to provide a 0.1 to 110-MHz output. This signal is coupled through a lowpass filter to a video amplifier. Then, the video-amplifier output is applied to a calibrated attenuator that enables the leveled signal at the 50-Ω rf-output terminal to be varied from +20 to −110 dBm.

Next, let us observe the alc feedback loop in Fig. 18.23. The video amplifier contains a detector that samples the signal power at the video-amplifier output. The detector output is a voltage level which is proportional to the signal power. This output from the detector is applied to the automatic level-control circuit, where it is compared to a reference voltage that is determined by the setting of the output-level vernier control. In turn, the output from the level control is stepped up by an alc amplifier and coupled back to the 200-MHz amplifier/modulator output in the primary signal path. The three modulation functions—rf blanking, frequency markers, and amplitude modulation—are also applied to the alc level-control and amplifier circuits.

With reference to the frequency-control feedback loop in Fig. 18.23, this network ensures high-frequency accuracy, stability, and linearity by controlling the 200.1- to 310-MHz VTO. This loop response corrects for any 100-MHz oscillator drift and keeps the VTO output frequency independent of the VTO's nonlinear tuning characteristic. That is, the loop locks the rf output frequency to the linear output tuning voltage of the sweep generator. We observe that the 200-MHz amplifier/modulator output is mixed with the 200.1- to 310-MHz VTO output in the loop mixer. In turn, the 0.1- to-110-MHz dif-

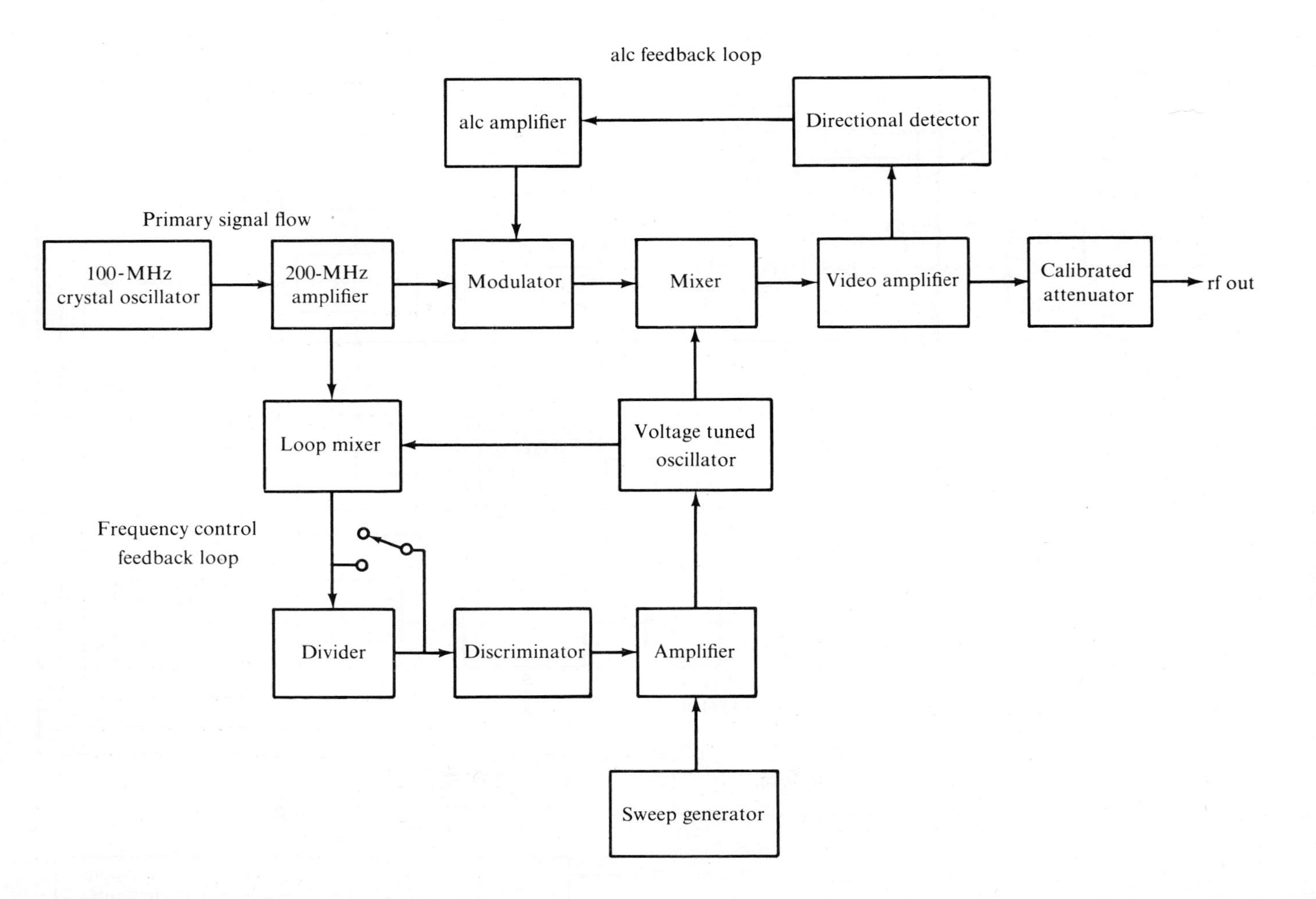

FIG. 18.22 *Simplified block diagram of a sweep generator.*

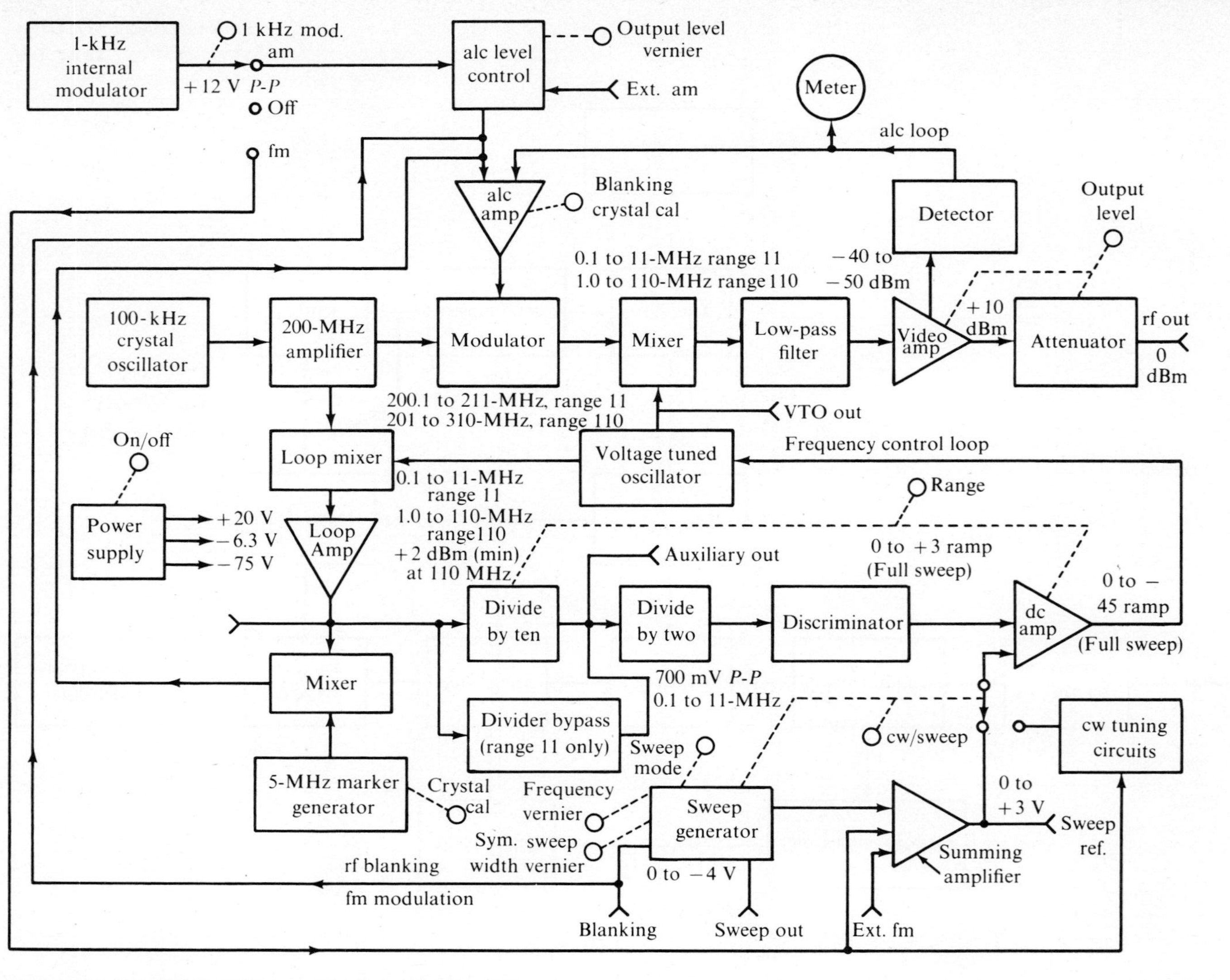

FIG. 18.23 *Detailed block diagram of a solid-state sweep generator.*

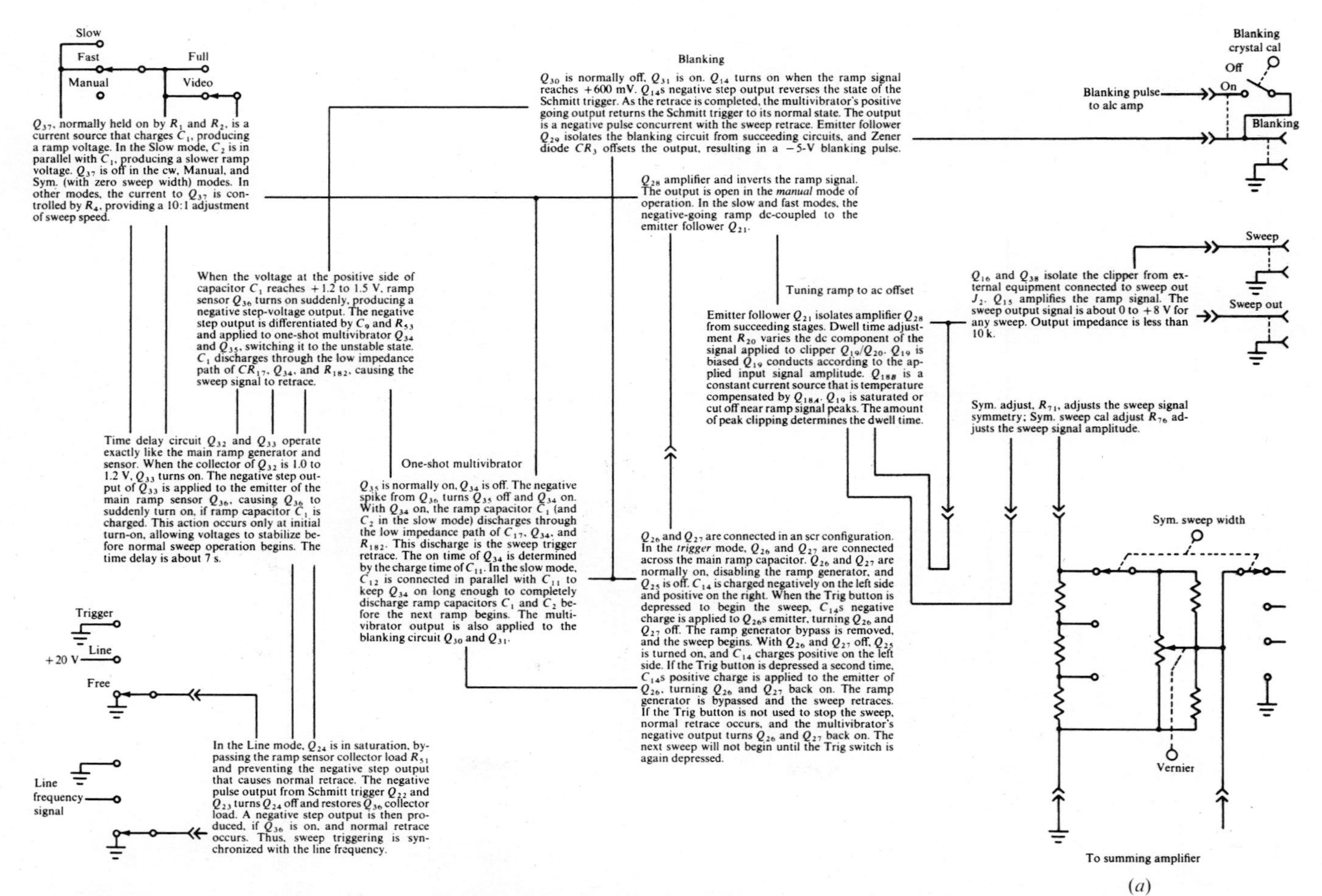

FIG. 18.24 *Solid-state sweep generator. (a) Sectional summary; (b) circuit diagram.*

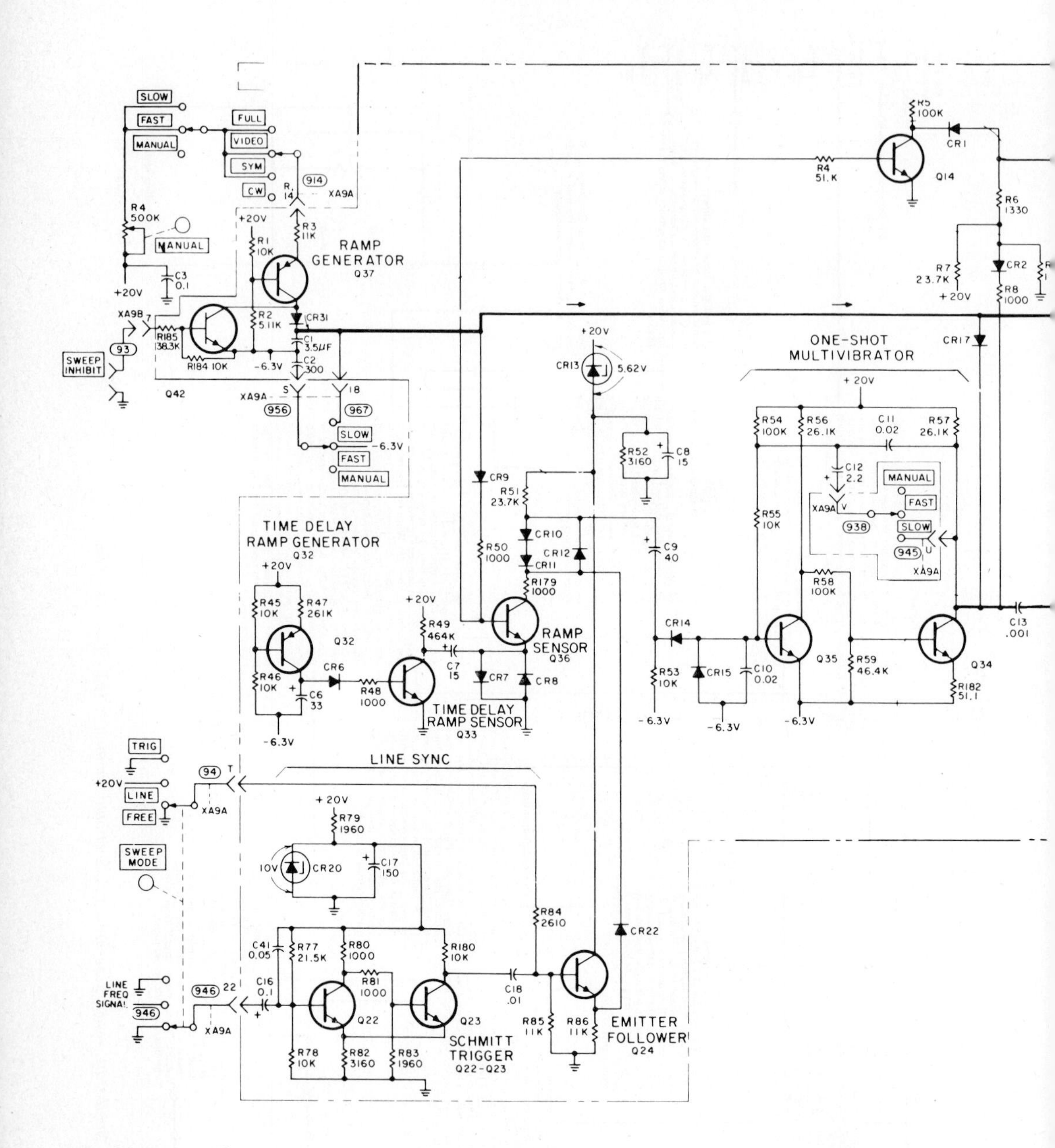

FIG. 18.24 *(cont'd)*

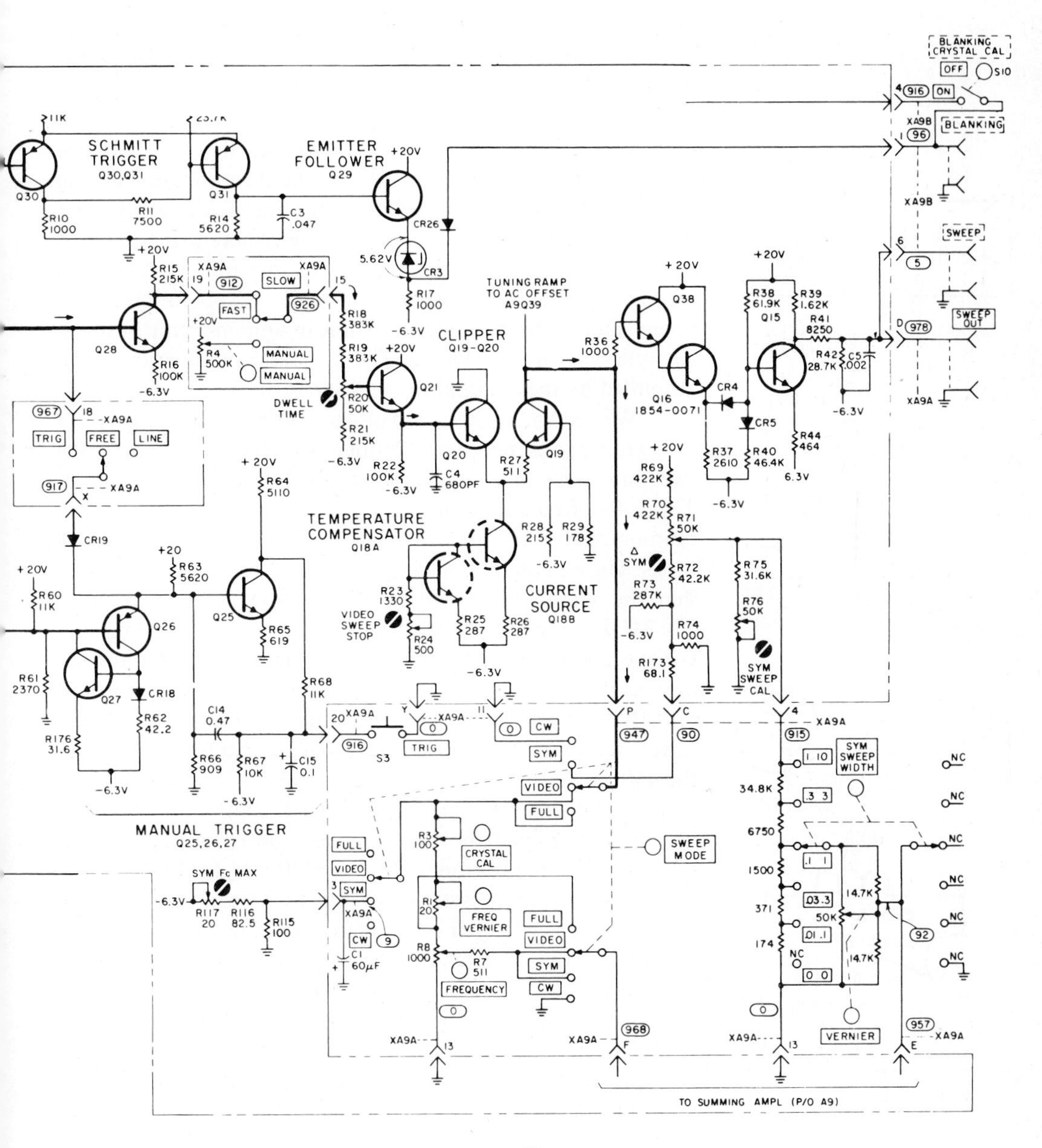
SCHMITT TRIGGER
Q30,Q31
EMITTER FOLLOWER
Q29
CLIPPER
Q19-Q20
TUNING RAMP TO AC OFFSET A9Q39
TEMPERATURE COMPENSATOR
Q18A
CURRENT SOURCE
Q18B
MANUAL TRIGGER
Q25,26,27
DWELL TIME
VIDEO SWEEP STOP
SYM SWEEP CAL
SYM Fc MAX
BLANKING CRYSTAL CAL
BLANKING
SWEEP
SWEEP OUT
SWEEP MODE
CRYSTAL CAL
FREQ VERNIER
FREQUENCY
SYM SWEEP WIDTH
VERNIER
TO SUMMING AMPL (P/O A9)

(b)

ference frequency from the mixer is applied to a frequency discriminator. This is the first step in the process of tuning the VFO.

Because the discriminator frequency range is limited, a "divide by 20" frequency divider precedes the discriminator during 1.0- to 110-MHz operation. The discriminator output is a voltage level proportional to the input frequency. This voltage level is applied to a dc amplifier where it is compared to a voltage reference determined by the frequency-control setting and the sweep generator mode of operation. (In a sweep mode, the reference is a linear voltage ramp; for cw, the reference is a dc value). The difference-voltage output is amplified by the dc amplifier and tunes the VTO.

We recognize that the voltage comparison in Fig. 18.23 is a continuous process at either a cw frequency or when sweeping over any portion of the 0.1- to 110-MHz range. Thus, the frequency-control feedback loop keeps the discriminator output equal to the reference tuning voltage at all times. The dc amplifier also contains a search circuit to ensure that the VTO tuning voltage tunes the VTO upward from 200 MHz. Note that the sweep generator produces the tuning-voltage reference for the dc amplifier in both sweep and cw modes. The sweep generator tuning-ramp output, when applied directly to the dc amplifier, causes the VTO to sweep the full range: 0.1 to 11 MHz in the 11 range, or 1.0 to 110 MHz in the 110 range.

In the Video mode (Fig. 18.23), the tuning ramp's upper limit is determined by the frequency-control setting, so the sweep is from the bottom of the band (0.1 or 1.0 MHz) to the particular frequency setting. In the Symmetrical mode, the tuning ramp is centered on a cw frequency determined by the setting of the frequency control. Internal and external frequency-modulation signals are superimposed on the sweep generator tuning voltage by a summing amplifier in the Symmetrical mode of operation. Circuit details for the sweep generator are shown in Fig. 18.24*b*, with sectional summaries given in Fig. 18.24*a*.

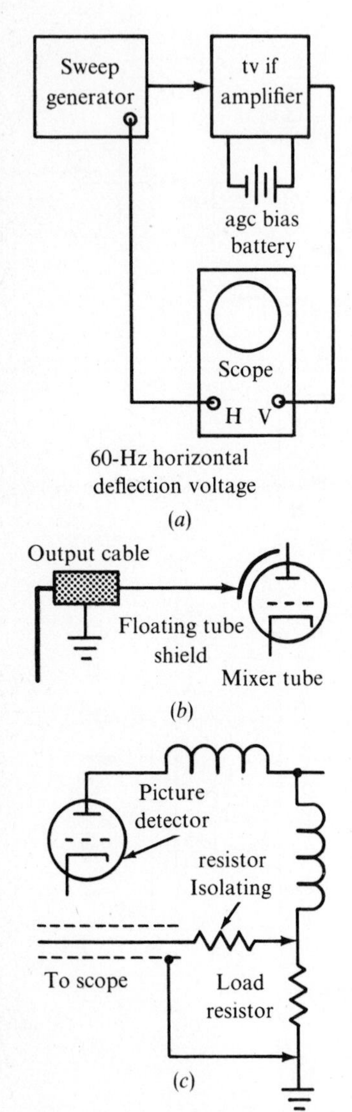

FIG. 18.25 *Display of tv if response curve. (a) Plan of test arrangement; (b) scope connection to output of picture detector.*

18.10 APPLICATIONS

One of the more basic sweep generator applications entails display of the if response curve for a television receiver. With reference to Fig. 18.25, the output from the sweep generator is capacitively coupled to the mixer section; this is commonly done by placing an ungrounded (floating) tube shield over the mixer tube. This metal shield has capacitance to the plate of the mixer tube, thereby coupling the if sweep signal into the mixer circuit. The output signal from the picture detector is taken across the detector load resistor. As shown in the diagram, an isolating resistor of approximately 30 kΩ is connected at the end of the scope input cable. This isolating resistor has a lowpass filter function which is useful in obtaining sharp frequency-marker displays.

An incidental consideration in obtaining tv if curves is is the stabilization of if gain by clamping the agc line. That is, battery bias is commonly provided for the if stages during alignment procedures. The sweep generator is set to the center frequency of the if section (such as 42 MHz), and with an appropriate sweep width (such as 6 MHz). The sweep output level is adjusted to avoid overload of the if amplifier, and the scope controls are adjusted to display a response curve such as illustrated in Fig. 18.26. If the sweep generator contains a built-in marker generator, frequency points can be located on the response curve by applying a marker frequency. Some sweep generators provide dual markers, which appear on the response curve as shown in Fig. 18.27.

Another basic sweep generator application consists in the display of the overall rf/if response curve for a television receiver. Figure 18.28*a* shows the plan of the test arrangement. It is similar to that for the if curve display, except that the sweep signal is applied at the antenna input terminals of the receiver. The sweep generator is operated at a center frequency corresponding to that of the channel under test, and with a sweep width of approximately 15 MHz. A typical overall rf/if response curve is illustrated in Fig. 18.28*b*.

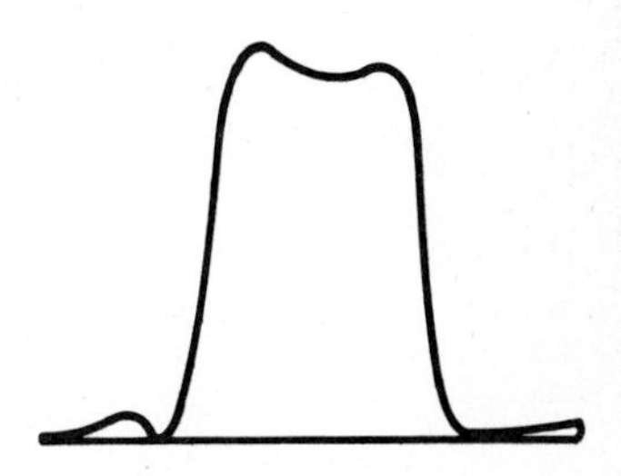

FIG. 18.26 *Typical tv if response curve.*

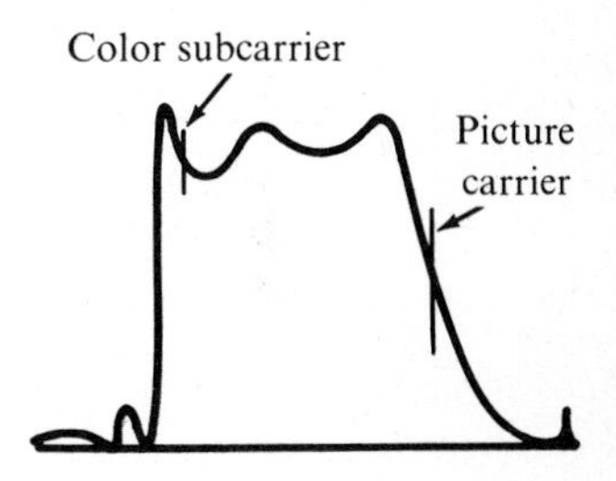

FIG. 18.27 *Tv if response curve for a color receiver; display includes two frequency markers.*

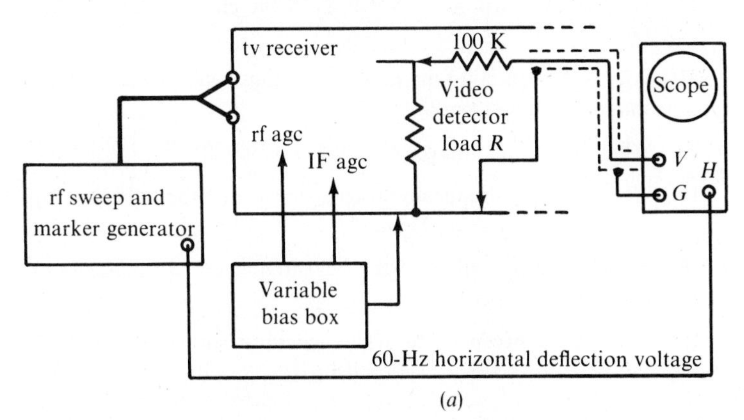

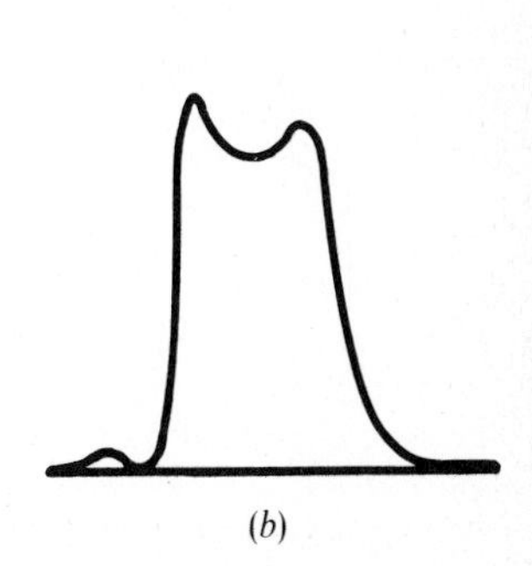

FIG. 18.28 *Display of overall rf/if response curve. (a) Test arrangement; (b) typical overall response curve.*

QUESTIONS AND PROBLEMS

1. What are sweep frequency generators?
2. What is the distinction between a cw and a frequency-modulated wave?
3. How would the frequency-response curve of the sound section of a tv receiver be displayed on an oscilloscope?
4. Why must the output level of a sweep generator be constant?
5. Draw a diagram of the basic arrangement for displaying a frequency-response curve on an oscilloscope.
6. What is the purpose of the 60-Hz sine-wave deflection voltage in the diagram shown in Fig. 18.2?
7. Why do designers commonly employ a 60-Hz frequency-modulating voltage in sweep generators?
8. Why is it necessary to blank the return trace on the crt when a 60-Hz sine wave is used as the deflection voltage?
9. What are the advantages of blanking the retrace to form a zero reference line as compared to allowing the two curves to be superimposed?
10. Explain the operation of the electromechanical fm modulator shown in Fig. 18.7.
11. Why is the arrangement depicted in Fig. 18.8 limited to a frequency deviation of less than 12 MHz on the vhf frequency range?
12. Explain the operation of the semiconductor frequency modulator shown in Fig. 18.9.
13. What is the advantage of the push-pull oscillator configuration over the single-ended oscillator?
14. What is the purpose of switch S_2 in the circuit shown in Fig. 18.10?
15. Why is it essential to maintain a constant value of output from an rf sweep generator?
16. What are the advantages of resistive padding?
17. Why are beat-frequency oscillators mandatory to develop video-frequency sweep signals?
18. Explain the operation of the beat-frequency sweep oscillator shown in Fig. 18.12.
19. What is the sweep range of the generator shown in Fig. 18.17 and how is this range accomplished?

20. What is the function of each vacuum tube in the sweep generator depicted in Fig. 18.17?
21. What is the purpose of the signal output at J_2 in the sweep generator depicted in Fig. 18.17?
22. How is the beat marker developed on the frequency response curve shown in Fig. 18.18?
23. Why is filter action necessary for the development of a clear marker on the frequency response curve on a crt?
24. Explain how absorption markers are produced.

SQUARE-WAVE AND PULSE GENERATORS

19.1 COMPARISON OF SQUARE-WAVE AND PULSE WAVEFORMS

Square-wave and pulse generators are closely related instruments; in some cases, the same instrument can be used to provide either a square-wave or a pulse output. However, a pulse generator operates at much lower average power and for this reason separate instruments are commonly employed. A pulse waveform is also termed a rectangular waveform; a square wave is a special case of a rectangular waveform. With reference to Fig. 19.1, a square wave can be analyzed in terms of a fundamental and old harmonics, whereas a pulse is analyzed in terms of a fundamental and both even and odd harmonics.

If a Fourier analysis is made of a square waveform and the resulting terms are plotted, we obtain the sinusoidal relations depicted in Fig. 19.2. This treatment illustrates the necessity for wideband circuitry in the design of square-wave generators that provide fast rise. If the higher harmonics of the waveform are attenuated or suppressed by narrowband generator circuitry, the output waveform will have slow rise. We recall the rule-of-thumb formula that relates rise time to the cutoff frequency of a circuit:

$$T_r = \frac{1}{3f_c} \tag{19.1}$$

where T_r is the rise time of the wavefront and f_c corresponds to the high-frequency cutoff point for the circuit.

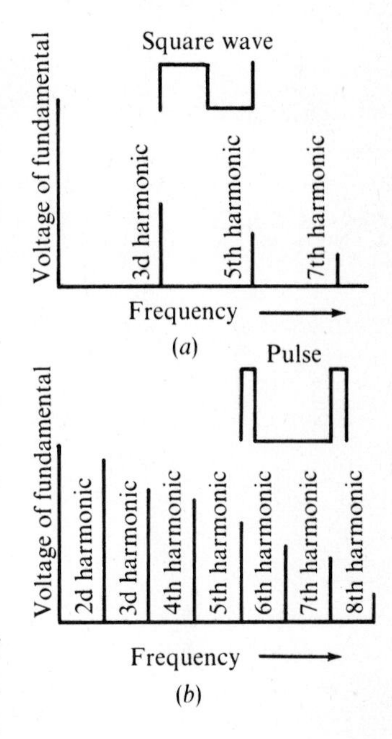

FIG. 19.1 *Harmonic analysis of a square wave vs. a pulse.*

19.2 METHODS OF SQUARE-WAVE GENERATION

Square waveforms can be generated in various ways; one of the simplest methods employs a sine-wave oscillator followed by clipper circuits. If a sine wave is clipped, amplified, and then clipped again, a semisquare wave is obtained. The rise time can be improved by successive amplification and clipping processes. This type of instrument utilizes an *RC* audio oscillator, as discussed in Chap. 18. A waveshaping circuit follows, which clips either the positive peak or the negative peak from the sine wave, as seen in Fig. 19.3. This initial clipping process may be accomplished with a series or a parallel configuration. However, when fast-rise waveforms are to be clipped, the series clipper circuit is more effective, in that the bandwidth of the parallel configuration is less owing to stray capacitance associated with the series resistor.

We observe that the series clipper arrangement in Fig. 19.3*a* is suitable for driving either a high-impedance or a low-impedance load, inasmuch as the diode action simulates an on-off switch. However, the parallel clipper arrangements must operate into a high-impedance load; otherwise, the voltage divider action of the shunt diode will be impaired, and poor clipping action will be obtained.

Next, let us observe the action of a biased clipper circuit, as exemplified in Fig. 19.4. Since the diodes are reverse biased, they cannot conduct until the applied ac voltage cancels the bias voltage. Accordingly, the clipped output waveform has positive and negative peak voltages equal to the bias voltage.

Figure 19.5 shows a cascade clipper arrangement that provides faster rise than a single clipper stage. It is evident that the rise time can be shortened progressively by successive amplification and clipping processes. A practical limit is imposed by the bandwidths of the clipper and amplifier stages. In other words, the rise time of a waveform cannot exceed the rise time of a particular circuit through which it passes. We find that triodes are also used as clippers; a cathode follower will clip the negative peak of a waveform when the tube is overdriven. This type of clipper is commonly used in

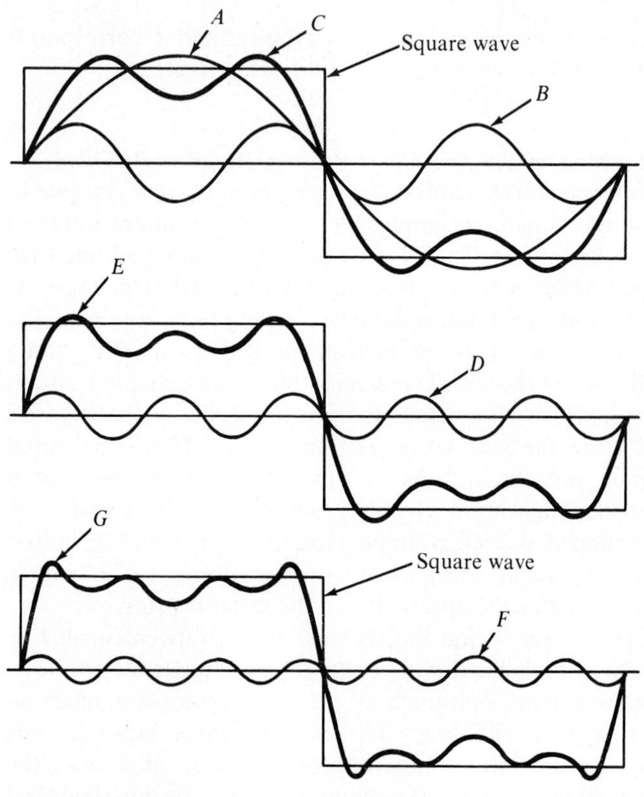

A: Fundamental
B: 3d harmonic
C: Fundamental plus 3d harmonic
D: 5th harmonic
E: Fundamental plus 3d and 5th harmonics
F: 7th harmonic
G: Fundamental plus 3d, 5th, and 7th harmonics

FIG. 19.2 *Sinusoidal components of a square waveform.*

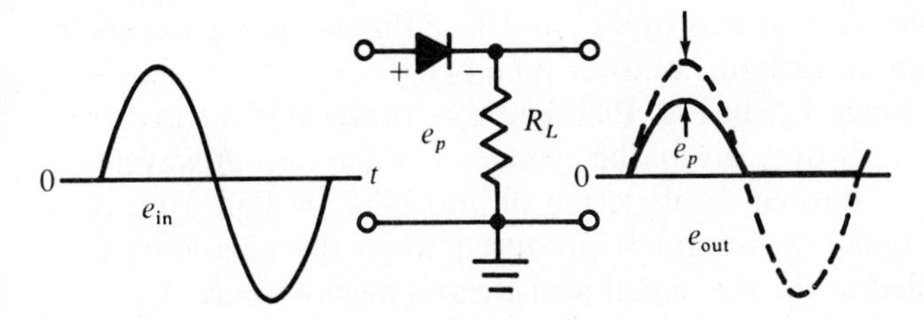

(*a*) Negative series clipper

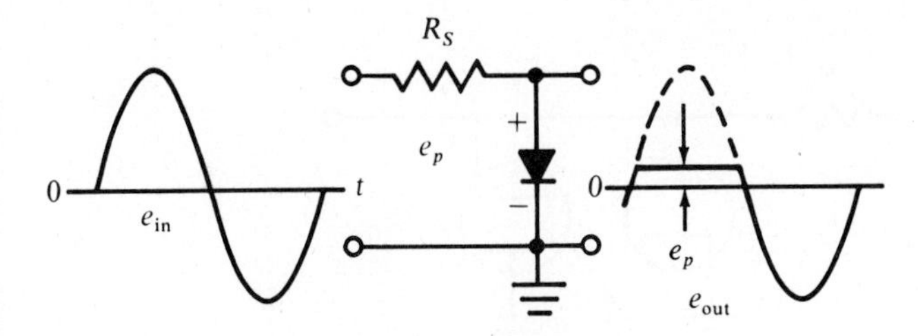

(*b*) Positive parallel clipper

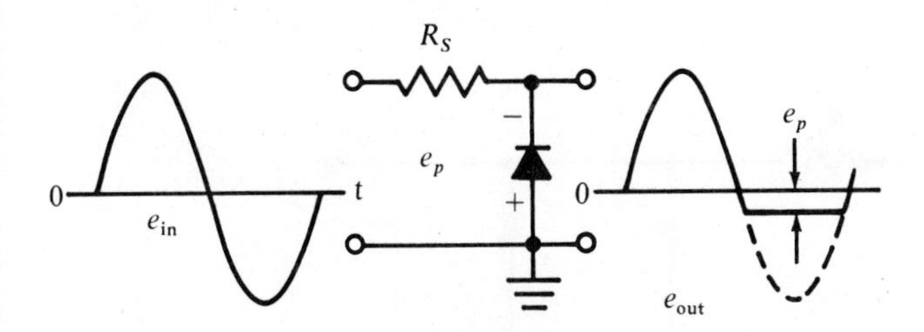

(*c*) Negative parallel clipper

FIG. 19.3 *Basic clipper circuitry.*

the output stage of a generator. It also does double duty as a buffer, when isolation is desirable between an *RC* oscillator and the subsequent waveshaping section.

A configuration for a typical service-type square-wave generator is shown in Fig. 19.6. This instrument provides a square-wave repetition rate from 20 Hz to 1 MHz. The square waveform has a rise time of less than 0.15 μs. A maximum output voltage of 10 V peak-to-peak is available. Tubes V_1 and V_2 operate in a bridged-T *RC* oscillator configuration; the sine-wave output from this oscillator is fed to cathode follower V_3, and is available from the associated attenuator. The sine-wave signal is also fed to V_{4A}, which operates as a buffer and cathode follower clipper. Subsequent amplification and clipping is provided by V_{5A} and V_{5B}. Final clipping occurs in the output cathode follower tube V_{4B}.

Since V_{4A} in Fig. 19.6 develops a trapezoidal waveform, and V_{5A} is overdriven, the symmetry of the output waveform can be adjusted by variation of grid bias for V_{5A}. That is, a rectangular waveform is produced when the bias voltage is adjusted to provide equal positive and negative periods. Note that the plate of V_{5A} is coupled through an RC circuit to the grid of V_{5B}; the 10-pF capacitor is a compensating capacitor

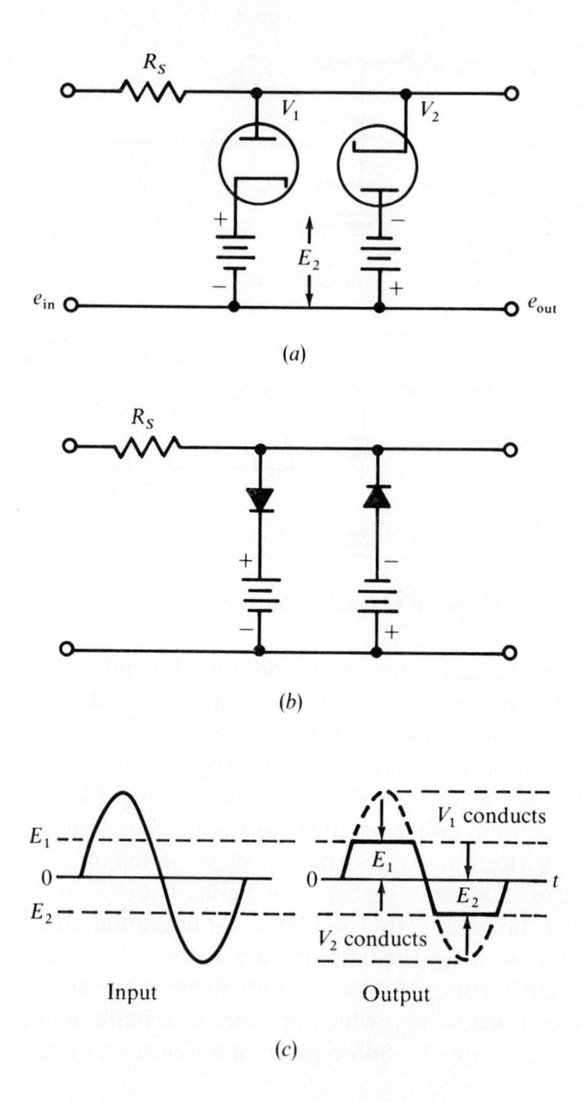

FIG. 19.4 *Biased clippers. (a) Vacuum-tube diode circuit; (b) semiconductor-diode circuit; (c) waveshaping action of clippers.*

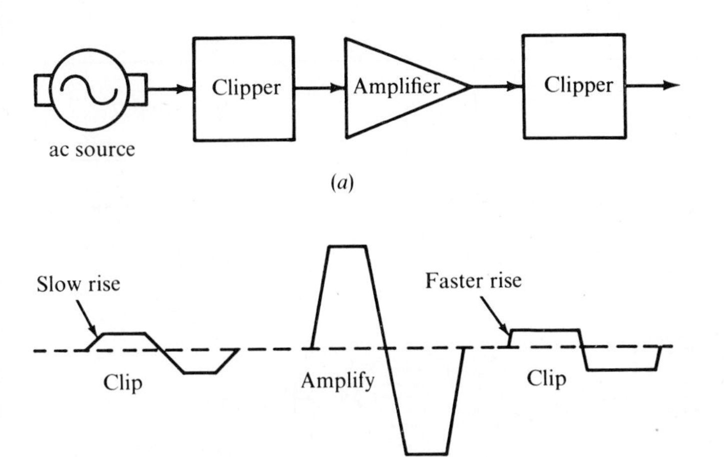

FIG. 19.5 *Cascade clipper arrangement. (a) Block diagram; (b) waveshaping action.*

to permit operating at maximum bandwidth, and thereby optimizing the rise time of the waveform. Frequency compensation also minimizes tilt in the generated waveform. Because of tolerances on production components, the grid input capacitance of V_{5B} is made adjustable by means of the 5- to 25-pF trimmer capacitor connected from grid to ground.

19.3 MULTIVIBRATOR SQUARE-WAVE GENERATOR

When a generator is designed to provide square-wave output only, it usually employs a multivibrator instead of a sine-wave oscillator. A multivibrator oscillator affords an advantage in that its output waveform has a fast rise, which minimizes the number of amplifier and clipper stages that are required. This type of square-wave generator is exemplified in Fig. 19.7; the output waveform has a rise time of less than 0.07 μs, and its repetition rate is adjustable in steps from 50 Hz to 500 kHz. We observe that the semisquare waveform from duotriode V_1 is fed to the gated-beam clipper tube V_2. This type of tube has a comparatively sharp cutoff characteristic, and the rise time of the output waveform is minimized by operation at low plate and screen voltages.

Tube V_3 in Fig. 19.7 operates as an overdriven amplifier, and improves the rise time of the waveform by cutoff and saturation clipping action. Tube V_4 operates primarily as an electronic impedance transformer to provide a low-impedance output circuit. This cathode follower stage also provides

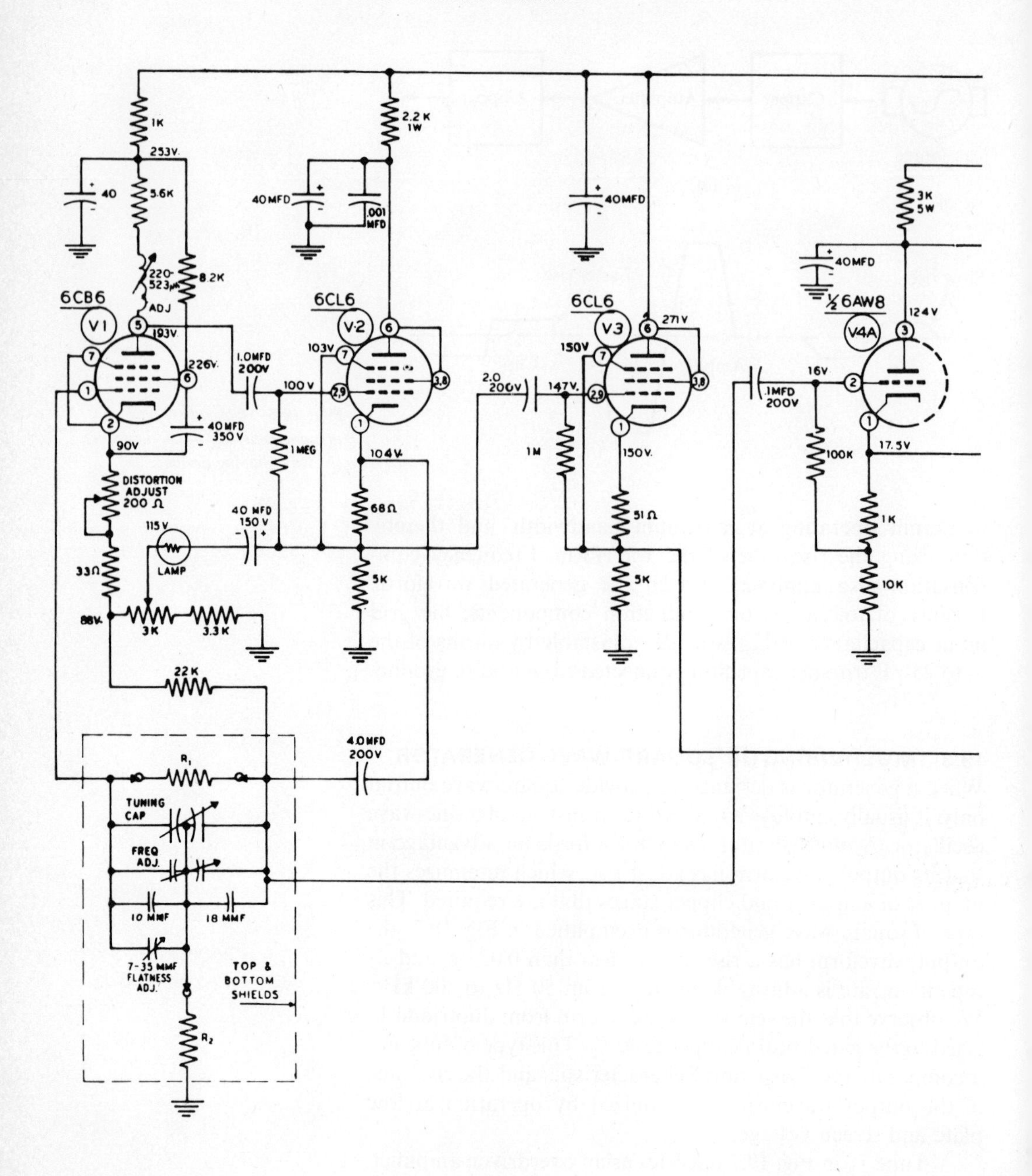

FIG. 19.6 *Configuration of a typical service-type square-wave generator. (Courtesy, Heath Co.)*

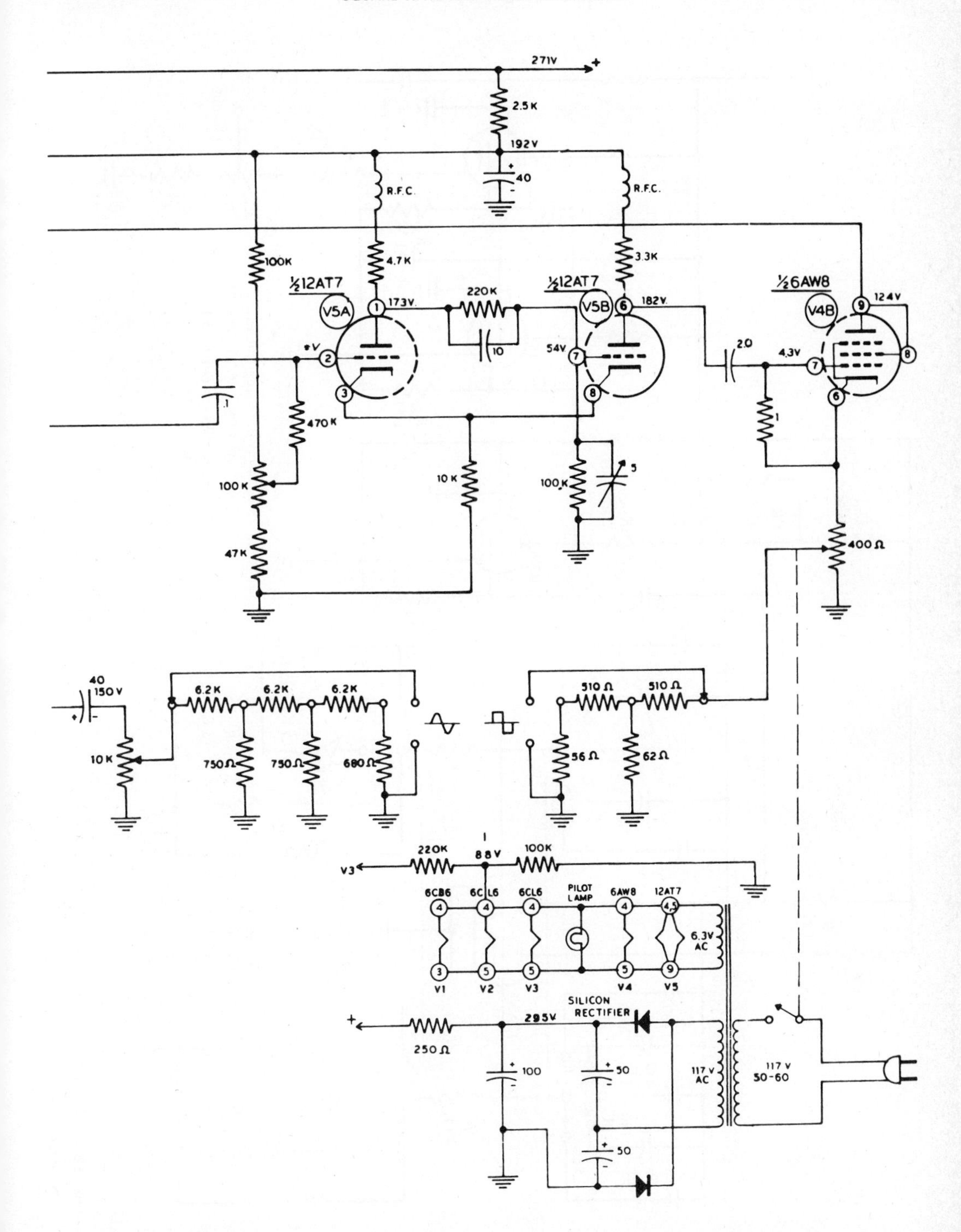
271V
2.5K
192V
40
R.F.C.
R.F.C.
100K
4.7K
3.3K
½12AT7
½12AT7
½6AW8
V5A
V5B
V4B
173V.
220K
182V.
124V
+V
10
54V
2.0
4.3V
.1
470K
1
5
10K
100K
100K
47K
400Ω
40
150V
10K
6.2K
6.2K
6.2K
750Ω
750Ω
680Ω
510Ω
510Ω
56Ω
62Ω
220K
88V
100K
V3
6CB6
6CL6
6CL6
PILOT LAMP
6AW8
12AT7
6.3V AC
V1
V2
V3
V4
V5
SILICON RECTIFIER
295V
250Ω
100
50
50
117V AC
117V 50-60

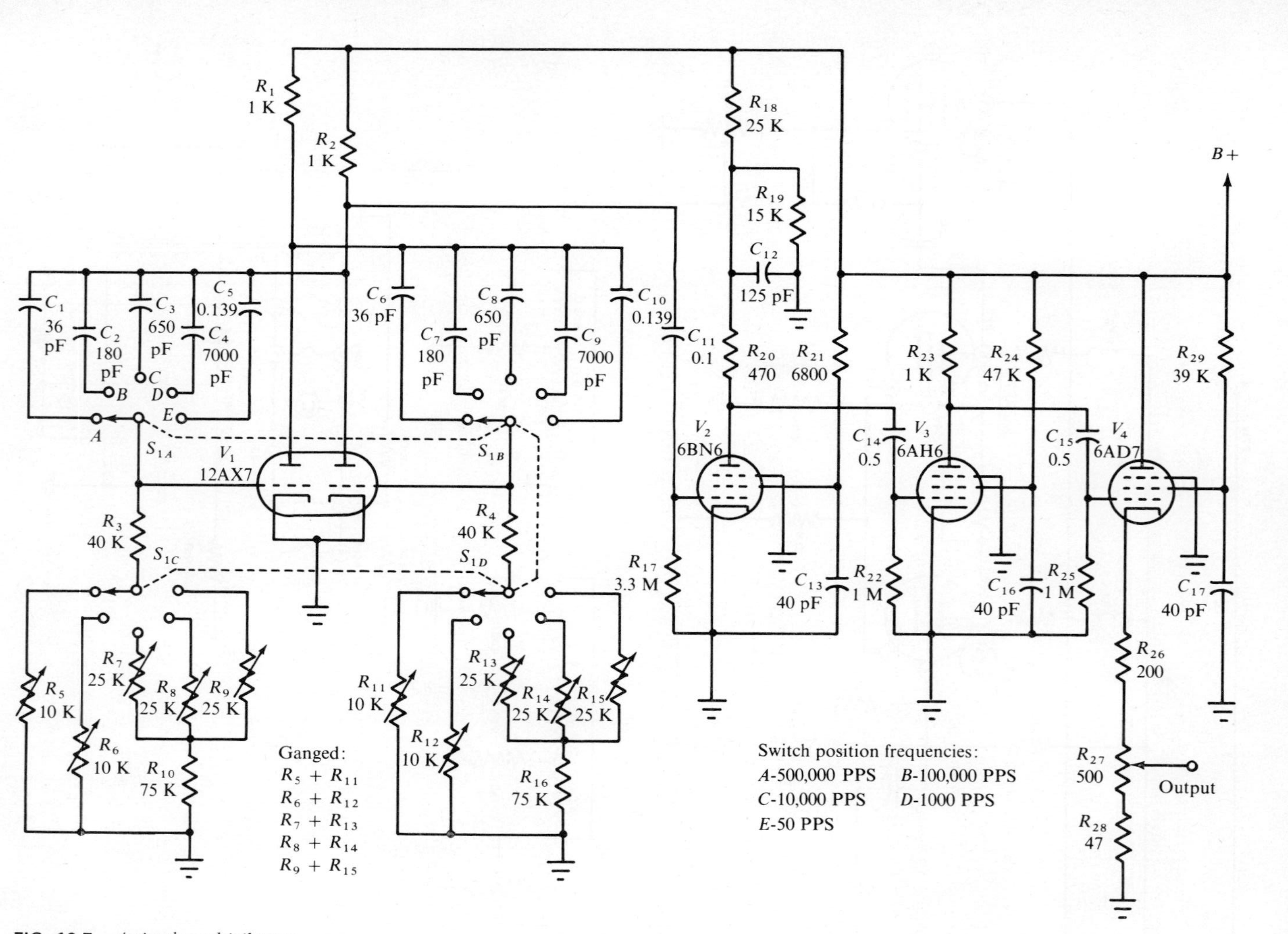

FIG. 19.7 *A simple multivibrator-type square-wave generator.*

negative-peak clipping. R_{26} is a protective bias resistor; if the output lead were short-circuited to ground, a protective bias would nevertheless develop for V_4 by means of the voltage drop across V_4. Resistor R_{27} provides an output voltage range of 0.8 to 8 V. If a lower output limit were provided, the output network would have to be elaborated to maintain fast rise at low amplitudes. Stray capacitance imposes a design limitation in the simple arrangement utilized in Fig. 19.7.

A comparatively large bandwidth can be obtained in amplifier-clipper stages by means of low-value plate load resistors. This simple approach, however, entails a limited output amplitude and thus designers usually employ substantially higher values of plate load resistance combined with high-frequency peaking. Peaking coils are comparatively difficult design components because they tend to introduce overshoot and ringing in fast-rise circuits. In contending with this source of distortion, the designer may prefer to utilize less efficient methods of high-frequency compensation that are easier to work with. For example, C_{13}, C_{16}, and C_{17} are partial screen-bypass capacitors in Fig. 19.7; the tubes develop full gain at high frequencies, but operate at reduced gain at low frequencies.

Let us consider the configuration for a basic lab-type square-wave generator. With reference to Fig. 19.8, a multivibrator oscillator is utilized, followed by two limiter-amplifier stages which drive the output stage. Special circuitry is employed to obtain a rise time of less than 20 ns with no overshoot in the waveform. Direct coupling is used between stages, high-frequency compensation is provided, high-transconductance low-capacitance tubes assist in developing fast rise, and the output stage operates as an inverted amplifier with the plate of the tube connected to ground through the plate load resistor.

The configuration of the generator in this example is shown in Fig. 19.9. Tubes V_1 and V_2 are connected in a symmetrical multivibrator circuit, with two repetition-rate controls. A switching system is used to change the oscillator time-constant in steps, and a potentiometer is employed for vernier control between steps. Note that R_4, R_5, R_{27}, and R_{28} are parasitic-suppression resistors. This multivibrator provides a repetition rate from 25 Hz to 1 MHz. Because of tolerances on tubes and components, the output waveform will not necessarily be symmetrical from one switch setting to another. Therefore, a symmetry control (R_{25}) is included to vary the screen grid voltages of the oscillator tubes and thereby permit adjustment of waveform symmetry.

High-frequency compensation in Fig. 19.9 is provided by C_1 and C_{13}. At high repetition rates, supplementary high-frequency compensation is required; this is the function of $C_{1.1}$ and $C_{11.1}$. These are screen-feedback capacitors, and their comparatively small values cause their effect to be significant only at high frequencies. The semisquare output waveform is taken from across cathode resistor R_{29}. This is a low-impedance circuit point and in turn is not loaded significantly by the drive requirements of the following stage. Tube V_3 is a shaping amplifier which clips the applied semisquare wave. Cathode resistor R_{32} develops protective bias for V_3 in case of multivibrator failure. High-frequency compensation in this stage is provided by a peaking coil in the plate circuit.

Output from the shaping amplifier is coupled to the grids of the driver amplifier tubes V_4 and V_5. These tubes are connected in parallel to obtain adequate power output. Parasitic oscillation is suppressed by resistors R_{35} through

FIG. 19.8 *Block diagram of a laboratory-type square-wave generator. (Courtesy, Tektronix Inc.)*

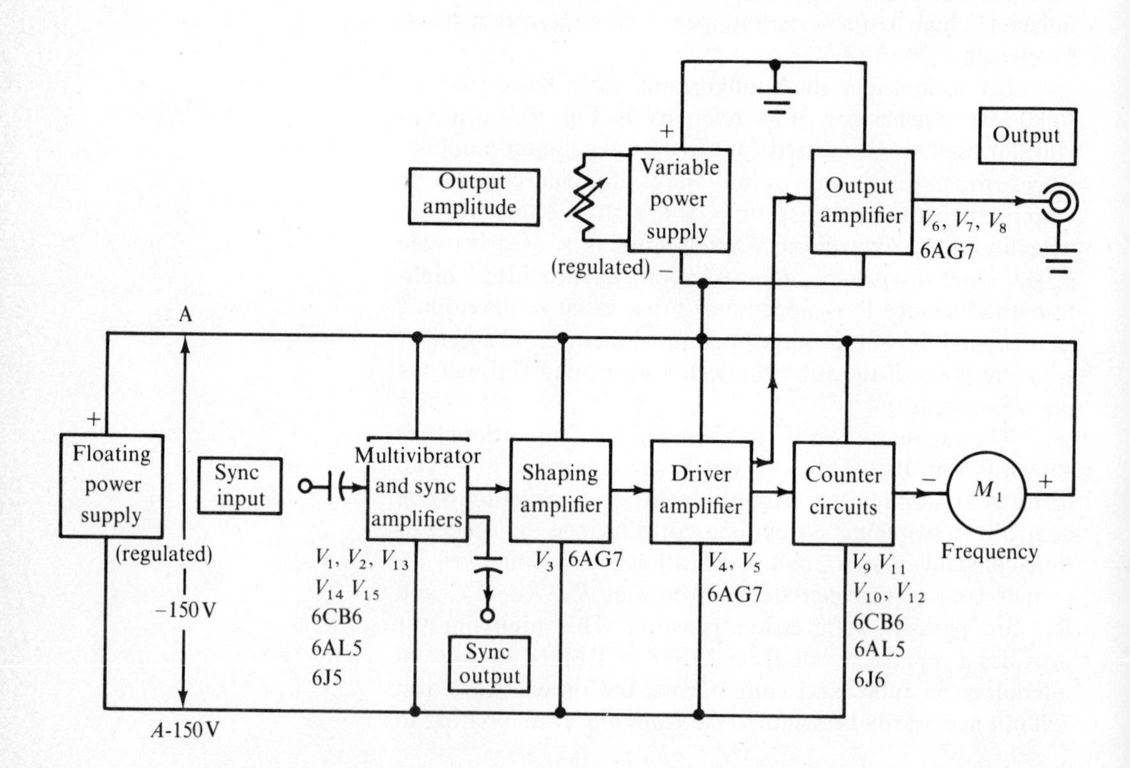

R_{40}. Tubes V_4 and V_5 are driven between cutoff and zero, thereby clipping the waveform and improving its rise time. High-frequency compensation is provided by peaking coil L_2. The output from the driver amplifier is directly coupled to the grids of the output amplifier tubes. A sample of the driver output is also fed to the meter amplifier tube V_9. This tube energizes a frequency meter, as explained subsequently.

Note that the output amplifier in Fig. 19.9 comprises three tubes, V_6, V_7, and V_8, connected in parallel. Parasitic suppression is provided by resistors R_{44} through R_{51}. The output amplifier tubes are driven between cutoff and zero, and the output waveform is developed across R_{54}. In application, R_{54} is also shunted by the circuit or device under test. To vary the amplitude of the output waveform, a variable power supply is employed to apply an adjustable voltage to the cathodes of V_6, V_7, and V_8. The output amplitude range is from 10 to 100 V peak-to-peak across R_{54}. If R_{54} is shunted by a terminated 93-Ω output cable, the output amplitude range is from 1.5 to 15 V peak-to-peak.

We now consider the frequency-meter section depicted in Fig. 19.9, This meter circuit provides a direct readout of the square-wave repetition rate. A sample of the generated square wave is taken from across R_{42} and is amplified by V_9. Peaking coil L_2 serves to maintain fast rise. One section of V_{10} (pins 2 and 5) operates as a clamp diode to prevent the plate of V_9 from rising above the supply voltage. This voltage rise stems from the collapsing field of L_3. Note that the other section of V_{10}—called a plate-catcher diode—establishes the level at which the negative excursion will be fixed. V_{11} is a cathode follower voltage-regulator tube which determines the voltage at which the plate of V_9 will be caught by V_{10}. That is, the meter-adjust potentiometer R_{63}, in conjunction with limit resistors R_{62} and R_{64}, permits adjustment of cathode voltage on V_{11} so that V_9 will constantly generate a square wave of 65 V peak-to-peak amplitude.

Note that a capacitor, selected from C_{23} to C_{31} in Fig. 19.9 by the range-switch sections $SW_{1\text{-}G}$ and $SW_{1\text{-}H}$, is charged on each positive excursion of the square wave because the meter diode V_{12} (pins 2 and 5) clamps the switch arm $SW_{1\text{-}H}$ at "common ground" potential (actually to the 160 to −320 V bus plus a small potential drop across R_{66}). The negative excursion of the square-wave cycle then discharges the selected capacitor through meter M_1 (resistance approximately 5000 Ω). A shunt resistor is connected in parallel with the meter

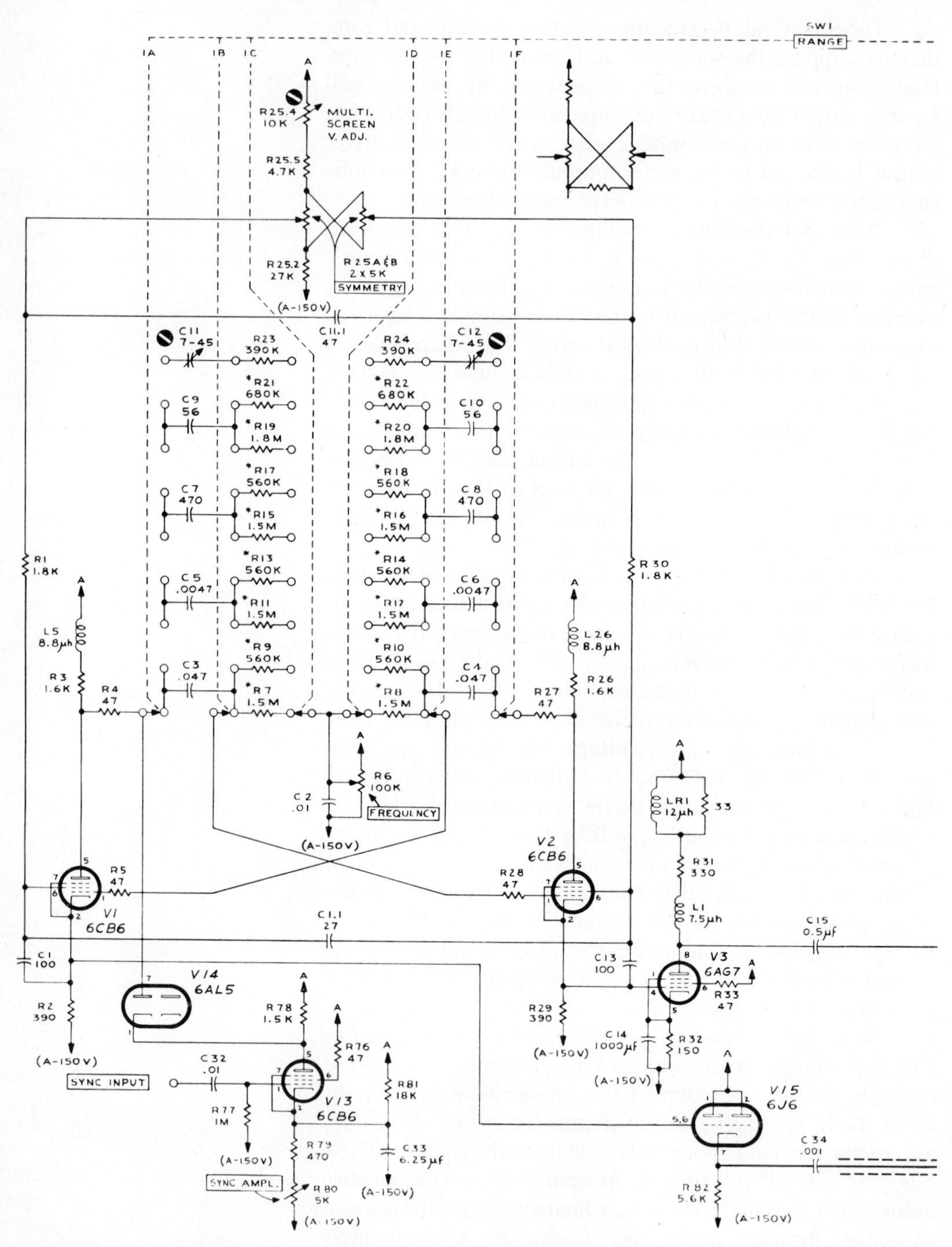

FIG. 19.9 *Configuration of a square-wave generator.* (*Courtesy, Tektronix Inc.*)

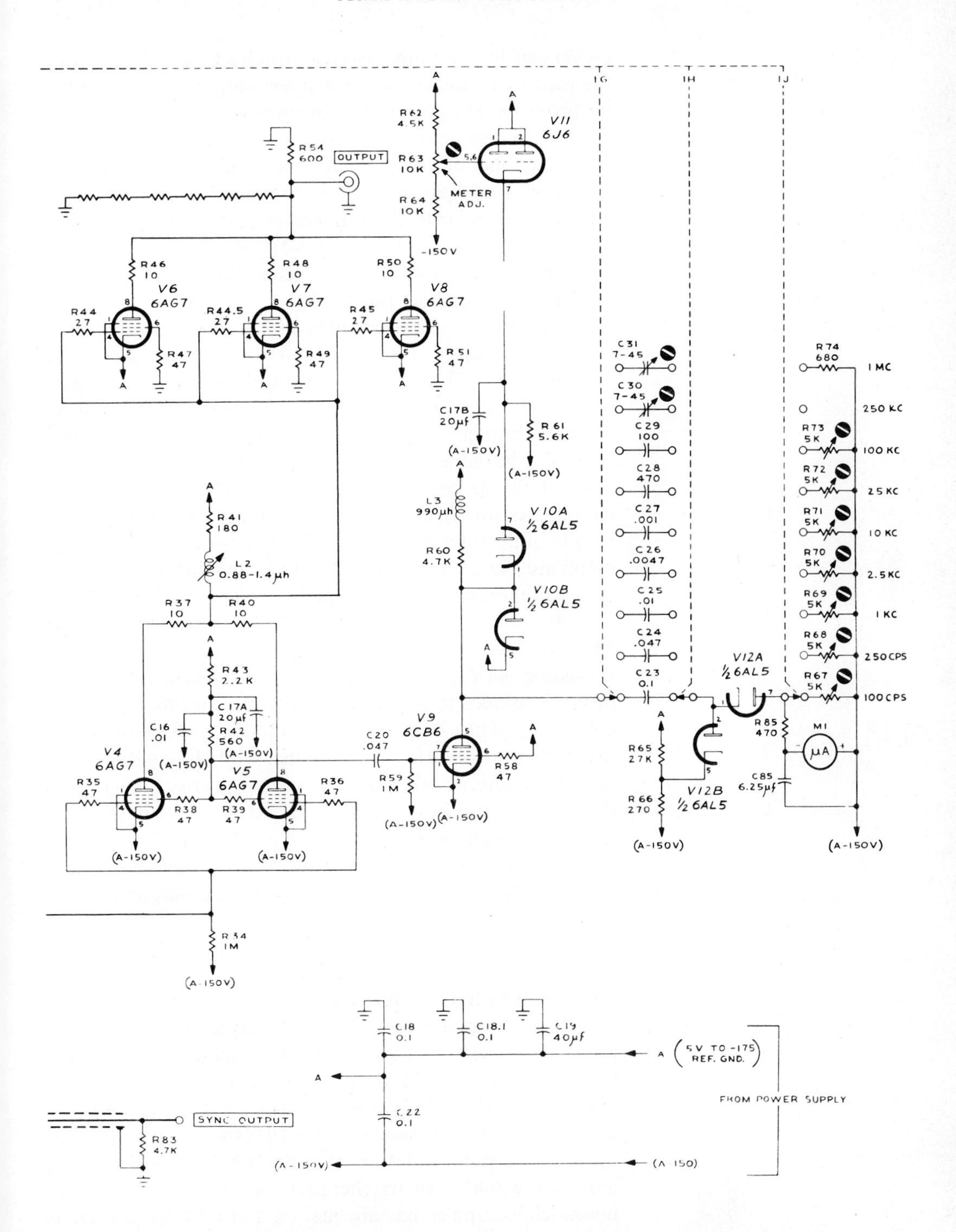
OUTPUT
R54 600
R62 4.5K
R63 10K
R64 10K
METER ADJ.
-150V
V11 6J6
1G
1H
1J
R46 10
R48 10
R50 10
V6 6AG7
V7 6AG7
V8 6AG7
R44 27
R44.5 27
R45 27
R47 47
R49 47
R51 47
C17B 20μf
R61 5.6K
(A-150V)
L3 990μh
R60 4.7K
V10A ½ 6AL5
V10B ½ 6AL5
R41 180
L2 0.88-1.4μh
R37 10
R40 10
R43 2.2K
C17A 20μf
R42 560
C16 .01
V4 6AG7
V5 6AG7
R35 47
R38 47
R39 47
R36 47
C20 .047
V9 6CB6
R59 1M
R58 47
C31 7-45
C30 7-45
C29 100
C28 470
C27 .001
C26 .0047
C25 .01
C24 .047
C23 0.1
V12A ½ 6AL5
V12B ½ 6AL5
R65 27K
R66 270
R85 470
C85 6.25μf
M1
μA
R74 680
R73 5K
R72 5K
R71 5K
R70 5K
R69 5K
R68 5K
R67 5K
1 MC
250 KC
100 KC
25 KC
10 KC
2.5 KC
1 KC
250 CPS
100 CPS
R34 1M
C18 0.1
C18.1 0.1
C19 40μf
C22 0.1
A (5V TO -175 REF. GND.)
FROM POWER SUPPLY
(A-150)
SYNC OUTPUT
R83 4.7K

movement by the range-selector switch section $SW_{1\text{-}1}$. Either the resistor or the capacitor is made variable for each range for provision of calibration adjustments.

We know that the charge in a capacitor represents a certain number of electrons (coulombs). Since each cycle of the square wave drives one charge through the meter, the scale reading will indicate the number of charges per second, or the repetition rate of the square wave. Resistors R_{65} and R_{66} develop a bias voltage on the meter diode tube (pin 5) for cancellation of contact potential.

Since a multivibrator can be synchronized by means of an injected frequency, a sync section is also provided in the configuration of Fig. 19.9 to increase the utility of the instrument. Sync signals fed to the sync input binding post are coupled through capacitor C_{32} to the grid of sync amplifier tube V_{13}. Variable grid bias for this tube is provided by the sync amplitude control, R_{80}, which operates in the cathode circuit. Resistor R_{79} develops a minimum fixed bias for protection of the tube. We observe that bias voltages greater than cutoff are available because of current flow through R_{81}; thus, large-amplitude sync signals can be accommodated.

In operation, high-frequency compensation is obtained in the cathode circuit of V_{13} (Fig. 19.9) by means of partial cathode bypassing via C_{33}. The amplified sync signal from the plate of V_{13} is coupled to the plate of multivibrator tube V_1 through diode V_{14}. This is a disconnect diode; it disconnects the multivibrator from the sync amplifier while V_1 conducts. This action prevents another sync impulse from reaching V_1 until the multivibrator has completed its cycle. Note that an output signal for synchronizing external equipment is developed by cathode follower V_{15}. The square wave is differentiated by C_{34} and R_{83}, so that a spike-shaped sync output signal is developed.

19.4 PULSE GENERATION

Pulse generators are basically similar to square-wave generators. However, a pulse generator supplies rectangular waveforms instead of a symmetrical square waveform. If R_{25} were provided with a considerably greater range in the multivibrator section shown in Fig. 19.9, the instrument would operate as a pulse generator. On the other hand, the output amplifier would then be energized out of the tube ratings, inasmuch as a pulse has unequal on and off intervals. Other things being equal, the power tubes in a pulse generator must have a higher power rating than in a square-wave generator. It is usually desired to have a choice of positive or negative pulses, with provision for single-pulse generation. Accordingly, the plan of a pulse generator is often more elaborate than that of a square-wave generator.

FIG. 19.10 *Appearance of a typical pulse generator. (Courtesy, Simpson Electric Co.)*

A typical pulse generator is illustrated in Fig. 19.10. This instrument provides a pulse repetition rate from 10 Hz to 100,000 Hz with a rise time of less than 20 ns. The pulse width is variable from 0.1 to 1000 μs. Meters are provided to indicate the pulse repetition rate and the pulse duration in microseconds. Either positive or negative pulses can be supplied, at an amplitude up to 10 V peak-to-peak. The amplitude is controlled by means of a step attenuator and a continuous attenuator; toggle switches are utilized in the step-attenuation section. As would be anticipated, pulse-generating circuitry is much the same as square-wave circuitry, except that rectangular waveforms are processed instead of square waveforms.

All pulse waveforms necessarily have at least a slight amount of distortion in practice. Figure 19.11 depicts the chief

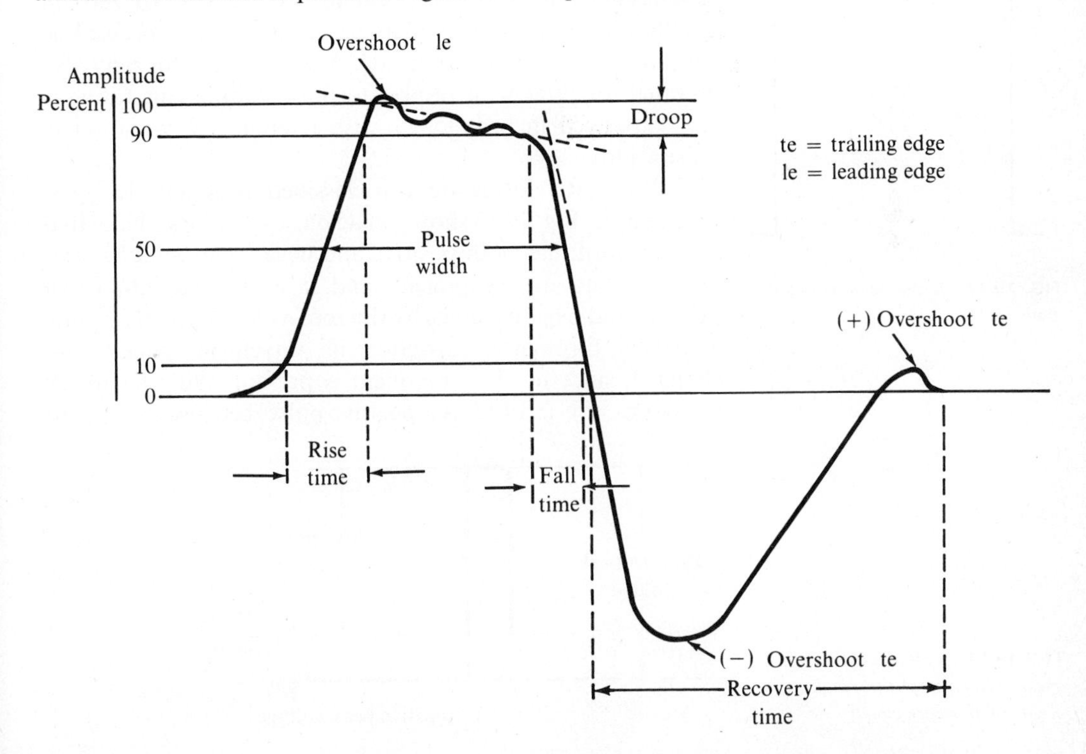

FIG. 19.11 *Pulse characteristics.*

characteristics of concern to the engineer:

1. Rise time
2. Fall time
3. Leading-edge overshoot
4. Droop (also called tilt)
5. Pulse width
6. Trailing-edge negative overshoot
7. Trailing-edge positive overshoot
8. Recovery time

An engineer may also be concerned with the damping time of the oscillatory interval following an overshoot. The damping time is measured in microseconds, as depicted in Fig. 19.12, and is the time required for the damped sine wave to decay to zero. Another characteristic that may be of concern is called jitter; this is defined as an irregular random variation in the pulse repetition rate. Although jitter is very slight in a well-designed instrument, it cannot be reduced to zero in practice. Jitter is measured in percentage and expressed as variation in pulse repetition rate. A pulse generator may also be rated for jitter with respect to pulse width. This rating is also expressed in percentage, with reference to variation in pulse width.

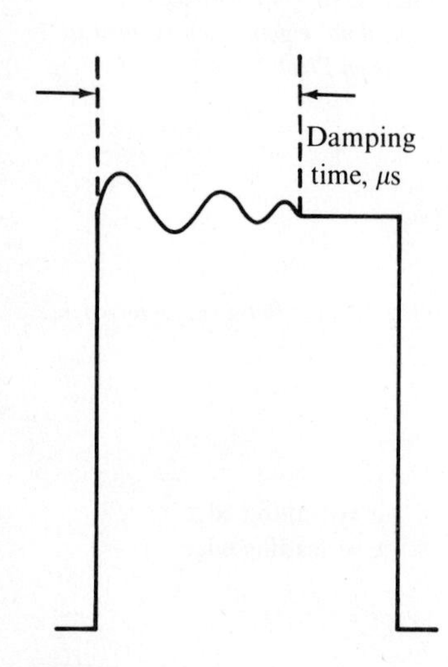

FIG. 19.12 *Example of damping time.*

Pulse waveforms are also classified as ac and dc types; for example, Fig. 19.13 shows an ac pulse waveform. Note that the waveform has both positive and negative excursions; this pulse has no dc component, and it is divided into equal positive and negative areas by the zero-volt axis. If an ac pulse is passed through a capacitor, its waveform remains unchanged, since no dc component is present. We refer to the waveform in Fig. 19.13 as a positive pulse, because its greatest

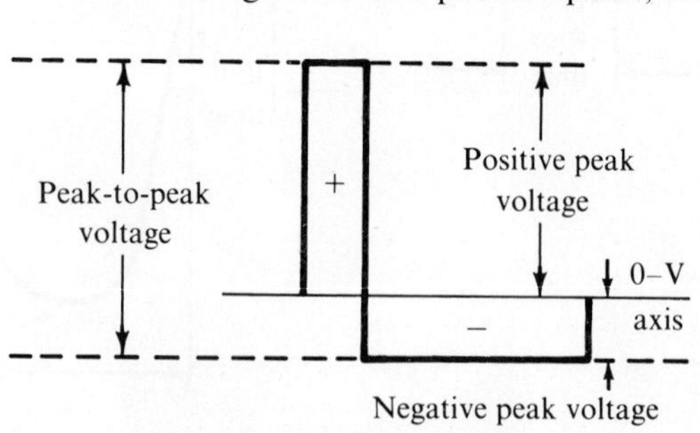

FIG. 19.13 *Voltages in an ac pulse waveform (positive and negative areas are equal).*

excursion has positive polarity. If this waveform is passed through a conventional amplifier stage, it will become a negative pulse, because its greatest excursion will then have a negative polarity.

With reference to Fig. 19.14, we observe the appearance of positive and negative dc pulses. A dc pulse starts at the zero-volt level and has one polarity only; in other words, it has a dc component that causes the waveform to rest on the zero-volt level. If a dc pulse is passed through a capacitor, it is changed into an ac pulse, for the dc component is removed by the coupling capacitor, and the zero-volt axis in the ac pulse then appears at the level of the dc component in the dc pulse. This shift in the zero-volt location is easily demonstrated by applying a dc pulse to an oscilloscope, and switching the scope from dc response to ac response.

Pulse generators are greatly elaborated for special applications. For example, a sophisticated generator can provide rectangular pulses as described above, pulse bursts such as illustrated in Fig. 19.15, binary pulse words up to 112 bits, doublet pulses, pulses superimposed on wider pulses or square waves, staircase waveforms, step functions as depicted in Fig. 19.16, and both triangular and trapezoidal waveforms. Students who are interested in the design of sophisticated pulse generators are referred to specialized texts on this subject.

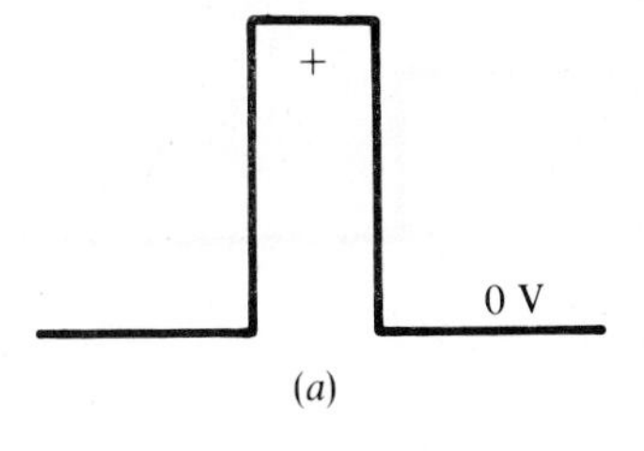

(*a*)

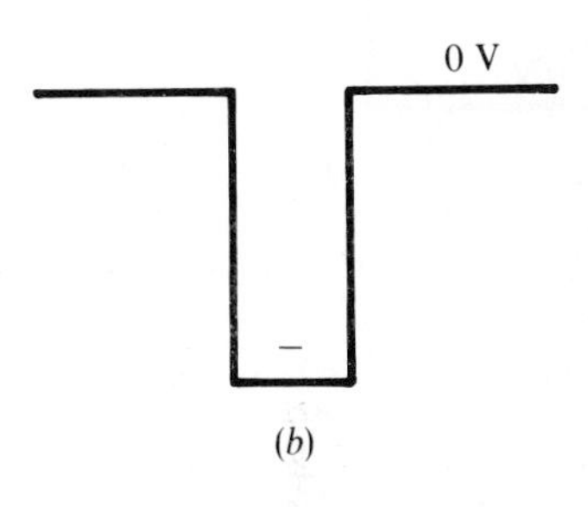

(*b*)

FIG. 19.14 *Pulses. (a) Positive dc; (b) negative dc.*

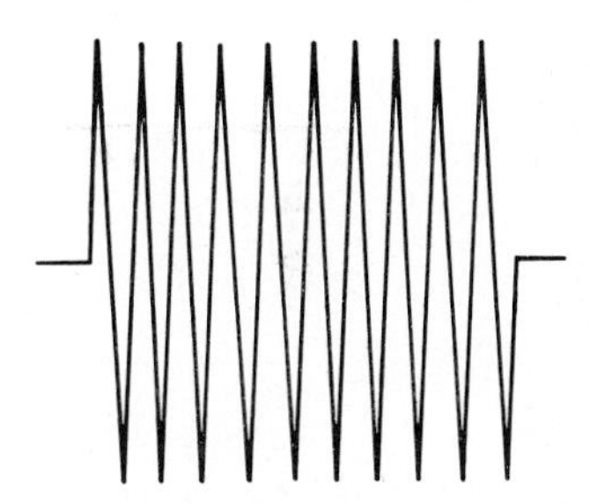

FIG. 19.15 *A pulse-burst waveform.*

19.5 APPLICATIONS

Pulse or square-wave tests are useful for determining the approximate Q value and bandwidth of a tuned circuit. Figure 19.17 shows a commonly used test arrangement. A generator with a fast rise is used, and the output signal is coupled into the tuned circuit by means of a "gimmick" consisting of a loop or a few turns of wire. In turn, the LC circuit is shock excited and undergoes transient oscillations termed *ringing*. The ringing waveform is evaluated for the Q value of the circuit as shown in Fig. 19.18. We count the peaks in the waveform from its maximum point to its 37-percent-of-maximum-amplitude point. The number is then multiplied by pi (3.1416), to calculate the approximate Q value. In the example shown, there are 10 peaks from the 100 percent point to the 37 percent point and therefore the Q value is approximately 31.4.

If a triggered-sweep scope is used, the ringing frequency may be determined from the settings of the time-base controls

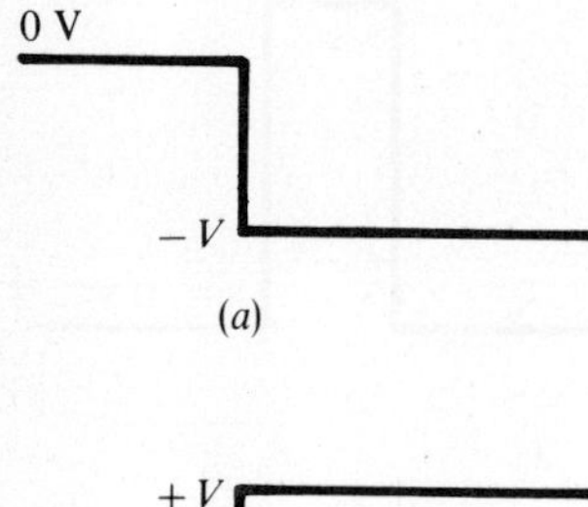

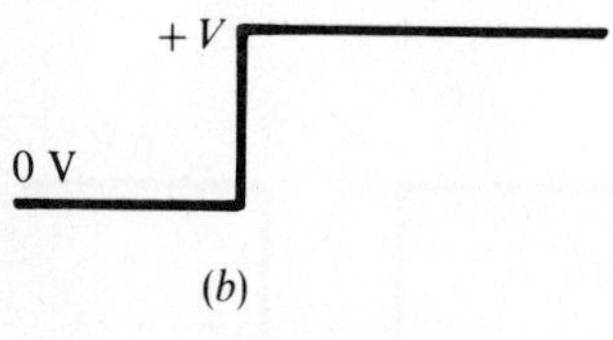

FIG. 19.16 *Step functions. (a) Negative; (b) positive.*

in Fig. 19.17. In turn, the bandwidth of the *LC* circuit is easily calculated after the Q value has been determined. That is, the circuit bandwidth is equal to approximately f_o/Q, where f_o is the ringing frequency, and Q is the value measured as described above. Note that although this is an approximate measurement, its accuracy is adequate for most engineering purposes. The accuracy of the method is very good for high Q values, and is generally adequate for moderate Q values. It should not be used for circuits that have quite low Q values.

A somewhat similar ringing test can be made on a tuned-if transformer as shown in Fig. 19.19. When the primary and the secondary are tuned to the same center frequency, the ringing pattern illustrated in Fig. 19.20 is normally displayed. This pattern can be evaluated to determine each of the parameters involved in the transformer action. Comparative tests are usually made; for example, pulse tests provide a quick and efficient method of incoming inspection for if transformers in factories.

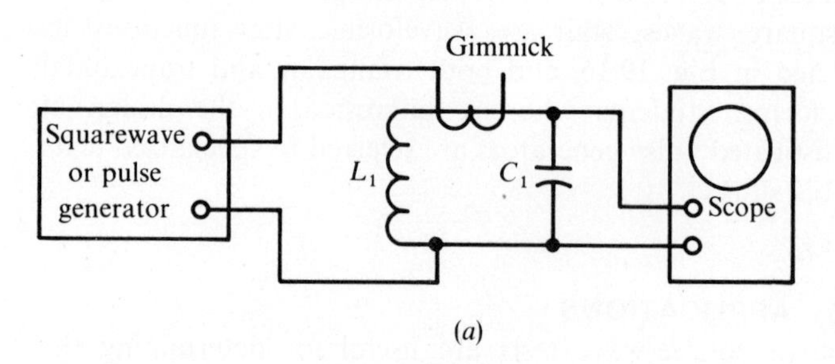

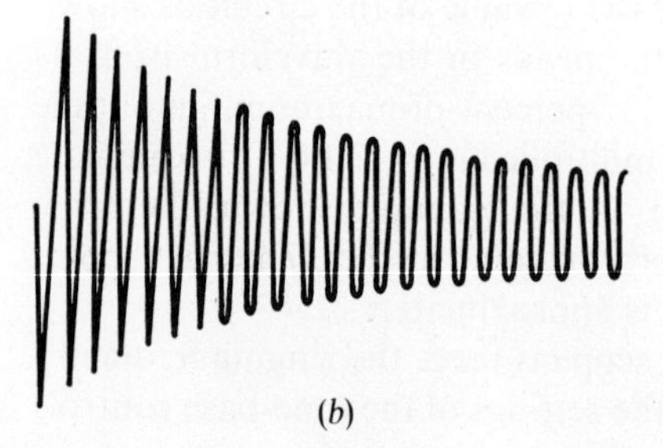

FIG. 19.17 *Test arrangement for ringing an LC circuit. (a) Test setup; (b) ringing waveform.*

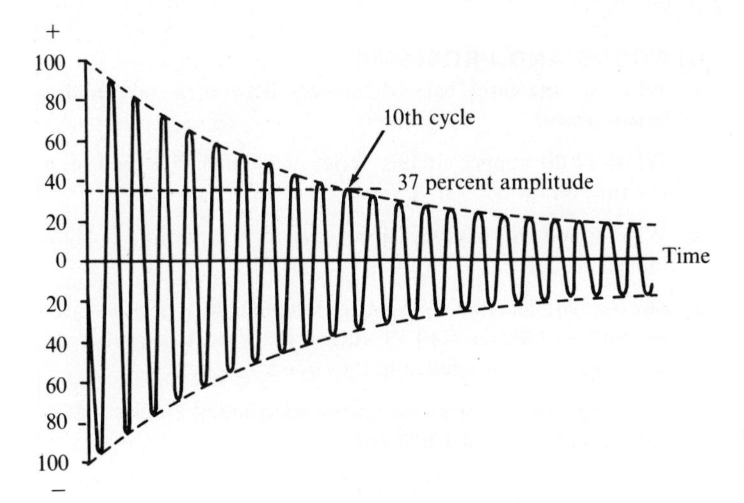

FIG. 19.18 *Example showing Q equals 10π or 31.4.*

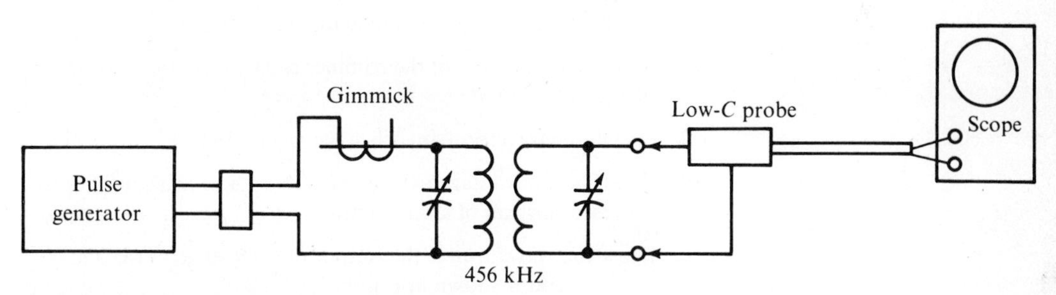

FIG. 19.19 *Test arrangement for ringing test of if transformer.*

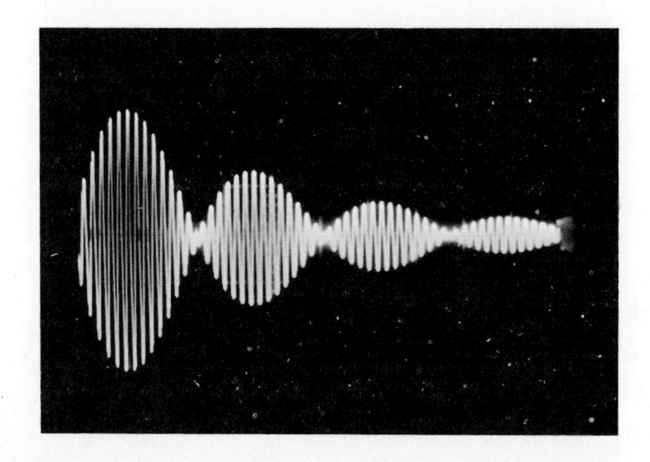

FIG. 19.20 *Typical ringing patterns produced by a tuned if transformer.*

QUESTIONS AND PROBLEMS

1. What are the three basic differences between a pulse and a square wave?
2. Which of the clipper circuits, series or parallel, give the fastest rise time and why?
3. Why should the parallel clipper circuit operate into a high-impedance load?
4. Draw a parallel clipper circuit to develop a 20 V peak-to-peak output (+10 V and −10 V) when a 100 V peak-to-peak sinusoidal waveform is applied to the circuit.
5. What limitation is imposed on the rise time of a pulse that is passed through a particular circuit?
6. What is the function of the 10-μF capacitor connected between V_{5A} and V_{5B} in Fig. 19.6?
7. When used as an output stage of a square-wave generator, what two functions does the cathode follower perform?
8. What is the purpose of the trimmer capacitor in the grid circuit of V_{5B} in Fig. 19.6?
9. Sketch the waveform at each stage in Fig. 19.6.
10. What is the advantage of using a multivibrator in a square-wave generator instead of an oscillator?
11. Draw the waveform at the output of each stage in the square-wave generator shown in Fig. 19.7.
12. Why is the gated-beam tube, in the square-wave generator depicted in Fig. 19.7, operated with low values of screen-grid and plate voltages?
13. What is the purpose of each of the stages in the square-wave generator depicted in Fig. 19.7?
14. How do capacitors C_{13}, C_{16}, and C_{17} increase the frequency response of the amplifier V_4 shown in Fig. 19.7?
15. What is the disadvantage of using low-value plate load resistors to increase the bandwidth of an amplifier?
16. List three reasons for a lack of symmetry in the output waveform of the square-wave generator shown in Fig. 19.9.
17. What is the purpose of capacitors $C_{1.1}$ and $C_{11.1}$ in the square-wave generator shown in Fig. 19.9?
18. What is the purpose of resistor R_{32} in the circuit shown in Fig. 19.9?

19. Parasitic oscillations are suppressed by fixed resistors. Five resistors are used to reduce parasitic oscillations in tubes V_4 and V_5 in the circuit shown in Fig. 19.9; what are the numbers of these resistors?

20. What is the maximum value of the output signal from the generator shown in Fig. 19.9?

21. Explain the operation of the frequency-meter section of the circuit shown in Fig. 19.10.

22. What is the function of V_{11} in the circuit shown in Fig. 19.9?

23. What are the requirements for a sync signal that is to synchronize the multivibrator circuit in Fig. 19.9?

24. Explain the operation of the sync section of the square-wave generator in Fig. 19.9?

25. Why are the circuits for pulse generators usually more elaborate than the circuits for square-wave generators?

26. What characteristics of square waves are of chief concern?

27. Define the term *jitter* as it applies to a square wave.

28. What is the difference between an ac and a dc type of pulse?

TUBE AND SEMICONDUCTOR DEVICE TESTERS

TUBE TESTERS

20.1 BASIC TESTS

Tube testers are necessary in factories and laboratories to check the characteristics of electron tubes. Table 20.1 exemplifies the design parameters of a television receiving tube. It is evident that comparatively elaborate and sophisticated test equipment must be employed to check certain parameters, such as the plate characteristics in the regions where maximum average current and power ratings are exceeded. This is accomplished by means of pulse techniques.

Tube testers are necessary in service shops to determine whether a receiver trouble symptom is being caused by a defective tube or by a circuit fault. Although the most reliable test of a tube is a performance check in a normally operating circuit, this method is often unavailable. An extremely large inventory of tube types is stocked by most service shops.

The basic requirements for a service-type tube tester

include tests for interelectrode short circuits, leakage, cathode emission, transconductance, power output, and grid current caused by gas or grid emission. Certain types of tubes should be checked for the value of grid-bias voltage that produces plate current cutoff, and others need to be checked for the value of plate voltage at the screen-grid knee point. For example, a twin triode will not oscillate in a multivibrator circuit if its plate cutoff point is too remote. A horizontal-output tube will produce subnormal beam deflection if its screen-grid knee point occurs at a comparatively high value of plate voltage. (See also Table 20.1.)

20.2 BASIC TUBE-TESTING REQUIREMENTS

With the exception of simple heater-continuity testers, all tube testers have facilities for adjusting some or all of the test voltages applied to a tube. Adjustment may be provided by switches, punched cards, or a multiplicity of tube sockets, as explained in greater detail subsequently. Tubes often become defective because of interelectrode leakage or short circuits. For example, the grid in a high-transconductance receiving tube may be mounted very close to the cathode. If the structure warps slightly due to thermal cycling, the grid wires may contact the cathode. Heater-to-cathode leakage or short

TABLE 20.1 *Pentode characteristics for 44-MHz IF amplifier design (tubes operated under rated conditions that give highest G_m)**

TYPE	R_{input}, Ω	R_k UNBYPASSED FOR MINIMUM ΔC_{in} FOR ΔBIAS, Ω	C_{in} UNSHIELDED RATED pF	C_{out} UNSHIELDED RATED pF	ΔC_{in} COLD TO HOT	RATED g_m μmho	EFFECTIVE g_m WHEN UNBYPASSED R_k IS USED μmhos
6BJ6	1160	150*	4.5	5.0	1.8	3800	2100§
6CB6	10900	68	6.3	1.9	1.8	6200	3900
6BH6	11300	150*	5.4	4.4	2.3	4600	2300§
6AG5	15000	82	6.5	1.8	2.3	5100	3800
6AK5	35800	68	4.0‡	2.8‡	1.2	5100	3500
6BC5	12400	68	6.5	1.8	2.0	6100	4000
6AH6	5300	68	10.0‡	3.6‡	3.2	11000	5000
6AU6	7900	140†	5.5	5.0	2.6	5200	2500§
6BA6	9500	130†	5.5	5.0	2.5	4400	2500§

*Data courtesy of Sylvania.
†This value greater than that required for correct bias.
‡With shield—has little effect on C_{in}.
§Will be less actually due to excess bias, unless the grid is returned to a positive potential. See footnote †.

circuits can develop due to thermal cycling and abrasion of insulation. Leakage from grid to plate may be caused by overloading of a tube, with resulting production of vapor deposits.

Since short circuits and leakage are common defects, service-type tube testers are designed to make an initial test of a tube as depicted in Fig. 20.1. If there is appreciable leakage between the electrodes under test, the neon bulb will glow in response to leakage current flow. In such case, the tube is rejected without making other tests. Note that some tube testers may be damaged if a merit test is attempted on a short-circuited tube. Interelectrode leakage tests are always made with the heater energized, because a leakage condition will sometimes develop only while the tube is at operating temperature. Note that the configuration in Fig. 20.1 is basically a continuity indicator. This principle is also employed in the simplest types of tube testers that check only for heater continuity, such as illustrated in Fig. 20.2.

The simplest merit test of a tube is an emission test, which indicates the ability of the cathode to emit electrons. Figure 20.3*a* shows a basic circuit for making an emission test. Note that all the electrodes in the tube, except the cathode, are connected to the plate. The filament or heater is operated at rated voltage, and a low positive voltage is applied to the plate circuit; the resulting emission current is indicated by a meter. Readings that are well below the average for a particular tube type indicate that the total number of available electrons has been reduced to a point that the tube is unable to function properly. In service-type testers, an ac voltage is utilized in the plate circuit, as depicted in Fig. 20.3*b*, and the dc meter is deflected by the resulting rectified half-sine waves.

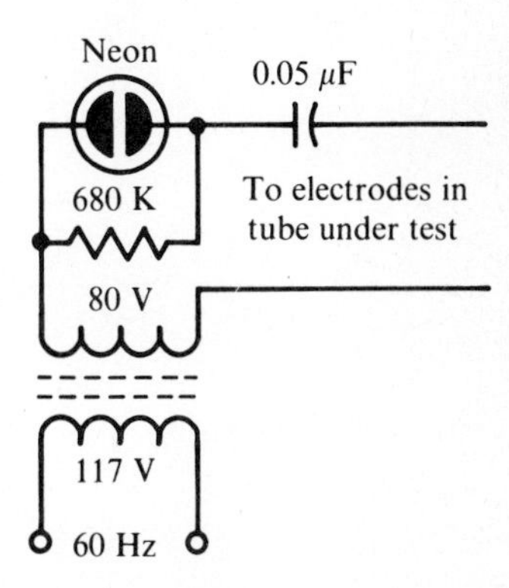

FIG. 20.1 *Interelectrode short and leakage circuit.*

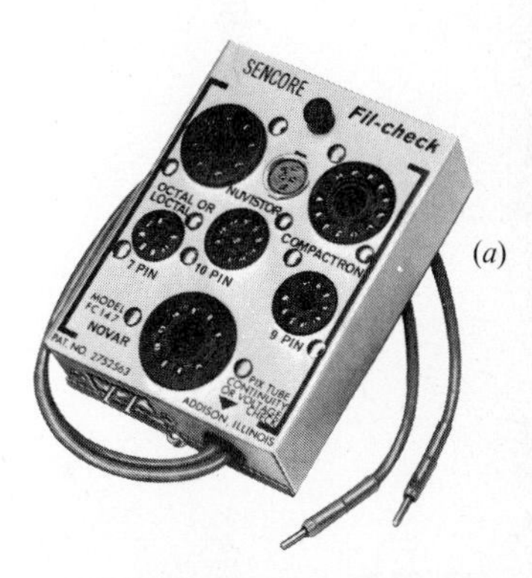

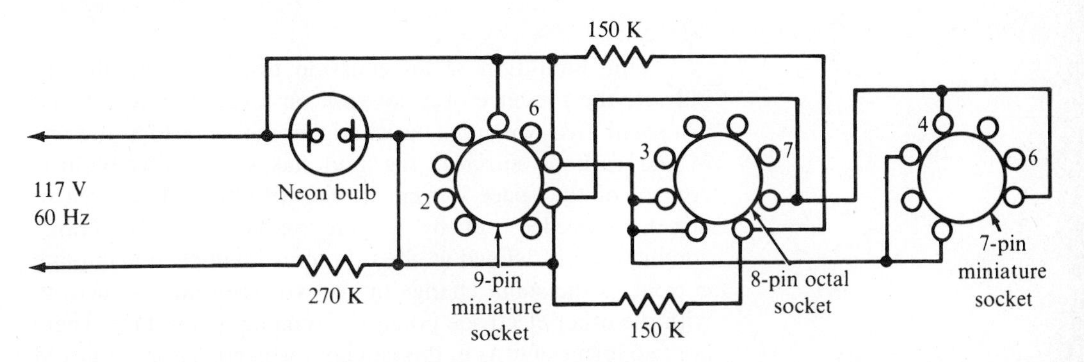

FIG. 20.2 *A heater continuity checker. (a) Appearance; (b) circuit diagram. (Courtesy, Sencore)*

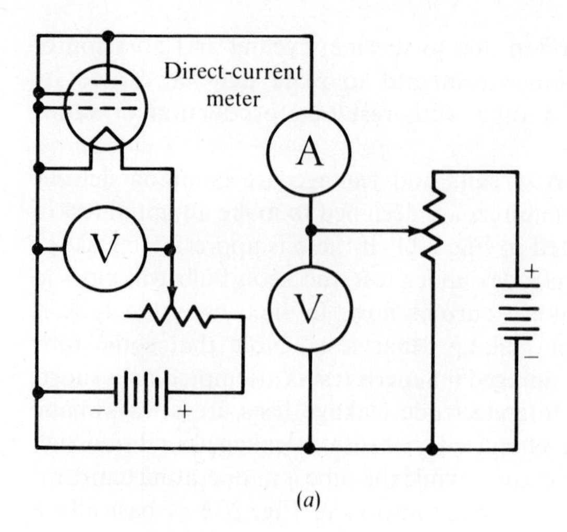

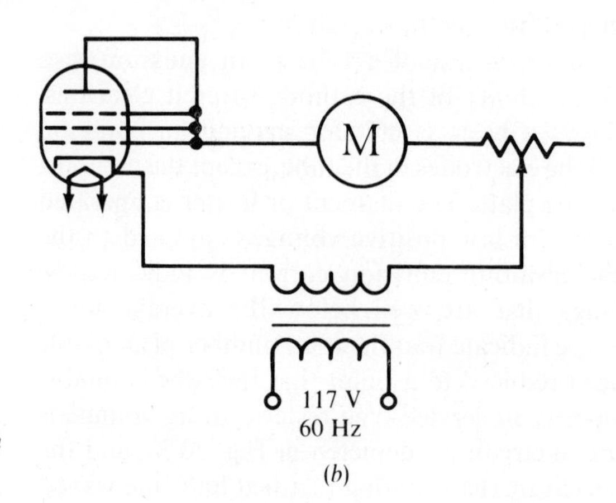

FIG. 20.3 *Emission test circuit. (a) With dc supply voltage; (b) with ac supply voltage.*

One limitation of an emission test is its inability to indicate the presence of a hot spot on a cathode. A *hot spot* is a small area which has much higher emission than the rest of the cathode surface; the grid has comparatively little control of the space current in this situation. This type of defect is weeded out by a transconductance test. Transconductance is defined as the ratio of a small change in plate current to the small change in grid voltage that produces it, with the other electrode potentials remaining constant. There are two forms of tests in this category which may be employed in tube testers. With reference to Fig. 20.4*a*, one of the basic test circuits is termed a *grid-shift* configuration. Let us consider the basis of this test.

We observe in Fig. 20.4*a* that appropriate operating voltages are applied to the tube electrodes. In turn, the meter indicates an associated value of plate current. The grid bias is then shifted by a small amount, such as 1 V, and another plate current reading results. The difference between these two plate current values is directly related to the transconductance value of the tube. Accordingly, the scale of the meter may be calibrated in micromhos. It is evident that transconductance values are indicative of control grid action. The chief limitation of the grid-shift method is that the tube is checked under dc conditions, whereas a tube usually operates in the audio-frequency, radio-frequency, or very-high-frequency range. Therefore, the more elaborate transconductance testers utilize an ac grid drive voltage.

With reference to Fig. 20.4*b*, the grid of the tube under test is driven by a small ac voltage, such as 1 V which preferably has a frequency in the normal operating range of the tube. Note that a dc meter is not used in the plate circuit; instead, an ac meter of the dynamometer type is employed. If a 1-V rms signal is applied to the grid, the reading of the plate current meter multiplied by 1000 gives the transconductance value in micromhos. Of course, the meter scale may be calibrated to read micromho values directly. A dynamometer type of meter is preferred because it indicates true rms values, regardless of any waveform distortion that might be introduced by the tube. Moreover, this type of meter has a very low internal resistance, which minimizes the change in effective plate resistance due to the presence of the meter in the plate circuit.

Figure 20.4*c* shows a block diagram for a typical service-type dynamic transconductance tube tester. This instrument employs a 5-kHz test signal; hence it is basically designed for checking audio amplifier tubes. Diodes, of course, must be accommodated by an emission test section (not shown in the diagram). Tube manufacturers tend to prefer another type of test, called the power-output test, when only one method of checking miscellaneous types of tubes is provided by the instrument. For example, when a voltage-amplifier tube is tested, a power output value is indicative of the amplification and output voltage obtainable from the tube. When a power output tube is tested, a power output value gives an accurate

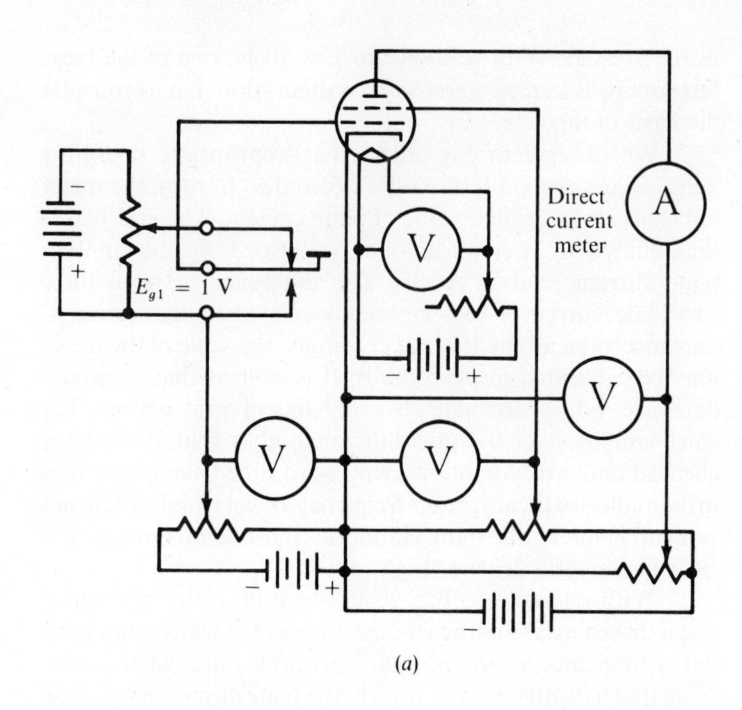
Direct current meter
A
V
V
V
V
E_{g1} = 1 V

(a)

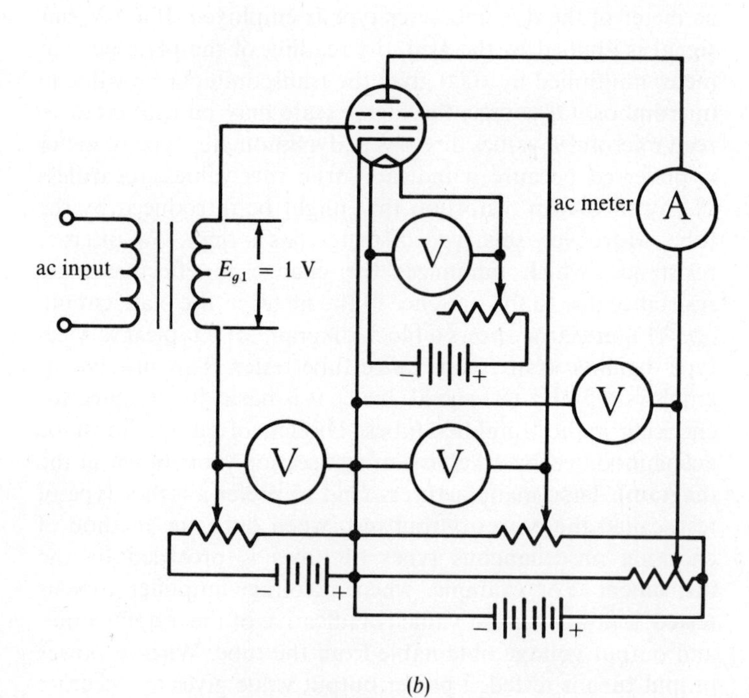
ac meter
A
ac input
E_{g1} = 1 V
V
V
V
V

(b)

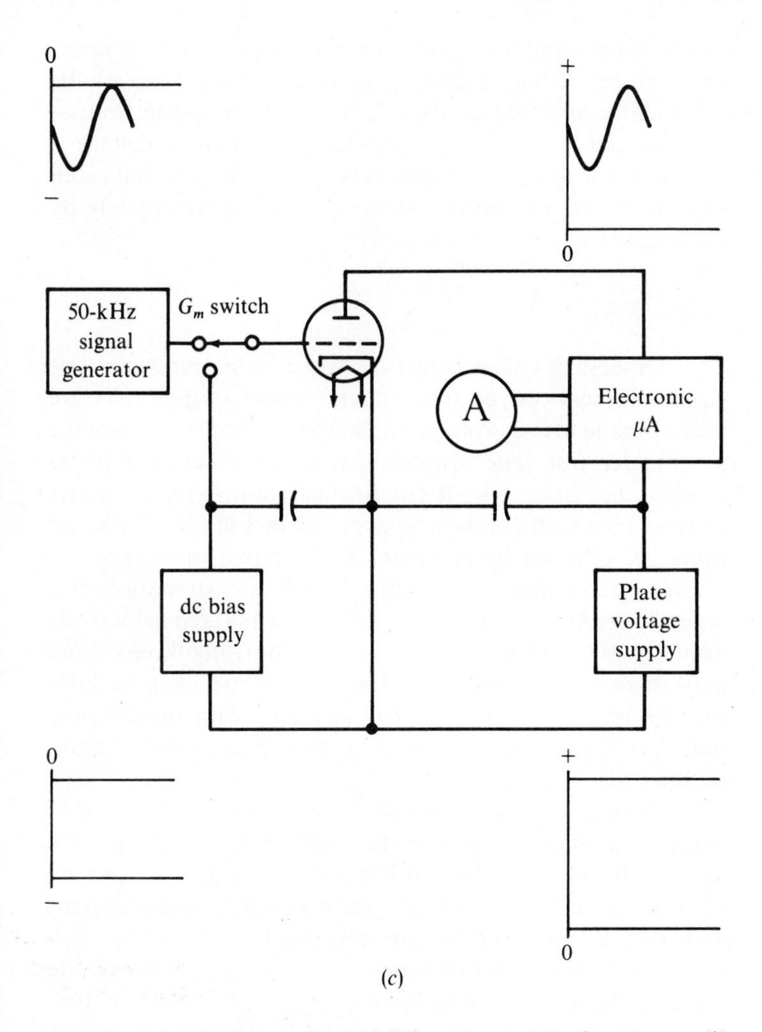

FIG. 20.4 *Basic tube tester. (a) Grid-shift or static transconductance test; (b) ac drive or dynamic transconductance test; (c) block diagram of a service-type dynamic transconductance tube tester.*

indication of tube performance. Let us consider the features of this test method.

With reference to Fig. 20.5*a*, the basic circuit for a power output test of tubes in Class *A* operation is depicted. Note that the ac output voltage is developed across the plate load impedance *L*, and is indicated by the ac meter. The dc plate current is blocked from the meter circuit by capacitor *C*. Power output values can be calculated from the indicated current values and the value of the load resistor; or, the meter scale can be calibrated directly in power values. When a tube is normally operated in a Class *B* circuit, the power output test depicted in Fig. 20.5*b* is preferred. An ac drive voltage is applied to the grid, and the plate current that flows is indicated by a dc meter. The power output is given approximately by the formula:

$$P_o = \frac{I_b^2 \times R_L}{0.405} \text{ W} \tag{20.1}$$

Diodes, of course, must be tested by an emission-type circuit. In service-type tube testers, power output tests are usually made with ac voltage applied to all the electrodes of the tube under test. The arrangement is often called a plate-conductance or a cathode-conductance test. A typical plate-conductance configuration is depicted in Fig. 20.5*c*. Picture tubes are checked by emission. The external appearance of a tube tester is illustrated in Fig. 20.6. It is customary in this type of instrument to provide a single scale marked in intervals from 0 to 100, and an "English" scale comprising three sectors marked *Replace*, *?*, and *Good*. The internal switching facilities provide test conditions whereby any tube with rated power output will produce a meter indication in the Good section on the scale.

Picture tubes can be tested for emission or for beam current with most service-type tube testers. An extension cable such as illustrated in Fig. 20.7 is used to connect the picture tube to the tester. The distinction between emission current and beam current in a picture tube can be seen in Fig. 20.8. Only the electrons emitted from the central area of the cathode proceed through the grid aperture to form the beam that travels to the screen. The remainder of the cathode area is largely ineffective in formation of the electron beam. Accordingly, a picture tube will become defective if the emission from the central area of the cathode is weak, although the remainder of the cathode area might have normal emission.

Let us now consider a widely used transconductance test circuit, the basic arrangement of which is seen in Fig. 20.9. The power supply employs a mercury-vapor rectifier tube to produce plate supply voltage; in turn, substantial current demands can be met. A vacuum-type rectifier tube is used to produce the bias and screen supply voltages. Since the rectifier outputs are unfiltered, the tube under test (T_a) is operated

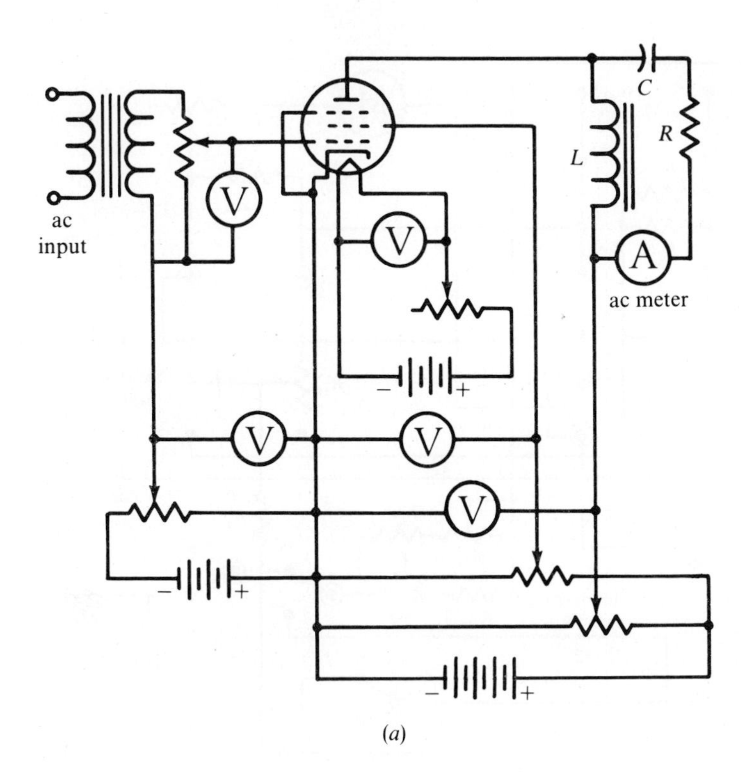

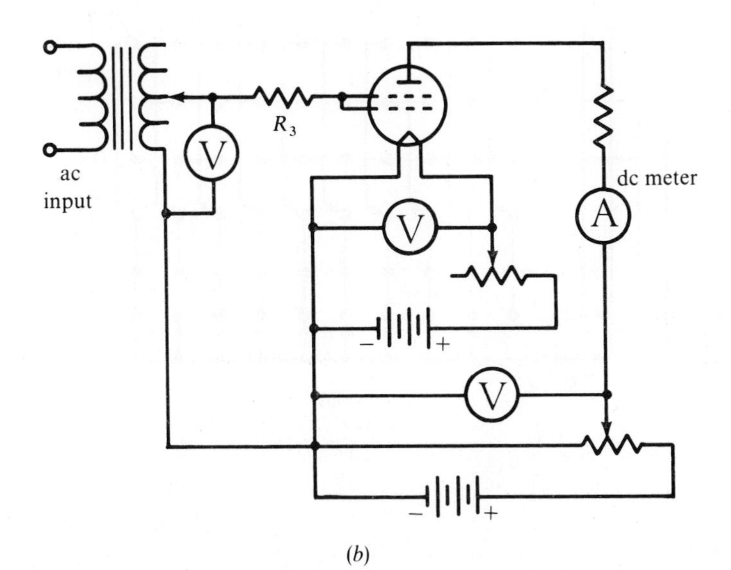

FIG. 20.5 *Tube tester circuits. (Courtesy, Triplett)*

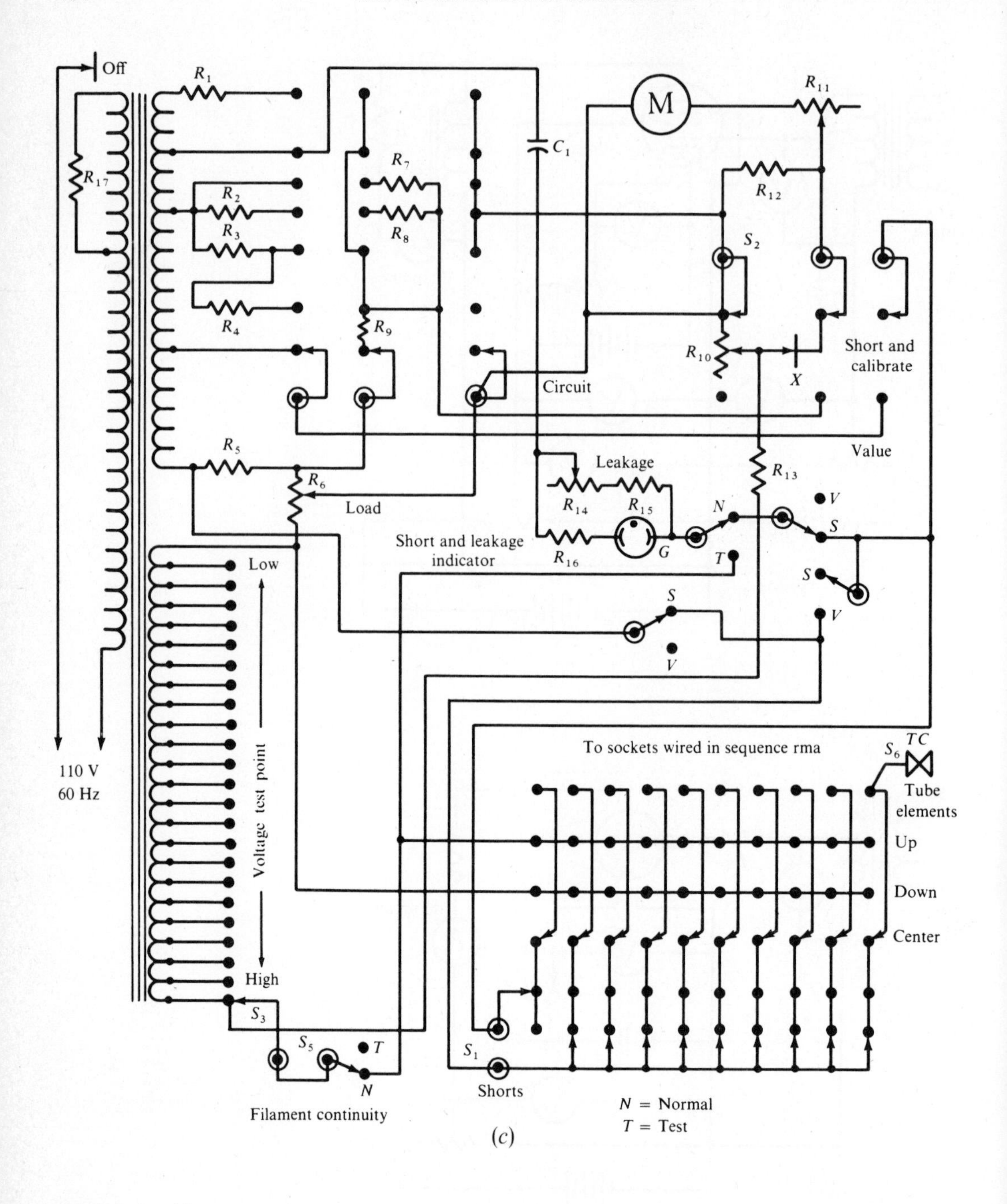

(c)

FIG 20.5 *(cont'd)*

with pulsating dc voltage. Note that voltages E_1, E_2, and E_3-E_4 are all in phase. Anodes P_1 and P_2 are alternately driven positive and negative. It is helpful to begin an analysis of the circuit action with the assumption that the grid signal voltage E_5 is not varied.

When anode P_1 is positive in Fig. 20.9, and S_1 is closed, electrons flow from K_2 to P_1, through E_1 to E, through R_1 to D, B, K_1, P_a, and S_1. A portion of this current flows through meter M and resistor R_2, tending to deflect the pointer to the left. However, anode P_1 is negative on the next half cycle, and anode P_1 is positive. Therefore, the pointer is subjected to equal and opposite forces, and does not deflect. Next, when the grid signal voltage E_5 is varied by the combined action of E_5 and T_2, the grid voltage becomes more negative while P_1 is positive than while P_2 is positive. In turn, the opposing currents through M are unbalanced, and the pointer is deflected. It can be shown that the magnitude of pointer deflection is proportional to the mutual conductance value of T_a.

Note that two grid-drive voltages are provided in Fig. 20.9, with values of 1 and 5 V rms. The 5-V signal is used in checking power-type tubes in order to provide a meter indication that is indicative of power output capability. Switch S_4 is closed during merit tests of a tube; it is opened to check for grid current. If grid current is present, due to gas or grid emission, the grid becomes slightly more positive when S_4 is opened, thus producing a meter deflection. Diodes are tested in an emission configuration, not shown in Fig. 20.9.

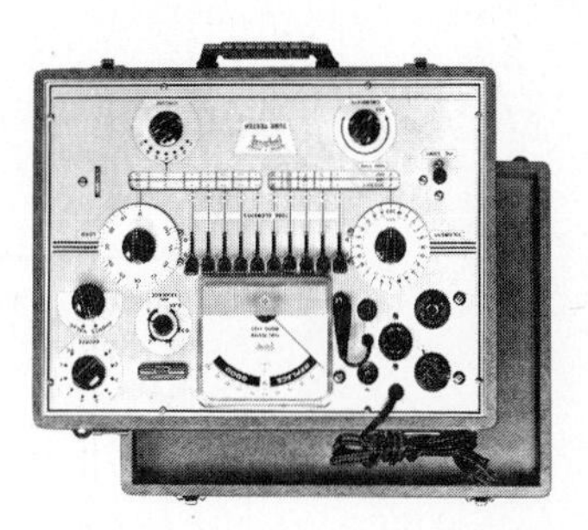

FIG. 20.6 *A service-type tester that provides a power output indication. (Courtesy, Triplett)*

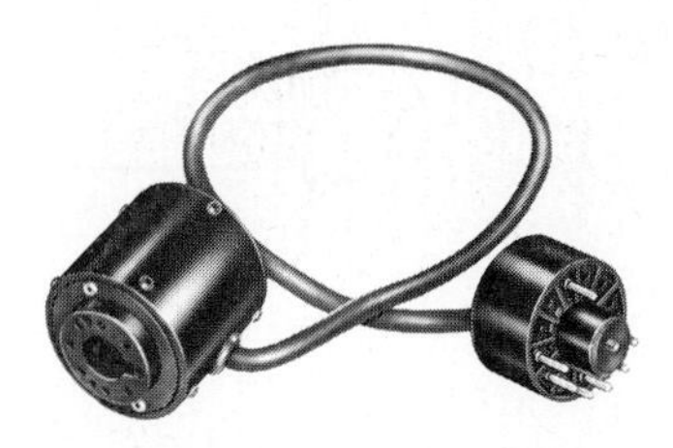

FIG. 20.7 *An extension cable for testing picture tubes. (Courtesy, Precision Apparatus Co.)*

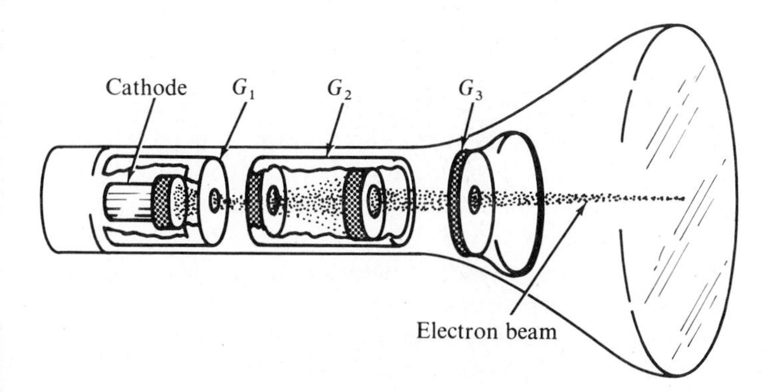

FIG. 20.8 *Central area of the cathode contributing to beam formation.*

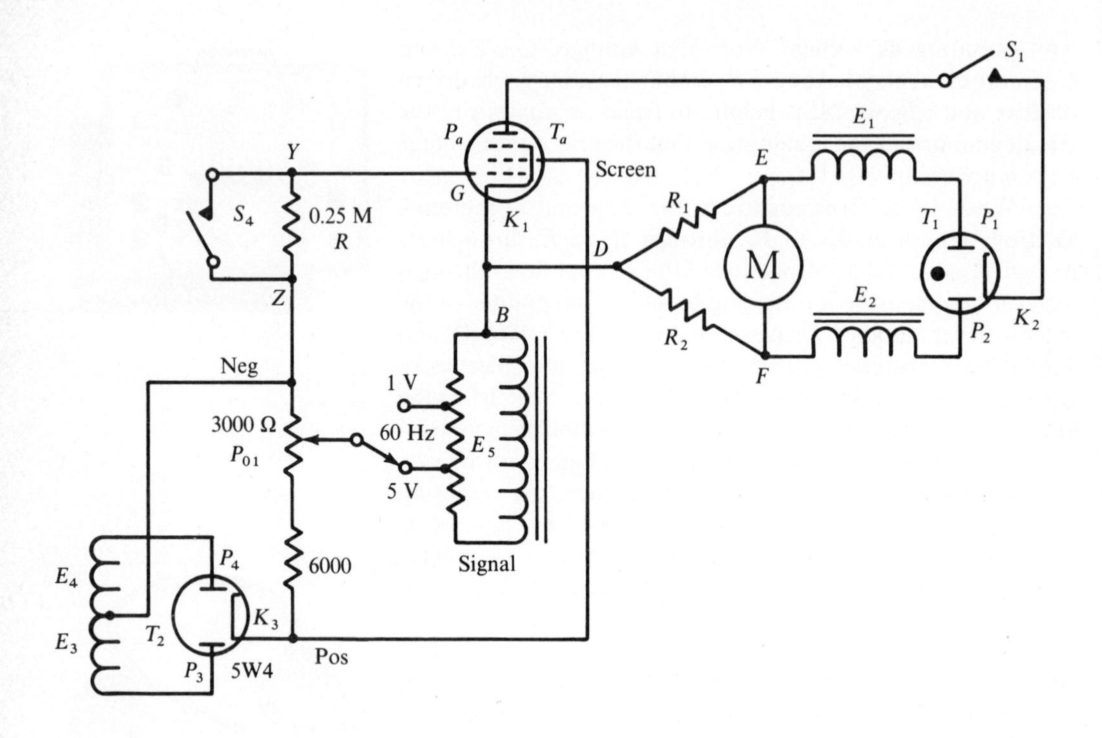

FIG. 20.9 *A transconductance test configuration that employs ac voltages.*

20.3 LABORATORY-TYPE TUBE TESTERS

Tube testers used in development and research work are highly specialized compared with service-type instruments. Similarly, tube testers employed in factory quality-control procedures have specialized functions. A lab-type tube tester is typically assembled from subunits and various instruments in order to provide accurate measurement of a particular tube parameter. Tubes are rated for input capacitance values, and a measurement of input capacitance requires comparatively sophisticated techniques. At higher operating frequencies, such as utilized in tv if amplifiers, a tube has a significant input resistance as well as an input capacitance value. Therefore, the test setup that is used must be designed to measure capacitance in the presence of shunt resistance.

It should be observed that a designer of a tv if amplifier is concerned with the value of unbypassed cathode resistance to be used in order to compensate for the input capacitance change of a pentode tube when its control grid bias is varied. The required value of cathode resistance depends both upon the input resistance and upon the input and output capacitances of the tube. Modern tv if amplifiers operate at approximately 44 MHz. With reference to Table 20.1, we observe these parameters for various types of tubes at a frequency of 44 MHz. The input resistance values, and the change in input capacitance values from ambient to normal operating temperature, are measured with the typical configuration shown in Fig. 20.10.

A measurement of input resistance is made first. Since the circuit resistance must be taken into account in the measurement, the engineer proceeds as follows: With the tube unplugged from its socket, the *LC* circuit in Fig. 20.10 is tuned to 44 MHz, the signal generator is set to 44 MHz, and the generator output is adjusted for an indication of 1 V rms on the VTVM. Accurately calibrated resistors are then connected between the grid and cathode terminals of the socket to provide meter readings from which the circuit resistance can be calculated. With a given value of resistance connected between the grid and cathode terminals, the *LC* circuit is retuned for maximum indication on the VTVM, and the

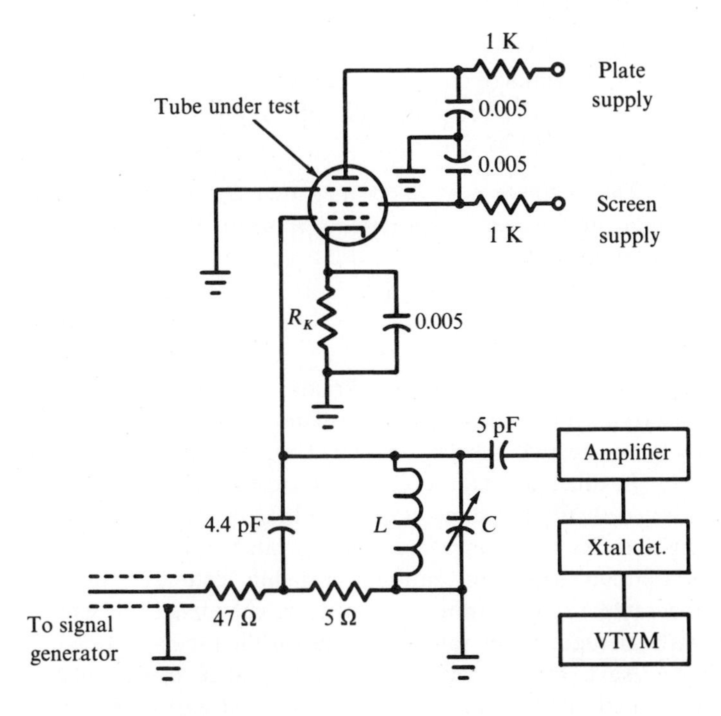

FIG. 20.10 *Test arrangement for R_{in} and ΔC measurements.*

circuit resistance value is calculated from the formula:

$$R_c = R\frac{E_c}{E_r} - 1 \tag{20.2}$$

where R_c = circuit resistance in ohms

R = calibrating resistance in ohms
E_c = VTVM reading with R disconnected
E_r = VTVM reading with R connected between grid and cathode terminals

To ensure accuracy, several values of R are tested, and the average of the calculations in accordance with formula (20.2) is taken as the most probable value of R_c. Next, the tube to be tested is plugged into the socket, and rated supply voltages are applied. No calibrating resistor is used in this part of the procedure. The LC circuit is retuned for maximum reading on the VTVM, and the input resistance of the tube is then calculated from the formula:

$$R_t = \frac{R_c}{(E_c/E_t) - 1} \tag{20.3}$$

where R_t = input resistance of tube in ohms
R_c = value of circuit resistance
E_c = VTVM reading with tube unplugged
E_t = VTVM reading with tube plugged into its socket

The engineer next takes the rated G_m of the tube into account, and corrects the foregoing value for R_t in accordance with the formula:

$$R_{\text{in}} = \frac{K}{G_m} \tag{20.4}$$

where K is a constant (more strictly, a parameter) equal to the product of the measured value of input resistance multiplied by the associated measured value of G_m in the test circuit.

In other words, the G_m of the tube in the test circuit is not necessarily the same as the rated G_m of the tube. Accordingly, the engineer measures the G_m value of the tube in the test circuit, using conventional grid-shift technique. In turn, the rated value of input resistance is calculated and stated with reference to the rated G_m value for the tube.

Next, the change in input capacitance of the tube is measured with the supply voltages connected and disconnected.

This is accomplished by noting the difference in values of C in Fig. 20.10 required to resonate the tuned circuit under the two conditions of test. Note that this difference is very nearly the same as the capacitance change that occurs between maximum rated plate current and plate current cutoff. In if amplifiers that are agc controlled, the input capacitance of a tube changes in response to a change in agc voltage because the G_m of the tube is being changed. This results in an undesirable detuning of the associated if circuit.

To stabilize the input capacitance of a tube over the normal range of agc voltage variation, a suitable value of unbypassed cathode resistance may be employed. The required value for the cathode resistor is given by the formula:

$$R_k = \frac{I_p \Delta C}{G_m L C_{glk}} \tag{20.5}$$

where R_k = required value of unbypassed cathode resistance
ΔC = measured change of input capacitance
I_p = plate current in the foregoing measurement
C_{glk} = grid-cathode capacitance of the tube with supply voltages disconnected

We find that employment of an unbypassed cathode resistor reduces the effective G_m of a tube by the ratio:

$$\frac{G_{m_1}}{G_{m_2}} = \frac{1}{1 + R_k C_{gk}} \tag{20.6}$$

in which $C_{gk} = G_m \dfrac{I_k}{I_p}$ (20.7)

Thus, the design engineer sacrifices stage gain to obtain stabilization of input capacitance. Note in passing that a widely used figure of merit for a high-frequency pentode is its gain-bandwidth product ($G \cdot BW$). This figure of merit is a practically constant value (parameter), and is given by the formula:

$$G \cdot BW = \frac{G_m}{2\pi C_T} \tag{20.8}$$

where C_T is the total input capacitance that was measured in the foregoing test procedure.

Although only a brief sampling of laboratory tube-testing procedures can be covered in this chapter, the techniques that are explained are illustrative of the sophistication that is

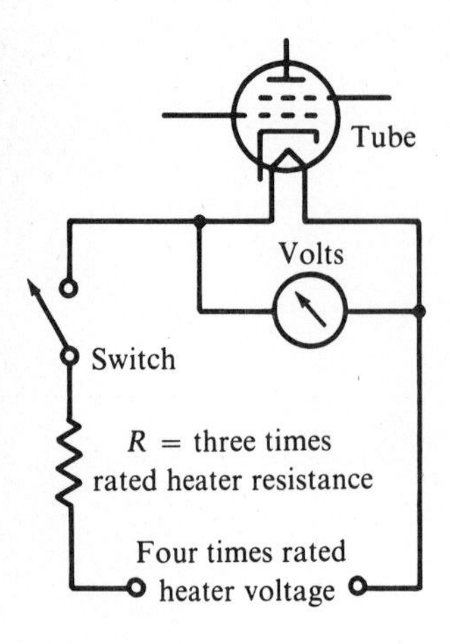

$R = \frac{3E}{1}$ where E and I are the rated heater voltage and current

FIG. 20.11 *Test setup for checking heater warm-up time.*

required of the technician or engineer who works in this area. We conclude this topic with a discussion of heater warm-up time measurement. Since there is necessarily a thermal lag when line voltage is suddenly applied across a series heater string, it is desirable that the tubes have a reasonably uniform temperature rise. If the heaters increase in temperature at about the same rate, undesirable inequalities in voltage distribution will be minimized. Engineers define warm-up time as the amount of time required for the voltage across the heater of a tube to reach 80 percent of rated value in the test circuit shown in Fig. 20.11. We observe that the heater is energized through a series resistance that has three times the rated value of the heater resistance. A source voltage is employed that has four times the value of the rated heater voltage.

Warm-up time is measured in Fig. 20.11 from the time that the switch is closed to the time that the voltmeter reads 80 percent of its final value. A tube manual specifies warm-up time as measured with this arrangement. For example, a 4CY5 tube has a rated warm-up time of approximately 11 s. If tubes with widely different warm-up times are connected in a series string, fast warm-up tubes will glow with excessive brilliance during the warm-up period. This tends to damage heater structures and shorten the life of a tube. If tubes with widely different warm-up times must be connected in a series string, the receiver design engineer can avoid difficulty by including a current slow-down device in the string, such as a Surgistor, or similar relay arrangement.

20.4 APPLICATIONS

Tube testers, both bench and portable models, are widely used in radio and television service work. Technicians generally take a portable tube tester along on house calls, because defective tubes are the most common cause of receiver failure. Simplified types of tube testers are often installed in supermarkets, drugstores, and other retail establishments for the convenience of the general public. These tube testers are sufficiently simple that they can be operated by nontechnical persons. Thus, radio and television tubes are merchandised by stores that may not stock any other electronic items. As noted previously, tube testers used in factories and laboratories are often highly complex instruments that require the attention of highly trained technical personnel.

QUESTIONS AND PROBLEMS

1. How are the plate characteristics of a vacuum tube tested in the regions where maximum average current and power ratings are exceeded?
2. What is the most reliable test of a vacuum tube?
3. What are the basic requirements of a service-type tube tester?
4. What are some of the defects that occur in vacuum tubes and what causes these defects?
5. When testing a vacuum tube, why should a short test always precede the merit test?
6. How is leakage current indicated by the circuit shown in Fig. 20.1?
7. What are the limitations of the tester depicted in Fig. 20.2?
8. Why is it necessary that the filament voltage be adjusted to the rated value when making an emission test for a vacuum tube?
9. What is one limitation of an emission test and how is this limitation overcome?
10. Define transconductance and explain how this test is accomplished on a vacuum tube.
11. What is the limitation of the grid-shift method of determining the transconductance of a tube?
12. What are the advantages of using a dynamometer-type instrument for indicating the transconductance of a vacuum tube?
13. Explain the power output test for a vacuum tube and state the advantages of this type of test.
14. How are diodes usually tested with a tube tester?
15. What is the English scale on the indicating meter of a tube tester?
16. How are picture tubes usually tested with a service-type tester?
17. What is the distinction between emission current and beam current in a picture tube?
18. Explain the circuit operation of the transconductance tester depicted in Fig. 20.9.
19. Why are two values of grid voltage available in the tester depicted in Fig. 20.9?
20. What difficulty is encountered in measuring the input capacitance of a vacuum tube?

21. Give a brief description of the procedure for measuring the input capacitance of a vacuum tube.
22. How does the use of an unbypassed cathode resistor affect the G_m of a vacuum tube?
23. How do engineers define the warm-up time of vacuum tubes?
24. How is damage to fast warm-up time tubes prevented when these tubes are connected in a series string?

SEMICONDUCTOR DEVICE TESTERS

21.1 BASIC TRANSISTOR PARAMETERS

Semiconductor device testers can be fundamentally classified into service-type and lab-type instruments. From a generalized viewpoint, this distinction can be compared with that of service-type and lab-type testers. We know that transistors are operated as amplifiers, oscillators, switches, mixers, detectors, and so on. These functions may be performed at low frequencies or at very high frequencies. A transistor may operate at low temperature in a particular application, and at a high temperature in another application. The problem of transistor testing is also complicated by the fact that the device draws control current. For example, a collector family of characteristic curves provides incomplete data unless supplemented by emitter or base families.

It is helpful to briefly review some of the more basic transistor parameters. The three fundamental configurations are depicted in Fig. 21.1. As indicated in the diagram, some of the basic parameters of practical concern are input resistance, output resistance, current gain, voltage gain, and power gain. A transistor, like a tube, has an upper frequency limit, which is called its alpha cutoff frequency, or its beta cutoff frequency, as will be explained subsequently. A transistor also has maximum voltage, current, and power ratings, which are temperature dependent. We can also compare the junction capacitances of a transistor with the interelectrode capacitances of a tube.

We recall that the current gain of a transistor in the common-base configuration is called *alpha*. This is an inherent parameter of the transistor and is defined as the ratio of the small change in collector current caused by a small change in emitter current. Thus, we write

$$\alpha = \frac{\Delta I_c}{\Delta I_e} \qquad (21.1)$$

where α = current gain
Δ = a small change
I_c = collector current
I_e = emitter current

The alpha of a junction transistor is always less than unity, although it may approach unity rather closely. Next, we recall that the current gain of a transistor in the common-emitter configuration is called *beta*. Again, this is an inherent parameter of the transistor and is independent of circuit values. Beta is defined as the ratio of the small change in collector current caused by a small change in base current. In turn, we write

$$\beta = \frac{\Delta I_c}{\Delta I_b} \qquad (21.2)$$

Note in passing that if we measure an alpha value, we can calculate the corresponding beta value, or vice versa, in accordance with the formulas:

$$\alpha = \frac{\beta}{1 + \beta} \qquad (21.3)$$

$$\beta = \frac{\alpha}{1 - \alpha} \qquad (21.4)$$

The simpler transistor testers are designed to measure beta values at an average or typical bias condition, and to measure the collector leakage current. On the other hand, an elaborate transistor tester will display complete families of characteristic curves on an oscilloscope screen. Most service-type transistor testers also provide for checking the more common types of semiconductor diodes. Conventional diodes are tested for front-to-back ratio; zener diodes are tested for knee point; tunnel diodes are checked for switching action. Service-type testers can also be classified as in-circuit and out-of-circuit types, as explained subsequently.

21.2 BASIC TRANSISTOR TEST METHODS

A semiconductor diode can be checked for its front-to-back ratio with an ohmmeter, as depicted in Fig. 21.2. The resistance values that are measured depend considerably upon the test voltage that is applied, since junction resistance is nonlinear. An ohmmeter test can also be made of each junction of a transistor, as shown in Fig. 21.3. In a typical test situation, the

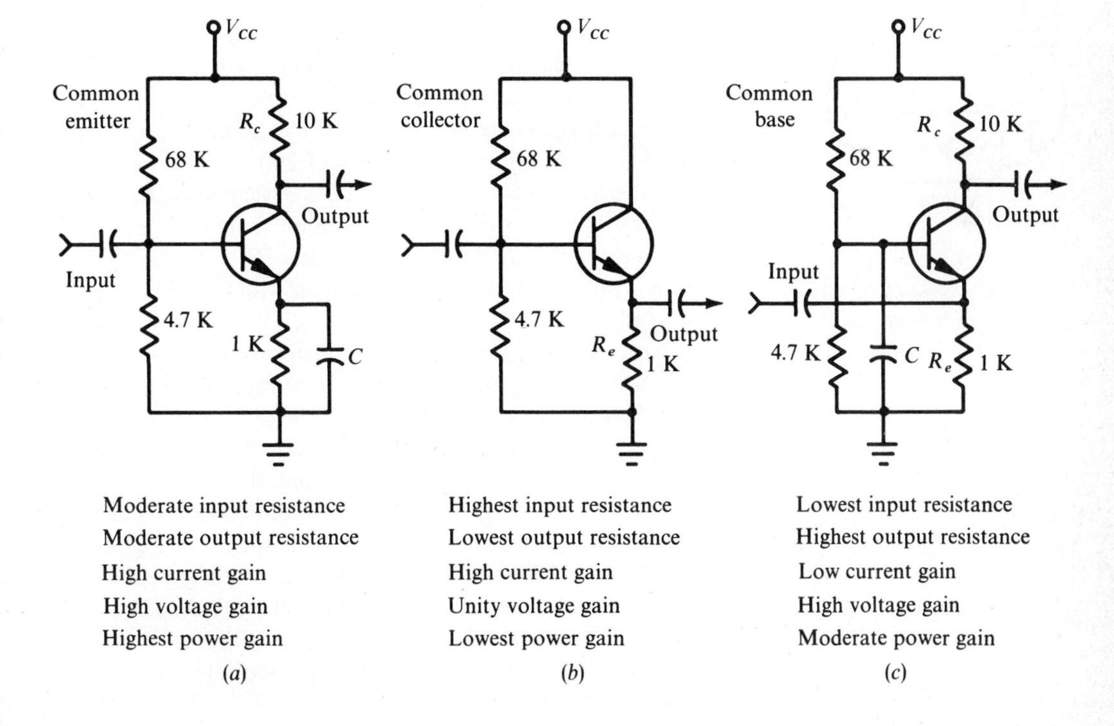

FIG. 21.1 *The three basic transistor configurations.*

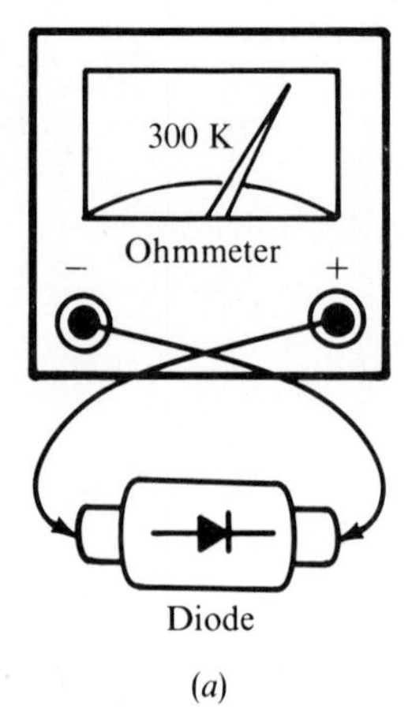

(*a*)

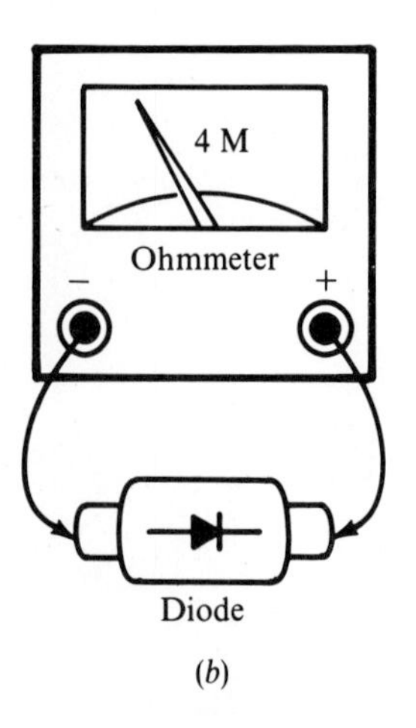

(*b*)

FIG. 21.2 *Ohmmeter test for the front-to-back ratio of a diode.*

resistance between emitter and base of a low-power transistor measures more than 10,000 Ω in the reverse direction and less than 100 Ω in the forward direction. Similar resistance values are measured for the collector-base junction. Note that the resistance between the collector and emitter leads depends considerably on the design of the transistor; it is possible that this front-to-back ratio might measure unity.

In general, it is good practice to avoid the use of the $R \times 1$ range of an ohmmeter in the foregoing tests, because it is possible that the maximum rated current for the transistor might be exceeded. An ohmmeter test will show whether a junction is open, shorted, or has a poor front-to-back ratio. Next, let us observe the slightly elaborated ohmmeter test for a transistor depicted in Fig. 21.4. The ohmmeter is used both as an indicator and as a voltage source in a control-action test. Three resistors and a switch are employed in this arrangement. When the switch is set to its first position, the ohmmeter normally indicates practically an open circuit. On the other hand, when the switch is thrown to its second position, the ohmmeter normally indicates a resistance in the order of 10,000 Ω.

A simple commercial transistor and diode tester that provides a proportional beta measurement employs the configuration shown in Fig. 21.5. Either a *PNP* or an *NPN* type of transistor can be tested. With the switch open, the collector leakage current is indicated by the meter; when the switch is closed, a forward current of 100 μA flows into the base of the transistor. The beta value is indicated on the meter scale. Jack J_1 is provided for checking the front-to-back ratios of conventional semiconductor diodes. This basic type of tester is available in many variations; a widely used design is illustrated in Fig. 21.6.

Note in Fig. 21.6 that switch positions are provided for measurement of I_{ceo} and I_{cbo}. The test configuration shown in Fig. 21.5 measures I_{ceo}, since the base lead is open while the collector leakage current is measured. An I_{cbo} test is made as depicted in Fig. 21.7*a*. We observe that the emitter lead is open-circuited during measurement of collector-junction leakage current. A basic I_{ceo} test circuit is shown in Fig. 21.7*b*. In the first analysis, an I_{ceo} test gives a reading beta times higher than an I_{cbo} test, because the collector leakage current is amplified by the transistor in the I_{ceo} configuration. Sometimes an I_{ces} test is provided, and this is essentially the same as an I_{cbo} test inasmuch as the base terminal of the transistor is short-circuited to the emitter terminal.

Note that a meter shunt is provided in the circuit of Fig. 21.7*a*. Since power-type transistors generally have larger collector leakage currents than small transistors, two or three current ranges are usually provided. We recall that saturation current is practically independent of the test voltage that is employed, although saturation current is temperature dependent. True leakage current obeys Ohm's law, and this component of I_{cbo} or I_{ceo} is proportional to the applied test voltage. Transistor manuals specify ratings for leakage current (total reverse current) at normal temperature, and a transistor is rejected if the rated value is exceeded. Similarly, a transistor is rated for beta value, and is rejected if its current gain is subnormal.

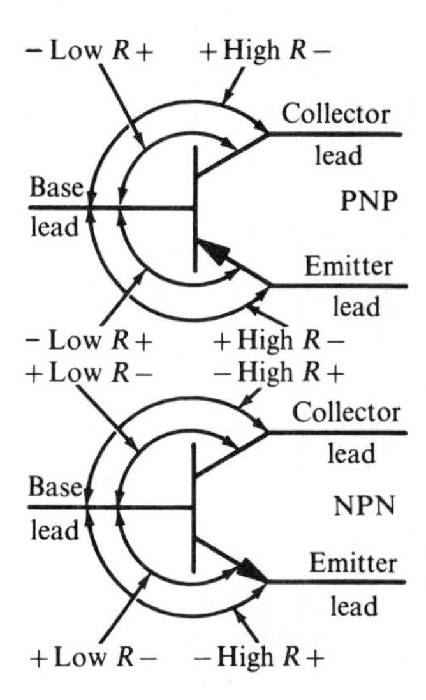

FIG. 21.3 *Comparative front and back resistances of transistor junctions.*

The more elaborate types of transistor testers also provide a measurement of $V_{ce(\mathrm{sat})}$ values. For example, see Fig. 21.8. This parameter is defined as the value of collector voltage at which the transistor goes into collector saturation. The value of $V_{ce(\mathrm{sat})}$ is measured with a test circuit such as shown in Fig. 21.9*a*. Note that a meter is connected in series with a collector load resistor to give a reading proportional to the collector current value. With fixed forward bias applied to the base, the collector voltage is increased until the collector current levels off. Next, the collector-emitter voltage is measured as shown in Fig. 21.9*b*, and this is the value of $V_{ce(\mathrm{sat})}$.

An elaborate transistor tester provides for dc beta measurements and ac beta measurements. A dc beta measurement is made by biasing the transistor to its normal operating point, and then shifting the bias current slightly. The corresponding change in collector current is accurately proportional to beta, and the scale can accordingly be calibrated in beta values. An ac beta measurement is usually made at 1 kHz in a service-type tester. A known small value of ac current is fed into the base, and the corresponding ac collector current is indicated by a meter in the collector circuit, with a scale calibrated in beta values.

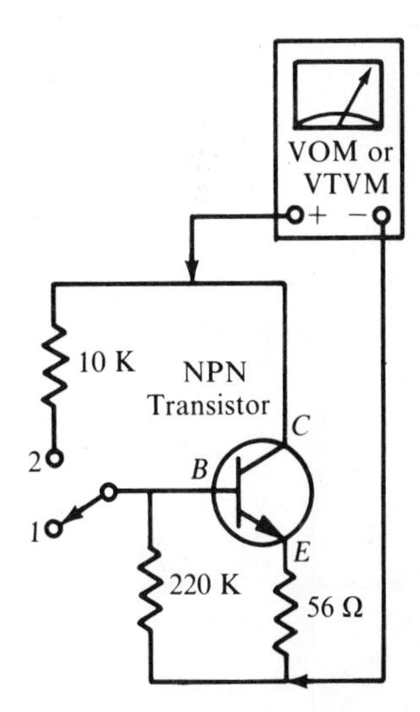

FIG. 21.4 *Control-action test with an ohmmeter.*

21.3 TUNNEL DIODE TESTS

Service-type semiconductor testers commonly provide a switching test for checking tunnel diodes. The basic configura-

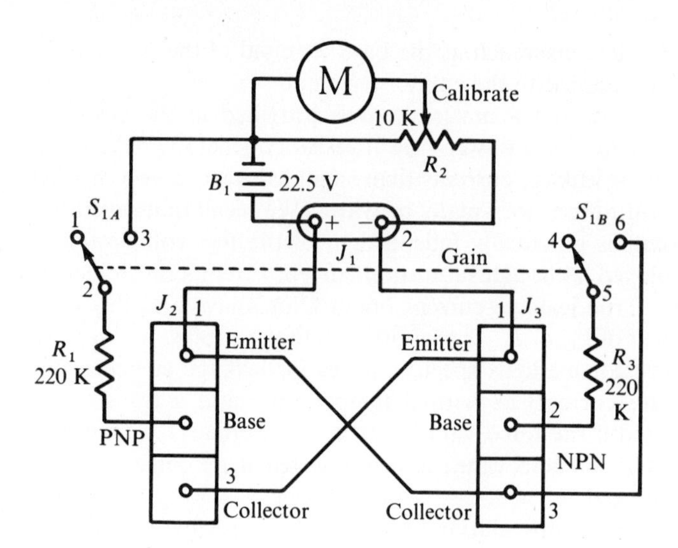

FIG. 21.5 *A simple beta and leakage measuring circuit.*

tion is shown in Fig. 21.10*a*. As the applied voltage is gradually increased, the voltmeter indicates a practically constant small value until a critical potential is reached. At this point, the voltmeter reading suddenly increases to a substantial fraction of 1 V. Next, as the applied voltage is decreased, a critical value is reached at which the reading suddenly drops to its former small value. That is, the tunnel diode has switched through a complete cycle of operation. A defective diode will fail to switch.

With reference to Fig. 21.10*b*, we observe that a tunnel diode will be in a stable state and exhibit positive resistance from zero to P_3. Conversely, the diode is unstable and exhibits negative resistance from P_3 to P_2. Again, the diode is stable and exhibits positive resistance from P_2 to P_4. Thus, the load line may fall either at (*A*) or (*B*) in the circuit of Fig. 21.10*a*. However, if the applied voltage is changed so that the load line is forced to enter the negative-resistance interval, however slightly, the circuit suddenly switches and the load line jumps to its other position. Since this jump is accompanied by a sudden change in voltage drop across the diode, the meter indicates which of the switching modes the diode has been placed into.

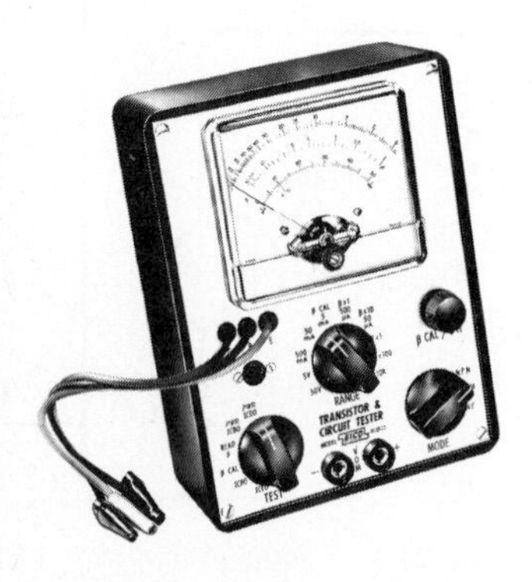

FIG. 21.6 *Appearance of a small transistor tester.* (*Courtesy, EICO*)

21.4 ZENER DIODE TEST CONFIGURATION

A zener diode permits only slight reverse-current flow until a critical reverse-voltage value is reached. Then, the reverse current increases very rapidly for a small increase in reverse voltage, as seen from the voltage-current characteristic in Fig. 21.11*a*. Accordingly, a basic test configuration employs a variable voltage source, a current meter, and a voltmeter, as depicted in Fig. 21.11*b*. A 1-K series resistor is utilized in this example as a current limiter, so that the current does not increase too abruptly when the potentiometer is advanced to the knee point on the characteristic.

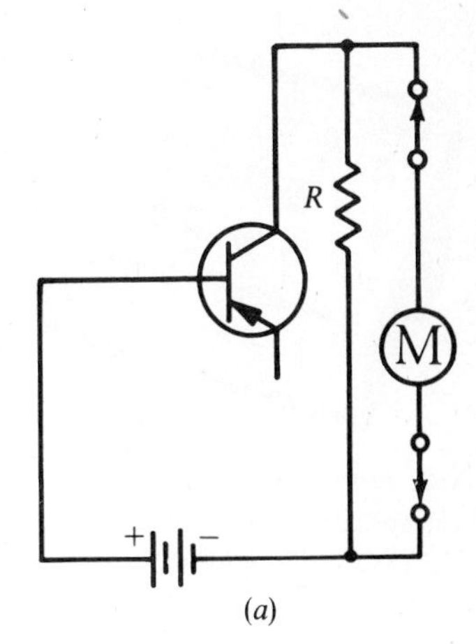

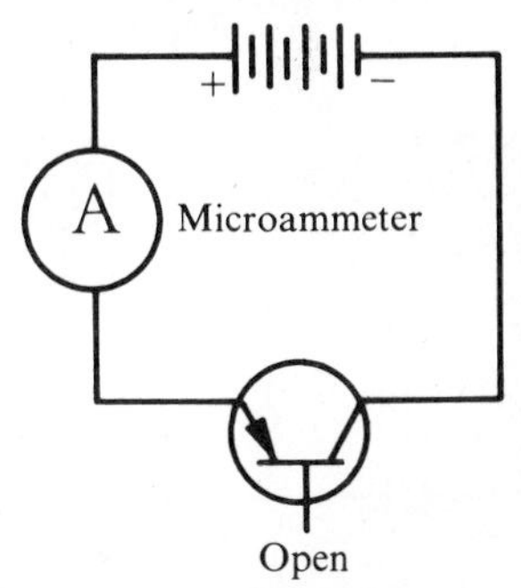

FIG. 21.7 *Transistor leakage test. (a) Basic I_{cbo} test configuration; (b) basic I_{ceo} test circuit.*

21.5 LABORATORY-TYPE SEMICONDUCTOR TESTERS

Laboratory-type semiconductor testers are comparatively specialized, and are assembled from various subunits. As an example, a semiconductor curve tracer may be used in combination with a staircase generator and an oscilloscope. This arrangement provides an accurate display of the collector family for a transistor, but gives no information concerning switching characteristics. To measure rise time, storage time, and fall time, and to check incidental waveform distortion, a suitable switch-circuit arrangement is employed in combination with a pulse generator and oscilloscope. Again, hybrid parameters are accurately measured in highly specialized test circuits that provide no data concerning collector families or switching characteristics.

Let us consider the chief features of a semiconductor curve tracer. A basic arrangement is depicted in Fig. 21.12*a*. The base is biased at some chosen value, and the collector is energized by half-sine waves. These half-sine waves are obtained by passing a 6.3-V 60-Hz current through a rectifier. A scope is then connected to display collector-current variation with respect to collector supply-voltage variation. That is, the voltage drop across the 100-Ω collector load resistor is applied to the vertical input terminals of the scope, and the supply voltage is applied to the horizontal input terminals. Note that the 20-K resistor is used to establish a moderate horizontal input impedance for the scope so that stray hum pickup will not pose operating problems.

A typical collector curve is shown in Fig. 21.12*b*. This type of single-curve presentation is informative, but excessive time would be required to change the base current manually in small steps and thereby to develop a family of collector curves. That is, it is necessary from a practical viewpoint to employ automatic stepping of the base current in order to display a complete family of collector curves simultaneously. Figure

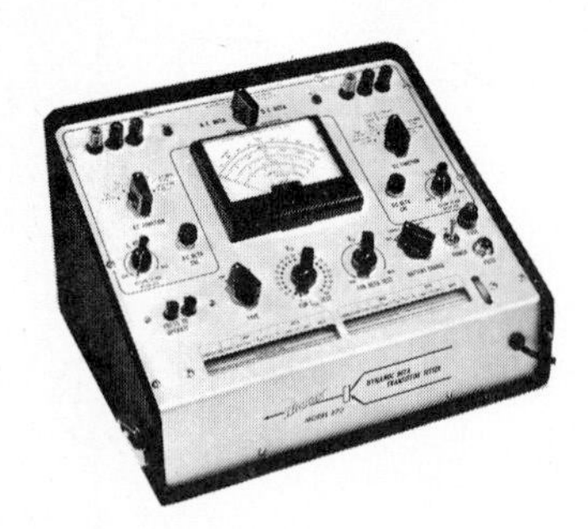

FIG. 21.8 *A comparatively elaborate transistor tester. (Courtesy, Hickok Instruments Co.)*

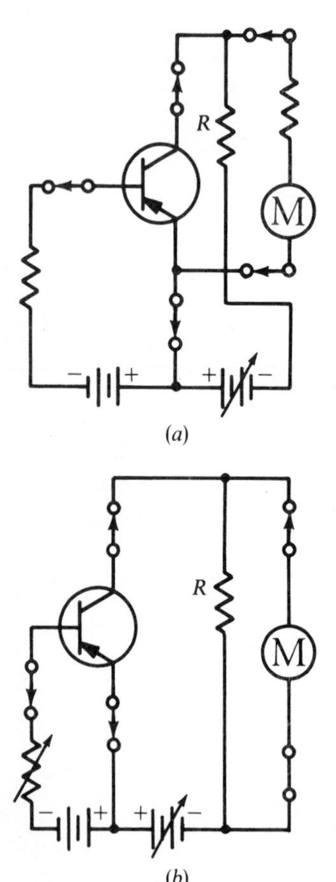

FIG. 21.9 *Transistor test circuits. (a) Collector saturation-current test; (b) measuring $V_{ce(sat)}$ values.*

21.13 shows an example of a collector family published by a transistor manufacturer. Note the maximum power dissipation hyperbola in the diagram. Since a transistor will be damaged if operated in the "forbidden region" of excessive power dissipation, the designer of a curve tracer must observe this requirement. Note also that the power dissipation curve is not given by a transistor curve tracer.

Figure 21.14 shows a simple block diagram for a curve tracer with a staircase generator in the base circuit of the transistor to provide incremental steps of base current. Various types of circuitry are used in the design of staircase generators. The modern trend is to utilize semiconductor counter arrangements with sequential decoders to provide a switching sequence. An illustration of the output waveform from a typical staircase generator is seen in Fig. 21.15. Students interested in specialized staircase generators are referred to pertinent engineering design handbooks.

In the elaboration of the basic test circuit shown in Fig. 21.14, it is evident that the polarities of the supply voltages, as well as their amplitudes, must be subject to suitable control. For example, the supply voltages must be reversed in polarity when an *NPN* transistor is to be tested instead of a *PNP* transistor. Again, *P*- or *N*-channel fet devices require appropriate polarities of test voltages. When a diode is being tested, it is usually desirable to apply a suitable sequence of reverse and forward bias voltages in order to display both the reverse and forward intervals of the diode characteristic. The basic relations are depicted in Fig. 21.16.

Next, let us consider a switching test of a transistor. The test is made with a fast-rise pulse, and the resulting output waveform has the general form depicted in Fig. 21.17. T_r is the rise time of the leading edge, T_s is the storage time, and T_f is the fall time. These distortions are produced by the physical properties of junctions and the semiconductor bulk substance. Basic switching circuits are fairly simple, as exemplified in Fig. 21.18. The transistor is considered to be "on" when its output waveform is above the 90-percent-of-maximum amplitude level, and is said to be "off" when the output waveform is below the 10-percent level. Note that we also measure the turn-on delay time (t_d) which is the elapsed time from application of the driving pulse to the 10-percent rise point on the output waveform.

The storage time (t_s) in Fig. 21.18 is also called the turn-off delay time. Note that the fall time is generally somewhat

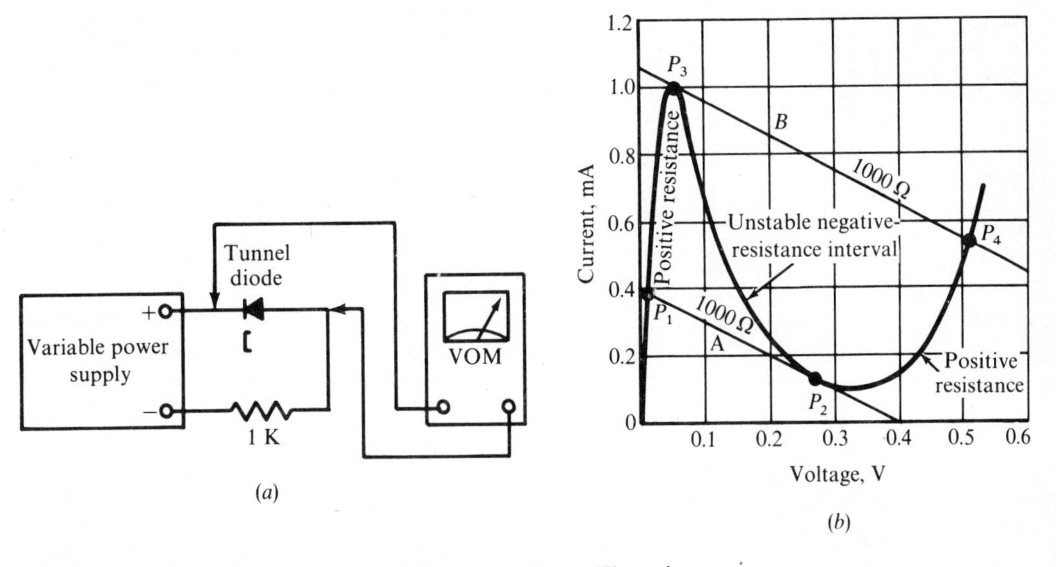

FIG. 21.10 *Tunnel diode test. (a) Test circuit; (b) diode switching characteristics.*

different from the rise time in a switching transistor. There is a general relation between the beta cutoff frequency of a transistor and its switching speed. The beta cutoff frequency is the upper frequency limit of the device in the ce configuration, and occurs at the point on the frequency characteristic where the output is 3 d*B* down. A rule of thumb states that the rise time of a switching output waveform is equal to one-third of the period corresponding to its beta cutoff frequency.

Another important class of laboratory test arrangements is employed to measure hybrid parameters of transistors. These

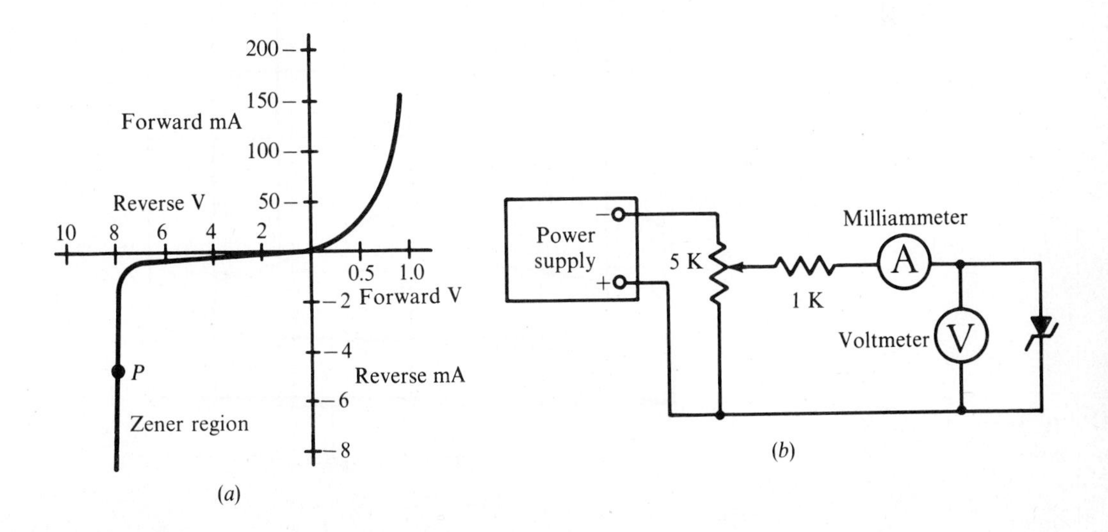

FIG. 21.11 *Zener diode test. (a) E-I characteristics; (b) basic test circuit.*

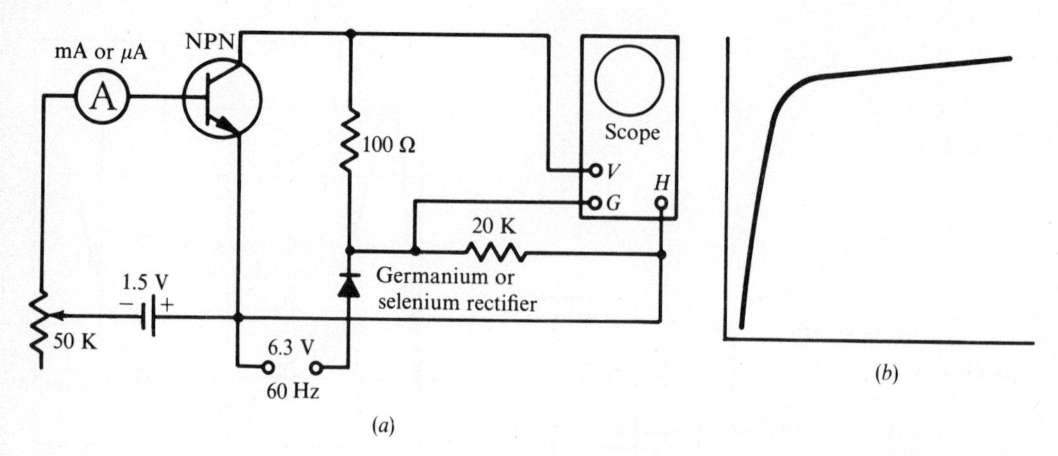

FIG. 21.12 *Transistor test. (a) Basic arrangement for displaying collector curves; (b) oscilloscope presentation.*

parameters are measured on the basis of a conventional four-terminal "black box," called a hybrid matrix. Figure 21.19 depicts the black box concept; its input terminals are identified as 1, 1, and its output terminals are identified as 2, 2. The *h*

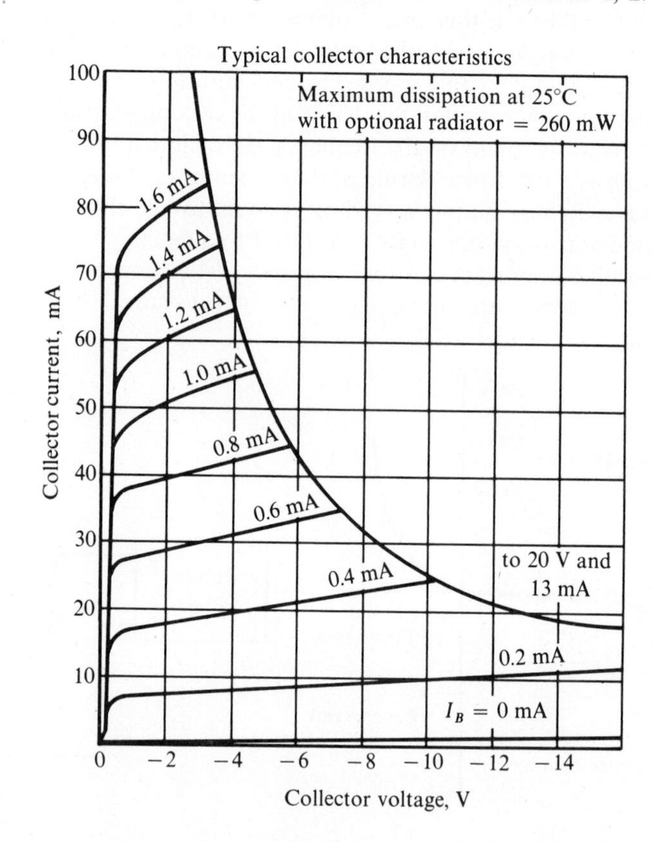

FIG. 21.13 *Typical collector family.*

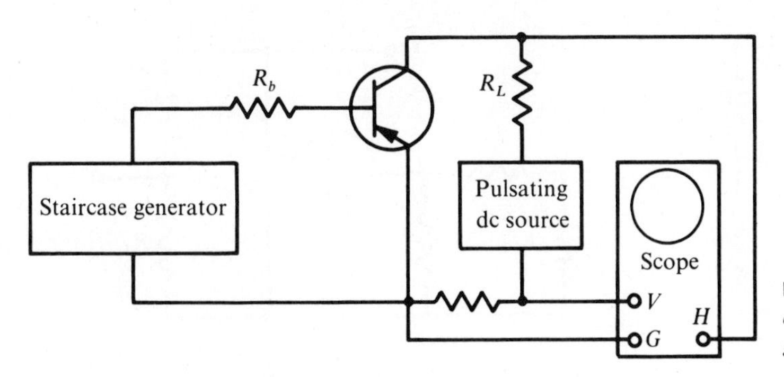

FIG. 21.14 *Basic curve-tracer arrangement, with base staircase generator.*

parameters for a transistor are defined as follows:

$$h_{11} = \frac{e_1}{i_1} = \text{input impedance with output short-circuited} \tag{21.5}$$

$$h_{12} = \frac{e_1}{e_2} = \text{reverse voltage ratio with input open-circuited} \tag{21.6}$$

$$h_{21} = \frac{i_2}{i_1} = \text{forward current gain with output short-circuited} \tag{21.7}$$

$$h_{22} = \frac{i_2}{e_2} = \text{output admittance with input open-circuited} \tag{21.8}$$

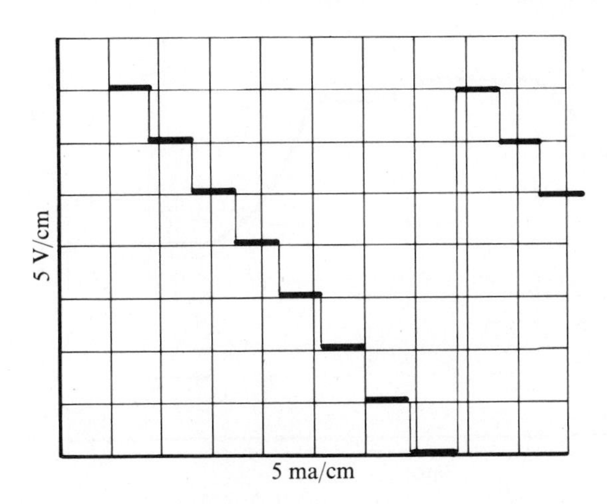

FIG. 21.15 *Output waveform from a typical staircase generator.*

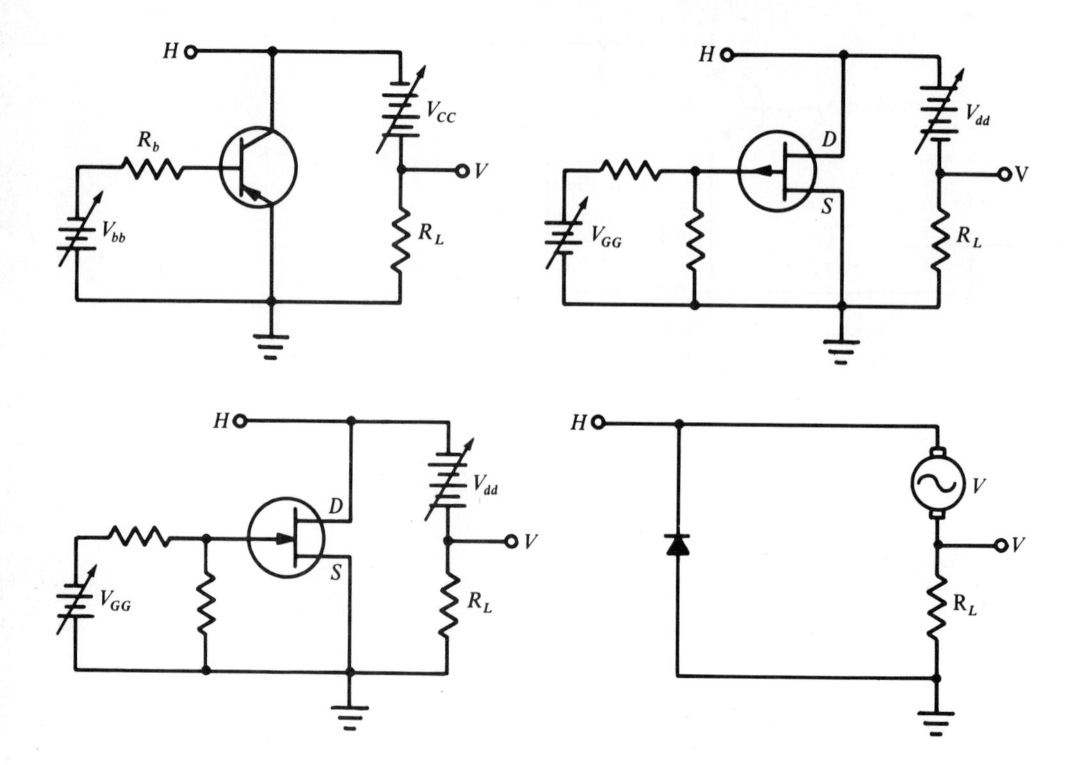

FIG. 21.16 *Polarity relations of power supply voltages for various semiconductor devices.*

Figure 21.20 shows the black box concept elaborated by the equivalent circuit for a transistor, and with connections of source and load. If we employ a common-base test circuit, this fact is indicated by writing h_{11b} or h_{ib}. Figure 21.21*a* shows a test circuit for measurement of h_{11b}. The test frequency is

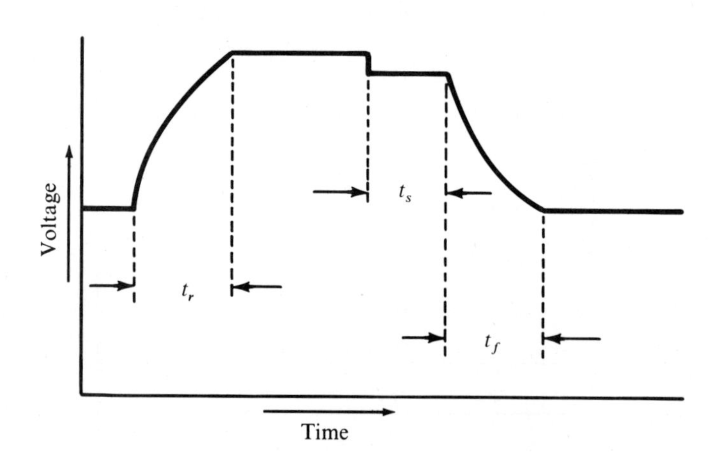

FIG. 21.17 *Pulse distortion by a switching transistor.*

commonly 1 kHz. We measure the values of e_g and i_g, and calculate the value of h_{11b} as follows:

$$h_{11b} = \frac{e_g}{i_g} \tag{21.9}$$

The transistor is biased for Class A operation in the test circuit, and a small ac voltage is employed to ensure essentially linear operation. The bypass capacitor C places the collector at ac ground potential. In turn, Formula (21.9) expresses the voltage-current quotient with the output of the transistor short-circuited. Although a noise voltage is present, it is generally small and can be ignored. Next, to measure h_{12b} or h_{rb}, we utilize the test circuit shown in Fig. 21.21*b*. Note that the generator signal is coupled into the collector circuit. We measure the signal voltage from collector to base (e_g), and from emitter to base (e_1). In turn, we write

$$h_{12b} = \frac{e_1}{e_g} \tag{21.10}$$

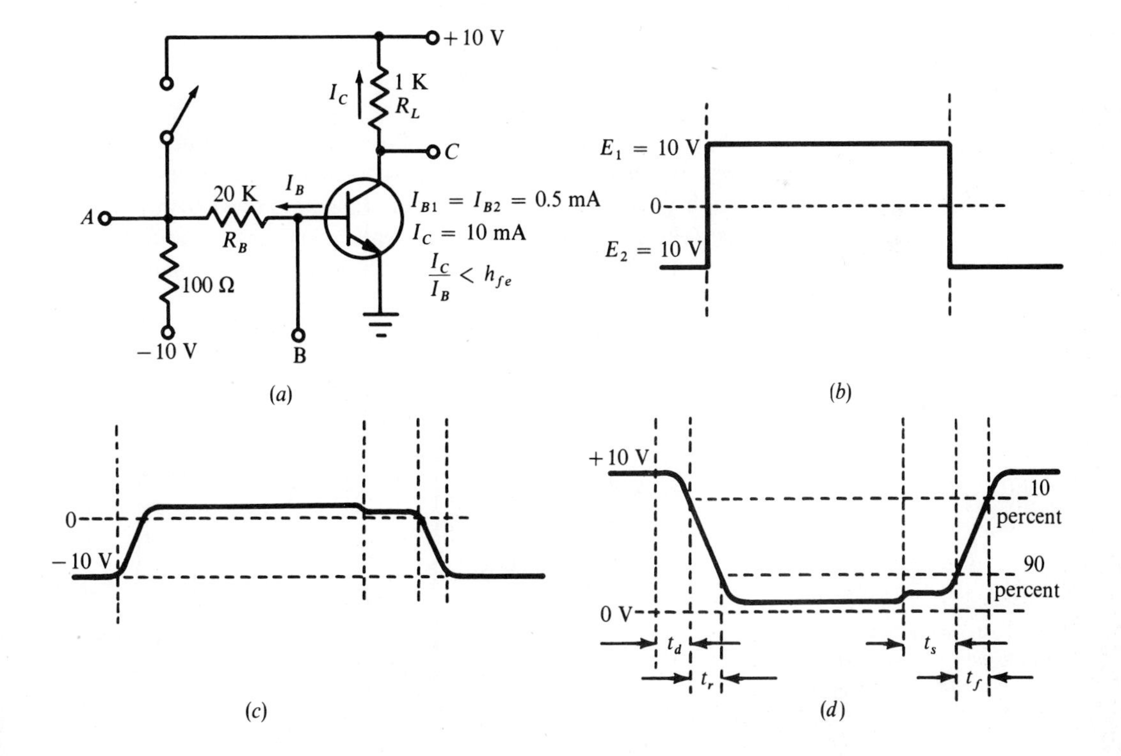

FIG. 21.18 *Basic switching test circuit and associated waveforms.*

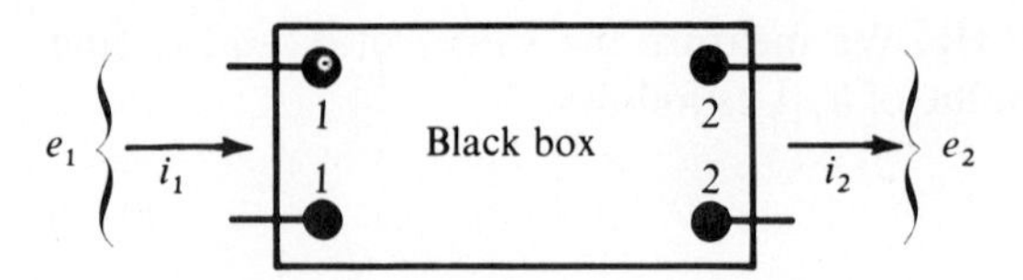

FIG. 21.19 *The black box concept.*

Since the VTVM has a very high impedance, it can be regarded as an open circuit in this test. As before, the noise voltage is ignored since its value is very small in normal circumstances. Next, to measure h_{21b} or h_{fb}, we employ the circuit shown in Fig. 21.21*c*. This is the alpha value, and is formulated:

$$h_{21b} = \frac{i_2}{i_g} \tag{21.11}$$

The arrangement shown in Fig. 21.21*d* is used to measure h_{22b} or h_{ob}, which is the output admittance value of the transistor. Thus

$$h_{22b} = \frac{i_g}{e_g} \tag{21.12}$$

Note that we measure i_g in terms of e_v/R_L in the foregoing test. This is justified because i_e is very small.

Now, let us proceed to the common-emitter configurations depicted in Fig. 21.22. To measure the value of h_{11e} or h_{ie}, we use the test circuit shown in (*a*). We observe that a high Q parallel-resonant circuit z_p is employed to provide ac isolation of the base electrode, while providing a low-resistance dc path. With a small ac signal applied by the generator, we measure e_g and i_g, and formulate h_{11e} as follows:

$$h_{11e} = \frac{e_g}{i_g} \tag{21.13}$$

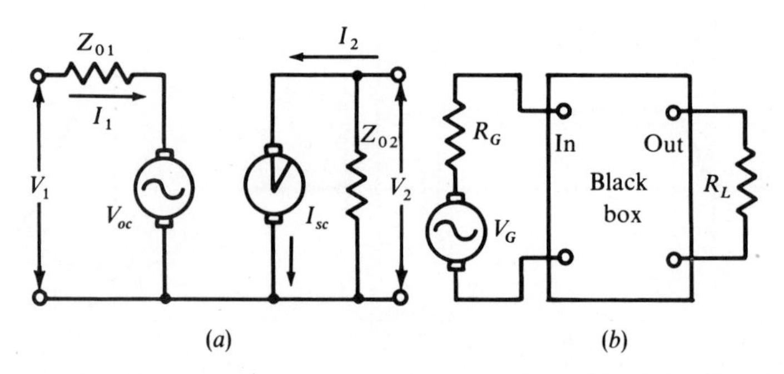

FIG. 21.20 *Black box configuration. (a) Equivalent circuit for a transistor; (b) source and load connections.*

Next, to measure h_{12e} or h_{re}, we use the circuit shown in Fig. 21.22*b*. In turn, the value of h_{12e} is formulated:

$$h_{12e} = \frac{e_1}{e_g} \tag{21.14}$$

To measure the value of h_{21e} or h_{fe}, also called beta, we employ the configuration depicted in Fig. 21.22*c*. In turn, we write

$$h_{21e} = \frac{i_2}{i_g} \tag{21.15}$$

FIG. 21.21 *Test circuits for measurement of h parameters in the common-base configuration.*

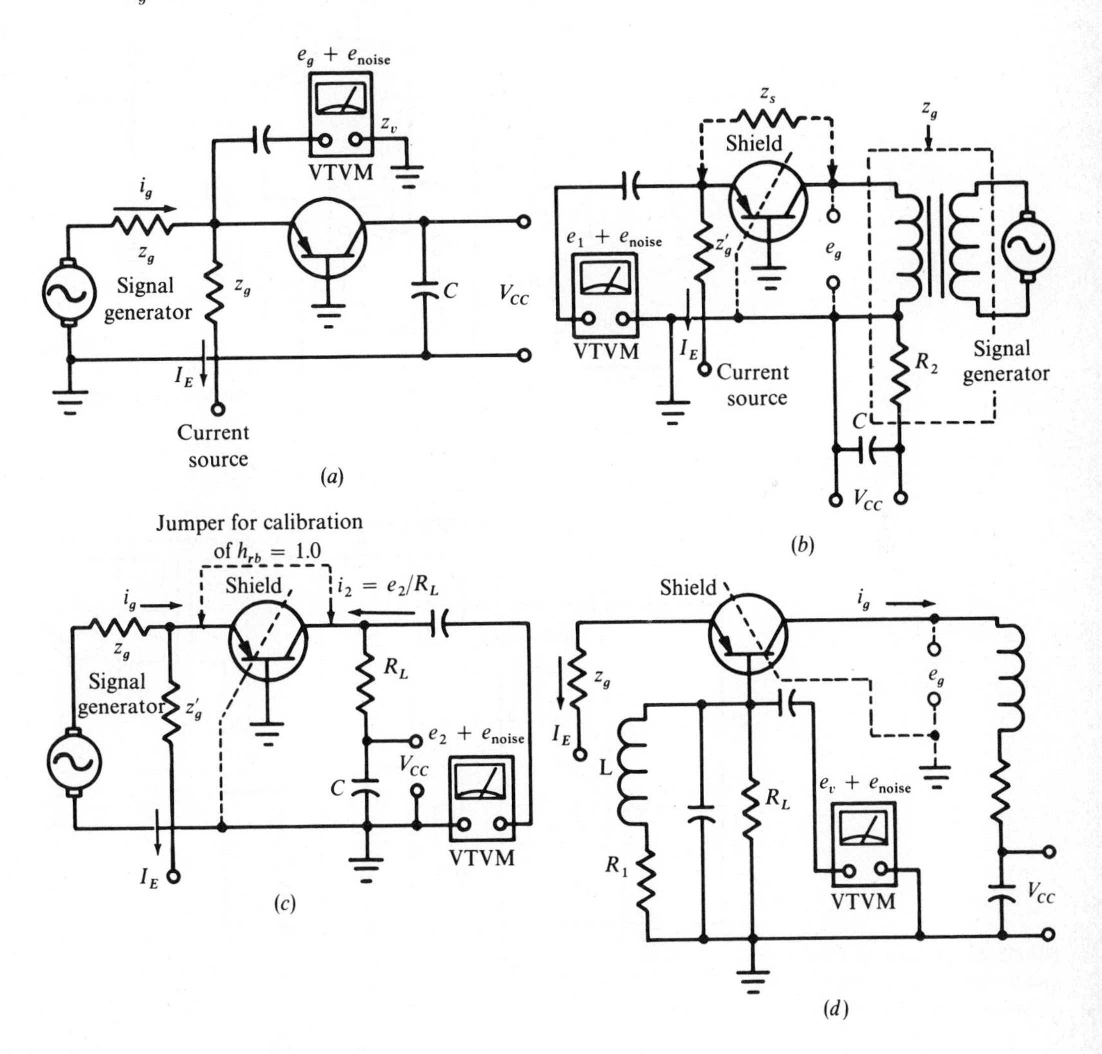

Finally, to measure the value of h_{22e} or h_{oe}, we employ the test circuit shown in Fig. 21.22*d*. The value of h_{22e} is then formulated:

$$h_{22e} = \frac{i_g}{e_g} \tag{21.16}$$

We observe in the foregoing relation that i_g is equal to e_o/R_L for all practical purposes. Of course, it is unnecessary to measure *h* parameters in the *cc* configuration, because these

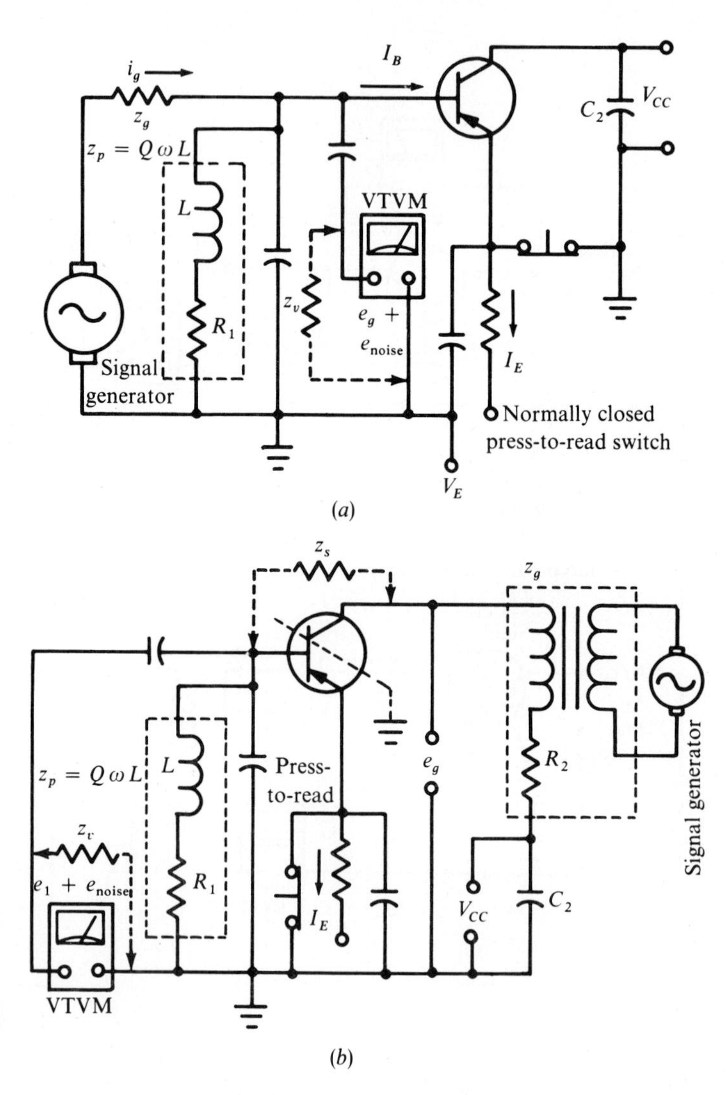

FIG. 21.22 *Test circuits for measurement of h parameters in the common-emitter configuration.*

are easily calculated from the parameters measured in the *cb* and *ce* configurations. For practical purposes, h_{21c} or h_{fc} is almost equal to h_{1e}. Also, h_{21c} is practically equal to 1. Only the low-frequency *h* parameters have been considered in the foregoing discussion. Test circuits for high-frequency measurements are comparatively elaborate, and the interested student is referred to specialized engineering texts and handbooks for pertinent information.

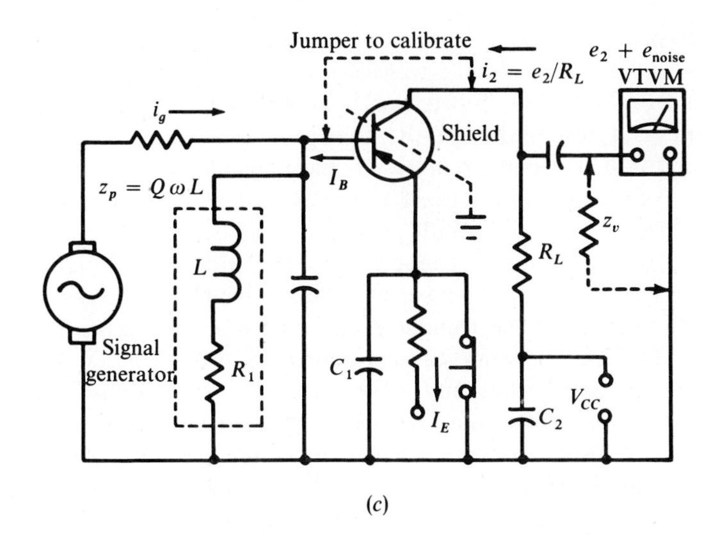

(*c*)

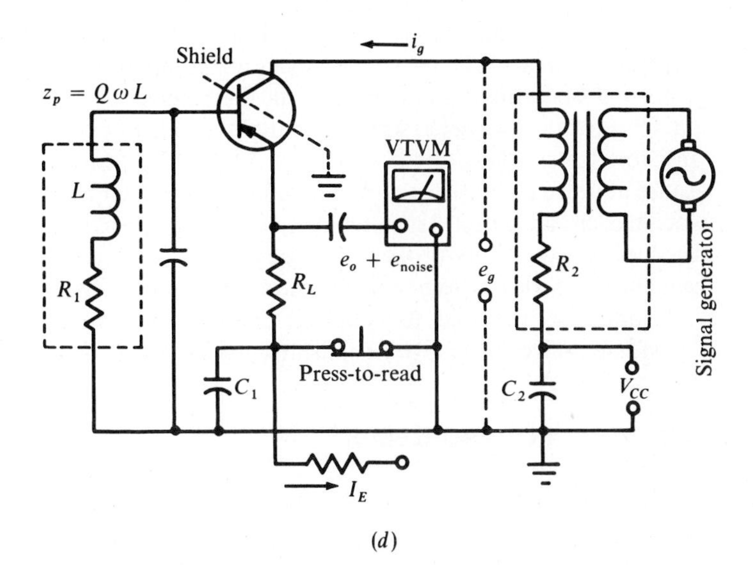

(*d*)

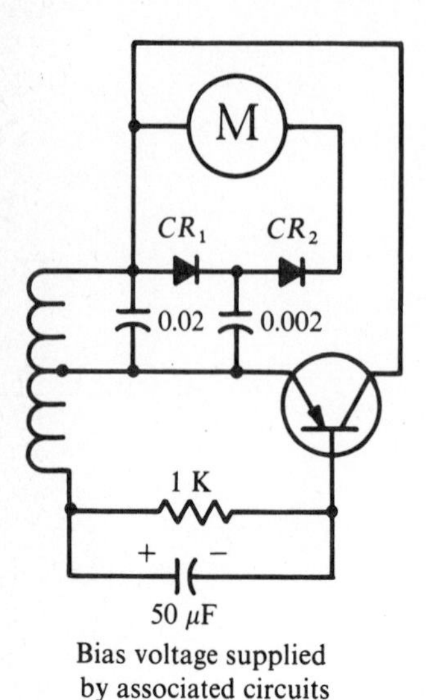

FIG. 21.23 *Circuit configuration of an oscillator-type in-circuit transistor tester.*

21.6 APPLICATIONS

Since small transistors are usually soldered into printed-circuit boards, it is impractical to remove a transistor for test. Accordingly, there has been a marked trend toward the development of in-circuit transistor testers. It is comparatively difficult to obtain an accurate test of transistor condition without removing it from its circuit, because there are many circuit parameters to contend with. For this reason, most in-circuit transistor testers are of the go/no go type; that is, the instrument merely serves to indicate whether a transistor is dead or workable. If a transistor is connected into a circuit that has very low impedance, it might check "dead" on an in-circuit test, although the transistor is actually in good condition. Thus, in-circuit testers have definite limitations.

One type of in-circuit transistor tester operates on the oscillator principle. With reference to Fig. 21.23, an *LC* feedback network is connected to the transistor, with dc voltage applied by the receiver circuit to which the transistor is connected. If the transistor has a normal beta value, the *LC* feedback network causes oscillation, provided that the receiver circuit does not have unusually low impedances. Oscillation causes an ac voltage to be generated in the tank circuit; this ac voltage is fed to a rectifier network and is indicated by a meter in the tester. No leakage test is made of the transistor, as this is not a practical in-circuit procedure.

Evaluation of transistor condition in-circuit can often be inferred by means of dc voltage measurements at the transistor terminals. The measured values are compared with the normal operating values specified in the receiver service data. Thus, a dc voltmeter serves essentially as an in-circuit transistor tester. This method is not always conclusive, because abnormal transistor terminal voltages can result from associated circuit defects, such as leaky coupling capacitors, off-value resistors, or broken printed-circuit conductors. Frequently, a series of tests and measurements, plus a chain of reasoning, is required to conclude definitely whether a transistor is indeed defective as indicated by in-circuit tests.

QUESTIONS AND PROBLEMS

1. Define the term *alpha* as it applies to a transistor.
2. Define the term *beta* as it applies to a transistor.

3. What is the relationship between the alpha and beta of a transistor?
4. Explain how a diode can be tested with an ohmmeter.
5. What is a precaution that must be taken when testing a diode with an ohmmeter?
6. Draw diagrams and explain the checks that may be accomplished on a transistor with an ohmmeter.
7. What is the importance of the collector leakage test of a transistor?
8. When is the $V_{cec(sat)}$ value of a transistor important?
9. Explain the difference between the dc and ac beta measurements for a transistor.
10. Explain the operation of the test setup in Fig. 21.10 as it is used to test a tunnel diode.
11. Explain the operation of the test setup in Fig. 21.11 as it is used to test a zener diode.
12. Explain how the circuit depicted in Fig. 12.12*a* could be used to develop a family of curves for a transistor.
13. How are the conditions "off" and "on" defined for the output pulse from a transistor circuit?
14. How is the turn-on time defined for a switching transistor?
15. Define the four *h* parameters for a junction transistor.
16. Why is a small value of ac voltage used when measuring the *h* parameters of a transistor?

ELECTRONIC COUNTERS AND FREQUENCY METERS

ELECTRONIC COUNTERS AND DIGITAL VOLTMETERS

22.1 BASIC REQUIREMENTS

Electronic counters are instruments that add up (count) the number of electrical pulses applied over an arbitrary period of time, indicating the sum in digital form, as seen in Fig. 22.1. For example, an electronic counter may be arranged to count the positive peaks of a sine wave; then, if a 1-MHz sine wave is applied for 1 ms, the counter will indicate 1000. If the frequency is known precisely, the counter serves as a time-interval measuring device. That is, we know that 1 ms has passed when the counter indicates 1000. On the other hand, if the time interval is known precisely, the counter serves as a frequency measuring device. In other words, we know that the applied frequency is 1 MHz when the counter indicates 1000 at the end of 1 ms.

FIG. 22.1 *An electronic counter. (Courtesy, Hewlett-Packard)*

An electronic counter employs some form of electronic switching, as shown in Fig. 22.2. Counters utilized in electronic instruments generally use transistor flip-flop circuitry. A flip-flop is also called a bistable multivibrator. It has two stable states: if the left-hand transistor is conducting, the right-hand transistor is cut off; if the left-hand transistor is driven into cutoff, the right-hand transistor is automatically driven into conduction. Basically, a bistable multivibrator employs the same circuitry as a free-running multivibrator, except that sufficient reverse bias is included such that the transistors cannot change state until a trigger pulse is applied.

A bistable circuit will remain at rest in either one of its stable states until the circuit is triggered by an input pulse. Then the circuit switches to its second stable state, where it remains until triggered by a subsequent pulse. A conventional bistable multivibrator operates as follows: With reference to Fig. 22.3, the circuit differs from that of a free-running (astable) multivibrator in the base-emitter voltage relations. As noted previously, if one transistor is conducting, the other is automatically biased beyond cutoff. With the initial application of dc—provided the transistors were exactly the same in every respect, the residual random noise voltages in the circuit would cause one of the transistors to momentarily conduct more current than the other. Owing to the feedback arrangement, this action is regenerative, with the result that one transistor is driven rapidly into conduction while the other is driven rapidly beyond cutoff.

With reference to Fig. 22.3, the application of a negative trigger pulse to the base of the nonconducting transistor, or a positive pulse to the base of the conducting transistor, will switch the state of the circuit. Collector triggering may be similarly accomplished, if desired. Two separate inputs are shown in Fig. 22.3. A trigger pulse at input *A* will change the state of the circuit. Once the state of the circuit is changed, an input pulse of the same polarity at input *B*, or an input of opposite polarity at input *A*, will again trigger the circuit. Other methods are also employed, but the end result is the same in any case.

22.2 BINARY COUNTING PROCESSES

Counters indicate the number of pulses that have been applied in terms of the familiar decimal system. On the other hand, the pulses are processed in terms of binary numbers. The reason

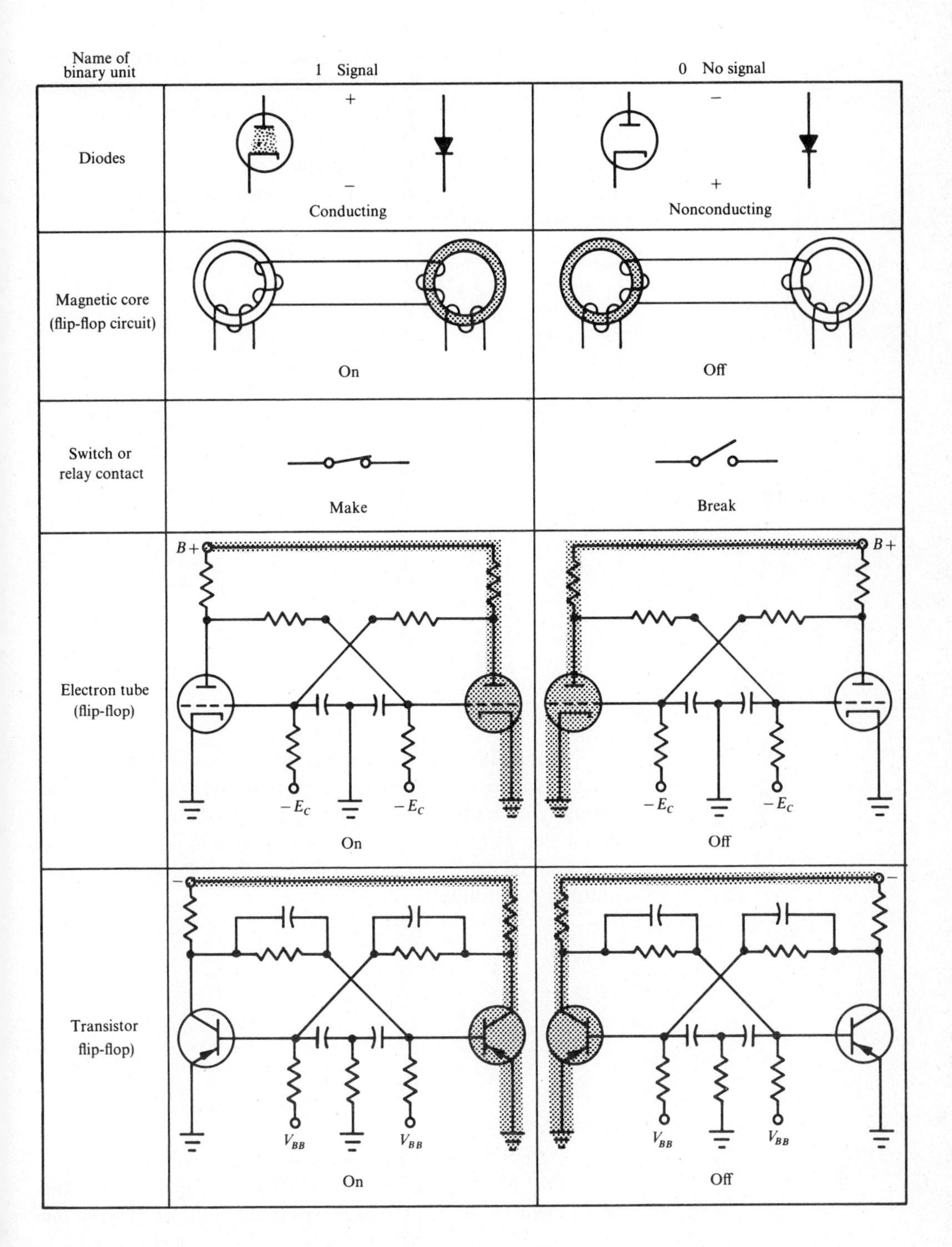

FIG. 22.2 *Some basic electronic switching devices.*

for basing the counting process on the binary number system is that a flip-flop has only two states: *on* and *off*. In this system, a number is either 1 or 0; it cannot be a fraction of 1, or more than 1, or negative. Since there are only two digits in the binary system (0 and 1), mathematicians state that the base of the binary system is 2. Let us observe some examples of binary numbers:

BINARY	DECIMAL
0	0
1	1
10	2
11	3
100	4
101	5
110	6
111	7
1000	8
1001	9
1010	10
1011	11
110	12
1101	13
1110	14
1111	15

Note that binary numbers have place values, just as decimal numbers have place values. In the decimal system, and similarly in the binary system, 1 is not the same as 10, and 10 is not the same as 100. In the binary system, it is helpful to note that place values correspond to exponents of the base number 2, as follows:

$1 = 2^0$

$2 = 2^1$

$4 = 2^2$

$8 = 2^3$

$16 = 2^4$, etc.

The binary number 0000010101 is equal to 21; the zeros prior to the first 1 are disregarded, and we observe the following values:

$$\begin{aligned}
\text{(binary)} \quad 10000 &= 16 \quad \text{(decimal)} \\
100 &= 4 \\
1 &= 1 \\
\text{or} \quad 16 + 4 + 1 &= 21 \quad \text{(decimal)}
\end{aligned}$$

Now, let us consider an electronic counter arrangement. Assume that four flip-flop circuits are connected in cascade as depicted in Fig. 22.3. This arrangement represents a four-

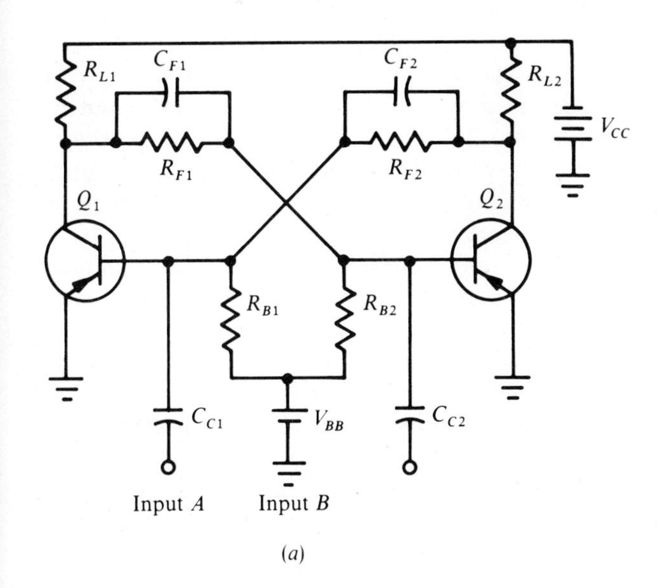

FIG. 22.3 *Conventional bistable multivibrator. (a) Schematic; (b) binary counter using bistable flip-flop.*

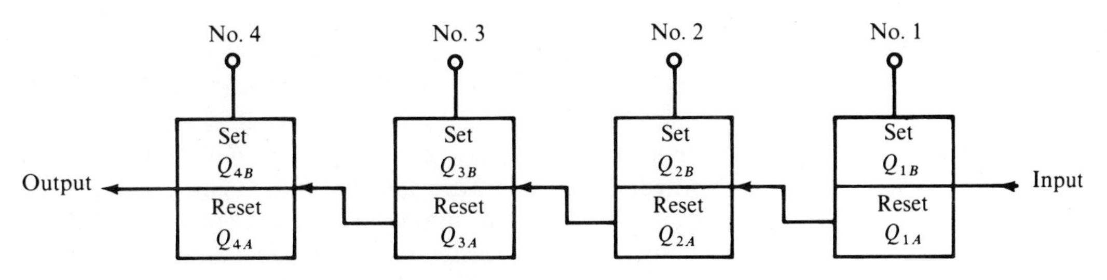

"1" or set = Upper transistors conducting, lower transistors off
"0" or reset = Lower transistors conducting, upper transistors off

(b)

digit binary counter. For convenience in representing binary numbers, we have shown the progression of the flip-flops from right to left. Neon-bulb indicators can be used to indicate a "1" or "set" condition when they are glowing; conversely, a "0" or "reset" condition is indicated by a neon bulb when it is not glowing. We read the indicators from left to right to determine the binary number that is represented. Assume to start with that all four flip-flops are in the zero (reset) condition (upper transistors off). If a negative pulse is now applied to the input, No. 1 ff_1 changes state, from reset to set, and its indicator goes on; however, No. 1 ff_1 produces no output, and the three other flip-flops continue to indicate zero. In this example, we read the four indicators as a binary number, 0001.

Now, let us consider the application of a second input pulse in Fig. 22.3. The second input pulse causes No. 1 ff_1 to return to zero, and also provides a negative-going input to No. 2 ff_2. In turn, No. 2 ff_2 changes to the 1 state but produces no output. The binary indication is now 0011 (decimal number 3). When a fourth input pulse is applied, No. 1 ff_1 returns to zero (from set to reset) and produces an output. In turn, No. 2 ff_2 returns to zero (from set to reset) and also produces an output. Accordingly, No. 3 ff_4 changes to the 1 state (reset to set) but produces no output. Thus the binary indication is now 0100 (decimal number 4).

After 15 input pulses have been applied, the neon bulbs will indicate the binary number 1111 (decimal number 15). On the sixteenth pulse, No. 1 ff_1 changes to zero and produces an output. No. 2 ff_2 therefore changes to zero and produces an output; No. 3ff_4 changes to zero and produces an output, which changes No. 4 ff_8 to zero. The counter is therefore in its original state (0000) after 16 input pulses have been applied. In other words, the maximum capacity of this counter is 15. If we wish to count above 15, more flip-flop stages must be included. To review briefly, let us start with all flip-flops in their "off" or reset condition, as depicted at *A* in Fig. 22.4. If three input pulses are now applied, the indicator lamps for flip-flops Nos. 1 and 2 are "on" but Nos. 3 and 4 are dark, as shown at *B*. In turn, the binary indication is 0011, which corresponds to the decimal number 3. Next, after five more input pulses are applied, the indicator lamp for flip-flop No. 4 is "on" but the other lamps are dark. Thus the binary number 1000 is indicated, which corresponds to the decimal number 8.

22.3 BINARY TO DECIMAL CONVERSION

Because instrument users are schooled in the decimal system and may have no knowledge of the binary system, commercial electronic counters are designed to indicate decimal numbers. This requires suitable elaboration of the binary counting process. Various conversion methods are utilized, some of which are comparatively complex. All conversion methods operate on the basis of auxiliary circuits which are triggered by pulses formed in the course of the binary counting process. One of the basic conversion methods is depicted in Fig. 22.5. A feedback path is provided from *D* to *B* and *C*. If no feedback were provided, the circuit waveforms would appear as depicted in Fig. 22.6. Sixteen pulses cause the counter to recycle. Note that after the eighth pulse, *A* and *B* are "off" and *D* is "on" for the first time. After 14 pulses, all stages are "on" except *A*.

The waveforms following the eighth and fourteenth pulses in Fig. 22.6 are employed in the decade conversion process, as shown in Fig. 22.7. Note that after the eighth pulse, *D* produces an output which is fed back to *B* and *C*. In turn, stages *B* and *C* are driven "on." The tenth pulse recycles all stages. Waveforms of the type shown in Fig. 22.7

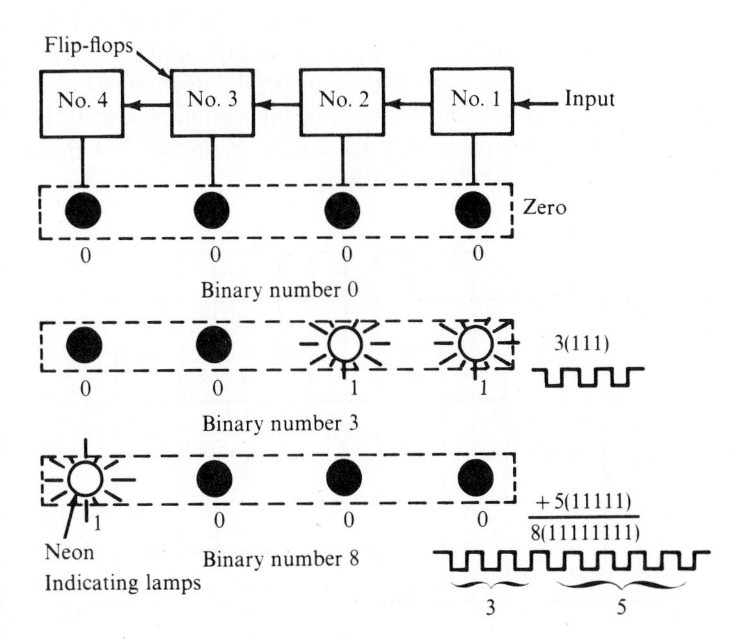

FIG. 22.4 *Addition by a flip-flop chain.*

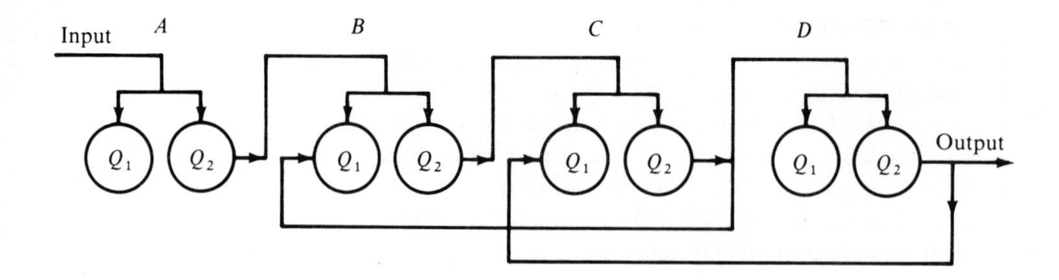

FIG. 22.5 *Block diagram of triggering connections whereby four bistable multivibrators can give a scale-of-10 or decimal circuit.*

permit an array of 10 glow lamps to be energized for representation of numbers from 0 to 9.

22.4 DIGITAL VOLTMETERS

A digital voltmeter (DVM) is illustrated in Fig. 22.8. This type of instrument displays a measured value in terms of discrete numerals, rather than as a pointer deflection on a continuous scale as used in analog instruments. Most designs of digital voltmeters fall into one of the following categories: (1) ramp, (2) staircase ramp, (3) dual ramp integrating, (4) integrating, (5) integrating and potentiometric, (6) successive approximation, and (7) continuous balance. In any case, the basic function that is performed is an *analog-to-digital* or (*A-to-D*) conversion. For example, a voltage value may be changed into a proportional time interval, which starts and stops an accurate oscillator. In turn, the oscillator output is applied to an electronic counter which is provided with a readout in terms of voltage values.

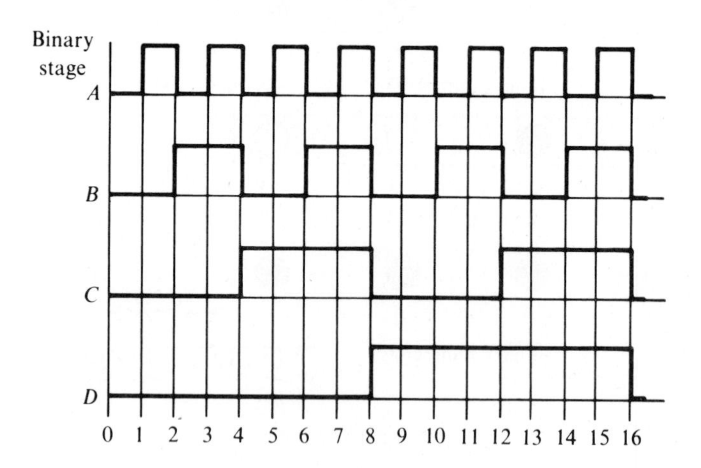

FIG. 22.6 *Operation of arrangement of the counter in Fig. 22.5 with no feedback.*

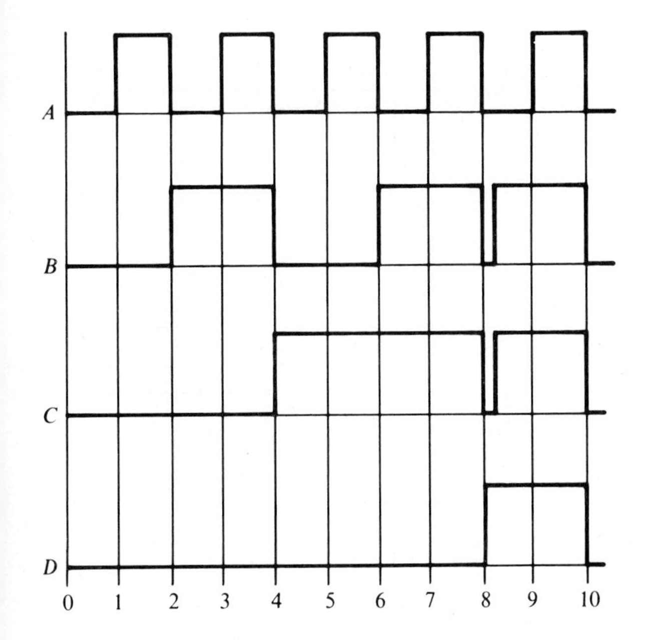

FIG. 22.7 *Operation with feedback.*

Digital readout has several advantages, compared with analog indication provided by a pointer on a scale. For example, an analog reading is subject to observational error, such as parallax and estimation errors. As explained previously, an analog scale usually has an increasing indication error toward its low end, unless the scale is hand-drawn. These sources of error are eliminated by digital readout. Operation is also simplified, because a DVM has a minimum of scales. Inexperienced personnel often tend to select a scale that does not correspond to the setting of the range switch, thus incurring a large operating error. Also, the rated accuracy of most DVMs is much greater than that of service-type voltmeters, and considerably greater than that of most laboratory-type analog voltmeters.

Let us consider the ramp-type digital voltmeter. Its operating principle is to measure the time that a linear ramp (linear rise or fall) takes to change from the input level to ground level (or vice versa). This time period is measured with an electronic time-interval counter and displayed on in-line indicating tubes. Conversion of a voltage value to a time interval is shown in the timing diagram of Fig. 22.9. At the start of the measurement, a ramp voltage is initiated. The ramp value is continuously compared with the voltage value

FIG. 22.8 *A digital voltmeter. (Courtesy, Hewlett-Packard)*

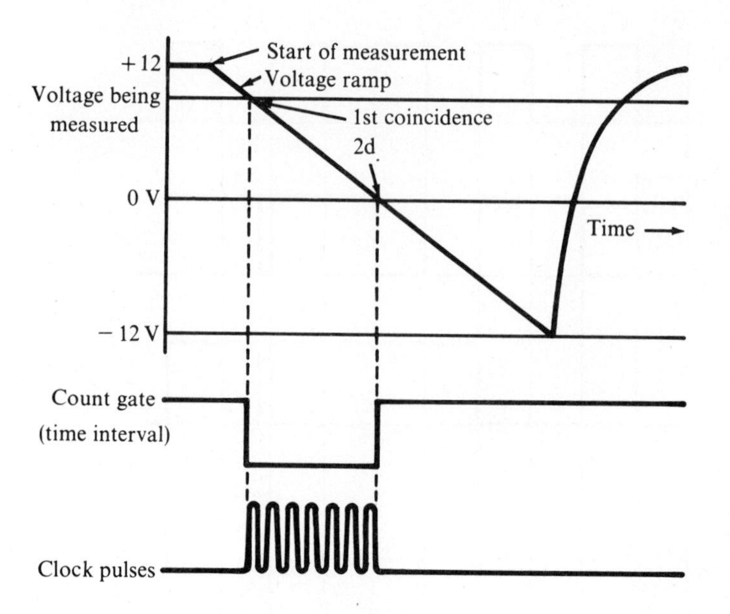

FIG. 22.9 *Voltage-to-time conversion.*

that is being measured. At the instant that these two values become equal, a coincidence circuit generates a pulse which opens a gate. The ramp continues until a second comparator circuit senses that the ramp has reached zero volts. The output pulse from this comparator closes the gate.

The time duration of the gate opening is proportional to the input voltage value. In turn, the gate allows pulses from a precision oscillator to pass to totalizing circuits (electronic counter), and the number of pulses counted during the gating interval is a measure of the voltage value. Figure 22.10 shows a typical block diagram. An instrument of this type has an accuracy of ± 0.05 percent of the reading, with reading rates up to 5 per second. An outstanding advantage of a digital instrument is the clarity of the digital display, which permits more accurate measurements to be made at a faster rate and with greater repeatability than is possible with analog instruments. Parallax problems are also eliminated by a digital display.

Any digital voltmeter has a fundamental cycle sequence which involves sampling, display, and reset sequences. In the instrument to be described next, application of an input voltage initiates the measurement cycle. With reference to Fig. 22.11, an oscillator is automatically switched into operation, and its output consists of pulses which are counted by an electronic counter. A *units* counter is first actuated; this units counter provides a carry pulse to the *tens* counter on every tenth input pulse. In turn, the tens counter provides its own carry pulse after it has counted 10 carry pulses from the units counter. This carry pulse is applied to the *hundreds* counter. If more than 10 carry pulses are applied to the hundreds counter, it provides its own carry pulse which switches on a warning lamp. This signifies to the operator that the next higher position should be employed on the range switch.

Each decade counter unit in a digital voltmeter is connected to a *D*-to-*A converter* (digital-to-analog converter). Outputs from the *D*-to-*A* converters are connected in parallel, which builds up a comparison voltage. At the instant that the comparator section senses that the input voltage and the comparison voltage are equal, it produces a trigger pulse that stops the oscillator. The *sample rate* function permits the DVM to "follow" a varying voltage. It is controlled by a simple relaxation oscillator, that triggers and resets the counters to zero every half second. The display circuits store each reading until a new sample value occurs; this eliminates blinking. Finally, when the input voltage is removed from the instrument, the reading automatically returns to zero and this completes the cycle sequence.

Next, let us consider the staircase ramp type of instru-

FIG. 22.10 *Block diagram of a ramp-type digital voltmeter.*

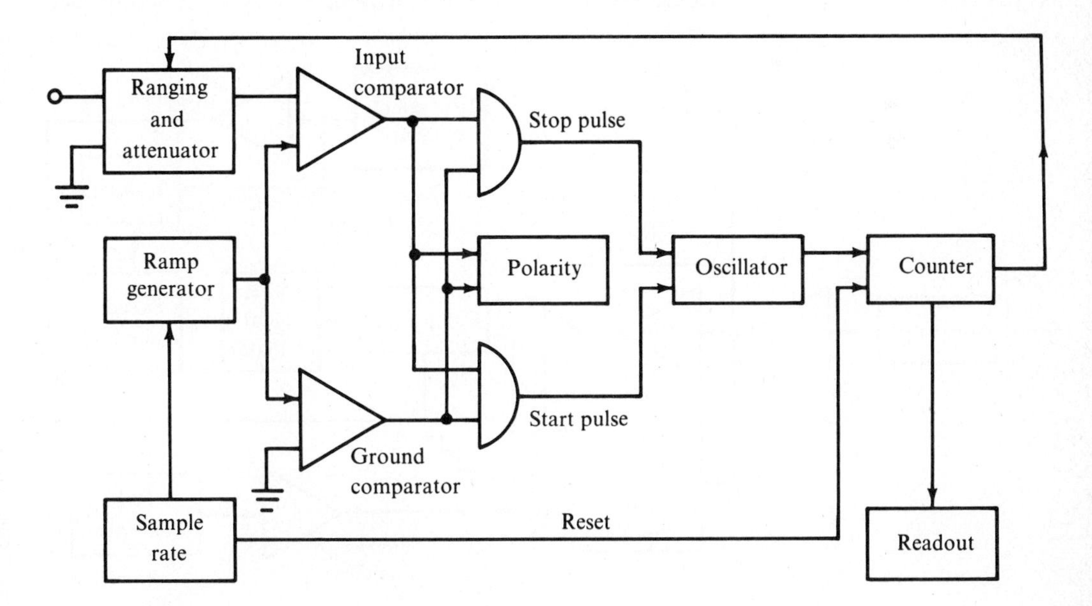

ment; a block diagram is depicted in Fig. 22.11. Voltage measurements are made by comparing the applied voltage value to an internally generated staircase ramp voltage. When the input and the staircase ramp voltages are equal, a comparator generates a signal to stop the ramp. The instrument will then display the number of steps that were required to make the staircase ramp value equal to the input voltage value. At the end of the sample, a reset pulse serves to reset the staircase to zero. The sampling rate in this example is fixed at 2 samples per second. The display circuits store each reading until a new reading is completed, thus eliminating the "blinking" that occurs with some types of digital voltmeters while a measurement is being made.

An integrating digital voltmeter is another basic type of instrument. It measures the true average of the input voltage over a fixed measuring period, in contrast to ramp types which measure the voltage at the end of the measuring interval. A widely used technique to accomplish integration is the use of a voltage-to-frequency converter, as depicted in Fig. 22.12. The circuitry functions as a feedback control system which governs the rate of pulse generation, making the average voltage of the rectangular pulse train equal to the dc input voltage value. The major advantage of this type of analog-to-digital conversion is its ability to measure accurately in the presence of large values of superimposed

FIG. 22.11 *Block diagram of a staircase digital voltmeter.*

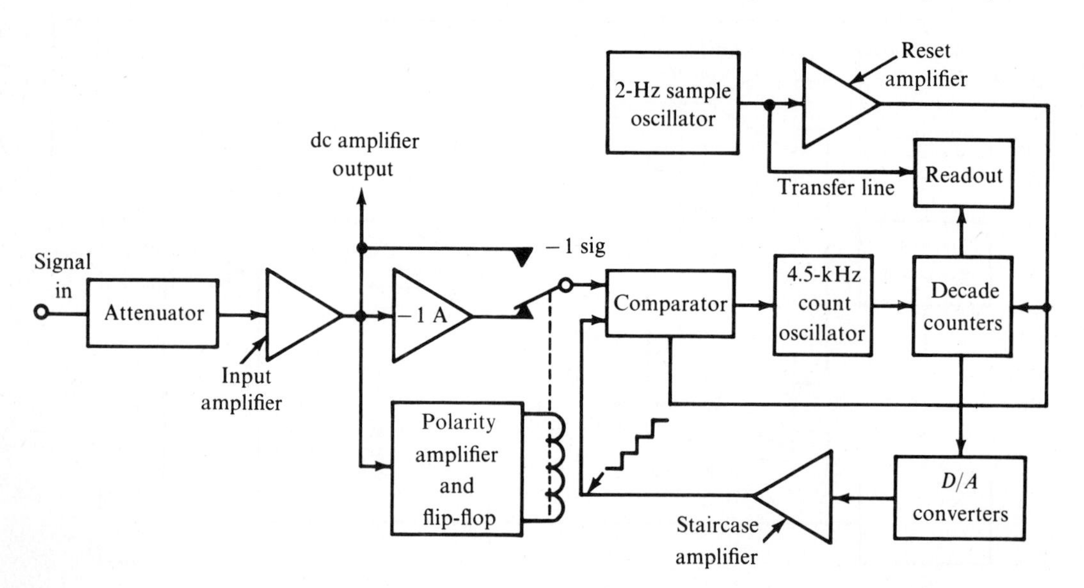

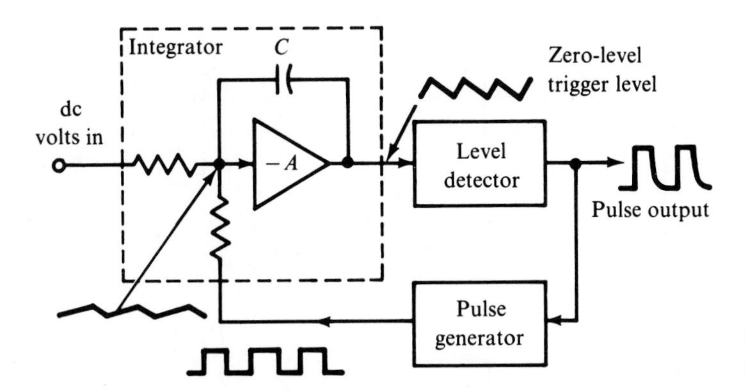

FIG. 22.12 *Voltage-to-frequency conversion.*

noise, because the input is integrated over the sampling interval. In turn, the reading represents a true average of the input voltage.

A block diagram for an integrating digital voltmeter is shown in Fig. 22.13. This type of instrument has an accuracy of approximately 0.01 percent. In this example, the average value of the applied voltage is measured over a 1/60-s sample period. It can make more than 43 separate 5-digit measurements in 1 s with a maximum resolution of 1 part in 130,000. Basically, the instrument consists of a voltage-to-frequency converter and a counter. A dc voltage applied to an integrating amplifier in the converter is changed to a pulse rate proportional to the applied voltage value. During the 1/60-s interval, the output from the voltage-to-frequency converter is applied to the 10^2 decade. An interpolation technique is used after the sampling period when pulses are entered into the 10^0 decade. These pulses are proportional to the charge remaining on the integrating capacitor after the 1/60-s sampling time. After the interpolation period, the counts present in all decades are displayed by in-line digital readout tubes.

Let us consider the integrating/potentiometric type of DVM shown in the block diagram of Fig. 22.14. It features (1) an integrating-type voltmeter which continually measures the true average of the input voltage, and (2) potentiometric action which provides high accuracy from precision resistance ratios and a stable reference voltage. The level comparisons are made on the basis of the basic null technique which is utilized in laboratory calibration of high-accuracy voltmeters. Details of this technique are explained in standard engineering texts on instrument calibration. This type of DVM provides an accuracy as high as ± 0.004 percent.

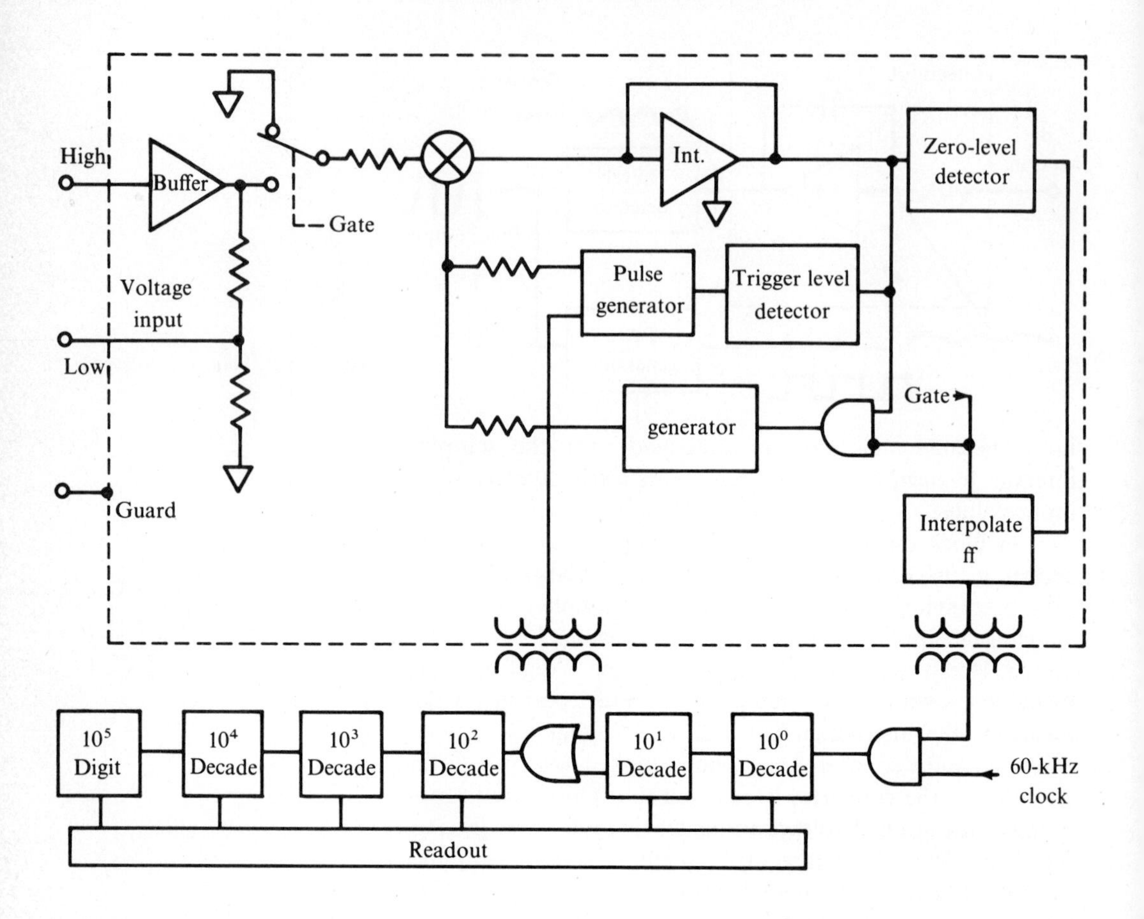

FIG. 22.13 *Block diagram of an integrating DVM.*

To measure ac voltage values with a DVM, an ac-dc converter is required. Figure 22.15 shows a typical arrangement for this unit. It produces an output voltage between 0 and 1 V dc, which is proportional to the average value of the applied ac volts in rms values. The principle of operation is evident from the diagram: an ac input voltage is stepped up by an amplifier and then rectified. A filter smooths the pulsating dc output, and pure dc voltage is made available for application to a DVM. As explained in a foregoing chapter, the dc voltage value is equal to the average value of the rectified ac voltage. Therefore, the instrument indicates correct ac voltage values for pure sine waveforms only.

Another type of DVM, the dual-slope integration design, is depicted in block form in Fig. 22.16. It measures dc voltages by means of an integrator, which produces a time interval proportional to the average value of the applied dc voltage. This time interval determines the gate time of the counter, and therefore determines the number of pulses totalized. Thus the number of pulses that are totalized is proportional to the average value of dc voltage under measurement.

During a precisely controlled time period of 1/10 or 1/60 of a second (selectable for optimum performance), the input signal is integrated, thereby producing an upward slope of voltage. This voltage, which is stored after integration, is proportional to the average of the dc input voltage. To start the downslope, a precise reference voltage of opposite polarity is switched to discharge the integrator. The zero crossing of the output voltage is detected by a zero-detect circuit. In turn, the counter is enabled to totalize pulses from a crystal oscil-

FIG. 22.14 *Block diagram of an integrating/potentiometric DVM.*

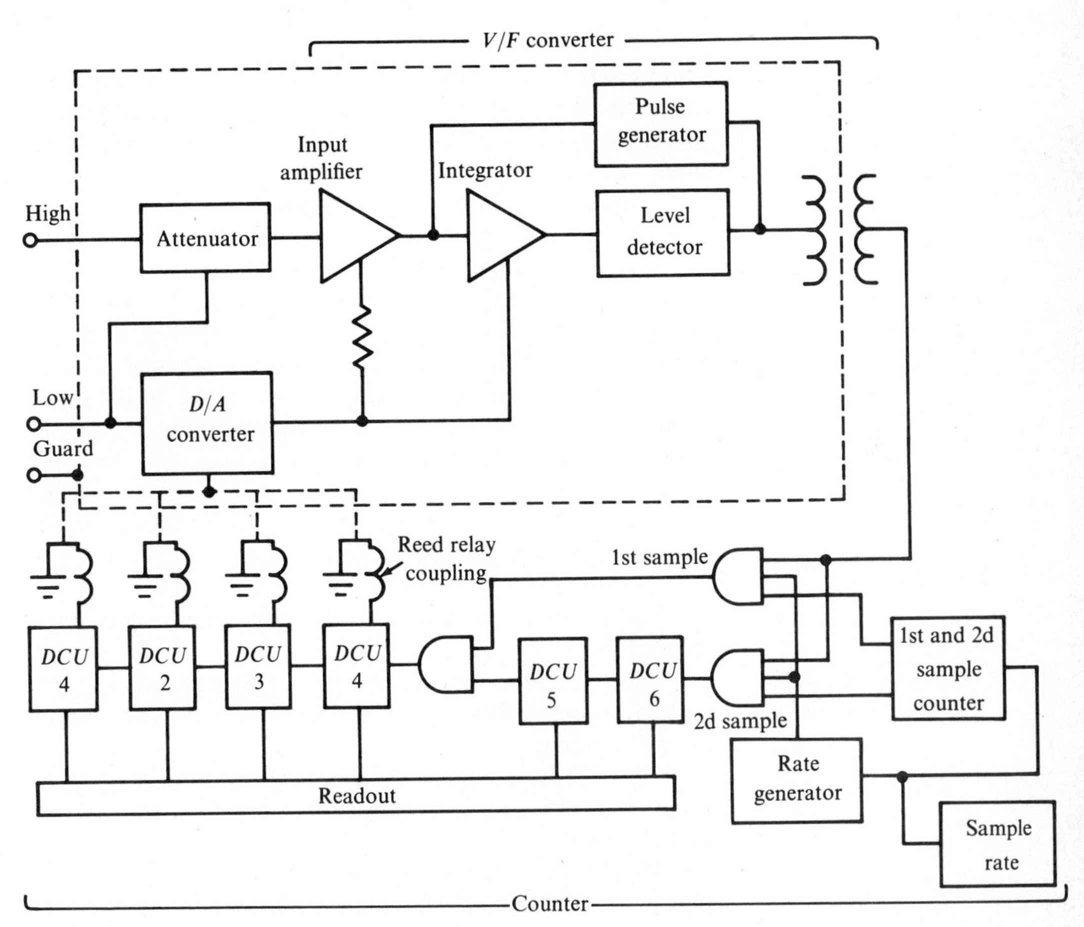

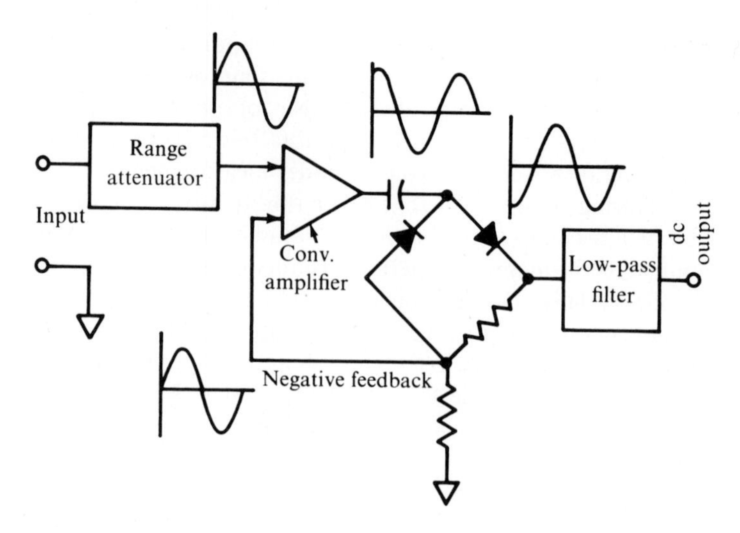

FIG. 22.15 *Typical ac/dc converter.*

lator during the discharge or downslope time of the integrator. Since the discharge time is proportional to the stored voltage value, the number of pulses totalized is proportional to the input voltage.

After completion of the integration cycle, the input amplifier is disconnected and automatically zeroed before the next measurement is taken. This autozeroing effectively compensates for any dc drift in the amplifier. To make a dc voltage measurement, the input voltage is applied to the X input terminals (Fig. 22.16). A dc ratio measurement is made by applying the source voltages to the X input and Y input terminals. The resulting ratio is indicated by the DVM and is equal to X/Y. This measurement is performed in the same manner as a dc voltage measurement, except that the downslope is not determined by the reference voltage but by the Y input voltage. The measurement sequence is as follows:

1. The Y input is measured to determine the proper range for Y.
2. This information is stored. The Y ranging is always performed automatically even if the instrument is switched to manual ranging.
3. The X input is applied to the integrator and the proper range is determined. (The range for X must be equal to or higher than that of Y.)
4. After the ranges for both inputs are determined, the X input will be enabled to charge the integrator.

5. The Y input will then be enabled (on the proper range) to discharge the integrator. In turn, the front-panel display is the ratio of X/Y.

The X and Y inputs are measured sequentially. The inputs are switched off and on in sequence. By switching both the high and the low of each input, complete isolation of X and Y and identical input impedances are achieved. As seen in Fig. 22.16, the instrument may be used with a true rms ac converter. The input circuitry, depicted in block form in Fig. 22.17, consists of an operational amplifier (an amplifier designed to perform a specified mathematical operation) with a gain that is accurately controlled to achieve attenuation of the input signal. An ac output from the input amplifier is fed to the modulator, and a second output is used as a trigger for the sync generator (nominally 5 Hz). The sync square-wave generator is used to synchronize the modulator and demodulator to the input signal.

A 1-kHz oscillator drives the dc-to-square wave converter which changes the dc output from the ac converter into a reference square wave. The amplitude of this square wave is proportional to the dc output value. The output from the modulator, at a nominal 5-Hz rate, consists of a composite signal: one-half input and one-half reference square wave. The agc amplifier controls the gain of the sampling amplifier and the integrator. This keeps the rms value of the signal applied to the thermocouple constant, and holds the gain of

FIG. 22.16 *Block diagram of a dual-slope integrating multifunction meter.*

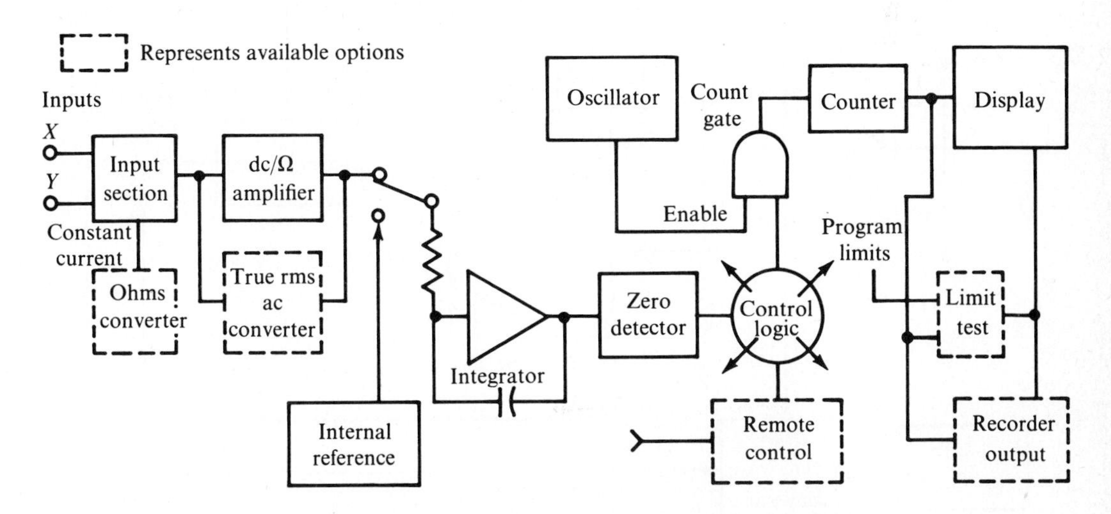

the system constant, regardless of the level of the input signal. The output from the thermocouple varies between two levels, reflecting a difference in the rms value of the input and reference signal. This error signal is amplified, and two signals 180° out of phase are fed to the demodulator.

The demodulator in Fig. 22.17 operates as a full-wave rectifier. The output pulses are amplified and integrated to develop the positive dc voltage output. This dc voltage is continuously corrected at nominally 5 times per second to ensure a dc voltage output proportional to the rms value of the input signal. From this true rms converter, the DVM is energized as explained previously for measurement of ac voltages and ac voltage ratios. Since the indication is fundamentally based on thermocouple action, the indicated rms value is independent of waveform. With respect to the ohms converter option shown in Fig. 22.16, refer to Fig. 22.18. This unit permits the DVM to be operated as an ohmmeter. Let us briefly consider how this is done.

In the configuration of Fig. 22.18, resistance and resistance-ratio measurements are made with a maximum current of 1 mA applied to the unknown resistor on the 10-

FIG. 22.17 *An ac/dc converter.*

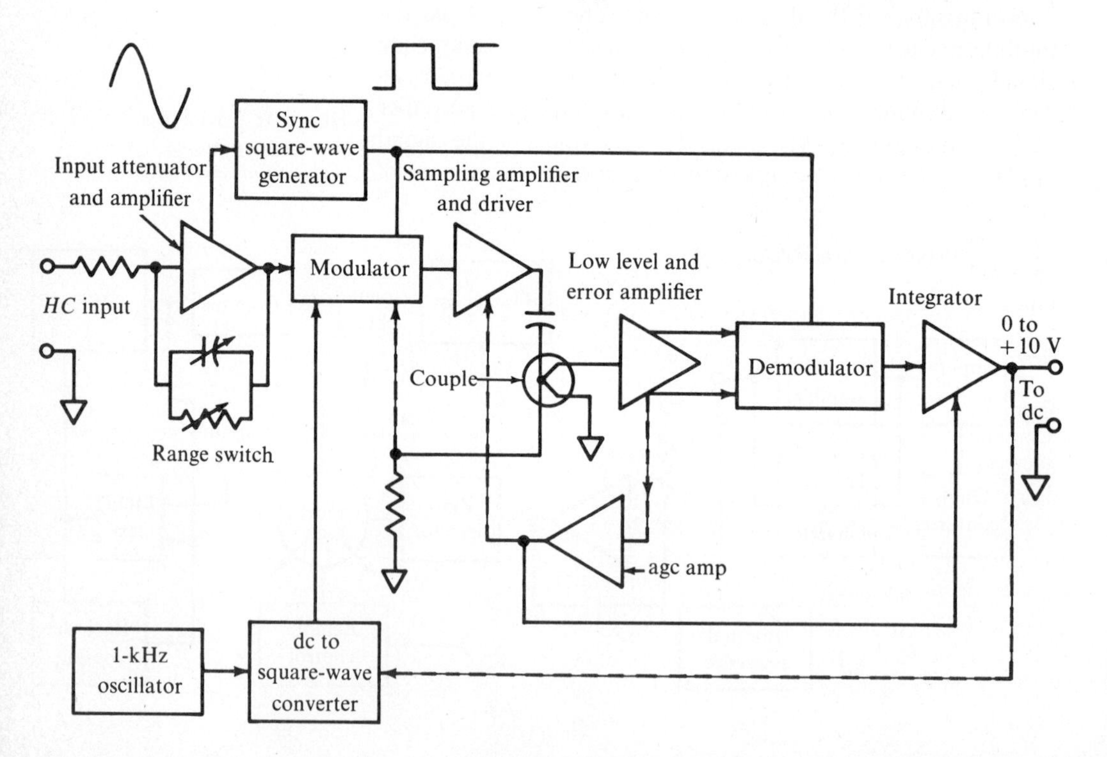

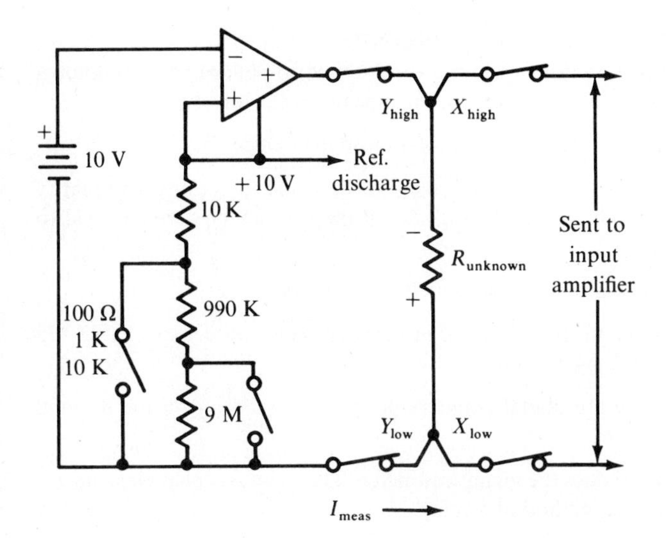

FIG. 22.18 *Ohms converter.*

kΩ range. This low current technique minimizes errors that could be caused by heating of the unknown resistor. A current source supplies three constant-current values of 1 mA, 10 μA, and 1 μA, with an open-loop voltage of 17 V maximum. In the resistance-measuring operation, the X input operates as the sensing terminals, and the Y input as the current terminals. A constant current is fed through the unknown resistor, and the resulting voltage drop is measured. This voltage drop is evidently proportional to the value of the unknown resistance.

22.5 APPLICATIONS

Digital voltmeters are used in the same manner as analog-type voltmeters, and are preferred in many cases because of the following features:

1. They are less costly where maximum indicating accuracy is required. (When only moderate accuracy is required, however, they are more costly than comparable analog instruments.)
2. They minimize the chance of gross operating error inasmuch as the observer is not required to choose a pertinent scale or to read a pointer position on a graduated scale.
3. They eliminate the possibility of parallax error. In a situation where unskilled personnel must make voltage measurements, a digital voltmeter may be preferred even when highly accurate indication is not required.

QUESTIONS AND PROBLEMS

1. Explain how a counter can be used to determine an unknown frequency when the count time interval is known.
2. Explain the operation of a flip-flop circuit.
3. What changes would have to be made in the trigger polarity for the flip-flop in Fig. 22.3 if the transistors were changed to *NPN*?
4. How are pulses processed in a counter?
5. Explain the operation of the four-stage counter in Fig. 22.3 in counting 15 pulses.
6. Why are digital counters designed to read out in the decimal system?
7. How does the digital voltmeter differ from an analog instrument in the method of display?
8. What are the advantages of digital readout instruments over analog instruments?
9. Discuss the operating principle of a ramp-type digital voltmeter.
10. What is the typical accuracy rating of a ramp-type digital voltmeter?
11. What is the method by which a digital voltmeter measures accurately in the presence of large values of superimposed noise?
12. What is the basis of the measuring method used in an integrating potentiometer type of DVM?
13. How does a DVM measure ac voltage?
14. How is the DVM in Fig. 22.15 limited in its ability to measure ac voltages?
15. How is dc drift compensated in the DVM in Fig. 22.16?
16. How is a dc ratio measurement made with the DVM in Fig. 22.16?
17. What is the purpose of an operational amplifier?
18. What unit permits the DVM in Fig. 22.18 to be used as an ohmmeter?

INDEX